Design and Synthesis of Bioactive Compounds for Therapeutic Applications

Design and Synthesis of Bioactive Compounds for Therapeutic Applications

Guest Editors

Gerald Guillaumet
Saïd El Kazzouli

MDPI

Basel • Beijing • Wuhan • Barcelona • Belgrade • Novi Sad • Cluj • Manchester

Guest Editors

Gerald Guillaumet
Institute of Organic and
Analytical Chemistry
Universite d'Orleans
Orleans
France

Saïd El Kazzouli
Euromed Research Center
Euromed University of Fes
Fes
Morocco

Editorial Office
MDPI AG
Grosspeteranlage 5
4052 Basel, Switzerland

This is a reprint of the Special Issue, published open access by the journal *Biomolecules* (ISSN 2218-273X), freely accessible at: https://www.mdpi.com/journal/biomolecules/special_issues/0AAJ31YT17.

For citation purposes, cite each article independently as indicated on the article page online and as indicated below:

Lastname, A.A.; Lastname, B.B. Article Title. *Journal Name* **Year**, *Volume Number*, Page Range.

ISBN 978-3-7258-7815-4 (Hbk)
ISBN 978-3-7258-7816-1 (PDF)
https://doi.org/10.3390/books978-3-7258-7816-1

Contents

About the Editors

Gerald Guillaumet

Gerald Guillaumet was born in France in 1946. He studied chemistry at the University of Clermont-Ferrand (France). He joined Prof. P. Caubère's group and received his Ph. D. in 1972 from the University of Nancy (France) in the field of arynic condensations. Working first as an assistant at the University of Clermont-Ferrand, he was appointed as Maître-Assistant then as Maître de Conferences at the University of Nancy. Nominated as a full professor in organic chemistry at the University of Orleans in 1983, he became the director of the Institute of Organic and Analytical Chemistry. He also served as the Chevalier of the Legion of Honour and President of University of Orleans from 19th November 2004 to 18th November 2009 and is the author of more than 430 scientific publications and 49 patents and has supervised 90 PhD students. His current research interests include heterocyclic chemistry (synthesis and methodologies including C-H functionalization), medicinal chemistry (drug discovery for CNS, metabolic and cardiovascular diseases, anticancer chemotherapy, chemical biology), and the enantioselective synthesis of natural and non-natural molecules.

Saïd El Kazzouli

Saïd El Kazzouli was born in Beni Mellal, Morocco, in 1975. He received his Master's degree (DEA) in 2000 from the Universities of Orleans, Paris V and Paris XI (training at CNRS of Orleans). He then received his Ph.D. in chemistry from the University of Orleans in 2004 under the supervision of Prof. G. Guillaumet and Prof. A. Mouaddib. He worked at the same University as a postdoctoral fellow with Prof. L. Agrofoglio and Prof. S. Berteina-Raboin from 2004 to 2006. In 2006, he joined the National Cancer Institute (NCI) at the National Institutes of Health (NIH) in the USA through a postdoctoral intramural research training award for 3 years with Dr V. E. Marquez. In 2009, he became a researcher (project leader) at INANOTECH, MAScIR Foundation, in Rabat. In 2013, he joined the Euro-Mediterranean University of Fes (Morocco) as Associate Professor, and in 2019 he was appointed as Full Professor at the same University. Here, he set up a Master's degree in environmental engineering, which he coordinated between 2013 and 2015. He then transformed the program into a Master's degree in environmental engineering and water management, which he coordinated from 2015 to 2020. From November 2022 to February 2025, he was the director of the School of Engineering in Biomedical and Biotechnology. He has also been the Director of the Laboratory of Medicinal Chemistry since 2013. His main research interests are drug design and discovery (anti-cancer, anti-virus, and anti-parasite), drug delivery by organic nanoparticles, and the development of macromolecules for MRI and biological needs. Saïd El Kazzouli has an h-index = 32 (Google Scholar, h-index = 30, Scopus), more than 100 publications, three book chapters and four patents.

Preface

The design and discovery of bioactive molecules has emerged as a powerful research area in medicinal chemistry for disease treatment and as a useful tool for various biological process investigations. Nowadays, small molecules as well as macromolecules are known as key building blocks in various medicinal applications, such as drugs for diagnostic devices. This reprint of *Biomolecules* explores the latest scientific advancements related to understanding and addressing challenges in medicinal chemistry. Recent research has uncovered complex interactions between chemistry, biochemistry, cheminformatics, and biology, revealing new therapeutic targets.

This reprint covers innovative medicinal chemistry approaches, including the design and synthesis of bioactive molecules; the study of various extracts; and the binding and signalling activity of some receptors involved in their antimicrobial and antiradical activity, cognitive function, oxidative stress, and neuronal plasticity.

By presenting cutting-edge knowledge on molecular mechanisms, this reprint aims to empower medicinal chemists, bioinformaticians, and biologists to develop evidence-based, multidisciplinary strategies.

This reprint also explores the pharmacological potential and mechanisms of bisbenzylisoquinoline alkaloids from lotus seed embryos; the potential of bioactive compounds in pancreatic cancer therapy; recent advances in design, synthesis, and biological activity studies of 1,3-selenazoles as well as advances in the synthesis and biological applications of enoxacin-based compounds.

This reprint aims to inspire further research and dialogue, ultimately contributing to a future that focuses on the design and synthesis of bioactive compounds for therapeutic applications and the design and synthesis of bioactive molecules for disease treatment.

Gerald Guillaumet and Saïd El Kazzouli
Guest Editors

biomolecules

MDPI

Review

Advances in the Synthesis and Biological Applications of Enoxacin-Based Compounds

Garba Suleiman [1], Nabil El Brahmi [1], Gérald Guillaumet [1,2,*] and Saïd El Kazzouli [1,*]

1 Euromed Research Center, School of Engineering in Biomedical and Biotechnology, Euromed University of Fes (UEMF), Fez 30000, Morocco; s.garba@ueuromed.org (G.S.); n.elbrahmi@ueuromed.org (N.E.B.)
2 Institut de Chimie Organique et Analytique, Université d'Orléans, UMR CNRS 7311, BP 6759, CEDEX 2, 45067 Orléans, France
* Correspondence: gerald.guillaumet@univ-orleans.fr (G.G.); s.elkazzouli@ueuromed.org (S.E.K.)

Abstract: A comprehensive review of advances in the synthesis and biological applications of enoxacin (1, referred to as ENX)-based compounds is presented. ENX, a second-generation fluoroquinolone (FQ), is a prominent 1,8-naphthyridine containing compounds studied in medicinal chemistry. Quinolones, a class of synthetic antibiotics, are crucial building blocks for designing multi-biological libraries due to their inhibitory properties against DNA replication. Chemical modifications at positions 3 and 7 of the quinolone structure can transform antibacterial FQs into anticancer analogs. ENX and its derivatives have been examined for various therapeutic applications, including anticancer, antiviral, and potential treatment against COVID-19. Several synthetic methodologies have been devised for the efficient and versatile synthesis of ENX and its derivatives. This review emphasizes all-inclusive developments in the synthesis of ENX derivatives, focusing on modifications at C3 (carboxylic acid, Part A), C7 (piperazinyl, Part B), and other modifications (Parts A and B). The reactions considered were chosen based on their reproducibility, ease of execution, accessibility, and the availability of the methodology reported in the literature. This review provides valuable insights into the medicinal properties of these compounds, highlighting their potential as therapeutic agents in various fields.

Keywords: antibacterial; biological activity; anticancer; enoxacin; fluoroquinolone; naphthyridine; synthesis; quinolone

1. Introduction

Quinolones, a class of synthetic antibiotics, are widely recognized as crucial building blocks for designing multi-biological libraries [1,2]. Their inhibitory properties against DNA replication make them effective against various pathogens, including mycoplasma, bacteria, and protozoa [3–5]. These synthetic antibacterial drugs belong to the broader class of fluoroquinolones (FQs) and act by targeting DNA gyrase, topoisomerase enzymes, and topoisomerase IV, which are involved in DNA replication and repair processes in bacteria [6–13].

The discovery of nalidixic acid in 1962 marked the beginning of the use of quinolone derivatives as antibacterial agents worldwide [13,14]. The subsequent development of FQs in the 1970s and the 1980s significantly expanded their coverage [15,16]. FQs exhibit diverse biological activities, including against infectious diseases such as malaria and parasitic, bacterial, and fungal diseases [3,17–19], as well as viral infections such as hepatitis, human immunodeficiency virus (HIV), and herpes [20]. They are highly effective against Gram-negative Pseudomonas infections and have been employed in treating pneumonia and intra-abdominal infections [21]. Additionally, they show promise in treating autoimmune diseases, organ transplantation, and rheumatoid arthritis with low toxicity [2,22–24]. FQs can impede tumor growth by inducing damage to type II human DNA topoisomerases, similar to specific chemotherapy drugs such as etoposide [25,26], making them noteworthy agents in infectious disease management and potential adjuncts in certain cancer treatment strategies.

The critical structural attributes of quinolones have been identified, with 4-oxo-quinolone-3-carboxylic acid being a significant substructure in numerous quinolone derivatives with outstanding biological activities [27,28]. Chemical modifications at position 7 transform antibacterial FQs into anticancer analogs, while the carboxylic group at position 3 plays a vital role in enzyme binding and functional group transformation, enhancing anticancer potential [27,29,30]. FQs such as levofloxacin and moxifloxacin are designated by the WHO as second-line drugs for treating tuberculosis due to their broad and potent spectrum of activities as well as oral administration [31–33]. The versatility of quinolones and FQs makes them valuable tools in medicinal research and therapeutic applications across different disciplines.

FQs with a 1,8-naphthyridine core are a specific subset of the fluoroquinolone class, where the quinolone nucleus is replaced by a naphthyridine structure. In the case of FQs with a 1,8-naphthyridine core, the compounds primarily differ at two key positions, N1 and C7, with modifications often occurring at C3 and C7. Figure 1 depicts the 1,8-naphthyridine core, clearly labeling N1 through N8 to emphasize these distinctions within the structure. To illustrate, enoxacin (1, referred to as ENX) is known for having a piperazinyl group at C7 and an ethyl group at N1. In contrast, gemifloxacin, while also featuring the 1,8-naphthyridine core, has an aminopyrrolidinyl group at C7 and a cyclopropyl group at N1. Other FQs with this core typically have a different group at the C7 and N1 positions, as illustrated in Figure 2.

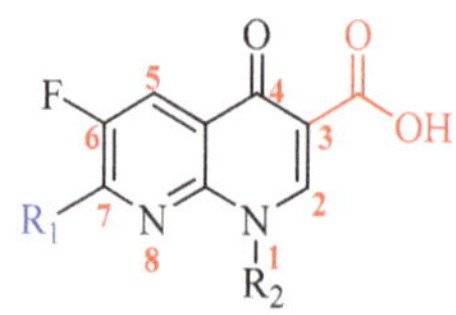

Figure 1. Labeled structural representation of FQs containing a 1,8-naphthyridine core. There are two important distinct positions, C7 and N1 (R_1 and R_2), with diverse substituents.

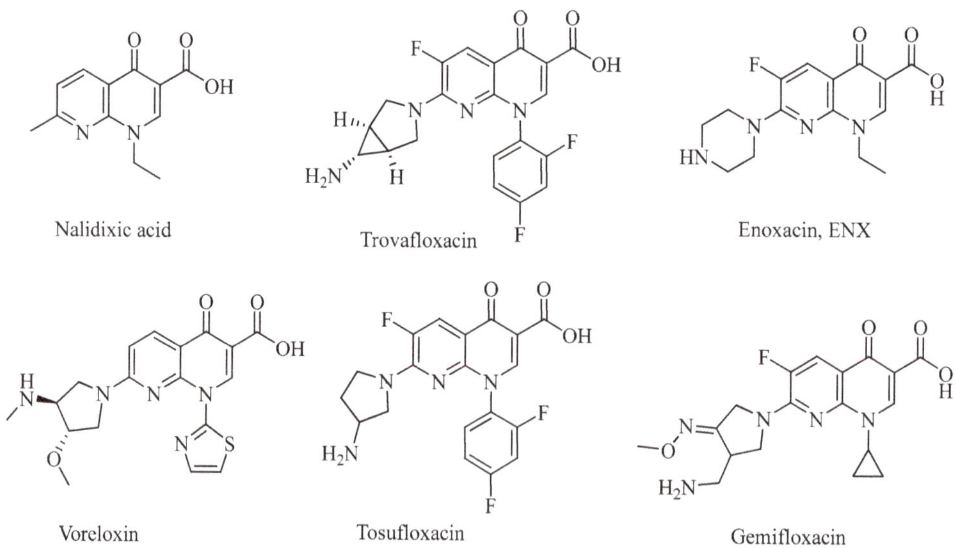

Figure 2. Fluorinated and non-fluorinated 1,8-naphthyridine-containing molecules.

In 1980, ENX, a 1,8-naphthyridine derivative of nalidixic acid, was discovered [34]. Although six distinct isomeric forms of naphthyridine exist, 1,8-naphthyridine derivatives have been extensively researched [35–37]. This unique skeleton has led to various bioactive compounds derived from natural sources, demonstrating significant biological applications [38–40]. ENX, a fluorinated antibacterial drug, and voreloxin, a non-fluorinated potential anticancer agent, are prominent 1,8-naphthyridines studied in medicinal chemistry [26,41]. Other important 1,8-naphthyridine-containing molecules with demonstrated biological activity include nalidixic acid, trovafloxacin, tosufloxacin, voreloxin, and gemifloxacin (Figure 2).

ENX, a second-generation fluoroquinolone, is known for its wide-spectrum antibacterial activity against both Gram-positive and Gram-negative bacteria [42–44]. Structurally, ENX comprises two fused six-membered rings with a 1,8-naphthyridine core as the parental structure (Figure 3) [45,46]. This drug is often well-tolerated and has a low frequency of side effects. It is typically delivered orally in the form of tablets. However, due to the development of resistance by many strains of bacteria, including *Escherichia coli* (*E. coli*) and *Pseudomonas aeruginosa* (*P. aeruginosa*), it is no longer considered a first-line treatment for bacterial infections [47]. Over the past few decades, scientists have examined the potential usage of ENX and its derivatives for several therapeutic applications [48–50]. In vitro tests have revealed that ENX exerts significant cytotoxicity in human cancer cells [48,51]. Moreover, it has also been reported to enhance the anticancer effects of other chemotherapeutic medications, including paclitaxel [51–53]. In addition, ENX possesses antiviral properties, making it effective against many different infections, including HIV and hepatitis C virus (HCV) [48,52].

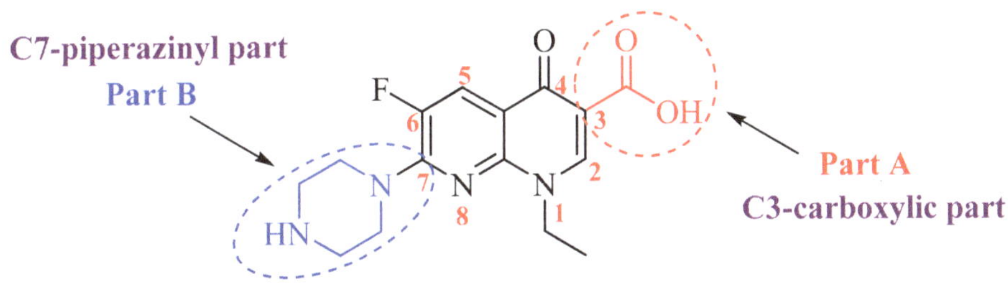

Figure 3. Main sites for the structural modification of ENX.

A recent study on repositioning FQs demonstrated the potential of repurposing ENX for its use as a potential treatment against COVID-19 (SARS-CoV-2) [54–56]. Although there are many motives for reviewing the chemical synthesis of ENX and its derivatives, some of the critical reasons are selectivity [57], repositionability [51], oral bioavailability [58], a better safety profile, pro-oxidative activity, and regulation of microRNA biogenesis [59]. ENX 's unique microRNA-interfering activity sets it apart from other FQs and topoisomerase II drugs [45].

Several synthetic methodologies have been devised and implemented and are known for their efficiency, versatility, and convenience [1,60,61]. However, no exhaustive review has exclusively presented the synthesis of ENX and its derivatives based on the current literature and understanding [62]. In this review, we highlight key developments in the synthesis of 4-quinolone-3-carboxylic acid derivatives with a 1,8-naphthyridine core, specifically focusing on ENX, and discuss its medicinal properties where relevant. Recent publications have discussed the expanded therapeutic potential of diverse heterocyclic molecules beyond their conventional applications [63–66].

The present analysis is structured into three distinct segments: part A focuses on the modification of the carboxylic acid at the C3 position, part B addresses the modification of the piperazinyl group at the C7 position, and the section on other modifications explores combined modifications involving both parts A and B. The reactions considered in this review were chosen based on their capacity for reproducibility, relative ease of execution, accessibility, and the availability of the methodology as reported in the literature. Below is the structural representation of ENX with labeled atom positions comprising the C3-carboxylic part, the C7-piperazinyl part, and the fluoroquinolone core (Figure 3).

2. Modifications of ENX-Based Compounds

2.1. C3 Modification of ENX (Part A)

In 2009, You and colleagues [67] designed and synthesized a novel series of quinolone and naphthyridine derivatives as potential topoisomerase I inhibitors by modifying the scaffold in three steps. The first step involved condensation of ENX with **2** in polyphosphoric acid (PPA) at 170–250 °C to obtain **3a–c** or **4** (Table 1). In the subsequent step, intermediate **3a–c** was nitrated in a mixture of concentrated sulfuric acid (H_2SO_4) and nitric acid (HNO_3) in an approximately equal ratio at 5 °C, followed by heating at 40–45 °C for 1–2 h, yielding **5a–c**. In the final step, the nitro-containing compound **5c** was subjected to hydrogenation over Pd/C in 1 N hydrochloric acid (HCl) solution to produce **6** (Scheme 1). All derivatives containing three kinds of heterocycles, benzoxazole, benzimidazole, and benzothiazole, at the C3 position were screened in vitro for their antiproliferative effects against oral epidermal carcinoma (KB), ovarian carcinoma (A270), and hepatocellular carcinoma (Bel-7402) cells using a 1-*N*-methyl-5-thiotetrazole (MTT)-based assay (Table 1). In summary, the 3-benzothiazolenaphthyridine skeleton **3c** showed the highest antiproliferative activity (IC_{50} = 2.4–2.7 μM) against three tumor cell lines. Conversely, nitro-containing 3-benzoxazolenaphthyridine scaffold **5b** displayed even better cytotoxic activity (IC_{50} = 31.8–3.0 μM). Surprisingly, reducing the nitro group in **5a** to **6** resulted in significantly diminished cytotoxicity. This reinforces the hypothesis that an electron-withdrawing group is essential for cytotoxic activity.

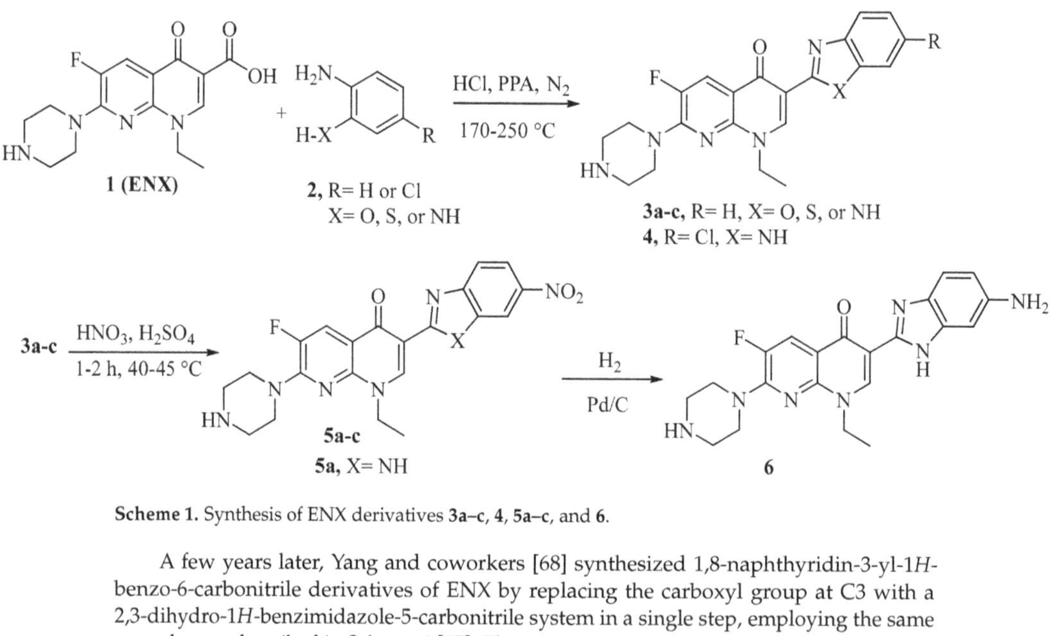

Scheme 1. Synthesis of ENX derivatives **3a–c**, **4**, **5a–c**, and **6**.

A few years later, Yang and coworkers [68] synthesized 1,8-naphthyridin-3-yl-1*H*-benzo-6-carbonitrile derivatives of ENX by replacing the carboxyl group at C3 with a 2,3-dihydro-1*H*-benzimidazole-5-carbonitrile system in a single step, employing the same procedure as described in Scheme 1 [67]. The target compound was realized by condensing ENX with **7** at 170–250 °C in PPA to yield product **8** (Scheme 2). Their studies were primarily centered around investigating the potential molecular mechanism by which it exhibits its antitumor activity against non-small-cell lung cancer (NSCLC). The results revealed that compound **8** exhibited significantly stronger inhibitory effects against NSCLC compared to its leading compound ENX, both in cultured cells and in a xenograft mice model. It also increases reactive oxygen species (ROS) generation and DNA damage response (DDR) in a dose-dependent manner. The ROS scavenger *N*-acetyl-cysteine (NAC) reduced DDR and apoptosis triggered by **8**, confirming that its antitumor actions are due to oxidative stress. Thus, **8** promotes oxidative stress and cell death by activating the mitochondrial and endoplasmic reticulum (ER) stress pathways [68].

Table 1. In vitro antiproliferative activity of compounds **3a–c**, **4**, **5a–c**, and **6**.

Compound	R	X	Antiproliferative Activity (IC_{50}, μM)		
			KB	A2780	Bel7402
3a	H	NH	2.0	4.8	4.1
3b	H	O	11.7	15.3	16.8
3c	H	S	2.4	2.7	2.4
4	Cl	NH	10.3	6.3	21.5
5a	NO_2	NH	22.4	12.4	10.8
5b	NO_2	O	1.8	ND	3.0
5c	NO_2	S	179.3	200.2	24.6
6	-	-	30.1	42.3	93.3

ND: not determined.

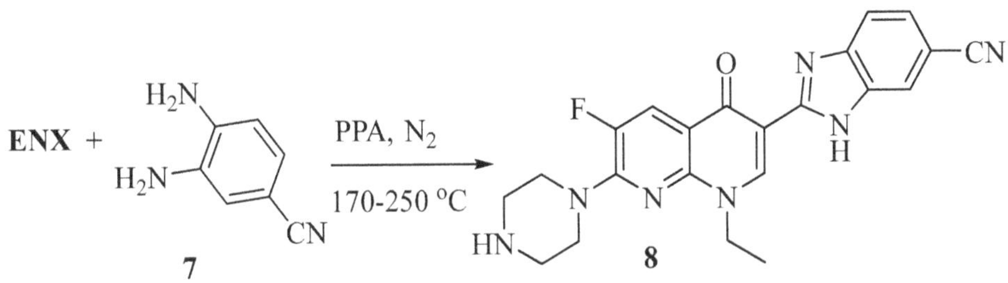

Scheme 2. Synthesis of 1,8-naphthyridin-3-yl-1*H*-benzo[d]imidazole-6-carbonitrile **8**.

In a study conducted by Arayne and colleagues [69], the synthesis of carboxy-substituted ENX analogs as antibacterial agents was documented. This synthesis involved the amidation of the 3-carboxylic acid group of ENX using aromatic amines (RNH_2) and phenyl hydrazine. Initially, an ENX ester, **9**, was prepared in methanol with a catalytic amount of H_2SO_4 at reflux for 7–8 h. The resulting intermediate **9** was further reacted with different aromatic amines, as well as phenyl hydrazine, under reflux for 2–3 h, yielding the desired carboxamides **10a–d** and the carbohydrazide **11** in moderate to good yields (Scheme 3). Compounds **10a–d** and **11** were tested against various bacteria, revealing remarkably improved antimicrobial effectiveness against Gram-negative strains. Furthermore, their potential to influence the immune response was assessed in a separate study [70]. To evaluate their immunomodulatory activity, the impact on the oxidative burst activity of phagocytes in whole blood, as well as macrophages and neutrophils, was investigated. Among the synthesized derivatives, compounds **10c** and **10d** exhibited the highest level of inhibition in whole blood (IC_{50} = 2.6 and 1.4 μg/mL), macrophages (IC_{50} = 3.2 and 1.4 μg/mL), and isolated neutrophils (IC_{50} = 0.8 and 1.4 μg/mL), respectively (Table 2).

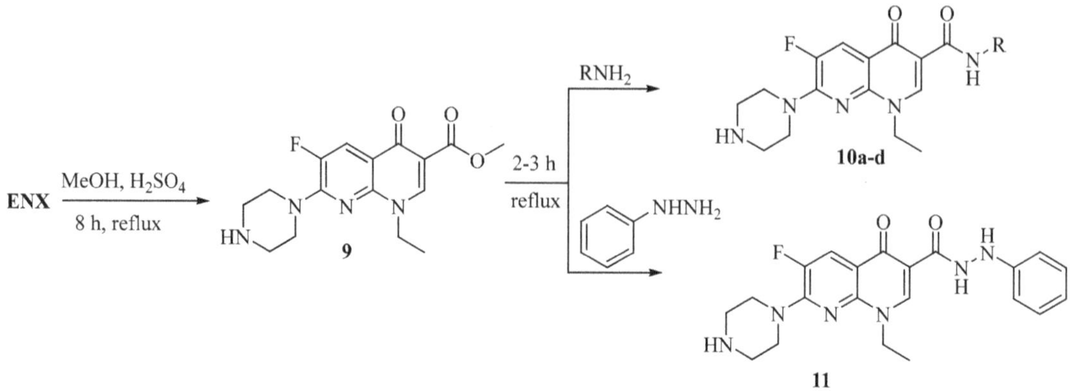

Scheme 3. Synthesis of aryl-substituted ENX carboxamides **10a–d** and carbohydrazide **11**.

Table 2. Immunomodulatory effect of ENX carboxamides **10a–d** and carbohydrazide **11** (Comparable effects of **10a–d** and **11** on the oxidative burst activity of whole blood phagocytes, neutrophils, and macrophages).

		Oxidative Burst Effects (IC_{50}, µg/mL)			
Compound	**R**	**Oxidative Burst of Whole Blood Using**	**Oxidative Burst of PMNs Using**	**Oxidative Burst of Macrophages Using**	**Oxidative Burst of Whole Blood Using**
		Luminol	**Luminol**	**Lucigenin**	**Luminol**
10a	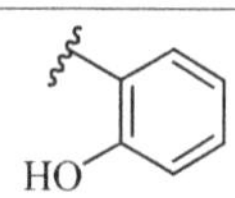	8.5	7.6	17.5	8.7
10b	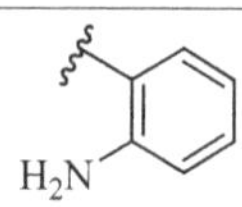	2.6	0.8	1.0	3.2
10c	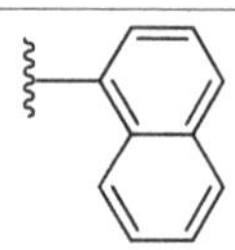	13.3	9.1	22.3	9.5
10d	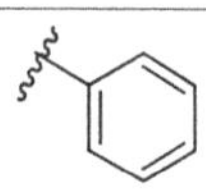	>25	>25	>25	>25
11	-	1.4	1.4	2.6	1.4
ENX	-	>25	>25	>25	>25

PMNs: Polymorphoneutrophils.

2.2. *C7 Modification of ENX (Part B)*

According to the literature, C7 piperazinyl quinolone modifications are effective not only against Gram-positive and Gram-negative pathogens [71] but also have numerous biological applications against cancer [72,73], inflammation [28], osteoclasts [74], viral infections [75], and other diseases [76,77]. As prospective osteo-adsorptive drugs, Herczegh and coworkers [78] developed a series of bisphosphonate FQ derivatives. The piperazinyl group of ENX was transformed with tetraethyl ethene-1,1-diylbis(phosphonate) **12**. In the first step, ENX was combined with **12** in the presence of triethylamine (Et_3N) in dichloromethane (DCM), under stirring at room temperature (rt), for 3 h. Afterwards, an aqueous work-up and recrystallization from toluene produced the bis-(diethoxy-phosphoryl)-ethyl ester **13**. The ester was then hydrolyzed with bromotrimethylsilane $(CH_3)_3SiBr$ in DCM at rt for 72 h, yielding **14** as a hydrobromide salt. Treatment of the salt with water (H_2O) at rt for 6 h, followed by agitation in DCM and subsequent ether washing, resulted in an average yield of the desired compound, bis-phosphonic-ENX derivative **14** (Scheme 4).

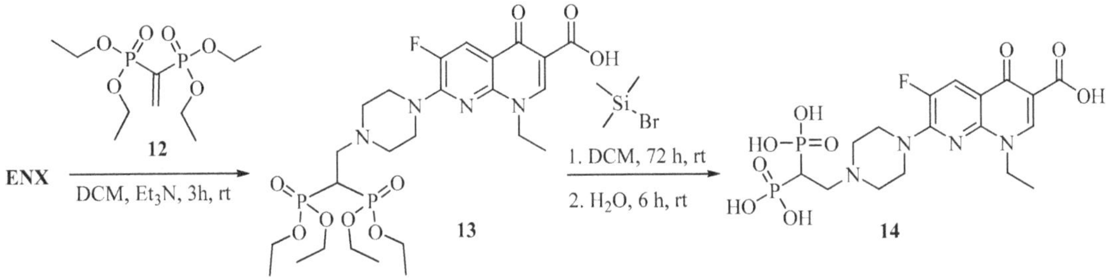

Scheme 4. Synthesis of bis-phosphonic-ENX **14**.

In another study, Vracar and colleagues [79] discovered that ENX and bis-phosphonic-ENX, **14**, have been found to induce the release of extracellular vesicles from 4T1 murine breast cancer cells, which possess inhibitory effects on osteoclastogenesis. Surprisingly, adding a bisphosphonate moiety boosted bone binding affinity. Moreover, bis-phosphonic-ENX, similar to ENX, displayed inhibitory effects on the binding of V-ATPase to microfilaments, as well as on bone resorption in vitro. In summary, bis-phosphonic-ENX offers multiple benefits beyond preventing bone mineral loss. It not only modifies the composition of bone glycoproteins, making them more resistant to fractures, but also completely suppresses osteoclast differentiation. Both ENX and bis-phosphonic-ENX demonstrate similar potency, with IC_{50} values around 10 μM, indicating their strong inhibitory effects on osteoclasts.

Darekhordi and colleagues [80] reported the synthesis of the medicinally important ENX derivative **16** under moderate conditions in a single-step approach. The synthesis involved reacting ENX with **15** using potassium carbonate (K_2CO_3) in dimethylformamide (DMF) at reflux for 24 h, yielding **16** in a reasonable yield (Scheme 5). In addition, the antibacterial efficacy of the synthesized conjugate was tested via the agar diffusion method and exhibited concentration-dependent improved activity against *E. coli*, *Klebsiella pneumoniae* (*K. pneumoniae*), and *Staphylococcus aureus* (*S. aureus*).

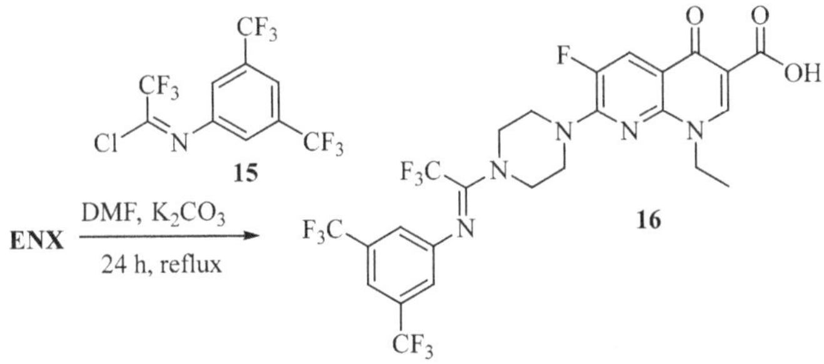

Scheme 5. Synthesis of ENX derivative **16**.

In their study, Xiao and colleagues [81] described the synthesis of FQ–flavonoid hybrids using a well-designed pharmacophore system, aiming to develop a multi-target bacterial topoisomerase inhibitor with potential as an efflux pump inhibitor. The synthesis involved the reaction of ENX with different flavonoids **(17)**, such as apigenin and naringenin, while including an ethylene linker in the process (Scheme 6). In the initial step,

17a–c was *o*-selectively alkylated with 1,2-dibromoethane in the presence of K_2CO_3 in DMSO at 70 °C for 15 h, yielding compounds **18a–c**. Then, compounds **18a–c** were reacted with ENX in DMSO in the presence of DMAP at 60 °C for 40–50 h, yielding FQ–flavonoid hybrids **19a–c** in reasonable yields (55–75%). The antibacterial efficacy of the hybrids was tested against different microorganisms, including Tetracycline-resistant *Bacillus subtilis* ATCC 6633 (*B. subtilis*), amphotericin B-resistant *Candida albicans* (*C. albicans*), multiple drug-resistant *E. coli* ATCC 35218, and methicillin-resistant *S. aureus* ATCC 25923. Some of these compounds displayed impressive antibacterial properties, particularly against drug-resistant strains. Remarkably, derivative **19a** exhibited outstanding activity against *B. subtilis* and *C. albicans* with minimum inhibitory concentrations (MICs) of 0.45 μg/mL and 2.60 μg/mL in comparison to the standard drug ciprofloxacin (CPX), which had MIC values of 2.70 μg/mL and 32.4 μg/mL for the respective microorganisms (Table 3).

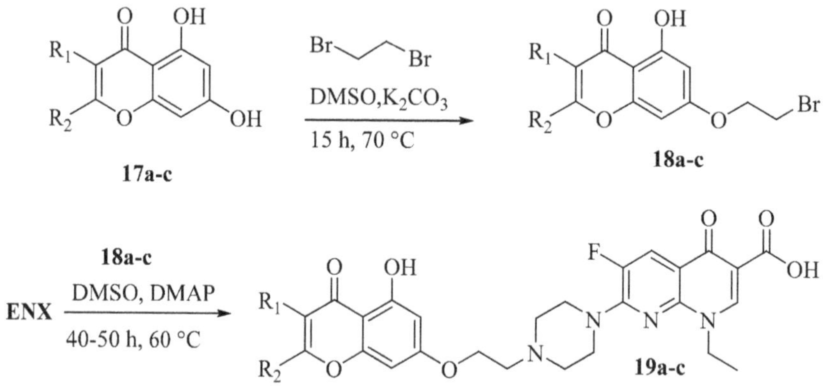

Scheme 6. Synthesis of ENX flavonoid-based analogs **19a–c**.

Table 3. In vitro antibacterial activity of **19a–c** against selected microbes.

			Antibacterial Activity (MIC, μg/mL)			
Compound	**R1**	**R2**	***E. coli***	***B. subtilis***	***S. aureus***	***C. albicans***
19a	H	H	46.3	0.45	21.5	2.60
19b	H	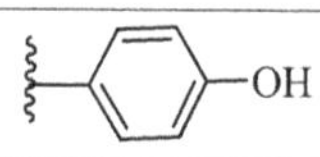	>50	16.1	33.6	17.5
19c	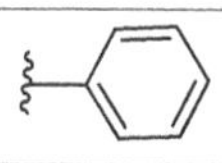	H	>50	>50	>50	>50
CPX	-	-	5.65	2.70	6.82	32.4

A methylene-bridged nitrofuran *N*-substituted piperazinylquinolone was designed and synthesized by Emami and colleagues [82]. ENX mixed with 2-(bromomethyl)-5-nitrofuran **20** in DMF in the presence of sodium hydrogen carbonate ($NaHCO_3$) as a base at rt for 120 h resulted in the formation of the desired compound **21** (Scheme 7) at a good yield (81%). The antibacterial assessment demonstrated that the efficacy of 7-piperazinylquinolones with (5-nitrofuran-2-yl) derivatives against diverse bacterial strains is contingent upon the nature of the substituents located at the N1 and C7 sites. Overall, the compound displayed noteworthy antibacterial efficacy against *Staphylococci* in a manner that was dependent on their concentration. Compound **21** showed the best inhibitory activity against *S. aureus* with a MIC of 0.39 μg/mL.

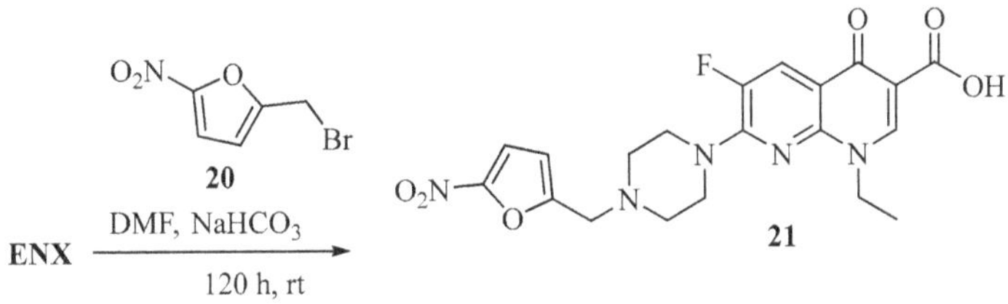

Scheme 7. Methylene-bridged nitrofuran *N*-substituted quinolone synthesis **21**.

In another report [83], four novel ENX derivatives were synthesized by introducing 2-(5-chlorothiophen-2-yl)ethyl into the piperazine ring. The synthesis was performed by reacting ENX with intermediates **22a–d** in DMF at rt, employing $NaHCO_3$ and yielding **23a–d** in 62–73% yields (Scheme 8). The introduction of 2-(5-chlorothiophen-2-yl)ethyl into the piperazine ring of ENX resulted in enhanced cytotoxicity against various cancer cell lines compared to the unmodified ENX [84]. Compound **23** exhibited varying modifications to the ethyl spacer structures. Regarding their cytotoxicity against cancer cell lines, including melanoma (SKMEL-3), breast (MCF-7), epidermoid (A431), bladder (EJ), colon (SW480), and KB cell lines, compounds **23b** and **23c** demonstrated the most significant impact. Specifically, **23b** displayed an IC_{50} range of 3 to 10 μM, while **23c** showed an IC_{50} range of 3 to 20 μM (Table 4). On the other hand, **23d** exhibited IC_{50} values of 2 to 14 μM for melanoma, epidermoid, cervical, and bladder cell lines. In summary, incorporating the 2-(5-chlorothio-phen-2-yl)ethyl group into the piperazinyl portion of ENX enhanced its cytotoxic properties compared to the parent ENX, although the extent of improvement depended on the structure of the spacer. By introducing an additional functionality, the antitumor effectiveness rose considerably (Table 4).

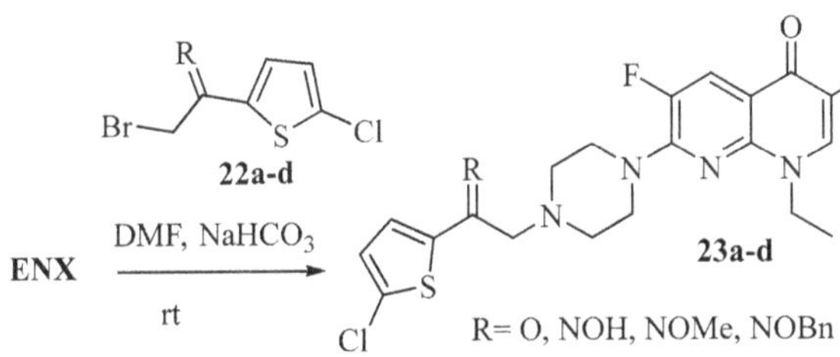

Scheme 8. Synthesis of ENX derivatives **23a–d**.

Table 4. In vitro cytotoxic evaluation of compounds **23a–d** against a panel of cell lines.

Compound	R	Anticancer Activity (IC_{50}, μM)					
		SKMEL-3	MCF-7	A431	EJ	SW480	KB
23a	O	106	106	131	66	100	117
23b	NOH	10.3	3.6	5.6	5.0	3.2	4.8
23c	NOMe	13	19	2.9	5.9	6.7	4.7
23d	NOBn	13.6	125	2.2	8.0	42	12
ENX	-	196	193	175	178	159	137

Synthesis and pre-formulation studies were conducted on a pharmacologically inactive precursor of ENX, resulting in the synthesis of **24** [85]. The synthesis involved reacting ENX with formaldehyde (CH_2O) in a solution of dichloromethane and methanol mixed in an equal ratio at rt for 3 h. The resulting compound was obtained in 89% yield (Scheme 9). The antimicrobial effectiveness of the prodrug was evaluated in comparison to ENX using the agar diffusion method, specifically targeting *E. coli*, *P. aureginosa*, and *S. aureus*. The most noteworthy outcome was observed against *E. coli*, where the MIC was determined to be 0.2 μg/mL.

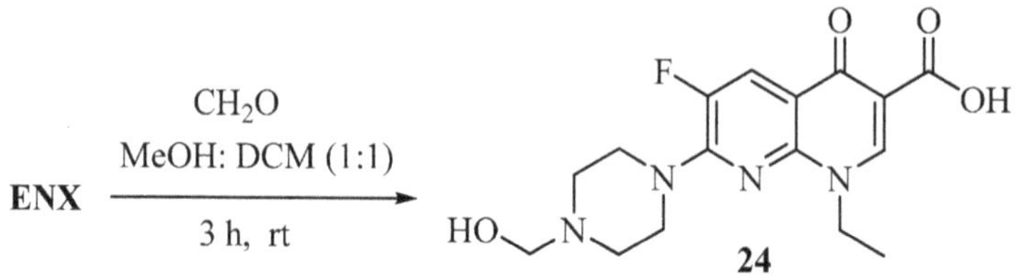

Scheme 9. Synthesis of *N*-piperazinyl-substituted ENX prodrug **24**.

N-substituted piperazinyl quinolone **26** was synthesized and examined for in vitro antibacterial activity against various strains of bacteria [86,87]. Through the reaction of ENX with **25** and $NaHCO_3$ in DMF at 85–90 °C for 12 h, **26** was obtained in satisfactory yield (Scheme 10). The antibacterial evaluation demonstrated that **26** exhibited potent and superior activity against the tested Gram-positive bacteria compared to reference FQs such as ENX. Compound **26** exhibited the highest activity against *B. subtilis*, with a MIC value of 0.008 μg/mL, surpassing the ENX value of 0.125 μg/mL.

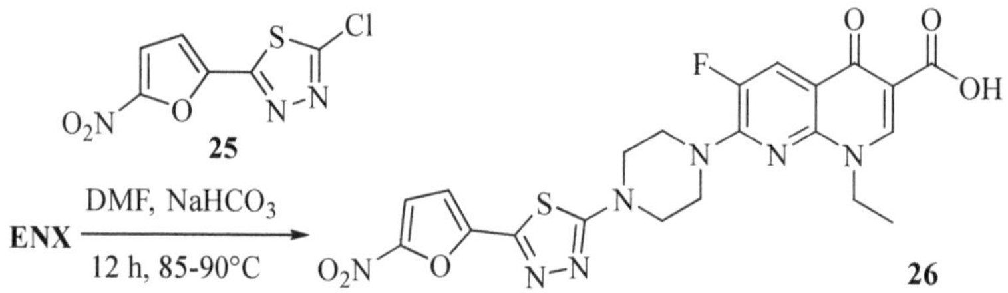

Scheme 10. Synthesis of *N*-substituted piperazinyl quinolone **26**.

Foroumadi et al. [88] reported a series of *N*-substituted piperazinyl quinolones using thiadiazole derivatives **27** with ENX and $NaHCO_3$ in DMF at 85–90 °C for 12 h (Scheme 11). This method successfully synthesized bioactive derivatives of *N*-[5-(chlorobenzylthio)-1,3,4-thiadiazol-2-yl] piperazinyl quinolones **28a–d** in moderate yields (62–67%). To evaluate the efficacy of the synthesized compounds, the agar dilution method was employed against a panel of bacteria including *S. aureus*, *Staphylococcus epidermidis* (*S. epidermidis*), *B. subtilis*, *E. coli*, *K. pneumoniae*, and *P. aeruginosa*. The results indicate that the obtained derivatives exhibited moderate antibacterial activity against the tested microorganisms (Table 5).

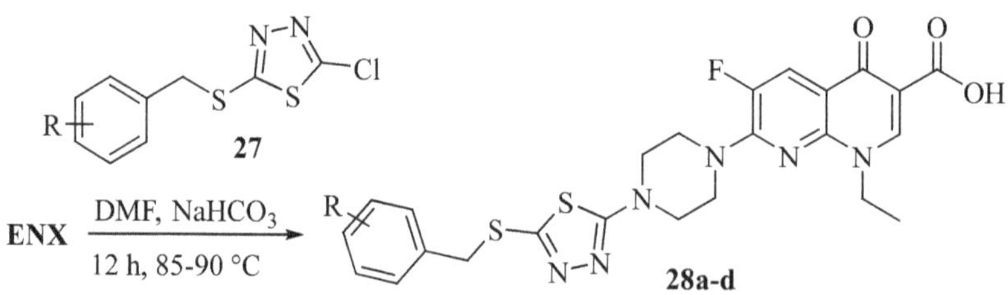

Scheme 11. Synthesis of *N*-[5-(chlorobenzylthio)-1,3,4-thiadiazol-2-yl] quinolones **28a–d**.

In a similar study, a variety of ENX-substituted derivatives **30a–g** were synthesized and tested for antibacterial activity in vitro by combining the ENX with appropriate intermediates **29a–g** [89]. The target derivatives were obtained through the *N*-alkylation of ENX with properly substituted intermediates **29a–g** by employing $NaHCO_3$ as a base in DMF as an appropriate solvent in good yields (76–79%) (Scheme 12). The in vitro antibacterial activity of **30a–g** against various bacterial strains revealed that compounds **30a–c** and **30g** demonstrate antibacterial activity similar to ENX against certain bacterial strains, particularly Gram-positive bacteria such as Staphylococci and Gram-negative bacteria such as *E. coli* and *Enterobacter cloacae* (*E. cloacae*). However, none of the derivatives consistently outperformed ENX across all the tested strains (Table 6).

Table 5. In vitro antibacterial activity of **28a–d** against different bacterial strains.

Compound	R	Antibacterial Activity (MIC, μg/mL)					
		S. aureus	*S. epidermidis*	*B. subtilis*	*E. coli*	*K. pneumoniae*	*P. aeruginosa*
28a	2-Cl	1	2	4	>4	>4	>4
28b	3-Cl	>4	>4	>4	>4	>4	>4
28c	4-Cl	>4	>4	>4	>4	>4	>4
28d	2,4-diCl	4	4	>4	>4	>4	>4
ENX	-	1	0.5	0.125	0.25	0.25	4

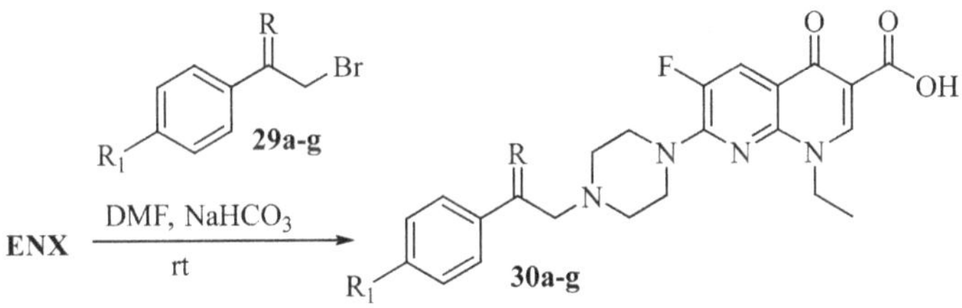

Scheme 12. Synthesis of ENX-substituted derivatives **30a–g**.

Table 6. In vitro antibacterial activity of **30a–g** against various bacterial strains.

Compound	R	Antibacterial Activity (MIC, μg/mL)						
		R_1	*S. aureus*	*S. epidermis*	*E. coli*	*K. pneumoniae*	*E. cloacae*	*P. aeruginosa*
30a	O	H	2	2	0.25	0.5	0.5	4
30b	O	F	4	2	0–5	2	1	8
30c	NOH	H	0.5	0.5	16	0.5	16	>64
30d	NOH	F	1	0.5	16	0.25	16	>64
30e	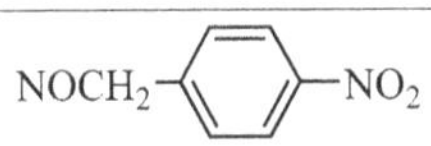	H	16	16	4	16	16	>64
30f	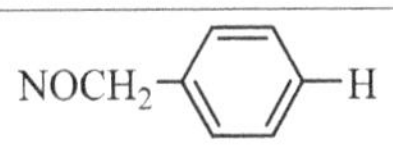	F	64	64	16	64	16	>64
30g	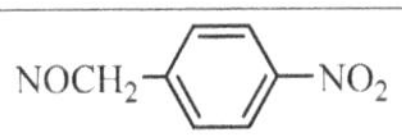	F	0.5	0.5	8	0.5	8	>64
ENX	-	-	1	0.5	0.13	0.5	0.13	4

Foroumadi et al. [90] described the synthesis and antibacterial activity evaluation of piperazinyl-substituted ENX analogs **32a–d.** The synthesis involved reacting **31** with ENX using $NaHCO_3$ in DMF at rt, resulting in the generation of ENX analogs **32a–d** in 45–72% yields (Scheme 13). The synthesized derivatives were evaluated against a variety of bacterial strains. All the tested derivatives show appreciable antibacterial activity against *B. subtilis*, with inhibitory concentrations ranging from 1.56 to 6.25 μg/mL. Although **32b** has consistently shown moderate activity across the tested strains, none of the compounds **32a–d** demonstrated potent antibacterial effects that were comparable to the reference drug ENX (Table 7).

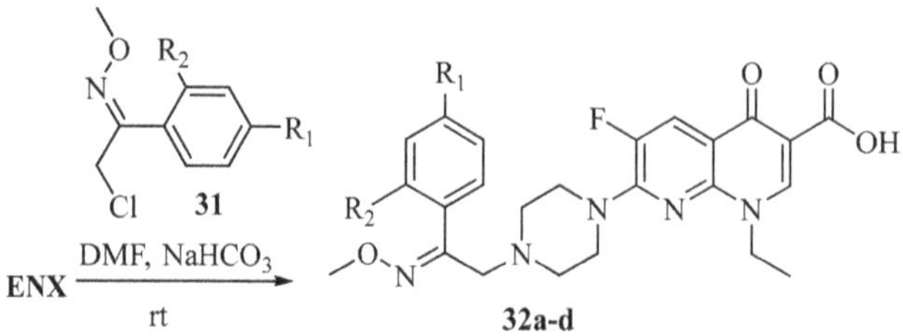

Scheme 13. Synthesis of ENX derivatives **32a–d**.

Table 7. In vitro antibacterial activity of **32a–d** against various bacterial strains.

Compound	R_1	R_2	Antibacterial Activity (MIC, μg/mL)					
			S. aureus	*S. epidermis*	*B. Subtilis*	*E. coli*	*K. pneumoniae*	*P. aeruginosa*
32a	H	H	25	25	6.25	12.5	12.5	>100
32b	F	H	12.5	6.25	6.25	6.25	1.56	100
32c	H	F	25	25	1.56	50	25	>100
32d	Cl	Cl	25	25	3.13	12.5	6.25	50
ENX	-	-	15.6	0.78	0.098	0.098	0.098	6.25

The same group [91] synthesized ENX furan-containing analogs from the furan-based intermediate **33**. *N*-[2-(furan-3-yl)-2-oxoethyl] or *N*-[2-(furan-3-yl)-2-oxyiminoethyl] **34a–d** was produced by treating ENX with **33** in the presence of $NaHCO_3$ at rt in moderate yields (41–59%) (Scheme 14). Evaluation of **34** against various bacterial strains revealed that **34a–c** exhibit comparable antibacterial activity to ciprofloxacin (CPX) against *S. aureus*, methicillin-resistant *S. aureus* (MRSA I and II), *S. epidermidis*, and *B. subtilis*. Specifically, compound **34a** has a MIC range of 0.39 to 0.78 μg/mL against these strains, which is similar to the MIC range of 0.19 to 0.39 μg/mL observed for CPX. Compound **34b** demonstrates a potency of 0.39 μM against *S. aureus*, MRSA, and *S. epidermidis*, closely matching the efficacy of CPX. Likewise, compound **34c** shows a MIC of 0.78 μg/mL against the same strains, again aligning with the antibacterial potency of CPX (Table 8).

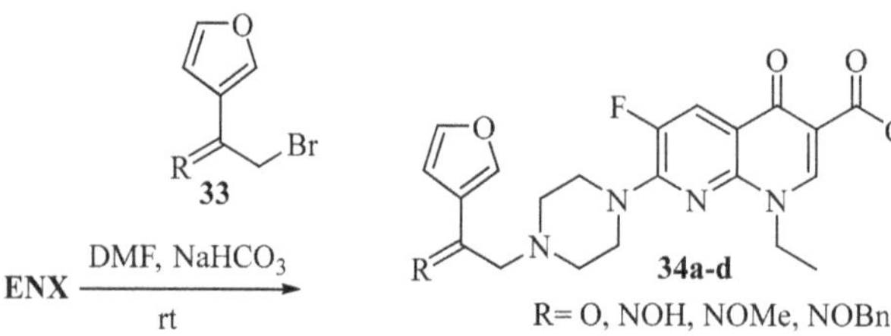

Scheme 14. Synthesis of ENX furan-containing analogs **34a–d**.

Emami et al. [92] reported the synthesis and antibacterial evaluation of ENX coumarin-derived analogs **36a–d**. The synthesis of the hybrids required the reaction of ENX with coumarin-based precursors **35** (Scheme 15). This reaction took place in DMF in the presence of $NaHCO_3$ at rt for 6–72 h, resulting in the desired analogs **36a–d** in moderate to excellent yields (57–91%). The antimicrobial efficacy of **36a–d** was assessed using the agar diffusion method. Compound **36a** exhibits the most potent antibacterial activity across all tested bacteria, including *S. aureus*, MRSA I, MRSA II, *S. epidermidis*, *B. subtilis*, *E. coli*, and *K. pneumoniae*, with MIC values ranging from 0.049 to 3.13 μg/mL. Notably, **36a** shows comparable or superior activity to the reference compound ENX against *S. aureus*, MRSA I,

MRSA II, *S. epidermidis*, *B. subtilis*, and *E. coli*. Compound **36b** also demonstrates significant antibacterial activity, with MIC values between 0.39 μg/mL and 12.5 μg/mL. However, **36b** is generally less potent compared to ENX. On the other hand, compounds **36c** and **36d** exhibit weaker antibacterial potency compared to both **36a** and **36b**, with MIC values that are generally higher than those of ENX (Table 9).

Table 8. In vitro antibacterial activity results of compounds **34a–d**.

Compound	R	Antibacterial Activity (MIC, μg/mL)							
		S. aureus	MRSA I	MRSA II	*S. epidermis*	*B. Subtilis*	*E. coli*	*K. pneumoniae*	*P. aeruginosa*
34a	O	0.78	0.78	0.78	0.78	0.39	0.39	0.19	12.5
34b	NOH	0.39	0.39	0.39	0.39	1.56	1.56	0.39	50
34c	NOMe	0.78	0.78	0.78	0.78	1.56	1.56	0.78	>100
34d	NOBn	25	12.5	12.5	12.5	3.13	3.13	1.56	>100
NOR	-	0.39	0.78	0.78	0.39	0.025	0.049	0.025	3.13
CPX	-	0.19	0.39	0.39	0.19	0.012	0.012	0.012	0.39

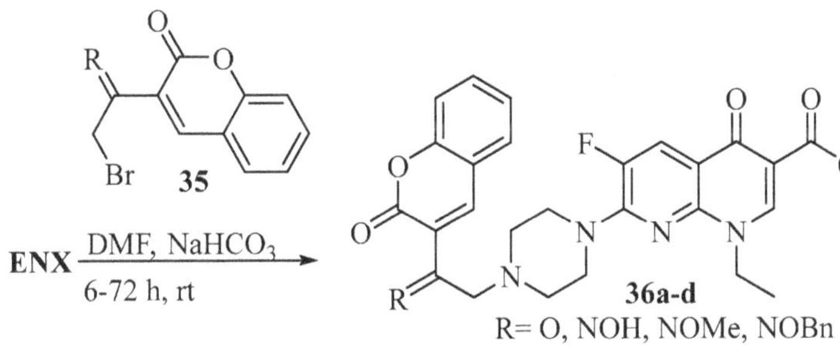

Scheme 15. Synthesis of ENX–coumarin hybrid **36a–d**.

Table 9. In vitro antibacterial activity of **36a–d** against various bacterial strains.

Compound	R	Antibacterial Activity (MIC, μg/mL)							
		S. aureus	MRSA I	MRSA II	*S. epidermis*	*B. Subtilis*	*E. coli*	*K. pneumoniae*	*P. aeruginosa*
36a	O	0.78	0.78	0.78	0.39	0.39	0.049	0.049	3.13
36b	NOH	3.13	3.13	3.13	1.56	0.78	0.78	0.39	12.5
36c	NOMe	3.13	3.13	3.13	6.25	0.78	6.25	1.56	>100
36d	NOBn	50	>100	>100	100	100	100	12.5	>100
ENX	-	0.39	0.78	0.78	0.098	0.19	0.098	0.049	1.56

Shafiee et al. [93] documented the synthesis and antibacterial activity of naphthyl-containing ENX analogs **38a–d**. The desired compounds were successfully synthesized using a versatile and efficient synthetic pathway (Scheme 16). This approach involved reacting ENX with **37** in the presence of $NaHCO_3$ in DMF at rt for 72 h. The resulting products were obtained in good yield (51–83%). The antibacterial evaluation of these derivatives demonstrated promising activity against the tested analogs. Compound **38a** displays comparable or superior antibacterial activity to ENX across all tested strains, with IC_{50} values ranging from 0.049 to 0.780 μg/mL. Similarly, **38b** shows superior activity compared to ENX, particularly against *B. subtilis* and *E. coli*, with IC_{50} values of 0.190 and 0.390 μg/mL, respectively. In contrast, compounds **38c** and **38d** generally exhibit weaker antibacterial activity compared to **38a** and **38b**, as well as the reference compound ENX (Table 10).

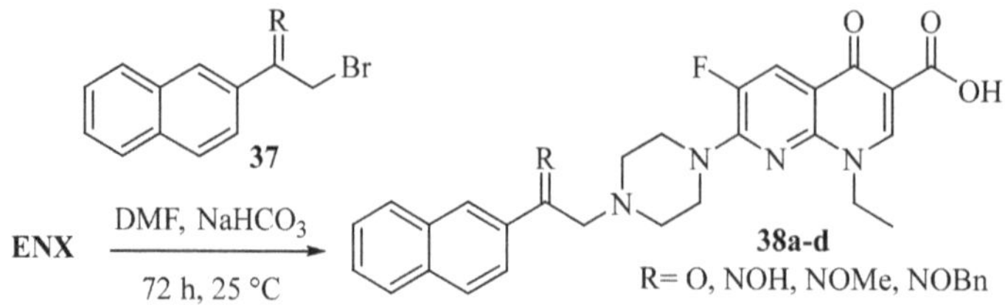

Scheme 16. Synthesis of naphthyl-based ENX analogs **38a–d**.

Table 10. In vitro antibacterial activity of **38a–d** against a panel of bacteria.

Compound	R	Antibacterial Activity (MIC, μg/mL)							
		S. aureus	MRSA I	MRSA II	*S. epidermis*	*B. Subtilis*	*E. coli*	*K. pneumoniae*	*P. aeruginosa*
38a	O	0.78	0.78	0.78	0.78	0.39	0.098	0.049	0.78
38b	NOH	0.78	0.78	0.78	0.78	0.19	3.13	0.39	>100
38c	NOMe	3.13	3.13	3.13	3.13	0.78	1.56	0.78	100
38d	NOBn	>100	>100	>100	100	100	100	25	>100
ENX	-	0.78	0.78	0.78	1.26	0.78	0.098	0.098	1.56

Ahmed and colleagues [94] conducted a groundbreaking study where they skillfully synthesized and screened new alternative molecules of ENX derivatives as potential antibacterial and antibiofilm agents (Scheme 17). ENX was acylated with acid chlorides **39** using Et_3N as a base in refluxing tetrahydrofuran (THF). The desired products **40a–e** were obtained with a moderate yield (49–64%). Evaluation of the antimicrobial potential of **40** against a panel of pathogens via the micro-broth dilution method revealed that all the synthesized derivatives were found to be active at low concentrations against MRSA, *K. pneumoniae, and Proteus mirabilis* (*P. mirabilis*) with MIC values in the range of 12.5 to 25 μg/mL compared to the parent molecule, ENX. Specifically, compounds **40b**, **40c**, and **40e** inhibited the growth of MRSA at a 1 μg/mL concentration better than the parent drug ENX. The antibiofilm inhibitory properties of the synthesized derivatives revealed that **40b**, **40c**, and **40e** inhibited MRSA biofilm formation in the concentration range of 0.5 to 1 μg/mL (Table 11).

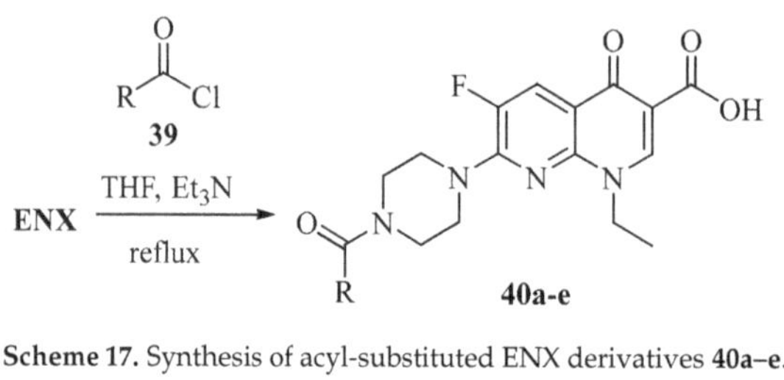

Scheme 17. Synthesis of acyl-substituted ENX derivatives **40a–e**.

Wang and coworkers [95] generated a library of 3-arylfuran-2(5*H*)-one-fluoroquinolone hybrids **46a–e**. Initially, substituted phenylacetic acids **41a–e** were converted to sodium phenylacetates **42a–e** in a dilute NaOH solution. Subsequent treatment of the intermediate salt with ethyl bromoacetate in DMSO at rt for 4 h resulted in the formation of phenylacetic acid ethyl esters **43a–e** in excellent yields (90–95%). Cyclization of **43a–e** was accomplished using sodium hydride (NaH) in THF at 0 °C to rt, leading to the formation of 4-hydroxy-3-phenylfuran-2(5*H*)-ones **44a–e.** The introduction of an ethyl linker was achieved by dissolving **44a–e** in acetone and adding 1,2-dibromoethane and Et_3N, followed by refluxing the mixture for 3–5 h, resulting in the formation of compounds **45a–e** in good yields. Finally, the target products **46a–e** were realized in moderate yields by combining ENX with **45a–e** in the presence of KI and DMAP in DMSO at 60 °C for 72 h (Scheme 18). The conjugated compounds were evaluated against a range of bacteria including tetracycline-resistant *B. subtilis*, *E. coli*, and *S. aureus*. Many of these analogs displayed antibacterial activity that was akin to the reference drug, CPX. Specifically, **46b** exhibited superior antibacterial efficacy across all the tested bacteria, with MIC_{50} values ranging from 1.6 to 2.6 μg/mL, which were significantly better than CPX with MIC_{50} values between 2.7 and 6.82 μg/mL (Table 12).

Table 11. In vitro antimicrobial/antibiofilm activity evaluation of **40a–e**.

Compound	R	Antimicrobial/Antibiofilm Activity (μg/mL)								
		K. pneumoniae			*Proteus mirabilis*			MRSA		
		MIC	MBC	MBIC	MIC	MBC	MBIC	MIC	MBC	MBIC
40a		25	25	6.25	12.5	50.0	25.0	6.4	12.1	4.0
40b	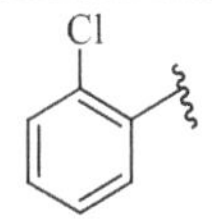	8.0	16.0	8.0	32.5	65.0	16.0	1.0	2.0	0.5
40c	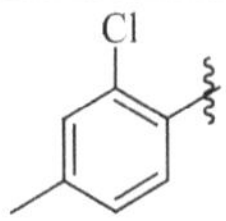	25.0	50.0	6.25	25.0	50.0	25.0	1.0	2.0	1.0
40d	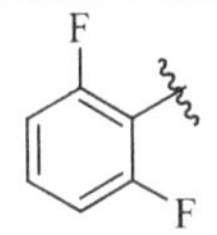	12.5	25.0	6.25	12.5	50.0	25.0	12.5	30.0	2.0
40e	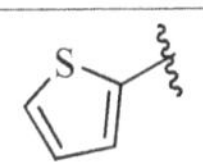	12.5	25	6.25	25.0	50.0	12.5	1.0	2.5	0.5

MBC: minimum bactericidal concentration; MBIC: minimum biofilm inhibitory concentration; MIC: minimum inhibitory concentration.

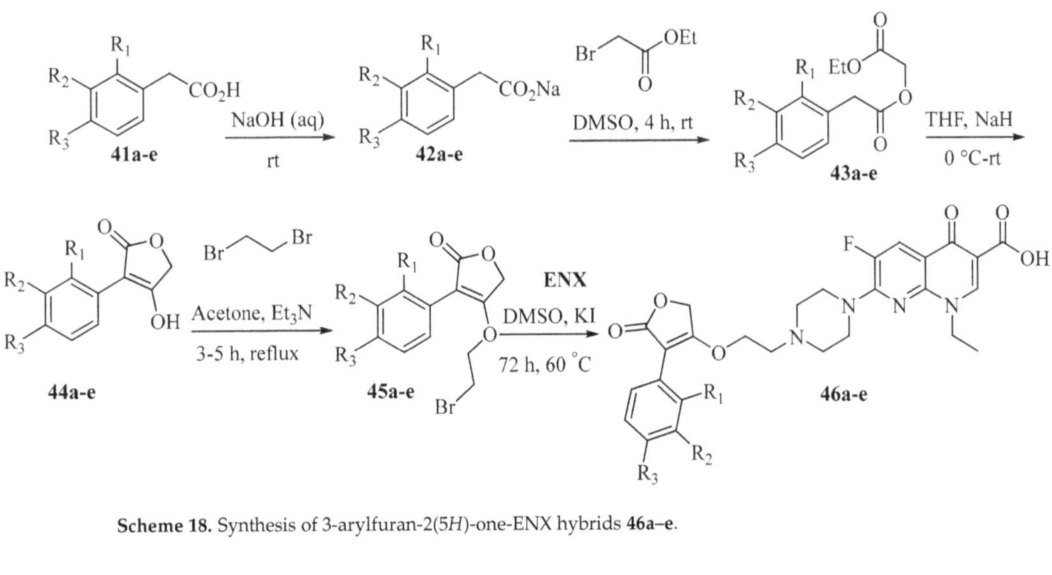

Scheme 18. Synthesis of 3-arylfuran-2(5*H*)-one-ENX hybrids **46a–e**.

Table 12. In vitro antibacterial activity of compounds **46a–e**.

Compound	R_1	R_2	R_3	Antibacterial Activity (MIC (μg/mL)		
				E. coli	*S. aureus*	[a] *B. subtilis*
46a	H	H	H	5.6	6.8	12.6
46b	F	H	H	2.6	2.6	1.6
46c	H	Cl	H	2.9	8.7	15.3
46d	H	H	Cl	9.6	24.9	13.2
46e	H	Br	H	12.2	13.1	4.7
CPX	-	-	-	5.65	6.82	2.70

[a] *B. subtilis*: tetracycline-resistant *Bacillus subtilis*.

Shaheen et al. [96] developed and produced a series of novel FQs that exhibit strong inhibitory effects on α-glucosidase (Scheme 19). The analogs were prepared by subjecting ENX to reflux conditions with various substituted benzyl chlorides **47a–g** in anhydrous acetone, in the presence of K_2CO_3, for 4–8 h. This process resulted in the desired monosubstituted compounds **48a–g** with satisfactory yields. The synthesized derivatives were then subjected to in vitro screening for α-glucosidase inhibition, along with in silico docking studies. The analogs **48a–g** demonstrated strong α-glucosidase inhibitory activity ranging from 48.7 to 74.5 μM, in comparison to the IC_{50} value of 425.6 μM observed for the reference α-glucosidase standard inhibitor drug, 1-deoxynojirimycin (Table 13). Docking studies of **48a–g** reveal that the molecular interactions of mono-benzylated derivatives align well with their inhibitory activity. These compounds were observed to form polar contacts with the active site of proteins, mainly involving residues such as Glu771, Asp392, Trp391, and Arg428.

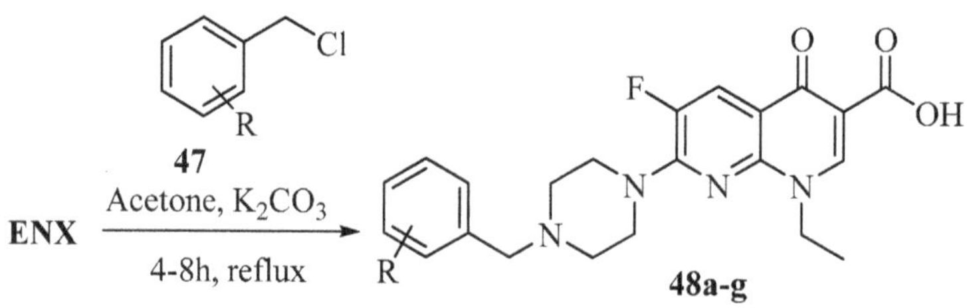

Scheme 19. Synthesis of piperazinyl mono-benzylated ENX derivatives **48a–g**.

Table 13. In vitro α-glucosidase inhibitory activity of compounds **48a–g**.

Compound	R	α-Glucosidase Inhibitory Effect (GIC, μM)
48a	-	57.8
48b	4-Me	69.8
48c	4-Cl	74.5
48d	2,4-diCl	63.8
48e	3,4-diCl	52.7
48f	2,6-diCl	74.2
48g	2,6-diF	48.7
DNJ	-	425.6

DNJ: 1-deoxynojirimycin (standard inhibitor α-glucosidase); GIC: α-glucosidase inhibitory concentration.

2.3. Other Modifications (Parts A and B)

This category encompasses modifications performed on both the **C3** and **C7** sites of the 1,8-naphthyridine core of ENX derivatives.

In the same report, Shaheen and colleagues [96] developed and produced novel disubstituted benzyl FQ derivatives with excellent α-glucosidase inhibitory effects (Scheme 20). The analogs were prepared as demonstrated in Scheme 19. However, in this case, the ENX was refluxed with various substituted benzyl bromides **49a–c** in the presence of K_2CO_3 for 4–8 h, resulting in the formation of disubstituted derivatives **50a–c**. The in vitro α-glucosidase inhibition screening showed that compound **50a** had the highest potency among all tested analogs, with an IC_{50} value of 45.8 μM. Other analogs in this series, **50b** and **50c**, also exhibited notable inhibitory activity, with IC_{50} values of 67.8 μM and 59.8 μM, respectively. These values are significantly lower than the IC_{50} of 425.6 μM for the reference α-glucosidase inhibitor. Interestingly, **50a** is not only more potent than the reference drug but also surpasses the parent compound, ENX, which has an IC_{50} of 58.9 μM. Specifically, **50a** is about 9.3-fold more potent than the reference drug, stressing its strong potential as a lead candidate for further development. Docking studies of compounds **50a–c** indicate that their molecular interactions are consistent with their observed inhibitory activity. These studies show that the di-benzylated derivatives form polar contacts with the active site of the enzyme, primarily interacting with residues such as Gly566, Glu771, Trp391, Asp508, Arg428, and Asp392 (Table 14).

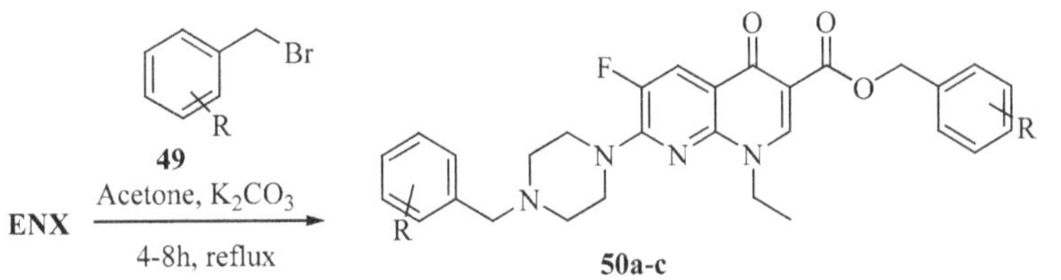

Scheme 20. Synthesis of piperazinyl di-benzylated ENX derivatives **50a–c**.

Table 14. In vitro α-glucosidase inhibitory activity of compounds **50a–c**.

Compound	R	α-Glucosidase Inhibitory Effect (GIC, μM)
50a	2-Br	45.8
50b	2-Cl,4-F	67.8
50c	4-NO_2	59.8
ENX	-	58.9
DNJ	-	425.6

3. Future Perspectives

The recent developments discussed in this review shed light on the synthesis of 4-quinolone-3-carboxylic acid derivatives, with a particular focus on scaffolds containing a 1,8-naphthyridine core reminiscent of ENX. These advancements pave the way for future exploration and innovation in this field. One promising avenue for future research is the further exploration of C3 modifications, as they have shown potential for generating diverse analogs with improved medicinal properties. By employing strategic modifications at the C3 position, researchers can fine-tune the pharmacological profile of these compounds, enhancing their efficacy and reducing potential side effects. Additionally, the C7 modification segment warrants further investigation, as it offers opportunities to optimize the physicochemical properties and biological activities of 4-quinolone-3-carboxylic acid derivatives. By carefully manipulating the C7 position, researchers can potentially enhance the bioavailability, target specificity, and overall therapeutic potential of these compounds.

Lastly, the approach that combines modifications from both parts A and B (other modifications) presents a promising direction for the design and synthesis of novel enoxacin derivatives with diverse pharmacological applications. Within this framework, researchers can explore a wide range of structural modifications in order to produce analogs with specialized features and unique biological activities. Overall, these prospects for the future emphasize the intriguing possibility for further breakthroughs in the synthesis and research of 4-quinolone-3-carboxylic acid derivatives.

4. Conclusions

In conclusion, this review provides a comprehensive analysis of developments in the synthesis of 4-quinolone-3-carboxylic acid derivatives, focusing on scaffolds containing a 1,8-naphthyridine core akin to ENX. The reviewed literature showcases various modifications at the C3 and C7 positions, as well as their combination, demonstrating their impact on the structural diversity, medicinal properties, and potential pharmacological applications of these compounds. The chosen reactions were selected based on their reproducibility, ease of execution, and the accessibility of the described methodologies. Researchers seeking to design and synthesize novel ENX derivatives with diverse pharmacological activities will find the insights presented in this review both valuable and insightful. This comprehensive analysis sets the stage for future investigations, where researchers can explore the untapped potential of 4-quinolone-3-carboxylic acids, specifically ENX derivatives, thereby opening new avenues for drug discovery and therapeutic interventions.

Author Contributions: Conceptualization, G.S., N.E.B., G.G. and S.E.K.; resources, N.E.B., G.G. and S.E.K.; writing—original draft preparation, G.S.; writing—Review and editing, G.S., N.E.B., G.G. and S.E.K.; supervision, N.E.B. and S.E.K.; project administration, N.E.B., G.G. and S.E.K. All authors have read and agreed to the published version of the manuscript.

Funding: This research received no external funding.

Institutional Review Board Statement: Not applicable.

Informed Consent Statement: Not applicable.

Data Availability Statement: Not applicable.

Acknowledgments: We would like to express our sincere gratitude to Euromed University of Fes (UEMF), Morocco, and the African Scientific, Research and Innovation Council (ASRIC) for their unwavering support and resources provided during the course of this project.

Conflicts of Interest: The authors declare no conflicts of interest.

References

1. Mugnaini, C.; Pasquini, S.; Corelli, F. The 4-quinolone-3-carboxylic acid motif as a multivalent scaffold in medicinal chemistry. *Curr. Med. Chem.* **2009**, *16*, 1746–1767. [CrossRef] [PubMed]
2. Naeem, A.; Badshah, S.L.; Muska, M.; Ahmad, N.; Khan, K. The current case of quinolones: Synthetic approaches and antibacterial activity. *Molecules* **2016**, *21*, 268. [CrossRef] [PubMed]
3. Sadeghi, M.; Sarvi, S.; Emami, S.; Khalilian, A.; Hosseini, S.A.; Montazeri, M.; Shahdin, S.; Nayeri, T.; Daryani, A. Evaluation of anti-parasitic activities of new quinolones containing nitrofuran moiety against *Toxoplasma gondii*. *Exp. Parasitol.* **2022**, *240*, 108344. [CrossRef] [PubMed]
4. Chu, D.T.; Plattner, J.J.; Katz, L. New directions in antibacterial research. *J. Med. Chem.* **1996**, *39*, 3853–3874. [CrossRef]
5. Greenwood, D.; Finch, R.; Davey, P.; Wilcox, M. *Antimicrobial Chemotherapy*; Oxford University Press: Oxford, UK, 2007.
6. Millanao, A.R.; Mora, A.Y.; Villagra, N.A.; Bucarey, S.A.; Hidalgo, A.A. Biological effects of quinolones: A family of broad-spectrum antimicrobial agents. *Molecules* **2021**, *26*, 7153. [CrossRef]
7. Robicsek, A.; Strahilevitz, J.; Jacoby, G.A.; Macielag, M.; Abbanat, D.; Park, C.H.; Bush, K.; Hooper, D.C. Fluoroquinolone-modifying enzyme: A new adaptation of a common aminoglycoside acetyltransferase. *Nat. Med.* **2006**, *12*, 83–88. [CrossRef]
8. Drlica, K.; Zhao, X. DNA gyrase, topoisomerase IV, and the 4-quinolones. *Microbiol. Mol. Biol. Rev.* **1997**, *61*, 377–392. [CrossRef]
9. Zechiedrich, E.L.; Cozzarelli, N.R. Roles of topoisomerase IV and DNA gyrase in DNA unlinking during replication in Escherichia coli. *Genes Dev.* **1995**, *9*, 2859–2869. [CrossRef]
10. Drlica, K.; Hiasa, H.; Kerns, R.; Malik, M.; Mustaev, A.; Zhao, X. Quinolones: Action and resistance updated. *Curr. Top Med. Chem.* **2009**, *9*, 981–998. [CrossRef]

11. Kaur, P.; Anuradha; Chandra, A.; Tanwar, T.; Sahu, S.K.; Mittal, A. Emerging quinoline- and quinolone-based antibiotics in the light of epidemics. *Chem. Biol. Drug Des.* **2022**, *100*, 765–785. [CrossRef]
12. McDermott, P.F.; Walker, R.D.; White, D.G. Antimicrobials: Modes of action and mechanisms of resistance. *Int. J. Toxicol.* **2003**, *22*, 135–143. [CrossRef] [PubMed]
13. Lesher, G.Y.; Froelich, E.J.; Gruett, M.D.; Bailey, J.H.; Brundage, R.P. 1,8-Naphthyridine derivatives. A new class of chemotherapeutic agents. *J. Med. Pharm. Chem.* **1962**, *5*, 1063–1065. [CrossRef] [PubMed]
14. Sanchez, J.P.; Domagala, J.M.; Hagen, S.E.; Heifetz, C.L.; Hutt, M.P.; Nichols, J.B.; Trehan, A.K. Quinolone antibacterial agents. Synthesis and structure-activity relationships of 8-substituted quinoline-3-carboxylic acids and 1,8-naphthyridine-3-carboxylic acids. *J. Med. Chem.* **1988**, *31*, 983–991. [CrossRef] [PubMed]
15. Emmerson, A.M.; Jones, A.M. The quinolones: Decades of development and use. *J. Antimicrob. Chemother.* **2003**, *51* (Suppl. S1), 13–20. [CrossRef]
16. Pham, T.D.; Ziora, Z.M.; Blaskovich, M.A.J.M. Quinolone antibiotics. *Med. Chem. Commun.* **2019**, *10*, 1719–1739. [CrossRef]
17. Hu, Y.Q.; Gao, C.; Zhang, S.; Xu, L.; Xu, Z.; Feng, L.S.; Wu, X.; Zhao, F. Quinoline hybrids and their antiplasmodial and antimalarial activities. *Eur. J. Med. Chem.* **2017**, *139*, 22–47. [CrossRef]
18. Andersson, M.I.; MacGowan, A.P. Development of the quinolones. *J. Antimicrob. Chemother.* **2003**, *51* (Suppl. S1), 1–11. [CrossRef]
19. Senerovic, L.; Opsenica, D.; Moric, I.; Aleksic, I.; Spasic, M.; Vasiljevic, B. Quinolines and quinolones as antibacterial, antifungal, anti-virulence, antiviral and anti-parasitic agents. *Adv. Exp. Med. Biol.* **2020**, *1282*, 37–69. [CrossRef]
20. Richter, S.; Parolin, C.; Palumbo, M.; Palu, G. Antiviral properties of quinolone-based drugs. *Curr. Drug Targets Infect. Disord.* **2004**, *4*, 111–116. [CrossRef]
21. Eckmann, C.; Dryden, M.; Montravers, P.; Kozlov, R.; Sganga, G. Antimicrobial treatment of "complicated" intra-abdominal infections and the new IDSA guidelines? A commentary and an alternative European approach according to clinical definitions. *Eur. J. Med. Res.* **2011**, *16*, 115–126. [CrossRef]
22. Platten, M.; Nollen, E.A.A.; Rohrig, U.F.; Fallarino, F.; Opitz, C.A. Tryptophan metabolism as a common therapeutic target in cancer, neurodegeneration and beyond. *Nat. Rev. Drug Discov.* **2019**, *18*, 379–401. [CrossRef] [PubMed]
23. Lin, L.; Zhang, J. Role of intestinal microbiota and metabolites on gut homeostasis and human diseases. *BMC Immunol.* **2017**, *18*, 2. [CrossRef] [PubMed]
24. Tegze, A.; Sági, G.; Kovács, K.; Homlok, R.; Tóth, T.; Mohácsi-Farkas, C.; Wojnárovits, L.; Takács, E. Degradation of fluoroquinolone antibiotics during ionizing radiation treatment and assessment of antibacterial activity, toxicity and biodegradability of the products. *Radiat. Phys. Chem.* **2018**, *147*, 101–105. [CrossRef]
25. Van Maanen, J.M.; Retèl, J.; De Vries, J.; Pinedo, H.J. Mechanism of action of antitumor drug etoposide: A review. *J. Natl. Cancer Inst.* **1988**, *80*, 1526–1533. [CrossRef]
26. Hawtin, R.E.; Stockett, D.E.; Byl, J.A.W.; McDowell, R.S.; Nguyen, T.; Arkin, M.R.; Conroy, A.; Yang, W.; Osheroff, N.; Fox, J.A.J. Voreloxin is an anticancer quinolone derivative that intercalates DNA and poisons topoisomerase II. *PLoS ONE* **2010**, *5*, e10186. [CrossRef]
27. Dine, I.; Mulugeta, E.; Melaku, Y.; Belete, M. Recent advances in the synthesis of pharmaceutically active 4-quinolone and its analogues: A review. *RSC Adv.* **2023**, *13*, 8657–8682. [CrossRef]
28. Daneshtalab, M.; Ahmed, A. Nonclassical biological activities of quinolone derivatives. *J. Pharm. Pharm. Sci.* **2012**, *15*, 52–72. [CrossRef]
29. Abdel-Aal, M.A.A.; Abdel-Aziz, S.A.; Shaykoon, M.S.A.; Abuo-Rahma, G.E.A. Towards anticancer fluoroquinolones: A review article. *Arch. Pharm.* **2019**, *352*, e1800376. [CrossRef]
30. Samir, M.; Ramadan, M.; Hamed, M.; Osman, M.; Abou=Rahma, G. Recent strategies in design of antitumor and antibacterial fluoroquinolones. *J. Adv. Med. Pharm. Sci.* **2021**, *4*, 134–151. [CrossRef]
31. Jang, J.G.; Chung, J.H. Diagnosis and treatment of multidrug-resistant tuberculosis. *Yeungnam Univ. J. Med.* **2020**, *37*, 277–285. [CrossRef]
32. Lee, J.; Lee, C.H.; Kim, D.K.; Yoon, H.I.; Kim, J.Y.; Lee, S.M.; Yang, S.C.; Lee, J.H.; Yoo, C.G.; Lee, C.T.; et al. Retrospective comparison of levofloxacin and moxifloxacin on multidrug-resistant tuberculosis treatment outcomes. *Korean J. Intern. Med.* **2011**, *26*, 153–159. [CrossRef]
33. Sidamo, T.; Shibeshi, W.; Yimer, G.; Aklillu, E.; Engidawork, E. Explorative analysis of treatment outcomes of levofloxacin- and moxifloxacin-based regimens and outcome predictors in ethiopian MDR-TB patients: A prospective observational cohort study. *Infect. Drug Resist.* **2021**, *14*, 5473–5489. [CrossRef]
34. Litvinov, V.P.; Roman, S.V.; Dyachenko, V.D. Naphthyridines. Structure, physicochemical properties and general methods of synthesis. *Russ. Chem. Rev.* **2000**, *69*, 201–220. [CrossRef]
35. Anwair, M.A.; Zeglam, T.H.; Fhid, O.N.; Elmezogi, J.S.; Ong, C.S. Synthesis of some 1, 8-naphthyridine derivatives with comparative studies of cyclization in two different acids. *Int. J. Pharm. Biol. Sci. Arch.* **2012**, *3*, 752–758.
36. Madaan, A.; Verma, R.; Kumar, V.; Singh, A.T.; Jain, S.K.; Jaggi, M. 1,8-Naphthyridine derivatives: A review of multiple biological activities. *Arch. Pharm.* **2015**, *348*, 837–860. [CrossRef]
37. Mithula, S.; Nandikolla, A.; Murugesan, S.; Kondapalli, V.G.J.F.M.C. 1, 8-naphthyridine derivatives: An updated review on recent advancements of their myriad biological activities. *Future Med. Chem.* **2021**, *13*, 1591–1618. [CrossRef]
38. Lee, K.H. Novel antitumor agents from higher plants. *Med. Res. Rev.* **1999**, *19*, 569–596. [CrossRef]

39. Lee, K.H. Anticancer drug design based on plant-derived natural products. *J. Biomed. Sci.* **1999**, *6*, 236–250. [CrossRef]
40. Dholwani, K.K.; Saluja, A.K.; Gupta, A.R.; Shah, D.R. A review on plant-derived natural products and their analogs with anti-tumor activity. *Indian J. Pharmacol.* **2008**, *40*, 49–58. [CrossRef]
41. Henwood, J.; Monk, J.J. Enoxacin. A review of its antibacterial activity, pharmacokinetic properties and therapeutic use. *Drugs* **1988**, *36*, 32–66. [CrossRef]
42. Wood, M.J. Tissue penetration and clinical efficacy of enoxacin in respiratory tract infections. *Clin. Pharmacokinet.* **1989**, *16* (Suppl. S1), 38–45. [CrossRef] [PubMed]
43. Chin, N.X.; Neu, H.C. In vitro activity of enoxacin, a quinolone carboxylic acid, compared with those of norfloxacin, new beta-lactams, aminoglycosides, and trimethoprim. *Antimicrob. Agents Chemother.* **1983**, *24*, 754–763. [CrossRef]
44. Paton, J.H.; Reeves, D.S. Fluoroquinolone antibiotics. Microbiology, pharmacokinetics and clinical use. *Drugs* **1988**, *36*, 193–228. [CrossRef]
45. Jalbrzykowska, K.; Chrzanowska, A.; Roszkowski, P.; Struga, M. The new face of a well-known antibiotic: A review of the anticancer activity of enoxacin and its derivatives. *Cancers* **2022**, *14*, 3056. [CrossRef]
46. Polishchuk, A.; Karaseva, É.; Emelina, T.; Nikolenko, Y.M.; Karasev, V.J.J.o.S.C. Electronic structure and specroscopic properties of norfloxacin, enoxacin, and nalidixic acid. *J. Struct. Chem.* **2009**, *50*, 434–438. [CrossRef]
47. Ching, C.; Orubu, E.S.F.; Sutradhar, I.; Wirtz, V.J.; Boucher, H.W.; Zaman, M.H. Bacterial antibiotic resistance development and mutagenesis following exposure to subinhibitory concentrations of fluoroquinolones in vitro: A systematic review of the literature. *JAC-Antimicrob. Resist.* **2020**, *2*, dlaa068. [CrossRef]
48. Pfab, C.; Schnobrich, L.; Eldnasoury, S.; Gessner, A.; El-Najjar, N. Repurposing of antimicrobial agents for cancer therapy: What do we know? *Cancers* **2021**, *13*, 3193. [CrossRef]
49. Xu, H.; Mao, M.; Zhao, R.; Zhao, Q. Enoxacin exerts anti-tumor effects against prostate cancer through inducing apoptosis. *Technol. Cancer Res. Treat.* **2021**, *20*, 1–8. [CrossRef]
50. Gioia, U.; Francia, S.; Cabrini, M.; Brambillasca, S.; Michelini, F.; Jones-Weinert, C.W.; d'Adda di Fagagna, F. Pharmacological boost of DNA damage response and repair by enhanced biogenesis of DNA damage response RNAs. *Sci. Rep.* **2019**, *9*, 6460. [CrossRef]
51. Yadav, V.; Talwar, P. Repositioning of fluoroquinolones from antibiotic to anti-cancer agents: An underestimated truth. *Biomed. Pharmacother.* **2019**, *111*, 934–946. [CrossRef]
52. Lyu, B.; Wang, C.; Bie, Y.; Kong, J.; Wang, A.; Jin, L.; Qiu, Y.; Zhou, X. Enoxacin shows broad-spectrum antiviral activity against diverse viruses by enhancing antiviral RNA interference in insects. *J. Virol.* **2022**, *96*, e0177821. [CrossRef] [PubMed]
53. Felicetti, T.; Cecchetti, V.; Manfroni, G. Modulating microRNA processing: Enoxacin, the progenitor of a new class of drugs. *J. Med. Chem.* **2020**, *63*, 12275–12289. [CrossRef] [PubMed]
54. Kumar, D.; Chandel, V.; Raj, S.; Rathi, B. In silico identification of potent FDA approved drugs against Coronavirus COVID-19 main protease: A drug repurposing approach. *Chem. Biol. Lett.* **2020**, *7*, 166–175.
55. Gutierrez-Villagomez, J.M.; Campos-Garcia, T.; Molina-Torres, J.; Lopez, M.G.; Vazquez-Martinez, J. Alkamides and piperamides as potential antivirals against the severe acute respiratory syndrome Coronavirus 2 (SARS-CoV-2). *J. Phys. Chem. Lett.* **2020**, *11*, 8008–8016. [CrossRef] [PubMed]
56. Yanez, O.; Osorio, M.I.; Uriarte, E.; Areche, C.; Tiznado, W.; Perez-Donoso, J.M.; Garcia-Beltran, O.; Gonzalez-Nilo, F. In silico study of coumarins and quinolines derivatives as potent inhibitors of SARS-CoV-2 main protease. *Front. Chem.* **2020**, *8*, 595097. [CrossRef]
57. Romero, I.C.; Saravia, N.G.; Walker, J. Selective action of fluoroquinolones against intracellular amastigotes of Leishmania (Viannia) panamensis in vitro. *J. Parasitol.* **2005**, *91*, 1474–1479. [CrossRef]
58. Hooper, D.C.; Wolfson, J.S. The fluoroquinolones: Pharmacology, clinical uses, and toxicities in humans. *Antimicrob. Agents Chemother.* **1985**, *28*, 716–721. [CrossRef]
59. Sousa, E.; Graca, I.; Baptista, T.; Vieira, F.Q.; Palmeira, C.; Henrique, R.; Jeronimo, C. Enoxacin inhibits growth of prostate cancer cells and effectively restores microRNA processing. *Epigenetics* **2013**, *8*, 548–558. [CrossRef]
60. Asahina, Y.; Araya, I.; Iwase, K.; Iinuma, F.; Hosaka, M.; Ishizaki, T. Synthesis and antibacterial activity of the 4-quinolone-3-carboxylic acid derivatives having a trifluoromethyl group as a novel N-1 substituent. *J. Med. Chem.* **2005**, *48*, 3443–3446. [CrossRef]
61. Pasquini, S.; De Rosa, M.; Pedani, V.; Mugnaini, C.; Guida, F.; Luongo, L.; De Chiaro, M.; Maione, S.; Dragoni, S.; Frosini, M.; et al. Investigations on the 4-quinolone-3-carboxylic acid motif. 4. Identification of new potent and selective ligands for the cannabinoid type 2 receptor with diverse substitution patterns and antihyperalgesic effects in mice. *J. Med. Chem.* **2011**, *54*, 5444–5453. [CrossRef]
62. Zahoor, A.F.; Yousaf, M.; Siddique, R.; Ahmad, S.; Naqvi, S.A.R.; Rizvi, S.M.A. Synthetic strategies toward the synthesis of enoxacin-, levofloxacin-, and gatifloxacin-based compounds: A review. *Synth. Commun.* **2017**, *47*, 1021–1039. [CrossRef]
63. Bourzikat, O.; El Abbouchi, A.; Ghammaz, H.; El Brahmi, N.; El Fahime, E.; Paris, A.; Daniellou, R.; Suzenet, F.; Guillaumet, G.; El Kazzouli, S. Synthesis, anticancer activities and molecular docking studies of a novel class of 2-phenyl-5,6,7,8-tetrahydroimidazo [1,2-b]pyridazine derivatives bearing sulfonamides. *Molecules* **2022**, *27*, 5238. [CrossRef] [PubMed]

64. El Abbouchi, A.; El Brahmi, N.; Hiebel, M.-A.; Bignon, J.; Guillaumet, G.; Suzenet, F.; El Kazzouli, S. Synthesis and biological evaluation of ethacrynic acid derivatives bearing sulfonamides as potent anti-cancer agents. *Bioorg. Med. Chem. Lett.* **2020**, *30*, 127426. [CrossRef] [PubMed]
65. El Abbouchi, A.; El Brahmi, N.; Hiebel, M.-A.; Bignon, J.; Guillaumet, G.; Suzenet, F.; El Kazzouli, S. Synthesis and evaluation of a novel class of ethacrynic acid derivatives containing triazoles as potent anticancer agents. *Bioorg. Chem.* **2021**, *115*, 105293. [CrossRef]
66. El Brahmi, N.; El Abbouchi, A.; El Kazzouli, S. An overview on the synthesis and anticancer properties of ethacrynic acid and their analogues. *Results Chem.* **2023**, *6*, 101117. [CrossRef]
67. You, Q.D.; Li, Z.Y.; Huang, C.H.; Yang, Q.; Wang, X.J.; Guo, Q.L.; Chen, X.G.; He, X.G.; Li, T.K.; Chern, J.W. Discovery of a novel series of quinolone and naphthyridine derivatives as potential topoisomerase I inhibitors by scaffold modification. *J. Med. Chem.* **2009**, *52*, 5649–5661. [CrossRef]
68. Yang, L.; Yuan, Y.; Fu, C.; Xu, X.; Zhou, J.; Wang, S.; Kong, L.; Li, Z.; Guo, Q.; Wei, L. LZ-106, a novel analog of enoxacin, inducing apoptosis via activation of ROS-dependent DNA damage response in NSCLCs. *Free Radic. Biol. Med.* **2016**, *95*, 155–168. [CrossRef]
69. Saeed Arayne, M.; Sultana, N.; Haroon, U.; Ahmed Mesaik, M.; Asif, M. Synthesis and biological evaluations of enoxacin carboxamide derivatives. *Arch. Pharm. Res.* **2009**, *32*, 967–974. [CrossRef]
70. Dalhoff, A. Immunomodulatory activities of fluoroquinolones. *Infection* **2005**, *33* (Suppl. S2), 55–70. [CrossRef]
71. Alovero, F.L.; Pan, X.S.; Morris, J.E.; Manzo, R.H.; Fisher, L.M. Engineering the specificity of antibacterial fluoroquinolones: Benzenesulfonamide modifications at C-7 of ciprofloxacin change its primary target in streptococcus pneumoniae from topoisomerase IV to gyrase. *Antimicrob. Agents Chemother.* **2000**, *44*, 320–325. [CrossRef]
72. Mohammed, H.H.H.; Abuo-Rahma, G.; Abbas, S.H.; Abdelhafez, E.M.N. Current trends and future directions of fluoroquinolones. *Curr. Med. Chem.* **2019**, *26*, 3132–3149. [CrossRef] [PubMed]
73. Sharma, P.C.; Chaudhary, M.; Sharma, A.; Piplani, M.; Rajak, H.; Prakash, O. Insight view on possible role of fluoroquinolones in cancer therapy. *Curr. Top. Med. Chem.* **2013**, *13*, 2076–2096. [CrossRef] [PubMed]
74. Swedan, H.K.; Kassab, A.E.; Gedawy, E.M.; Elmeligie, S.E. Topoisomerase II inhibitors design: Early studies and new perspectives. *Bioorg. Chem.* **2023**, *136*, 106548. [CrossRef] [PubMed]
75. Alaaeldin, R.; Mustafa, M.; Abuo-Rahma, G.E.A.; Fathy, M. In vitro inhibition and molecular docking of a new ciprofloxacin-chalcone against SARS-CoV-2 main protease. *Fundam. Clin. Pharmacol.* **2022**, *36*, 160–170. [CrossRef]
76. Ojha, M.; Yadav, D.; Kumar, A.; Dasgupta, S.; Yadav, R. 1, 8-naphthyridine derivatives: A privileged scaffold for versatile biological activities. *Mini Rev. Med. Chem.* **2021**, *21*, 586–601. [CrossRef]
77. Boschelli, D.H.; Wang, Y.D.; Ye, F.; Wu, B.; Zhang, N.; Dutia, M.; Powell, D.W.; Wissner, A.; Arndt, K.; Weber, J.M.; et al. Synthesis and src kinase inhibitory activity of a series of 4-phenylamino-3-quinolinecarbonitriles. *J. Med. Chem.* **2001**, *44*, 822–833. [CrossRef]
78. Herczegh, P.; Buxton, T.B.; McPherson, J.C., 3rd; Kovacs-Kulyassa, A.; Brewer, P.D.; Sztaricskai, F.; Stroebel, G.G.; Plowman, K.M.; Farcasiu, D.; Hartmann, J.F. Osteoadsorptive bisphosphonate derivatives of fluoroquinolone antibacterials. *J. Med. Chem.* **2002**, *45*, 2338–2341. [CrossRef]
79. Vracar, T.C.; Zuo, J.; Park, J.; Azer, D.; Mikhael, C.; Holliday, S.A.; Holsey, D.; Han, G.; VonMoss, L.; Neubert, J.K.; et al. Enoxacin and bis-enoxacin stimulate 4T1 murine breast cancer cells to release extracellular vesicles that inhibit osteoclastogenesis. *Sci. Rep.* **2018**, *8*, 16182. [CrossRef]
80. Darehkordi, A.; Javanmiri, M.; Ghazi, S.; Assar, S. Synthesis of N-aryl-2,2,2-trifluoroacetimidoyl piperazinylquinolone derivatives and their antibacterial evaluations. *J. Fluor. Chem.* **2011**, *132*, 263–268. [CrossRef]
81. Xiao, Z.P.; Wang, X.D.; Wang, P.F.; Zhou, Y.; Zhang, J.W.; Zhang, L.; Zhou, J.; Zhou, S.S.; Ouyang, H.; Lin, X.Y.; et al. Design, synthesis, and evaluation of novel fluoroquinolone-flavonoid hybrids as potent antibiotics against drug-resistant microorganisms. *Eur. J. Med. Chem.* **2014**, *80*, 92–100. [CrossRef]
82. Emami, S.; Shahrokhirad, N.; Foroumadi, A.; Faramarzi, M.A.; Samadi, N.; Soltani-Ghofrani, N. 7-Piperazinylquinolones with methylene-bridged nitrofuran scaffold as new antibacterial agents. *Med. Chem. Res.* **2013**, *22*, 5940–5947. [CrossRef]
83. Mirzaei, M.; Foroumadi, A. Synthesis and in-vitro antibacterial activity of N-piperazinyl quinolone derivatives with a 2-thienyl group. *Pharm. Pharmacol. Commun.* **2000**, *6*, 351–354. [CrossRef]
84. Foroumadi, A.; Emami, S.; Rajabalian, S.; Badinloo, M.; Mohammadhosseini, N.; Shafiee, A. N-Substituted piperazinyl quinolones as potential cytotoxic agents: Structure-activity relationships study. *Biomed. Pharmacother.* **2009**, *63*, 216–220. [CrossRef] [PubMed]
85. Chadha, D.; Agarwal, S. Synthesis and preformulation studies of a prodrug of enoxacin. *Indian J. Pharm. Sci.* **1997**, *59*, 101–104.
86. Foroumadi, A.; Soltani, F.; Moshafi, M.H.; Ashraf-Askari, R. Synthesis and in vitro antibacterial activity of some N-(5-aryl-1,3,4-thiadiazole-2-yl)piperazinyl quinolone derivatives. *Farmaco* **2003**, *58*, 1023–1028. [CrossRef]
87. Foroumadi, A.; Mansouri, S.; Kiani, Z.; Rahmani, A. Synthesis and in vitro antibacterial evaluation of N-[5-(5-nitro-2-thienyl)-1,3,4-thiadiazole-2-yl] piperazinyl quinolones. *Eur. J. Med. Chem.* **2003**, *38*, 851–854. [CrossRef]
88. Foroumadi, A.; Firoozpour, L.; Emami, S.; Mansouri, S.; Ebrahimabadi, A.H.; Asadipour, A.; Amini, M.; Saeid-Adeli, N.; Shafiee, A. Synthesis and antibacterial activity of N-[5-(chlorobenzylthio)-1,3,4-thiadiazol-2-yl] piperazinyl quinolone derivatives. *Arch. Pharm. Res.* **2007**, *30*, 138–145. [CrossRef]
89. Foroumadi, A.; Emami, S.; Davood, A.; Moshafi, M.; Sharifian, A.; Tabatabaie, M.; Farimani, H.T.; Sepehri, G.; Shafiee, A. Synthesis and in-vitro antibacterial activities of N-substituted piperazinyl quinolones. *Pharm. Pharmacol. Commun.* **1997**, *3*, 559–563. [CrossRef]

90. Foroumadi, A.; Ghodsi, S.; Emami, S.; Najjari, S.; Samadi, N.; Faramarzi, M.A.; Beikmohammadi, L.; Shirazi, F.H.; Shafiee, A. Synthesis and antibacterial activity of new fluoroquinolones containing a substituted N-(phenethyl)piperazine moiety. *Bioorg. Med. Chem. Lett.* **2006**, *16*, 3499–3503. [CrossRef]
91. Foroumadi, A.; Mohammadhosseini, N.; Emami, S.; Letafat, B.; Faramarzi, M.A.; Samadi, N.; Shafiee, A. Synthesis and antibacterial activity of new 7-piperazinyl-quinolones containing a functionalized 2-(furan-3-yl)ethyl moiety. *Arch. Pharm.* **2007**, *340*, 47–52. [CrossRef]
92. Emami, S.; Foroumadi, A.; Faramarzi, M.A.; Samadi, N. Synthesis and antibacterial activity of quinolone-based compounds containing a coumarin moiety. *Arch. Pharm. (Weinheim)* **2008**, *341*, 42–48. [CrossRef] [PubMed]
93. Shafiee, A.; Emami, S.; Ghodsi, S.; Najjari, S.; Sorkhi, M.; Samadi, N.; Faramarzi, M.A.; Foroumadi, A. Synthesis and antibacterial activity of N-[2-(2-naphthyl)ethyl]piperazinyl quinolones. *J. Iran. Chem. Soc.* **2009**, *6*, 325–333. [CrossRef]
94. Ahmed, A.; Khan, A.K.; Karim, A.; Siddiqui, R.; Rafique, R.; Ali, S.A.; Solangi, M.; Taha, M.; Khan, K.M. Enoxacin-based derivatives: Antimicrobial and antibiofilm agent: A biology-oriented drug synthesis (BIODS) approach. *Future Med. Chem.* **2022**, *14*, 947–962. [CrossRef] [PubMed]
95. Wang, X.D.; Wei, W.; Wang, P.F.; Tang, Y.T.; Deng, R.C.; Li, B.; Zhou, S.S.; Zhang, J.W.; Zhang, L.; Xiao, Z.P.; et al. Novel 3-arylfuran-2(5H)-one-fluoroquinolone hybrid: Design, synthesis and evaluation as antibacterial agent. *Bioorg. Med. Chem.* **2014**, *22*, 3620–3628. [CrossRef]
96. Shaheen, A.; Ashiq, U.; Jamal, R.A.; Khan, K.M.; Gul, S.; Yousuf, S.; Ali, S.T. Design and synthesis of fluoroquinolone derivatives as potent α-glucosidase inhibitors: In vitro inhibitory screening with in silico docking studies. *ChemistrySelect* **2021**, *6*, 2483–2491. [CrossRef]

biomolecules

Review

Recent Advances in Design, Synthesis, and Biological Activity Studies of 1,3-Selenazoles

Nataliya A. Makhaeva , Svetlana V. Amosova , Andrey S. Filippov, Vladimir A. Potapov and Maxim V. Musalov *

A. E. Favorsky Irkutsk Institute of Chemistry, Siberian Branch of the Russian Academy of Sciences, 1 Favorsky Str., Irkutsk 664033, Russia; ggm@irioch.irk.ru (N.A.M.); amosova@irioch.irk.ru (S.V.A.); filippov@irioch.irk.ru (A.S.F.); v.a.potapov@mail.ru (V.A.P.)

* Correspondence: musalov_maxim@irioch.irk.ru

Abstract: The review examines recent advances in the design and synthesis of 1,3-selenazole derivatives since 2000. Various synthetic approaches to 1,3-selenazoles and reaction conditions are discussed. The beneficial properties of 1,3-selenazoles, especially their biological activity, are emphasized. Compounds with antitumor, antiviral (HIV-1 and HIV-2), antibacterial, antifungal, antiproliferative, anticonvulsant, and antioxidant activity are highlighted.

Keywords: heterocycles; 1,3-selenazoles; synthesis; biological activity; 1,3-selenazole derivatives

1. Introduction

A powerful impulse for the development of organoselenium chemistry came with the discovery of important biological properties of selenium as an essential trace element for animals and humans [1–3]. An important role in the body is played by the glutathione peroxidise enzyme, which contains the selenium atom as an active center. A number of organoselenium compounds exhibit high glutathione-peroxidase-like activity [4–6].

Over the past two decades, a number of important areas of organoselenium chemistry, including the synthesis and properties of 1,3-selenazole derivatives, have been developing quite intensively. The 1,3-selenazoles derivatives exhibit various biological activities, including antiviral [7,8], antibacterial [9], anticancer [10–12], antiparasitic [13], neuroprotective [14], and inhibitory activity to xanthine oxidase [15].

Selenazofurin, one of the best known organoselenium compounds, is the 1,3-selenazole derivative that exhibits high antitumor and antiviral properties (Figure 1) [16–20]. Selenazofurin is active against a broad spectrum of DNA and RNA viruses, including both influenza A and B viruses, and it is significantly more potent than its sulfur analogue, the antiviral drug thiazofurin. Selenazofurin is considered to be the first potent selenium-containing antiviral agent that may find medical use [16].

Cyclopeptides containing 1,3-selenazole units are known as promising biologically and pharmaceutically active compounds (Figure 1) [21–24]. Selenazole derivatives are considered as potential detectors for DNA [25], nucleic acids [26], and proteins [27–30]. In particular, polymethine cyanine dyes based on 1,3-selenazole derivatives [31,32], as well as polymethine benzoselenazole dye with an electron-withdrawing group at the nitrogen atom of the heterocycles [33], form stabile DNA and dsDNA-complexes, which makes these dyes appropriate for the quantitation and visualization of nucleic acids.

The development of the synthesis and study of the properties of various novel organochalcogen heterocyclic compounds, especially Se,N-containing heterocycles, is within the field of our scientific interests [34–38]. The efficient syntheses of a number of condensed 1,3-selenazole derivatives have been developed [35–38]. Among these derivatives, compounds with high antibacterial activity have been found. The comparison of the antibacterial properties of condensed 1,3-selenazole derivatives with their sulfur analogs (condensed 1,3-thiazoles) has shown higher activity of the Se/N compounds [38].

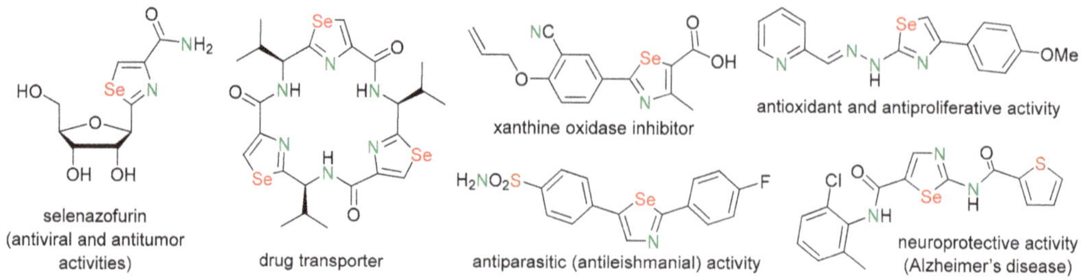

Figure 1. 1,3-Selenazoles with biological activity.

Several reviews have recently been published on the synthesis and biological activity of various Se,N-containing heterocyclic compounds [39–44].

This review covers design, synthesis, and biological activity studies on 1,3-selenazoles (Figure 2) and the application of these compounds as valuable intermediates in organic synthesis since 2000.

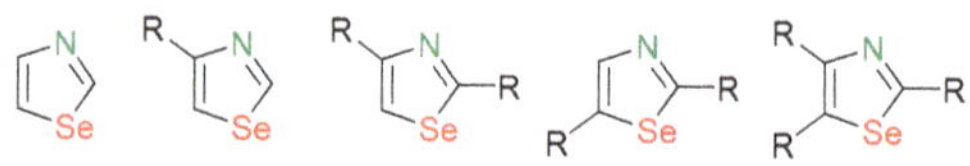

Figure 2. 1,3-Selenazoles considered in this review.

2. 1,3-Selenazole Derivatives

2.1. 1,3-Selenazole and 4-Substituted 1,3-Selenazoles

Parent 1,3-selenazole **1** was obtained in a 3% yield by the cyclization of selenoformamide with α-bromoacetaldehyde (Scheme 1) [45,46]. Selenoformamide was prepared from freshly prepared P_2Se_5 and formamide and used in crude form [47].

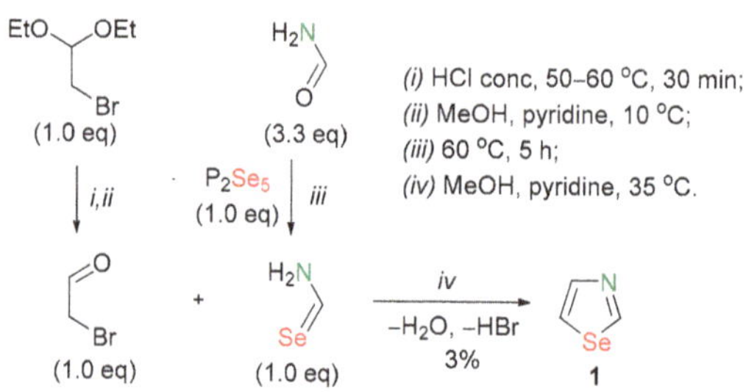

Scheme 1. Synthesis of parent 1,3-selenazoles **1**.

An effective method of the synthesis of 2-unsubstituted 1,3-selenazoles based on 2-benzoyl-1,3-selenazoles was developed (Scheme 2) [48,49]. The cyclization of α-bromoketones with selenoamides, prepared from arylacetonitriles and P_2Se_5 [47], afforded 2-aryl-1,3-selenazoles **2**, containing the benzyl-, *p*-chlorobenzyl-, and *p*-nitrobenzyl groups, in 60–99% yields. The oxidation of products with $K_2Cr_2O_7$ or with SeO_2 in 1,4-dioxane yielded 2-benzoyl-1,3-selenazoles **3** in 45–82% and 70–84% yields, respectively. The treatment of 2-benzoyl-1,3-selenazoles **3** with NaOH in ethanol afforded the desired selenazoles **4** in 60–90% yields [48,49].

The cyclization of α-bromoketones with selenoformamide provided a useful approach to producing 2-unsubstituted 1,3-selenazoles. For example, the reaction of phenacyl bromide with selenoformamide afforded 4-(4-nitrophenyl)-1,3-selenazole (**5**) in a 79% yield. In this case, only two rather than four synthetic stages were required. Obtained 1,3-selenazoles **5** have been shown to have considerable pharmacological significance (Scheme 3) [45,46,48].

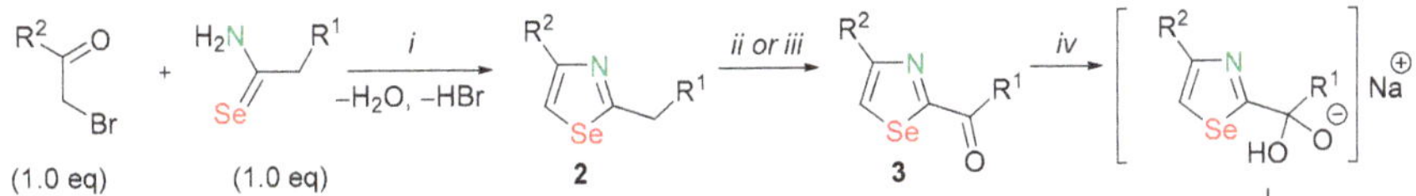

R^1	R^2	**2** (%)	**3** *(ii)* (%)	**3** *(iii)* (%)
Ph	Ph	99	64	70
Ph	4-Me-C_6H_4	77	–	74
Ph	4-Br-C_6H_4	99	60	80
Ph	4-NO_2-C_6H_4	88	76	70
4-Cl-C_6H_4	Ph	60	45	84
4-Cl-C_6H_4	4-Me-C_6H_4	71	79	84
4-Cl-C_6H_4	4-Br-C_6H_4	73	50	–
4-Cl-C_6H_4	4-NO_2-C_6H_4	67	62	72
4-NO_2-C_6H_4	Ph	78	53	72
4-NO_2-C_6H_4	4-Me-C_6H_4	86	42	83
4-NO_2-C_6H_4	4-Br-C_6H_4	86	57	60
4-NO_2-C_6H_4	4-NO_2-C_6H_4	81	82	58

R^2	**4** (%)
Ph	84
Ph	60
Ph	84
4-Me-C_6H_4	90
4-Br-C_6H_4	83
4-NO_2-C_6H_4	71

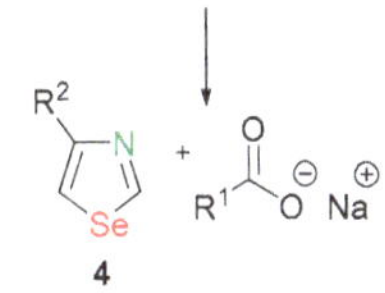

(i) EtOH, reflux, 5 min;
(ii) $K_2Cr_2O_7$ (1.0 eq), AcOH, 40 °C, 3 h;
(iii) SeO_2 (1.0 eq), 1,4-dioxane, 40 °C, 3 h;
(iv) NaOH (2.0 eq), EtOH, reflux, 1 h.

Scheme 2. Synthesis of 2-unsubstituted 1,3-selenazoles **4**.

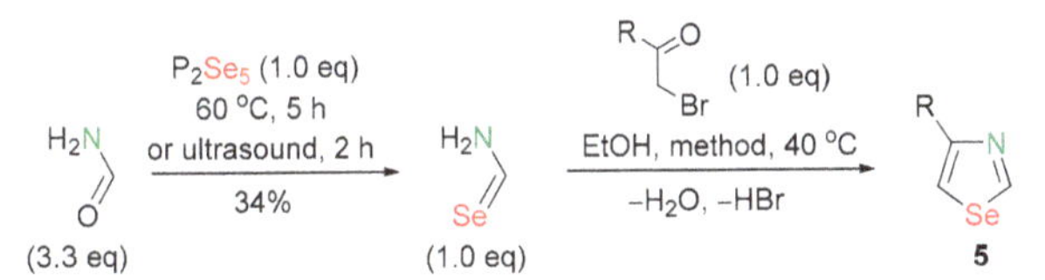

R	method	**5** (%)
4-Cl-C_6H_4	A	13
4-Cl-C_6H_4	B	22
4-Br-C_6H_4	A	9
4-F-C_6H_4	B	20
4-Me-C_6H_4	B	18
4-MeO-C_6H_4	B	6
4-NO_2-C_6H_4	A	28
4-NO_2-C_6H_4	B	79

A: Amberlite IR-120, B: pyridine (1.0 eq)

Scheme 3. Synthesis of 2-unsubstituted 1,3-selenazoles **5**.

2.2. 2,4-Substituted 1,3-Selenazoles

Various 1,3-selenazoles **7** were obtained on the basis of nitriles and freshly prepared P_2Se_5. The reaction of nitriles with P_2Se_5 in a refluxing EtOH/H_2O mixture afforded a chemoselective and convenient synthesis of a variety of selenoamides **6**, (18–84% yields) including a number of new functionalized derivatives. Selenoamides are useful building blocks for the synthesis of 1,3-selenazoles and other pharmacologically significant heterocycles containing nitrogen and selenium. The resulting selenoamides **6** were converted into 1,3-selenazoles **7** in 53–99% yields by the reaction with α-bromoketones in ethanol solution at room temperature (Scheme 4) [47].

The reaction of benzonitrile, cyclohexanecarbonitrile, 2-tolylnitrile, acetonitrile, diphenylacetonitrile, and 1-naphthonitrile with P_2Se_5 afforded the selenoamides **8** in 26–84% yields. At the same time, the slow adding of water to the reaction mixture resulted in the formation of small amounts of H_2Se, which subsequently added to the nitriles. The cyclization of selenoamides **8** with phenacyl bromides afforded the functionalized 1,3-selenazoles **9** in 34–98% yield. The yields of the products are generally higher for the aromatic substrates compared to aliphatic compounds (Scheme 5) [50].

2,4-Diaryl-1,3-selenazoles **13** were prepared by the two-component cyclization of the primary selenoamides with α-haloketones (Scheme 6) [51]. Selenoamides **10** were obtained in 86–98% yields by the reaction of Woollins' reagent with aromatic nitriles in refluxing toluene, followed by hydrolysis with water. A solution of α-haloketone in MeOH was added to a refluxing solution of arylselenocarboamide in MeOH and the reaction mixture was then refluxed for 1 h. A series of 2,4-diaryl-1,3-selenazoles **13** was prepared in 86–99% yields based on this approach. The reaction proceeded with the formation of intermediate **11** followed by the intramolecular cyclization and the dehydration process of intermediate **12**.

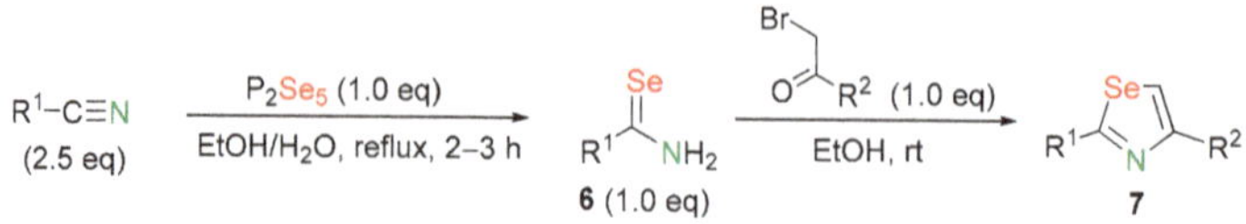

R¹	R²	**6** (%)	**7** (%)
Ph	Ph	84	87
2-Me-C_6H_4	Ph	30	93
Me	4-Br-C_6H_4	27	53
Et	Me	20	61
$PhCH_2$	Ph	40	99
2-Py	Ph	37	93
3-Py	4-Br-C_6H_4	18	61
4-Py	4-Br-C_6H_4	36	81
$NH_2C(O)CH_2$	Ph	47	95

R¹	R²	**6** (%)	**7** (%)
$EtOOCCH_2$	4-NO_2-C_6H_4	33	74
PhCH=CH	Me	42	61
1-naphthyl	Ph	35	80
4-NH_2-C_6H_4	Ph	22	87
4-Cl-C_6H_4-CH_2	4-Br-C_6H_4	73	73
4-NO_2-C_6H_4-CH_2	4-Me-C_6H_4	70	86
c-Hex	4-Br-C_6H_4	26	54
NH_2	Ph	37	87

Scheme 4. Synthesis of selenoamides **6** and 1,3-selenazoles **7**.

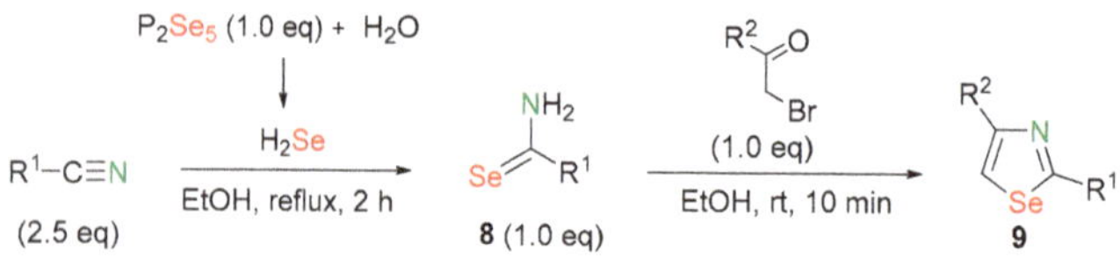

8 (%)	**9** (%)	R¹	R²
84	86	Ph	4-Me-C_6H_4
26	44	*c*-Hex	Ph
–	49	*c*-Hex	4-Me-C_6H_4
–	54	*c*-Hex	4-Br-C_6H_4
–	45	*c*-Hex	4-NO_2-C_6H_4
30	42	Me	4-Me-C_6H_4
–	45	Me	4-NO_2-C_6H_4
50	95	2-Me-C_6H_5	4-Br-C_6H_4
–	34	2-Me-C_6H_5	Et
81	96	Ph_2CH	Ph
–	98	Ph_2CH	4-NO_2-C_6H_4
35	73	1-naphthyl	4-Br-C_6H_4

Scheme 5. Synthesis of functionalized 1,3-selenazoles **9**.

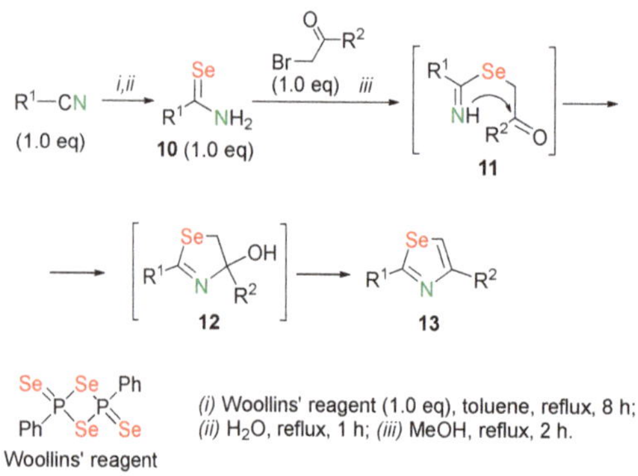

Y	R¹	R²	**10** (%)	**13** (%)
Br	4-Br-C_6H_4	4-Cl-C_6H_4	91	91
Br	Ph	4-Cl-C_6H_4	98	96
Br	3-Me-C_6H_4	4-Cl-C_6H_4	91	95
Br	3-Me-C_6H_4	4-Br-C_6H_4	–	93
Br	3,4-$(MeO)_2$-C_6H_3	4-Cl-C_6H_4	67	98
Br	3,4-$(MeO)_2$-C_6H_3	Ph	–	97
Cl	3,4-$(MeO)_2$-C_6H_3	4-Br-C_6H_4	–	86
Br	2,6-Cl_2-C_6H_3	4-NO_2-C_6H_4	86	97
Br	2,6-Cl_2-C_6H_3	4-Cl-C_6H_4	–	97
Br	4-F_3C-C_6H_4	4-Br-C_6H_4	90	83
Br	4-F_3C-C_6H_4	4-NO_2-C_6H_4	–	98
Cl	2-naphthyl	Ph	86	94
Br	2-naphthyl	4-Cl-C_6H_4	–	97
Br	1-naphthyl	4-Cl-C_6H_4	89	95
Br	1-naphthyl	3-NO_2-C_6H_4	–	93
Br	1-naphthyl	4-Me-C_6H_4	–	95
Br	2-naphthyl	4-Me-C_6H_4	–	99

Scheme 6. Synthesis of 2,4-diaryl-1,3-selenazoles **13**.

A versatile reagent for organic synthesis, [hydroxy(tosyloxy)iodo]benzene (HTIB), is especially useful in the reaction with ketones. The tosyloxylation of ketones with HTIB, followed by the treatment with primary selenoamides, provides a convenient method of synthesis for 1,3-selenazoles, avoiding the use of toxic and lachrymatory α-haloketones. Thus, 2,4-substituted 1,3-selenazoles **14** were prepared in 58-85% yields from the appropriate ketones and selenoamides under reflux in a solvent mixture MeCN/MeOH (Scheme 7) [52].

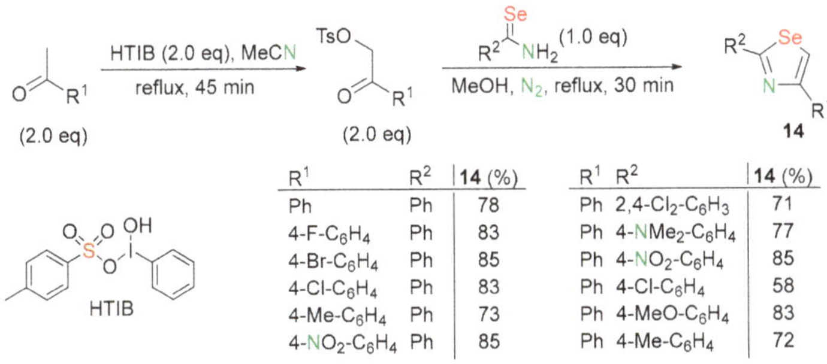

R^1	R^2	14 (%)
Ph	Ph	78
4-F-C_6H_4	Ph	83
4-Br-C_6H_4	Ph	85
4-Cl-C_6H_4	Ph	83
4-Me-C_6H_4	Ph	73
4-NO_2-C_6H_4	Ph	85
Ph	2,4-Cl_2-C_6H_3	71
Ph	4-NMe_2-C_6H_4	77
Ph	4-NO_2-C_6H_4	85
Ph	4-Cl-C_6H_4	58
Ph	4-MeO-C_6H_4	83
Ph	4-Me-C_6H_4	72

Scheme 7. Synthesis of 2,4-substituted 1,3-selenazoles **14** with using HTIB.

New 2-aryl-4-ferrocenyl-1,3-selenazoles **15** were synthesized in 43–72% yields by the reaction of aryl selenoamides with (2-bromoacetyl)ferrocene in refluxing ethanol. Aryl selenoamides were prepared by the treatment of aromatic nitriles with NaHSe in refluxing ethanol. 2-bromoacetyl ferrocene was prepared in 35% yield by the reaction of ferrocene with bromoacetyl bromide in the presence of $AlCl_3$ (Scheme 8) [53].

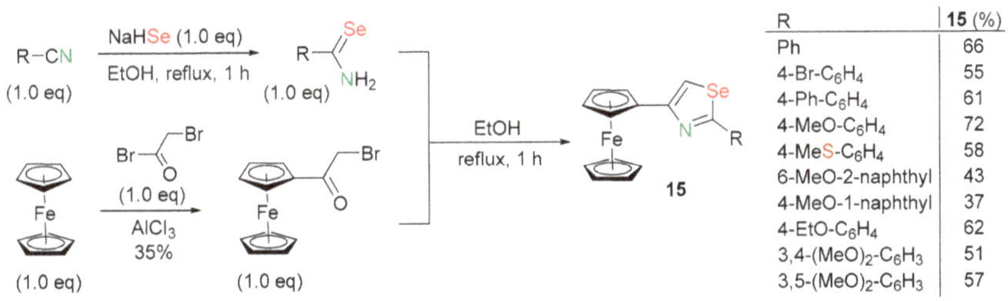

R	15 (%)
Ph	66
4-Br-C_6H_4	55
4-Ph-C_6H_4	61
4-MeO-C_6H_4	72
4-MeS-C_6H_4	58
6-MeO-2-naphthyl	43
4-MeO-1-naphthyl	37
4-EtO-C_6H_4	62
3,4-$(MeO)_2$-C_6H_3	51
3,5-$(MeO)_2$-C_6H_3	57

Scheme 8. Synthesis of 2-aryl-4-ferrocenyl-1,3-selenazoles **15**.

The obtained products **(15)** exhibit high antibacterial activity against *Escherichia coli, Staphylococcus aureus,* and *Pseudomonas aeruginosa* [53].

The reaction of cyanoacetamide with P_2Se_5 afforded malonic selenoamide **16** in 47% yield. The reaction of the product **16** with two equivalents of phenacyl bromide resulted in the formation of (4′-phenyloxazolyl-2′-yl)-4-phenylselenazolyl-2-methane (**17**) in 48% yield, containing the oxazole and selenazole rings (Scheme 9) [50].

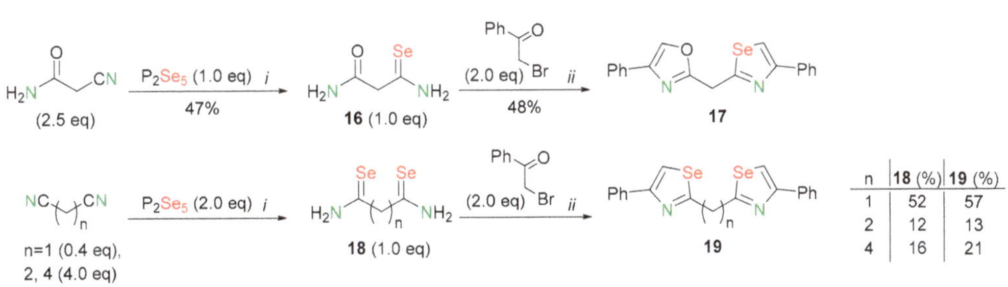

n	18 (%)	19 (%)
1	52	57
2	12	13
4	16	21

(i) **16:** NH_3 (pH 8), EtOH, H_2O, reflux, 2 h; **18** (n = 1,4): pyridine, H_2O, rt, 3 h; **18** (n = 2): EtOH, H_2O, reflux, 2.5 h; *(ii)* EtOH, rt, 20 min.

Scheme 9. Synthesis of functionalized 1,3-selenazoles **17** and **19**.

Moreover, the use of P_2Se_5 allowed researchers to carry out the convenient synthesis of selenocarboxylic diamides, which were converted into bis(selenazol-2-yl)alkanes. The reaction of P_2Se_5 with malonic dinitrile, 1,2-dicyanoethane, and 1,4-dicyanobutane yielded malonic diselenoamide, succinic diselenoamide, and adipic diselenoamide (**18**). The double cyclization of compounds (**18**) with phenacyl bromide in the EtOH/H_2O mixture afforded bis-selenazoles **19** (Scheme 9) [50].

Reactions of 1,3-dicyanobenzene with a potassium 4-methylbenzenecarboselenoate afforded 3-cyanobenzenecarboselenoamide **20** and 1,3-benzenedicarboselenoamide **21** in 49% and 24% yields, respectively.

The reaction was carried out under an argon atmosphere at 0 °C in THF in the presence of boron trifluoride ether complex. After workup, the residue was purified by flash chromatography. Prepared 3-cyanobenzenecarboselenoamide **20** and 1,3-benzenedicarboselenoamide **21** were used as substrates for the preparation of 1,3-selenazoles. Selenazole **22** and diselenazole **23** were obtained in 87% yields by the reaction of selenoamides **20** and **21** with phenacyl bromide in EtOH (Scheme 10) [54].

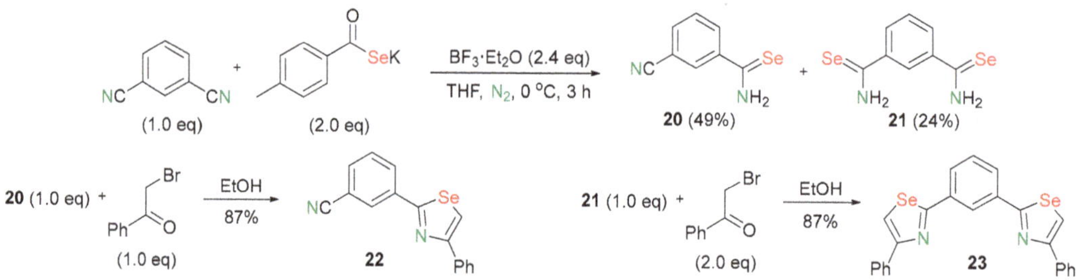

Scheme 10. Synthesis of 2,4-functionalized 1,3-selenazole **22** and bis-1,3-selenazole **23**.

A new cycloaddition reaction of the phenylselanylpropargyl alcohols with selenobenzamide in the presence of catalytic system $Sc(OTf)_3$-$MeNO_2$-H_2O-Bu_4NHSO_4 was developed (Scheme 11) [55]. The use of this effective catalytic system allowed researchers to generate α-selanylpropadienyl cations, which were involved into cycloaddition reactions to form cycloadducts **24**. 4-arylmethyl-2-phenyl-1,3-selenazoles **25** were obtained in 54–99% by the treatment of cycloadducts **24** with tributyltin hydride in the presence of azobisisobutyronitrile (AIBN) and methyllithium (Scheme 11) [55].

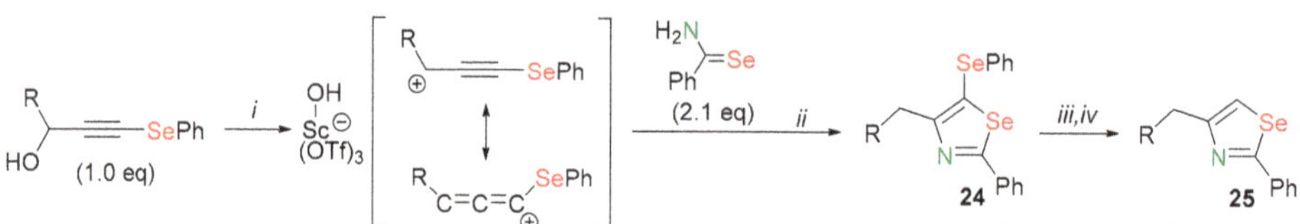

(i) $Sc(OTf)_3$ (5 mol %);
(ii) $MeNO_2/H_2O$ (v/v=10/1), Bu_4NHSO_4 (0.2 eq), reflux, 10 min;
(iii) Bu_3SnH (5.2 eq), AIBN (0.5 eq), toluene, reflux, 40 min;
(iv) MeLi (4.8 eq), THF, −78 °C, 10 min.

R	25 (%)
4-MeO-C_6H_4	99
2-thienyl	54
2,4,6-Me_3-C_6H_2	70
4-F-C_6H_4	78

Scheme 11. Synthesis of 4-arylmethyl-2-phenyl-1,3-selenazoles **25**.

A number of novel 2,4-diaryl-1,3-selenazoles **27** bearing pentafluorosulfanyl SF_5 functional groups were obtained by the two-component cyclization of selenoamide with α-haloketones (Scheme 12) [56]. The initial selenoamide **26** was obtained in an 87% yield from the reaction of Woollins' reagent with 4-pentafluorosulfanylbenzonitrile, followed by hydrolysis. The cyclization of selenoamide **26** with α-haloketones in refluxing ethanol made it possible to obtain a number of 1,3-selenazoles **27** in 89–99% yields (Scheme 12) [56].

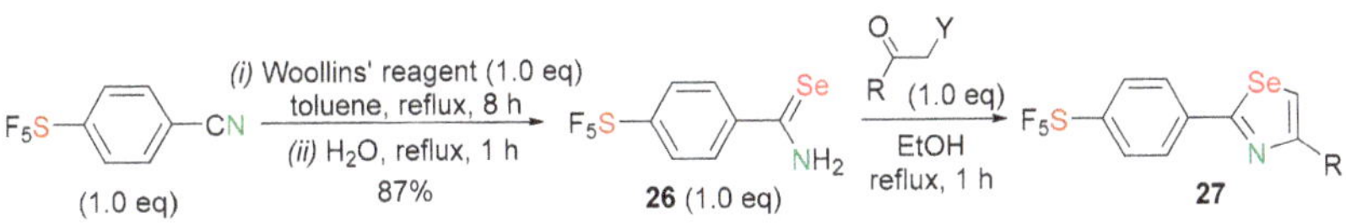

Y	R	27 (%)
Br	4-Cl-C_6H_4	99
Cl	2,4-Cl_2-C_6H_3	97
Br	3,4-Cl_2-C_6H_3	90
Br	4-Me-C_6H_4	94
Br	4-MeO-C_6H_4	95
Br	2,5-$(MeO)_2$-C_6H_3	89
Br	3-NO_2-C_6H_4	98
Br	4-NO_2-C_6H_4	97

Scheme 12. Synthesis of 2,4-diaryl-1,3-selenazoles **27**.

A method for the synthesis of 2,4-substituted 1,3-selenazoles **28** in 20–84% yields by the cyclocondensation of primary selenoamides and alkynyl(phenyl)iodonium salts was developed (Scheme 13) [57]. The synthesis was carried out in the presence of Et_3N in MeOH under a nitrogen atmosphere.

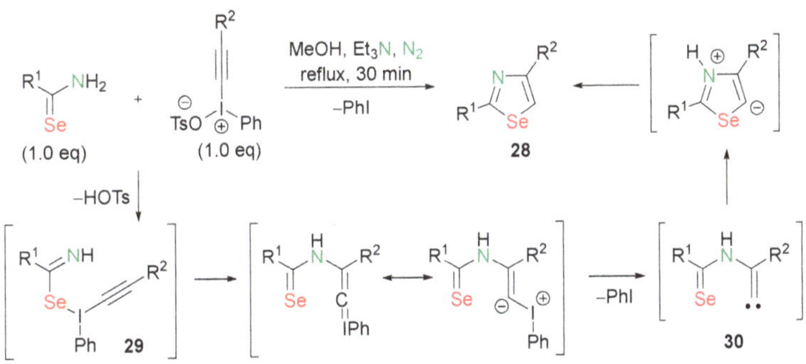

R^1	R^2	28 (%)
Ph	Ph	82
2,4-Cl_2-C_6H_3	Ph	74
4-NMe_2-C_6H_4	Ph	77
4-Cl-C_6H_4	Ph	56
4-Me-C_6H_4	Ph	83
4-MeO-C_6H_4	Ph	84
Ph	$MeOCH_2$	54
4-Cl-C_6H_4	$MeOCH_2$	59
4-Me-C_6H_4	$MeOCH_2$	70
Ph	n-C_6H_{13}	20

Scheme 13. Synthesis of 2,4-substituted 1,3-selenazoles **28** using alkynyl(phenyl)iodonium salts.

A probable mechanism for the formation of selenazoles **28** includes the selenophilic attack of the iodonium atom of alkynyl(phenyl)iodonium salts on the selenoamides with the formation of primary adducts **29**, followed by an uncommon polyhetero-Claisen rearrangement and 1,1-elimination of iodobenzene with the formation of carbene **30**. This synthetic method is simple and mild and provides good yields of the products (**28**) (Scheme 13) [57].

The cyclization of primary arylselenoamides **32** with α-bromoketones **31** in EtOH afforded a variety of 2,4-diaryl-1,3-selenazoles **33** in 21–77% yields (Scheme 14) [58]. The halogenation of the 2,4-diaryl-1,3-selenazoles **33** with SO_2Cl_2, bromine and iodine in diethyl ether yielded 1,1-dihalo-2,4-diaryl-1,3-selenazoles **34**–**36** in good yields (63–82%) while the reaction of compound **33** (R^1 = 4-EtO-C_6H_4, R^2 = 4-Br-C_6H_4) with ethyl iodide yielded selenonium salt **37** in an 81% yield (Scheme 14) [58].

The antiviral activity of 1,1-dihalo-2,4-diaryl-1,3-selenazoles **34** and **36** was investigated against the AIDS virus (HIV-1, HIV-2). It was found that these compounds exhibit antiviral activity against HIV-1 [58].

A practical strategy to access diversified 3-selenazole scaffolds was designed. Aryliodoazides were used as the versatile reagent; these can react with a wide range of selenoamides. The reaction of iodoazides with selenoamides proceeded efficiently using t-BuOK as the base in PEG-200 as a nontoxic solvent to obtain the target products (**38**) in 80–91% yields (Scheme 15). The special characteristics of this protocol are its ease of operation and high functional group tolerance, making it a practical and convenient route for the preparation of various libraries of 1,3-selenazoles (Scheme 15) [59].

R^1	R^2	33 (%)
Ph	4-Ph-C_6H_4	72
4-Br-C_6H_4	4-Br-C_6H_4	63
4-Br-C_6H_4	4-Ph-C_6H_4	77
4-OH-C_6H_4	4-Br-C_6H_4	52
4-MeO-C_6H_4	4-Br-C_6H_4	61
4-MeO-C_6H_4	4-Ph-C_6H_4	30
2-MeO-C_6H_4	4-Br-C_6H_4	69
2-MeO-C_6H_4	4-Ph-C_6H_4	55
4-EtO-C_6H_4	4-Br-C_6H_4	60
2,3-$(MeO)_2$-C_6H_3	4-Br-C_6H_4	48
2,3-$(MeO)_2$-C_6H_3	4-Ph-C_6H_4	73
3,5-$(MeO)_2$-C_6H_3	4-Br-C_6H_4	43
3,4-$(MeO)_2$-C_6H_3	4-Br-C_6H_4	66
3,4-$(MeO)_2$-C_6H_3	4-Ph-C_6H_4	73
4-MeS-C_6H_4	4-Br-C_6H_4	61
4-MeS-C_6H_4	4-Ph-C_6H_4	53
6-MeO-2-naphthyl	4-Ph-C_6H_4	59
6-MeO-2-naphthyl	4-Br-C_6H_4	62
6-MeO-2-naphthyl	Ph	70
4-MeO-1-naphthyl	Ph	40
4-Ph-C_6H_4	4-Br-C_6H_4	21
4-Ph-C_6H_4	4-Ph-C_6H_4	42
4-MeS-C_6H_4	4-MeS-C_6H_4	35
Ph	4-Br-C_6H_4	75

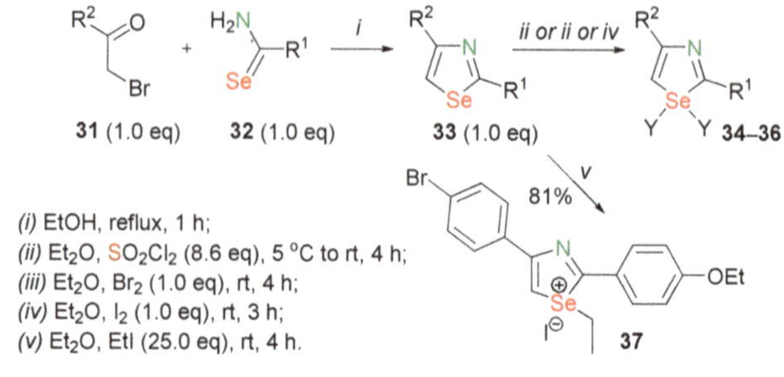

* anti-HIV-1 activity

R^1	R^2	34 (ii) (%) Y = Cl	35 (iii) (%) Y = Br	36 (iv) (%) Y = I
4-Br-C_6H_4	4-Br-C_6H_4	–	78	82
2-MeO-C_6H_4	4-Br-C_6H_4	73 *	69	78
4-EtO-C_6H_4	4-Br-C_6H_4	–	–	68 *
3,4-$(MeO)_2$-C_6H_3	4-Br-C_6H_4	–	75	79
3,4-$(MeO)_2$-C_6H_3	4-Ph-C_6H_4	70 *	74	67
4-MeS-C_6H_4	Ph	75	63	77
4-MeS-C_6H_4	4-Ph-C_6H_4	–	71	77 *
6-MeO-2-naphthyl	4-Br-C_6H_4	75	–	78

Scheme 14. Synthesis of 1,1-dihalo-2,4-diaryl-1,3-selenazoles **34–36** and selenonium salt **37**.

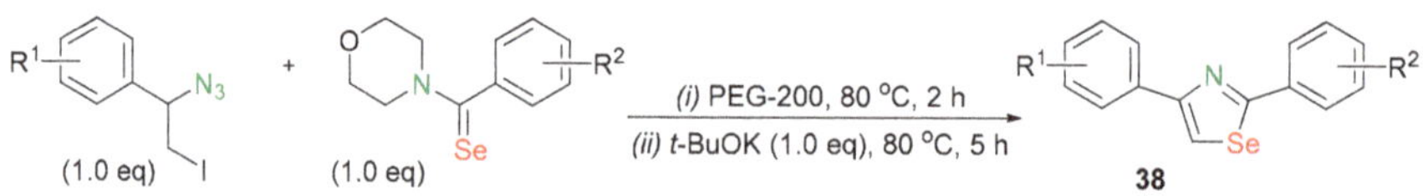

R^1	R^2	38 (%)
H	4-Me	89
H	4-MeO	91
H	2-Cl	81
H	3-Cl	82
H	3-Br	84
H	4-Br	85

R^1	R^2	38 (%)
4-MeO	4-Me	91
4-MeO	4-MeO	94
4-MeO	3-Cl	87
H	4- 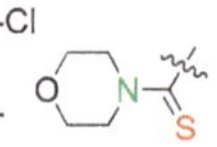	80

R^1	R^2	38 (%)
4-Cl	H	83
4-Cl	4-Me	84
4-Cl	4-MeO	90
4-Cl	2-Cl	80
4-Cl	3-Cl	81
4-Cl	3-Br	83
4-Cl	4-Br	85

Scheme 15. Synthesis of 2,4-aryl-substituted 1,3-selenazole derivatives **38**.

A number of 1,3,4-thiadiazole derivatives containing 1,3-selenazole with fragments of Schiff base were synthesized (Scheme 16) [60]. As a result of the interaction of benzonitrile with NaHSe, selenobenzamide was obtained, and this reacted with 1,3-dichloroacetone in acetone at room temperature to form compound **39**. Product **40** was obtained by the reaction of 1,3-selenazole **39** and 2-amino-5-mercapto-1,3,4-thiadiazole using PEG-400 as a catalyst under reflux. Finally, the condensation of compound **40** and substituted aromatic aldehydes in the presence of trimethyl orthoformate using benzyl triethylammonium chloride (BTEAC) as a catalyst made it possible to obtain the target compounds **41** under reflux. As a result of this reaction, a series of 1,3,4-thiadiazole derivatives containing 1,3-selenazole **41** with Schiff base fragments were prepared in 38–72% yields. These derivatives have an increased number of active groups in the molecule (Scheme 16) [60].

The majority of the obtained compounds exhibited high antiproliferative activity against mouse lymphocyte leukemia cells (L1210) and human breast cancer cells (MCF-7). In particular, compound **41** (R = 4-Cl-C_6H_4) demonstrated the most potent antiproliferative activity against MCF-7 cells compared to 2-phenyl-4-carboxyl-1,3-selenazole (PCS), which was used as a control compound [60].

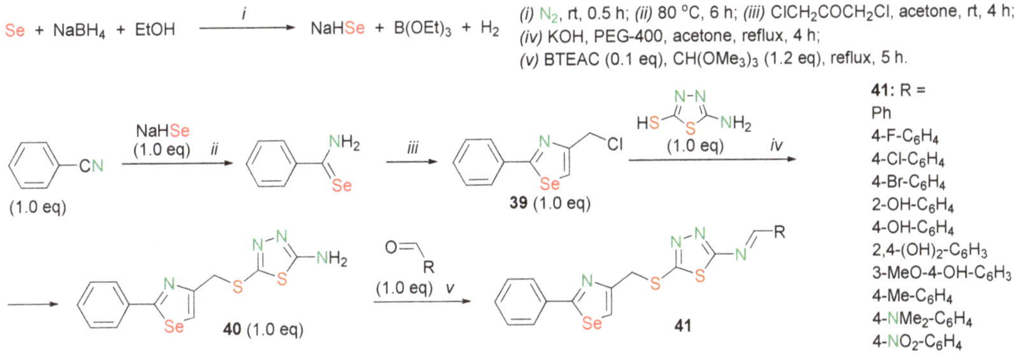

Scheme 16. Synthesis of 1,3-selenazole-containing 1,3,4-thiadiazole derivatives **41**.

The convenient synthesis of a chromenone derivative, bearing the 1,3-selenazole ring, was developed (Scheme 17) [61]. The condensation of salicylaldehyde with cyanoselenoacetamide under Knoevenagel reaction conditions led to the formation of intermediate **42**. Subsequent pyran ring closure proceeded to yield 2-oxo-2*H*-chromeno-3-carboselenamide **43** in a 75% yield. The product **43** entered easily into the Hantzsch-type condensation with phenacyl bromide in DMF at room temperature to form 3-selenazolyl-substituted 2-iminocoumarin **44**, which was rapidly hydrolyzed under the reaction conditions to form 3-(4-phenyl-1,3-selenazol-2-yl)-2*H*-chromen-2-one **45** in a 72% yield. Compound **45** was also obtained in a 75% yield by the condensation of salicylaldehyde with (4-phenyl-1,3-selenazol-2-yl)acetonitrile (Scheme 17) [61].

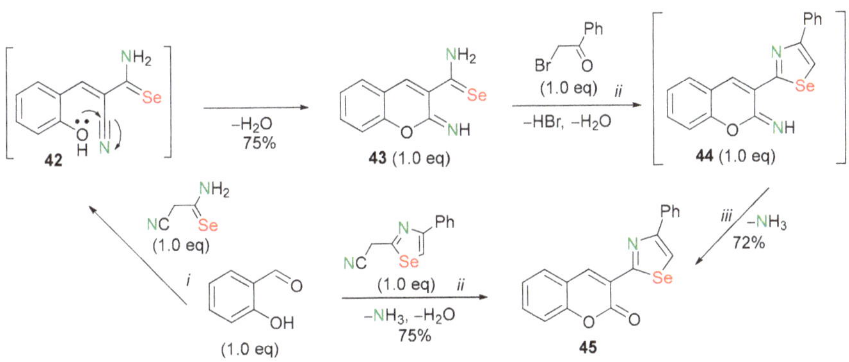

(i) (a) *N*-methylmorpholine (cat), EtOH, Ar, rt, 1 h; (b) rt, 24 h; *(ii)* DMF, Ar, rt, 3 h; *(iii)* H_2O/H^+ (1.0 eq), 48 h.

Scheme 17. Synthesis of 3-(4-phenyl-1,3-selenazol-2-yl)-2*H*-chromen-2-one **45**.

2,4-Subtituted 1,3-selenazoles were synthesized using new reagents, 3-amino-*N*-cyclopropyl-3-selanylidene-propanamide and methyl 3-amino-3-selanylidenepropanoate **47** (Scheme 18) [62]. The derivatives of 2-(carbamoselenoyl)acetic acid **47** were obtained by the reaction of 2-cyano-*N*-cyclopropylacetamide and methyl 2-cyanoacetate with H_2Se in diethyl ether at 0 °C under argon in the presence of triethylamine. Compounds **(47)** were converted to functionally substituted 1,3-selenazoles **50** by reaction with α-bromoketones in DMF at room temperature under argon (Scheme 18) [62].

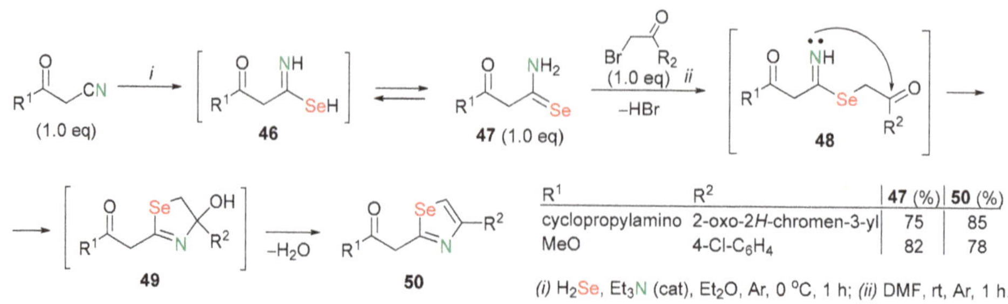

R^1	R^2	47 (%)	50 (%)
cyclopropylamino	2-oxo-2*H*-chromen-3-yl	75	85
MeO	4-Cl-C_6H_4	82	78

(*i*) H_2Se, Et_3N (cat), Et_2O, Ar, 0 °C, 1 h; (*ii*) DMF, rt, Ar, 1 h.

Scheme 18. Synthesis of functionally substituted 1,3-selenazoles **50**.

A plausible reaction pathway is proposed [62]. At the initial stage, the hydroselenide ion is generated from H_2Se and adds to the cyano group of starting cyanides to yield intermediate **46**, which is stabilized as selenoamides **47**. The reaction of selenoamides **47** with α-bromoketones leads to selenide **48**, which undergoes intramolecular cyclization to substituted dihydroselenazole **49**. The latter undergoes dehydration to afford final aromatic 1,3-selenazoles **50** (Scheme 18) [62].

A metal-free method was efficiently developed for the synthesis of sulfoximine tethered selenazoles. *N*-TBDPS-protected sulfoximine **51** was treated with *n*-BuLi followed by the addition of 2-bromo-*N*-methoxy-*N*-methylacetamide (Scheme 19) [63]. A key compound, α-bromoalkanone **52**, was obtained in an 84% yield. The target functionalized selenazoles **53** were synthesized in good yields (72–88%) by the reacting compound **52** and the selenoamides in 2-propanol at an ambient temperature (Scheme 19) [63]. The cyclization and the TBDPS deprotection processes occurred as a one-pot procedure. The developed metal-free approach to 2,4-substituted 1,3-selenazoles **53** can be considered a very useful contribution to the growing field of the synthesis and studies of sulfoximine compounds in medicinal chemistry.

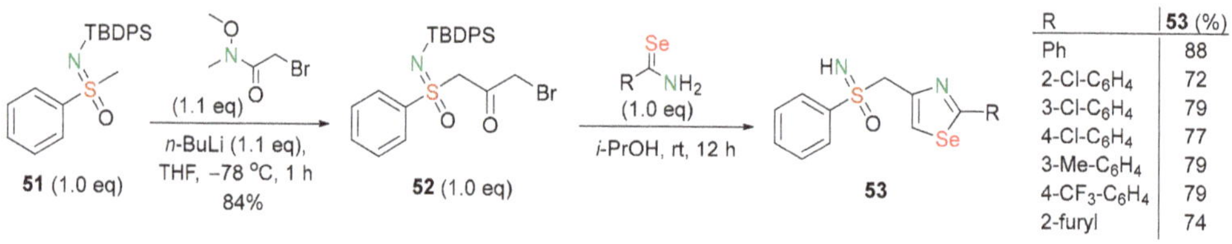

R	53 (%)
Ph	88
2-Cl-C_6H_4	72
3-Cl-C_6H_4	79
4-Cl-C_6H_4	77
3-Me-C_6H_4	79
4-CF_3-C_6H_4	79
2-furyl	74

Scheme 19. Synthesis of sulfoximine tethered 2,4-substituted 1,3-selenazoles **53**.

A novel method for the one-pot cascade synthesis of 2,4-disubstituted 1,3-selenazoles was developed (Scheme 20) [64]. Triethylamine, pentafluorophenyl diphenylphosphate (FDP), β-azido diselenide **56**, and triphenylphosphine were added to a solution of acid or anhydride in acetonitrile and the reaction mixture was stirred at 70 °C. This approach allowed obtaining selenazolines **57** in 54–77% yields. In turn, selenazolines **57** were oxidized by activated manganese oxide in 1,2-dichloroethane to yield selenazoles **58** in 89–95% yields (Scheme 20) [64]. Azido diselenide **56** was obtained in a 44% yield from commercially accessible *N*-Boc-*L*-serine methyl ester (**54**) in several stages with the formation of the intermediate *N*-Boc-protected homoselenocystine **55**. The valuable application of this method could be seen in the efficient total synthesis of selenazofurin, which showed high antitumor activities (Scheme 20) [64].

Sodium fluoride has been found to be a mild, simple, and efficient catalyst for the synthesis of 2,4-disubstituted 1,3-selenazoles using phenacyl bromides and 3-(2-bromoacetyl)-2*H*-chromen-2-one with selenourea in aqueous MeOH at an ambient temperature (Scheme 21) [65]. Presumably, the fluoride ion acts as a base in this reaction.

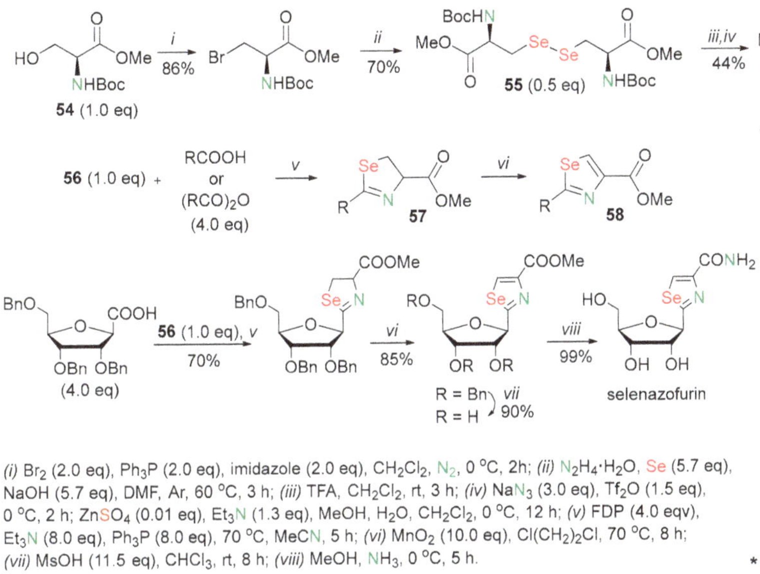

substrate	57 (%)	58 (%)
2-I-C_6H_4-COOH	60	89
4-Cl-C_6H_4-COOH	65	92
3-Br-C_6H_4-COOH	75	90
2-AcO-C_6H_4-COOH	65	93
3-Me-C_6H_4-COOH	71	92
n-BuCOOH	60	95
n-C_5H_{11}COOH	71	93
n-C_7H_{15}COOH	65*	90
(*n*-PrCO)$_2$O	54*	95
(*n*-C_5H_{11}CO)$_2$O	66*	93
(PhCO)$_2$O	76	94
Ph, COOH, *t*-Bu-Si-O, Ph	77	90
Ph, COOH, NHBoc	62	91

* coupling reagent FDP was not added

(*i*) Br_2 (2.0 eq), Ph_3P (2.0 eq), imidazole (2.0 eq), CH_2Cl_2, N_2, 0 °C, 2h; (*ii*) $N_2H_4{\cdot}H_2O$, Se (5.7 eq), NaOH (5.7 eq), DMF, Ar, 60 °C, 3 h; (*iii*) TFA, CH_2Cl_2, rt, 3 h; (*iv*) NaN_3 (3.0 eq), Tf_2O (1.5 eq), 0 °C, 2 h; $ZnSO_4$ (0.01 eq), Et_3N (1.3 eq), MeOH, H_2O, CH_2Cl_2, 0 °C, 12 h; (*v*) FDP (4.0 eqv), Et_3N (8.0 eq), Ph_3P (8.0 eq), 70 °C, MeCN, 5 h; (*vi*) MnO_2 (10.0 eq), $Cl(CH_2)_2Cl$, 70 °C, 8 h; (*vii*) MsOH (11.5 eq), $CHCl_3$, rt, 8 h; (*viii*) MeOH, NH_3, 0 °C, 5 h.

Scheme 20. Synthesis of 2,4-disubstituted 1,3-selenazoles **58** and selenazofurin.

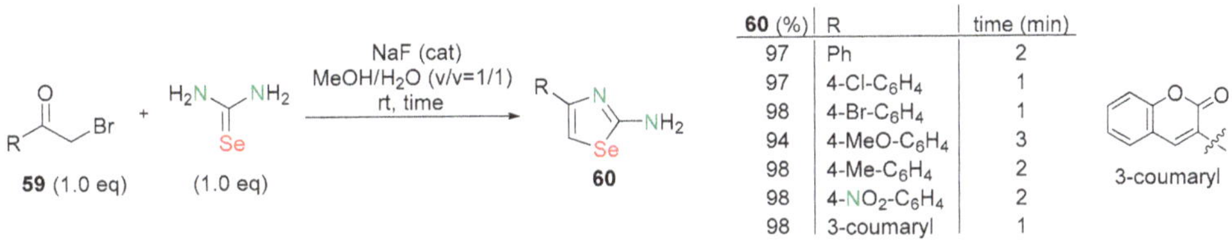

60 (%)	R	time (min)
97	Ph	2
97	4-Cl-C_6H_4	1
98	4-Br-C_6H_4	1
94	4-MeO-C_6H_4	3
98	4-Me-C_6H_4	2
98	4-NO_2-C_6H_4	2
98	3-coumaryl	1

Scheme 21. Synthesis of substituted 1,3-selenazole derivatives **60** catalyzed by NaF.

Sodium fluoride was added to the solution of the corresponding phenacylbromide or 3-(2-bromoacetyl)-2*H*-chromen-2-one **59** and selenourea in MeOH. After stirring the mixture at an ambient temperature for 1-3 min, water was added and analytically pure substituted 1,3-selenazole derivatives **60** were separated in excellent yields as solids (Scheme 21) [65].

A rapid and efficient procedure for the synthesis of 2,4-disubstituted-1,3-selenazoles **61,62** was developed based on the reaction of α-bromoketones with selenourea at room temperature in aqueous medium under ultrasonic irradiation. Analytically pure products **61,62** were formed in 91–99% yields for 10–60 s. The advantages of this methodology include its operational simplicity, ability to save time, use of mild reaction conditions, and lack of by-products (Scheme 22) [66].

2-amino-4-substituted-1,3-selenazoles **63** were synthesized in 86–95% yields from α-bromoketones and selenourea in the presence of β-cyclodextrin in water at 50 °C under atmospheric pressure. This methodology also avoids an inert atmosphere and high temperatures, making it more convenient to use (Scheme 23) [67].

Bromo acetyl benzene or 3-bromo acetyl coumarin derivatives and selenourea in the presence of $CuPy_2Cl_2$ reacted very efficiently in a mortar without any solvent. $CuPy_2Cl_2$ was added to a mixture of α-bromoketone and selenourea in a mortar and ground with a pestle at an ambient temperature. The reactions were completed within 30 min of milling, and the required products **64,65** were obtained in high yields (87–96%). The reaction under solid-phase conditions was environmentally safe, completed in higher yields, and had a more convenient workup (Scheme 24) [68].

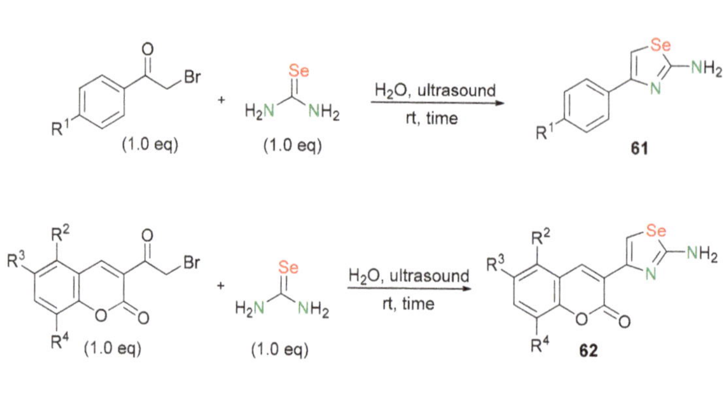

R^1	time (sec)	**61** (%)
H	10	97
F	15	95
Cl	20	97
Br	25	97
Me	15	92
MeO	10	97
NO_2	25	91
Ph	20	94

R^2	R^3	R^4	time (sec)	**62** (%)
H	H	H	40	96
H	Cl	H	50	99
H	Br	H	50	94
H	Br	Br	60	94
H	MeO	H	40	97
-benzo-		H	55	93

Scheme 22. Synthesis of 2,4-disubstituted-1,3-selenazoles **61,62** under ultrasonic irradiation.

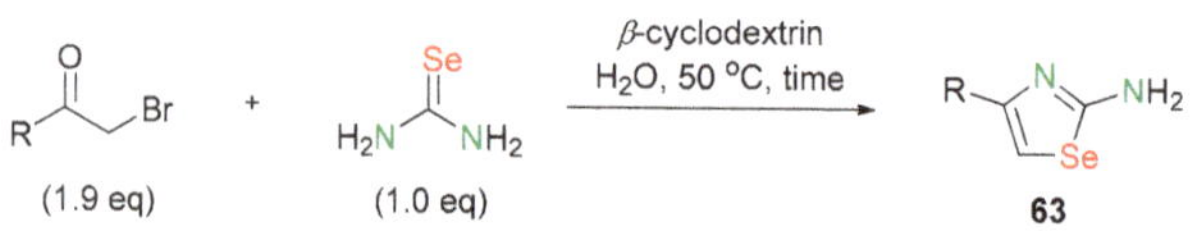

R	time (min)	**63** (%)
Ph	40	92
4-Me-C_6H_4	45	94
4-MeO-C_6H_4	40	95
4-Cl-C_6H_4	50	93
4-Br-C_6H_4	50	92
4-I-C_6H_4	50	90
4-NO_2-C_6H_4	35	94
t-Bu	60	86

Scheme 23. Synthesis of 2-amino-4-substituted-1,3-selenazoles **63**.

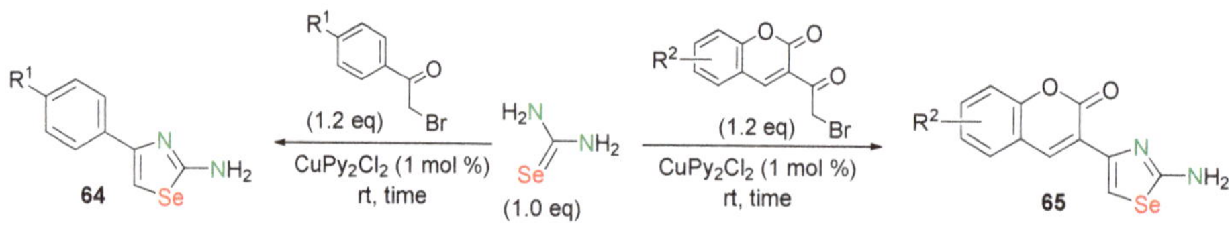

R^1	time (min)	**64** (%)
H	20	95
Me	18	95
Cl	20	92
Br	18	94
I	20	92
MeO	20	95
NO_2	20	96

R^2	time (min)	**65** (%)
H	25	96
8-Br	20	95
6,8-Cl_2	25	92
6,8-Br_2	23	96
8-Me	25	90
5,6-benzo	30	87

Scheme 24. Synthesis of 2-amino-1,3-selenazoles **64,65** in the presence of $CuPy_2Cl_2$.

A series of 1,3-selenazoles were synthesized via Ir-catalyzed sulfoxonium ylide insertion chemistry (Scheme 25) [69]. Sulfoxonium ylides **66** were synthesized as precursors to 1,3-selenazoles from trimethylsulfoxonium chloride and acid chloride substrate in THF using *t*-BuOK in 55–93% yields. Then, a mixture of *β*-ketosulfoxonium ylide **66** and selenourea in dichloroethane was heated to 80 °C and $[Ir(cod)Cl]_2$ was added to the mixture. The resulting mixture was stirred at 80 °C. After workup, 4-aryl-substituted 2-amino-1,3-selenazoles **67** were isolated in 30–48% yields. This was a useful, substrate-tolerant approach to 1,3-selenazoles that may find application in medicinal chemistry (Scheme 25) [69].

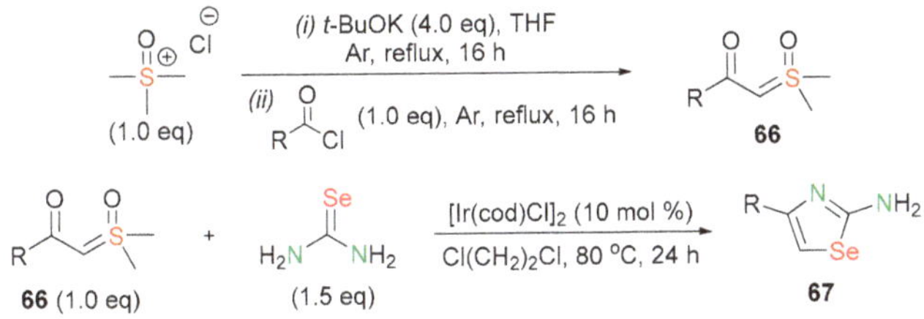

R	66 (%)	67 (%)
4-Me-C_6H_4	93	45
4-Cl-C_6H_4	73	34
4-Br-C_6H_4	73	41
4-NO_2-C_6H_4	55	30
4-MeO-C_6H_4	74	40
3-MeO-C_6H_4	70	48
2-thienyl	69	42

Scheme 25. Synthesis of 4-aryl-substituted 2-amino-1,3-selenazoles **67**.

An efficient and simple one-pot synthesis of Fmoc/Boc/Z-amino acid derived from 2-amino-1,3-selenazoles **69** was developed (Scheme 26) [70]. At first, *N*-Fmoc/Boc/Z-amino acid was transformed into the corresponding diazomethyl ketone. The conversion of the latter compound into bromomethyl ketone **68** was carried out in the presence of HBr within 10 min in a near-quantitative yield. Then, the 2-amino-1,3-selenazole analogues of Fmoc/Boc/Z-amino acids **69** were synthesized in 80-90% yields upon treating the *N*-protected bromomethyl ketones **68** with selenourea in acetone under ultrasonication (Scheme 26) [70].

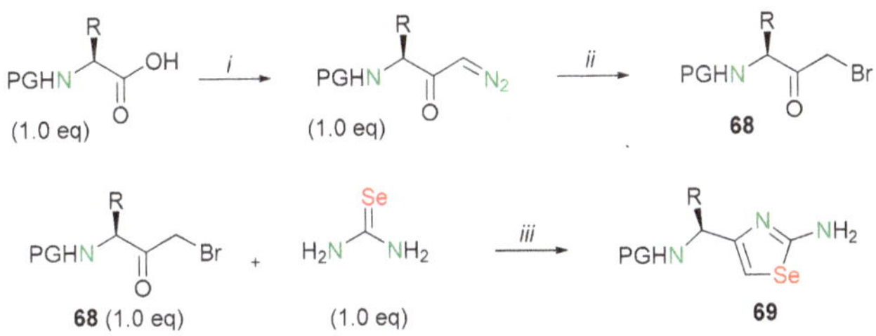

(i) Et_3N (1.0 eq), ECF (ethyl chloroformate) (1.0 eq), diazomethane, THF, Et_2O, −15 °C to 0 °C, 45 min; *(ii)* 45% HBr aq, rt, 5 min; *(iii)* acetone, ultrasound, rt, 10 min.

PG	R	68 (%)	69 (%)
Fmoc	Me	91	89
Fmoc	Bn	93	90
Fmoc	*i*-Pr	94	88
Fmoc	*i*-Bu	90	85
Fmoc	$BnOCH_2$	85	82
Fmoc	(*Z*)$NH(CH_2)_4$	86	83
Cbz	H	90	80
Cbz	Me	74	87
Cbz	Bn	87	84
Boc	*i*-Pr	89	81
Boc	(*Z*)$NH(CH_2)_4$	86	82

PG - protecting group

Scheme 26. Ultrasound mediated synthesis of 2-amino-1,3-selenazoles **69** obtained from Fmoc/Boc/Z-α-amino acids.

A simple and environmentally tolerant solvent- and catalyst-free protocol for the synthesis of 2-amino-1,3-selenazoles **70** through Hantzsch-type condensation was developed (Scheme 27) [71]. Powdered selenourea was added to 2-bromoacetophenone heated to the melting point. The reactions were completed within a few seconds, and the desired 4-aryl-1,3-selenazol-2-amine hydrobromides **70** were obtained in 42–93% yields (Scheme 27) [71]. This reaction procedure is an example of a promising approach that can be used in green chemistry.

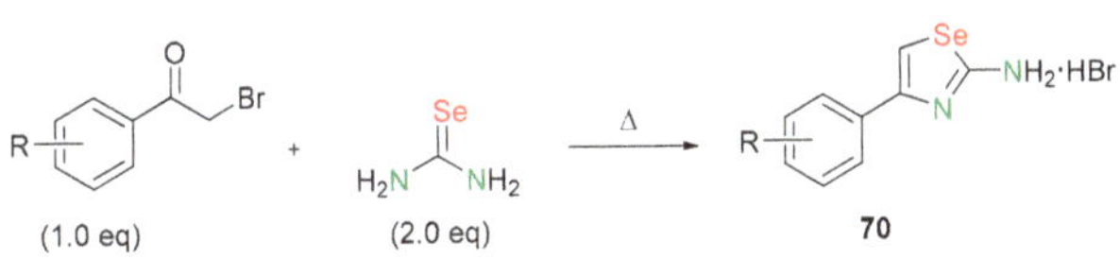

R	70 (%)
H	93
4-Cl	60
4-Br	42
4-F	76
4-Me	80
4-MeO	78
4-NO_2	93
4-CN	58

Scheme 27. Solvent-free synthesis of 4-aryl-1,3-selenazol-2-amine hydrobromides **70**.

A series of functionalized 2-arylamino-1,3-selenazoles **72** were synthesized (Scheme 28) [72]. The starting compounds, *N*-aryl-*N*′-benzoylselenoureas, were treated with an excess of hydrazine hydrate under solvent-free conditions. Upon the addition of cold water, the compounds **71** precipitated and were easily isolated (22–59% yields). The target 2-arylamino-1,3-selenazoles **72** were obtained in 32–89% yields from the cyclocondensation reaction of arylselenoureas **71** with 2-haloketones in the presence of Et_3N (Scheme 28) [72].

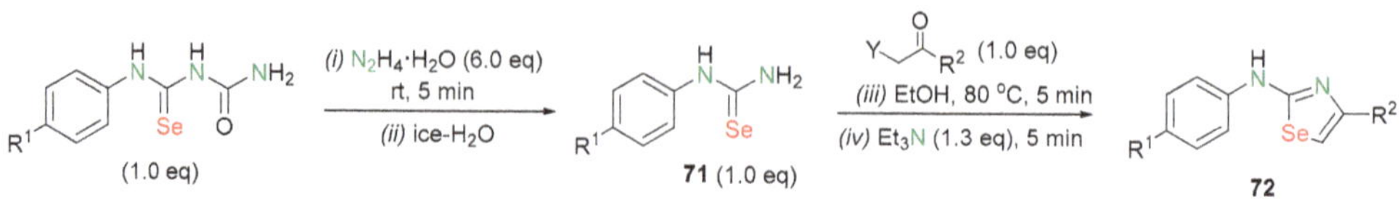

R¹	71 (%)
H	59
4-Me	54
4-MeO	22

Y	R¹	R²	72 (%)
Br	H	Ph	89
Br	H	4-Me-C_6H_4	69
Br	H	4-MeO-C_6H_4	71
Br	H	4-Ph-C_6H_4	74
Br	H	2-naphthyl	76
Br	H	*t*-Bu	52
Cl	H	Me	56
Cl	H	H	32

Y	R¹	R²	72 (%)
Br	Me	Ph	40
Br	Me	4-Me-C_6H_4	58
Br	Me	4-MeO-C_6H_4	48
Br	Me	4-Ph-C_6H_4	58
Br	Me	2-naphthyl	61
Br	Me	*t*-Bu	84
Cl	Me	Me	81
Cl	Me	H	60

Y	R¹	R²	72 (%)
Br	MeO	Ph	58
Br	MeO	4-Me-C_6H_4	58
Br	MeO	4-MeO-C_6H_4	65
Br	MeO	4-Ph-C_6H_4	68
Br	MeO	2-naphthyl	68
Br	MeO	*t*-Bu	35
Cl	MeO	Me	79
Cl	MeO	H	34

Scheme 28. Synthesis of functionalized 2-arylamino-1,3-selenazoles **72**.

The promising compounds (**72**, R^1 = H, R^2 = 4-MeO-C_6H_4; R^1 = H, R^2 = Me) were found, and these showed relatively low toxicity and high activity against the fungal strains *C. neoformans* and *C. albicans* [72].

2-aminoselenazoles **73** were obtained by two alternative methods. On the one hand, the reaction of P_2Se_5 with cyanamide yielded selenourea. The cyclization of selenourea with α-bromoketones afforded the expected primary 2-aminoselenazoles **73** in a 84–94% yield (Scheme 29) [50].

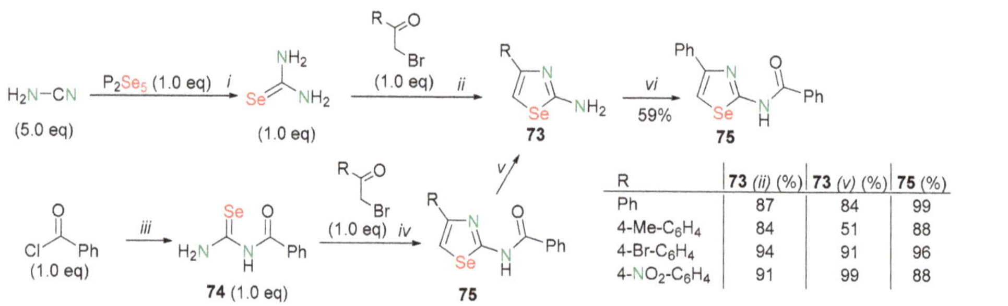

R	73 *(ii)* (%)	73 *(v)* (%)	75 (%)
Ph	87	84	99
4-Me-C_6H_4	84	51	88
4-Br-C_6H_4	94	91	96
4-NO_2-C_6H_4	91	99	88

(i) (a) H_2O, rt, 3h; (b) reflux, 1h; *(ii)* (a) EtOH, briefly reflux; (b) NH_3 aq, rt; *(iii)* KSeCN (1.0 eq), acetone, NH_3, rt; *(iv)* (a) acetone, reflux, 20 min; (b) NH_3 aq, rt; *(v)* (a) 68% H_2SO_4 aq, reflux, 30 min; (b) H_2O, 10% NaOH aq; *(vi)* PhCOCl (1.0 eq), pyridine, reflux, 15 min.

Scheme 29. Synthesis of 2-aminoselenazoles **73** and the protected 2-aminoselenazoles **75**.

However, another approach did not require the use of selenourea for the synthesis of 2-aminoselenazoles **73**. The stable benzoylselenourea (**74**) was obtained by the reaction of an acetone solution of potassium selenocyanate with benzoyl chloride and subsequent addition of ammonia. The cyclization of compound **74** with α-bromoketones afforded the protected 2-aminoselenazoles **75** (88–99% yields), which were deprotected by the treatment with sulfuric acid, providing products (**73**) in good yields (51–99%) (Scheme 29) [50].

In addition, 2-amino-5-phenyl-1,3-selenazole **73** (R = Ph) was transformed into 2-benzoylamino-4-phenyl-selenazole **75** (59% yield) by refluxing with benzoyl chloride in pyridine solution (Scheme 29) [49].

Fourteen biologically active 2,4-disubstituted 1,3-selenazoles **78** were synthesized (Scheme 30) [73]. At the first stage, selenosemicarbazones **76** and **77** were synthesized by the reaction of commercially available cyclohexanone and 2-methylcyclohexanone with KSeCN and hydrazine hydrate in the presence of HCl. The products **76** and **77** were obtained in 61% and 51% yields, respectively. Next, 1,3-selenazoles **78** were prepared in 44–70% yields and with high chemical and isomeric purity (100% *E*-isomer) by the Hantzsch-type cyclization reaction of cyclohexanone selenosemicarbazone **76** and racemic 2-methylcyclohexanone selenosemicarbazone **77** with appropriate para-substituted bromoacetophenone and 1-adamantyl bromomethyl ketone (Scheme 30) [73].

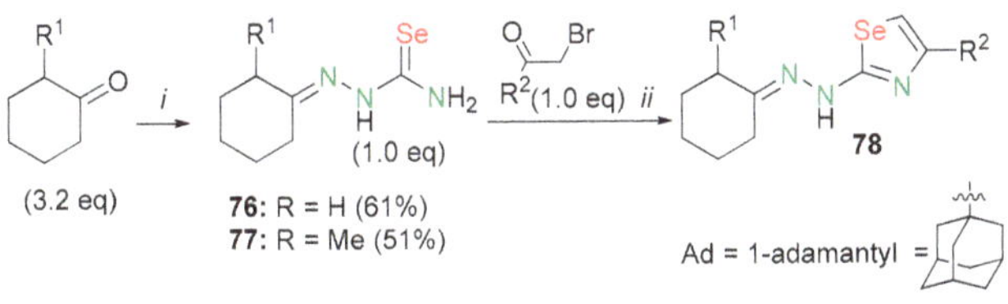

R^1	R^2	**78** (%)	R^1	R^2	**78** (%)
H	4-Br-C_6H_4	58	Me	4-Br-C_6H_4	67
H	4-F-C_6H_4	56	Me	4-F-C_6H_4	63
H	4-NO_2-C_6H_4	65	Me	4-NO_2-C_6H_4	70
H	4-CN-C_6H_4	62	Me	4-CN-C_6H_4	64
H	4-MeO-C_6H_4	47	Me	4-MeO-C_6H_4	53
H	3,4-Cl_2-C_6H_3	63	Me	3,4-Cl_2-C_6H_3	57
H	Ad	44	Me	Ad	66

(i) KSeCN (1.0 eq), 80% $N_2H_4 \cdot H_2O$ (4.9 eq), HCl conc, EtOH, H_2O, reflux, 2 h; *(ii)* EtOH, rt, 20 h.

Scheme 30. Synthesis of 2,4-disubstituted 1,3-selenazoles **78** with antimicrobial activity.

According to research results, the products with F, NO_2, CN, and MeO substituents and the cyclohexanone and 2-methylcyclohexanone derivatives exhibited antifungal activity. Compounds with 3,4-dichlorophenyl and adamantanyl substituents also showed antibacterial activity [73].

Nine 2,4-substituted 1,3-selenazoles based on dihydro-2*H*-thiopyran-4(3*H*)-one (**79**) were prepared. At first, 2-(tetrahydro-4*H*-thiopyran-4-ylidene)hydrazinecarboselenoamide (**80**) was prepared from compound **79** using KSeCN and hydrazine hydrate. The Hantzsch-type condensation of appropriate para-substituted bromoacetophenones with selenosemicarbazone **80** afforded the selenazoles **81** in good yields (67–96%) and with high chemical purity (Scheme 31) [74].

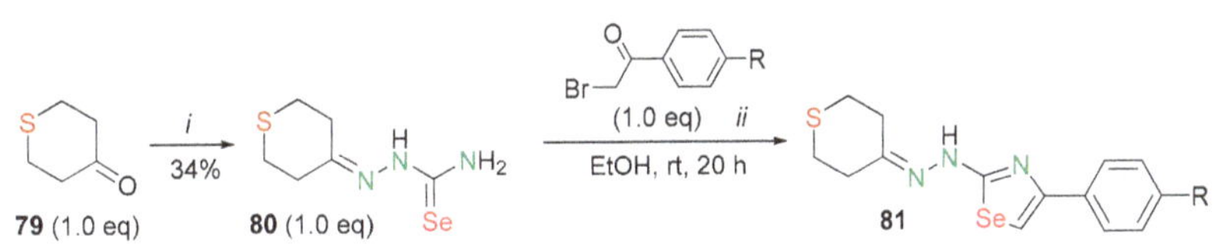

R	**81** (%)
F	94
Cl	95
Br	93
I	96
Me	91
MeO	80
CF_3	67
NO_2	95
CN	96

(i) KSeCN (1.0 eq), 80% $N_2H_4 \cdot H_2O$ (1.0 eq), HCl conc (1.0 eq), EtOH, H_2O, N_2, reflux, 2 h; *(ii)* EtOH, rt, 20 h.

Scheme 31. Synthesis of 2,4-disubstituted 1,3-selenazoles **81** based on dihydro-2*H*-thiopyran-4(3*H*)-one (**79**).

The synthesized compounds exhibited high antifungal activity against *Candida* spp and antibacterial activity against Gram-positive bacteria. In addition, some compounds showed significant anticonvulsant activity [74].

The synthesis of 1,3-selenazoles **83** with antimicrobial activities was developed (Scheme 32) [75]. At the first stage, the cyclohexanecarbaldehyde selenosemicarbazone **82** was synthesized from cyclohexanecarbaldehyde.

Then, the Hantzsch-type cyclization reaction of selenosemicarbazone **82** with appropriate para-substituted bromoacetophenones or ethyl 4-chloroacetoacetate under a nitrogen atmosphere afforded 2,4-substituted 1,3-selenazoles **83** in 18–52% yields (Scheme 32) [75].

Compounds (**83**) (except R = $EtOOCCH_2$) were found to be active against Gram-positive bacteria, including *Staphylococcus aureus* and pathogenic staphylococci [75].

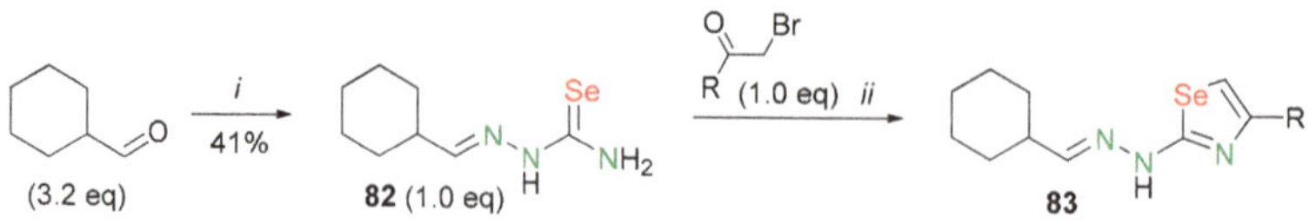

(i) KSeCN (1.0 eq), 80% $N_2H_4 \cdot H_2O$ (4.9 eq), HCl conc, EtOH, H_2O, N_2, reflux, 2.5 h; (ii) EtOH, rt, 20 h.

R	**83** (%)
4-Cl-C_6H_4	49
4-Br-C_6H_4	52
4-F-C_6H_4	41
4-Me-C_6H_4	52
4-NO_2-C_6H_4	42
4-CN-C_6H_4	42
$EtOOCCH_2$	18

Scheme 32. Synthesis of 2,4-disubstituted 1,3-selenazoles **83** based on cyclohexanecarbaldehyde.

Selenazolyl-hydrazones **84** were prepared by the reaction of selenosemicarbazones and α-bromocarbonyl derivatives (Scheme 33) [76]. Selenosemicarbazones were dissolved in a mixture of H_2O and EtOH (v/v=1/1) and α-bromocarbonyl derivatives were added. The reaction mixtures were stirred and refluxed. After the completion of the reactions, benzylidene-based (1,3-selenazol-2-yl)hydrazones **84** were prepared in 51–91% yields (Scheme 33) [76].

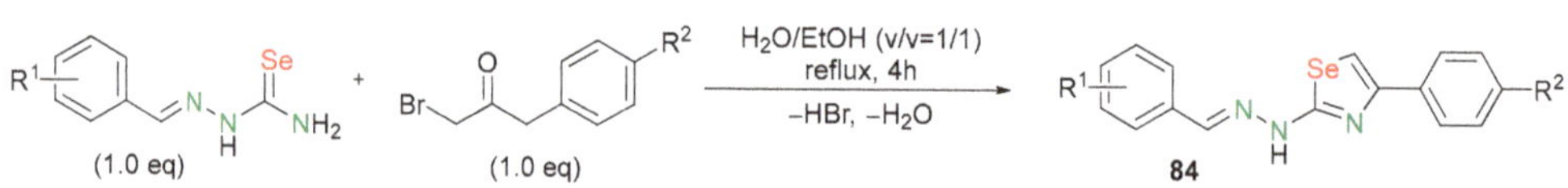

R^1	R^2	**84** (%)
H	H	77
2-NO_2	H	56
3-NO_2	H	73
4-NO_2	H	69

R^1	R^2	**84** (%)
H	MeO	91
2-NO_2	MeO	66
3-NO_2	MeO	70
4-NO_2	MeO	74

R^1	R^2	**84** (%)
H	Me	51
2-NO_2	Me	70
3-NO_2	Me	69
4-NO_2	Me	72

Scheme 33. Synthesis of benzylidene-based (1,3-selenazol-2-yl)hydrazones **84**.

The obtained selenazolyl-hydrazones **84** are selective monoamine oxidase inhibitors with antioxidant and antiproliferative activities [76].

Functionalized 1,3-selenazoles **86,** with an 4-oxo-4*H*-chromen-3-yl fragment, were constructed by the Hantzsch-type condensation reaction. This method is based on the cyclization of *N*-[(4-oxo-4*H*-chromen-3-yl)methylene]selenosemicarbazide (**85**) with bis-halogen and α-halocarbonyl compounds in DMF, containing a few drops of pyridine. Initial selenosemicarbazide **85** was prepared by the condensation of selenosemicarbazide and 4-oxo-4*H*-chromene-3-carboxaldehyde in absolute EtOH containing glacial acetic acid (Scheme 34) [77].

Compounds (**86**) (R = Ph, $ClCH_2$) demonstrated significant cytotoxic activity against the Hep-G2, HCT-116, MCF-7, and A549 cancer cell lines [77].

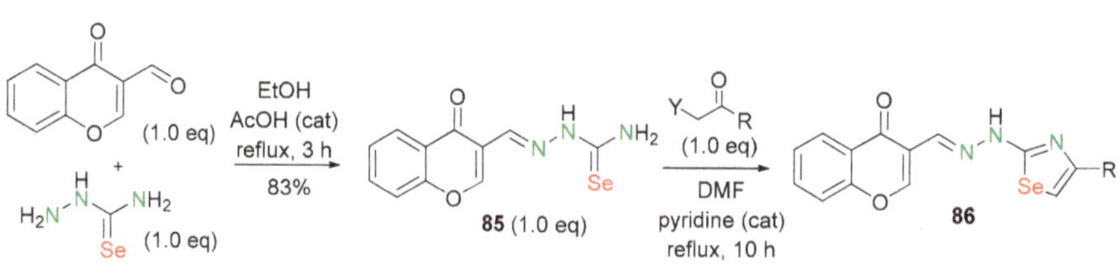

Y	R	**86** (%)
Cl	Ph	69
Br	Me	54
Cl	$ClCH_2$	65

Scheme 34. Synthesis of functionalized 1,3-selenazoles **86** with 4-oxo-4*H*-chromen-3-yl fragment.

A series of hydrazine functionalized 1,3-selenazoles was synthesized by the Hantzsch-type condensation reaction (Scheme 35) [78]. The synthesis of the 1,3-selenazoles used two precursors, selenosemicarbazones **87** and aroyl-selenosemicarbazide derivatives **90**, prepared by the condensation of selenosemicarbazide with aromatic carbaldehydes and acylation of selenosemicarbazide, respectively.

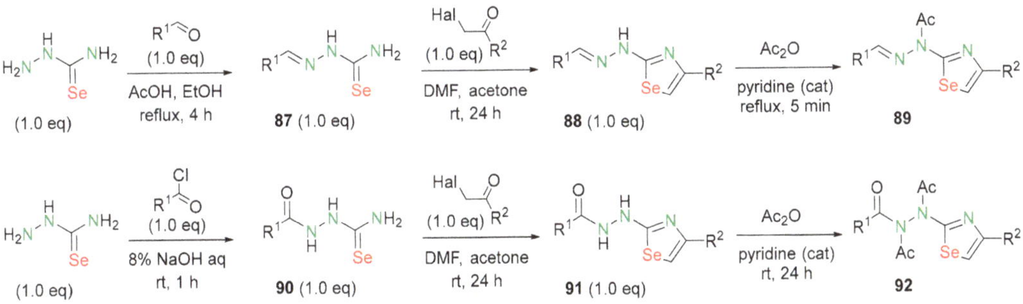

R^1	R^2	87 (%)	88 (%)	89 (%)
4-Cl-C_6H_4	Me	61	52	53
4-Cl-C_6H_4	$ClCH_2$	–	51	52
4-Cl-C_6H_4	Ph	–	56	68
4-MeO-C_6H_4	Me	62	56	56
4-MeO-C_6H_4	Ph	–	53	61
2-Ph-4-Th	$ClCH_2$	55	52	48

R^1	R^2	90 (%)	91 (%)	92 (%)
Ph	Me	58	59	51
Ph	Ph	–	63	53
4-Cl-C_6H_4	Me	65	67	52
4-Cl-C_6H_4	Ph	–	63	53

2-Ph-4-Th = 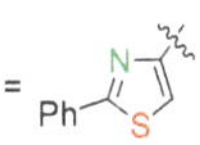

Scheme 35. Synthesis of hydrazine functionalized 1,3-selenazoles **88**,**89**,**91**,**92**.

The selenosemicarbazones **87** were further exposed to condensation using different α-halogenocarbonyl derivatives to yield a series of hydrazinoselenazoles **88** in 51–56% yields. The acetylation of the selenazoles **88** was readily accomplished using acetic anhydride in the presence of pyridine with the formation of selenazoles **89** in a 48–68% yield. The aroyl-hydrazinoselenazoles **91** and their diacetyl derivatives **92** were prepared in a similar way (Scheme 35) [78]. Compounds **88** and **89** (R^1 = 2-Ph-4-Th, R^2 = $ClCH_2$) exhibited significant cytotoxic activities against two human cancer cell lines, hepatocarcinoma cells (Hep-G2) and prostate cancer cells (DU-145) [78].

2-dialkylamino-1,3-selenazoles **94** were obtained in 19–90% yields by the reactions of *N*,*N*-unsubstituted selenoureas **93** with diketone derivatives in the presence of $FeCl_3$. (Scheme 36) [79].

The reaction of Woollins' reagent with *N*,*N*-dicyclohexylcyanamides followed by treatment with water led to the formation of 1,1-dicyclohexylselenourea **95** (Scheme 37) [80].

Selenourea **95** was converted into *N*,*N*-dicyclohexyl-4-substituted 1,3-selenazol-2-amine **98** in excellent yields (92–96%) by subsequent cyclization with four different α-haloketones. The reaction pathway included the formation of the intermediates **96** and **97**. The intermediate **96**, the addition product of selenourea **95** and α-halohekones, was involved in the cyclization reaction resulting in another intermediate **97**. This intermediate subsequently lost one molecule of water to form compounds **(98)** (Scheme 37) [80].

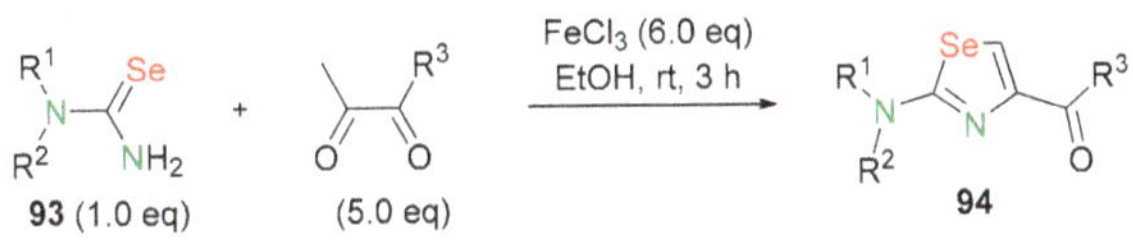

R^1	R^2	R^3	94 (%)
Me	Me	EtO	90
Et	Et	EtO	88
–$(CH_2)_2O(CH_2)_2$–		EtO	80
–$(CH_2)_5$–		EtO	87
–$(CH_2)_5$–		MeO	80
–$(CH_2)_5$–		Me	70
–$(CH_2)_5$–		Et	19

Scheme 36. Synthesis of 2-dialkylamino-1,3-selenazoles **94**.

The cyclization of α,α'-dichloroacetone with phenylselenourea afforded dihydroselenazole **99**, which was used in the efficient synthesis of selenazoles **100** and **101** (Scheme 38) [81]. Compound **99** was isolated by precipitation and used as a crude material for the preparation of selenazole **100**. The reaction of compound **99** with acetic anhydride led to the elimination of water and protection of the amino group by acetylation, yielding 4-chloromethyl-1,3-selenazole **100** in an 89% yield. The treatment of selenazole **100** with NaI afforded the highly reactive 4-iodomethyl-1,3-selenazole **101** in a 94% yield. The 4-halomethyl-2-amino-1,3-selenazoles **100** and **101** are the key intermediates for the synthesis of functionalized 1,3-selenazoles **102** (47–99% yields) (Scheme 38) [81].

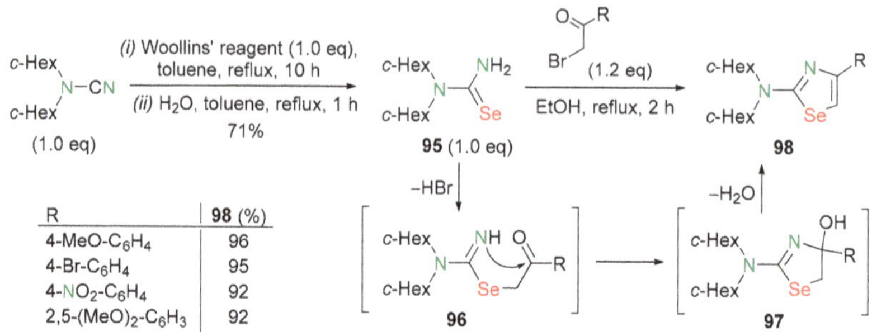

Scheme 37. Synthesis of *N,N*-dicyclohexyl-4-substirured 1,3-selenazol-2-amine **98**.

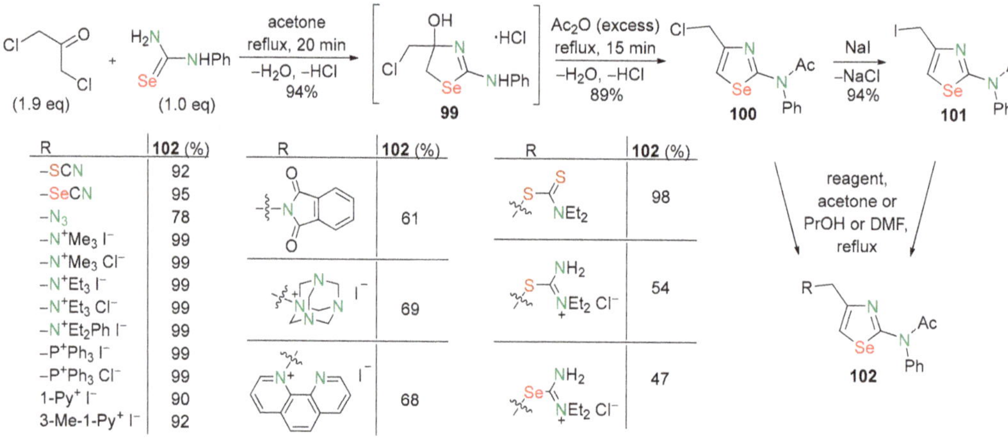

Scheme 38. Synthesis of 4-halomethyl-2-amino-1,3-selenazoles **100,101** as precursors of 2,4-functionalized 1,3-selenazoles **102**.

The tandem one-pot synthetic protocol for the synthesis of selenazoles under mild conditions from alkynes was developed (Scheme 39) [82]. The reaction of arylacetylenes with *N*-bromosuccinimide and water led to 2,2-dibromo-1-arylethanones, which further reacted with selenoureas to afford the desired products. The reaction was catalyzed by β-cyclodextrin in aqueous medium and resulted in 4-aryl-1,3-selenazol-2-amines **103** (61–72% yields) (Scheme 39) [82].

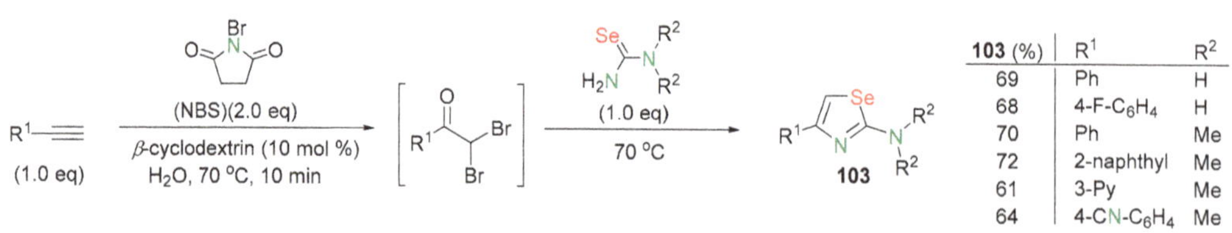

Scheme 39. Synthesis of 4-aryl-1,3-selenazol-2-amines **103**.

It is noteworthy that water is nontoxic, economically viable, and the most readily available reaction medium. The reaction, which was carried out in water using β-cyclodextrin as a catalyst, met the conditions of green chemistry [82].

An efficient and environmentally tolerant synthesis of 2-amino-1,3-selenazoles **104** in an ionic liquid/water solvent system was reported (Scheme 40) [83,84]. The synthesis of 2-amino-1,3-selenazoles **104** in 91–97% yields was developed by the condensation of various phenacyl bromides with selenourea in short reaction times (4–6 min) under ambient conditions. The addition of water to ionic liquid [Hbim]BF_4 (1-n-butylimidazolium tetrafluoroborate) in a 1:1 ratio enhanced the rate of cyclocondensation reaction and reduced the reaction time. This methodology has notable advantages such as the absence of any added catalyst, simple operation, and the recyclability of the ionic liquid. Moreover, acyl halides and phenacyl halides easily undergo condensation with selenourea in [Bmim]BF4 (1-butyl-3-methylimidazolium tetrafluoroborate) as ionic liquid by microwave irradiation (MW) to afford the desired products **104** in 95–98% yields (Scheme 40) [83,84].

R^1	R^2	time (min)	**104** *(i)* (%)
Ph	H	10	91
4-Me-C_6H_4	H	10	93
4-F-C_6H_4	H	15	94
4-Cl-C_6H_4	H	10	94
3-NO_2-C_6H_4	H	15	97
4-Ph-C_6H_4	H	20	96
β-$C_{10}H_7$	H	10	95

R^1	R^2	time (min)	**104** *(i)* (%)
Ph	Me	10	94
4-Me-C_6H_4	Me	10	95
4-F-C_6H_4	Me	10	96
3-NO_2-C_6H_4	Me	15	97
4-Ph-C_6H_4	Me	10	96
β-$C_{10}H_7$	Me	10	97

R^1	R^2	time (min)	**104** *(ii)* (%)
Ph	H	4	95
4-Me-C_6H_4	H	5	96
4-MeO-C_6H_4	H	6	96
4-Cl-C_6H_4	H	4	97
4-Br-C_6H_4	H	4	95
4-I-C_6H_4	H	6	95
3-NO_2-C_6H_4	H	6	98
t-Bu	H	4	96

Scheme 40. Synthesis of 2-amino-1,3-selenazoles **104** in ionic liquid/water solvent system or in ionic liquid under MW.

A series of new 4-substituted 2-(2-phenylethyl)amino 1,3-selenazoles **109** were synthesized based on (2-phenylethyl)amines (Scheme 41) [85]. Initial cyanamides **105** were prepared in almost quantitative yields from the reaction of BrCN with secondary amines in dry MeOH in the presence of anhydrous MeCOONa at room temperature. Two selenoureas **106** were obtained in 87–90% yields by the reaction of Woollins' reagent with cyanamides **105**, followed by post-treatment with water. The cyclization of selenoureas **106** with equimolar amounts of α-haloketones in refluxing ethanol solution yielded a series of 2-(2-phenylethyl)amino 1,3-selenazoles **109** in excellent yields (90–98%). The formation of products **109** occurred through intermediates **107** and **108** (Scheme 41) [85].

A simple protocol was developed for the synthesis of *N*-orthogonally protected dipeptidomimetics **113** linked with the selenazole ring by the condensation of *N*-amino selenoamides with α-bromomethyl ketones (Scheme 42) [86]. The synthesis of *N*-protected amino nitriles **110** was undertaken on the first stage. *N*-Methylmorpholine (NMM) and ethyl chloroformate (ECF) were added to a solution of *N*-protected amino acids in THF and stirred at −15 °C. After the treatment with aqueous ammonia solution, amino acid amides were obtained. Next, the solution of prepared amino acid amides in dry THF was treated with trifluoroacetic anhydride (TFAA) and pyridine and the reaction mixture was stirred at 0 °C to produce nitriles **110**. The reaction of nitriles **110** with P_2Se_5 yielded selenoamides **111**.

Bromomethyl ketones **112** were prepared from *N*-protected amino acid, ethyl chloroformate, diazomethane and HBr (Scheme 42) [86]. *N*-orthogonally protected dipeptidomimetics **113** were obtained in 75–83% yields by the condensation of *N*-amino selenoamides **111** with α-bromomethyl ketones **112** in refluxing acetone.

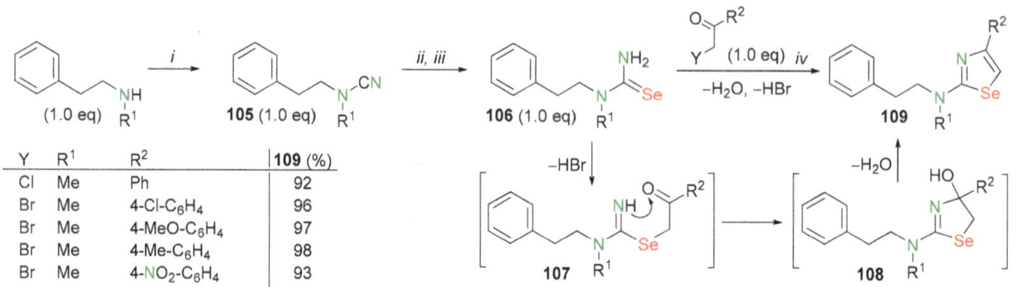

Y	R^1	R^2	109 (%)
Cl	Me	Ph	92
Br	Me	4-Cl-C_6H_4	96
Br	Me	4-MeO-C_6H_4	97
Br	Me	4-Me-C_6H_4	98
Br	Me	4-NO_2-C_6H_4	93
Br	Me	2,5-$(MeO)_2$-C_6H_3	90
Cl	Me	2,4-Cl_2-C_6H_3	91
Br	Me	4-Br-C_6H_4	95
Cl	EtOOC	Ph	96
Br	EtOOC	4-Cl-C_6H_4	96
Br	EtOOC	4-MeO-C_6H_4	95

R^1	105 (%)	106 (%)
Me	99	87
EtOOC	99	90

(i) BrCN (1.2 eq), AcONa (3.0 eq), MeOH, 0 °C to rt, 4 h
(ii) Woollins' reagent (0.5 eq), toluene, reflux, 4 h;
(iii) H_2O, toluene, reflux, 1 h;
(iv) EtOH, reflux, 1 h.

Scheme 41. Synthesis of 4-substituted-1,3-selenazol-2-amines **109**.

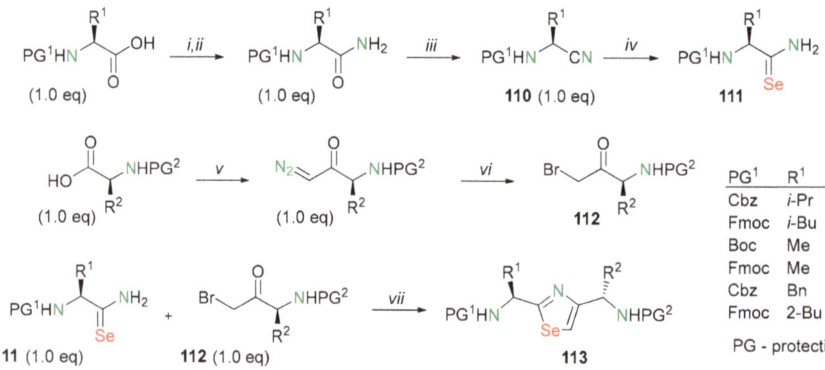

PG^1	R^1	PG^2	R^2	113 (%)
Cbz	*i*-Pr	Boc	*i*-Bu	82
Fmoc	*i*-Bu	Boc	Bn	78
Boc	Me	Cbz	Bn	81
Fmoc	Me	Boc	Bn	75
Cbz	Bn	Fmoc	Me	79
Fmoc	2-Bu	Boc	*i*-Pr	83

PG - protecting group

(i) NMM (1.1 eq), ECF (1.1 eq), THF, −15 °C, 20 min; *(ii)* NH_3 aq, 4 h; *(iii)* TFAA (1.2 eq), pyridine (2.0 eq), THF, 0 °C, 3 h; *(iv)* P_2Se_5 (2.0 eq), EtOH, H_2O, reflux, 4 h; *(v)* Et_3N (1.1 eq), ECF (1.1 eq), diazomethane, THF, Et_2O, −15 °C to 0 °C, 45 min; *(vi)* 45% HBr aq, rt, 5 min; *(vii)* acetone, reflux, 30 min.

Scheme 42. Synthesis of *N*-orthogonally protected dipeptidomimetics **113**.

An efficient protocol for the preparation of optically pure 1,3-selenazoles has been developed (Scheme 43) [22]. Aliphatic selenoamides **115** can be prepared by refluxing a mixture of the amino-acid-derived nitriles **114** and P_2Se_5 in an EtOH/H_2O mixture. However, in studies, these selenoamides **115** were unstable during purification. Therefore, after simple filtration and rapid concentration, the crude selenoamides **115** were then reacted with ethyl bromopyruvate (EBP) in the presence of excess $KHCO_3$ in DMF at −20 °C for 5 min. Dehydration was followed by adding trifluoroacetic anhydride (TFAA) to the reaction mixture to obtain 2,4-substituted selenazoles **116** in 32–74% yields. These reaction conditions appeared to be important for suppressing the racemization of the final products suitable for synthesis of optically pure peptides containing 1,3-selenazole fragments (Scheme 43) [22].

A new reagent, ethyl 3-amino-3-selenoxopropanoate **117** (82% yield), was obtained from ethyl cyanoacetate and H_2Se in the presence of Et_3N. The reaction of product **117** with phenacyl bromides involving the selenoamide group in DMF under an argon atmosphere led to the formation of the desired 1,3-selenazoles **118**. Upon heating selenazole **118** (R = H) with ammonia in ethanol solution, the ammonolysis of the ester group occurred to afford 2-thiazole-2-acetamide **119** in a 74% yield (Scheme 44) [87].

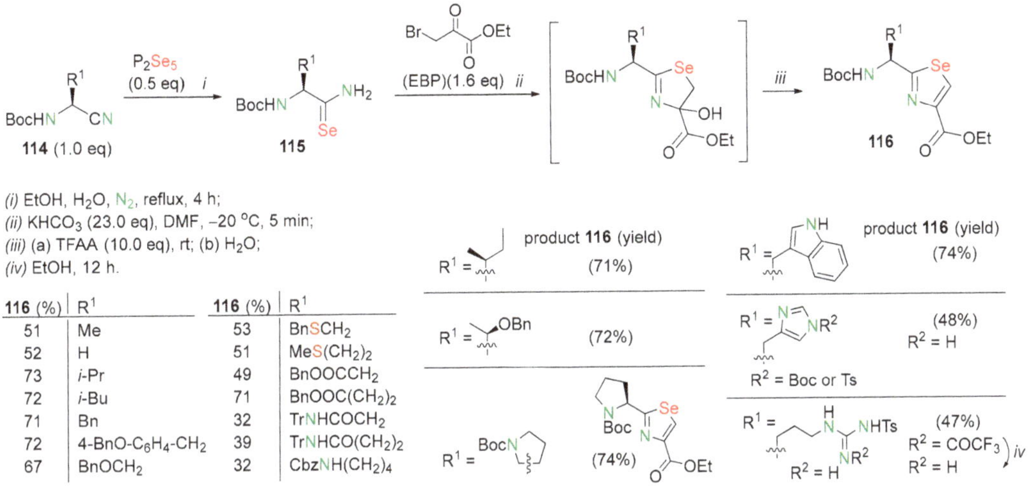

(i) EtOH, H_2O, N_2, reflux, 4 h;
(ii) $KHCO_3$ (23.0 eq), DMF, −20 °C, 5 min;
(iii) (a) TFAA (10.0 eq), rt; (b) H_2O;
(iv) EtOH, 12 h.

116 (%)	R^1	116 (%)	R^1
51	Me	53	$BnSCH_2$
52	H	51	$MeS(CH_2)_2$
73	i-Pr	49	$BnOOCCH_2$
72	i-Bu	71	$BnOOC(CH_2)_2$
71	Bn	32	$TrNHCOCH_2$
72	4-BnO-C_6H_4-CH_2	39	$TrNHCO(CH_2)_2$
67	$BnOCH_2$	32	$CbzNH(CH_2)_4$

Scheme 43. Synthesis of 2,4-substituted selenazoles **116**.

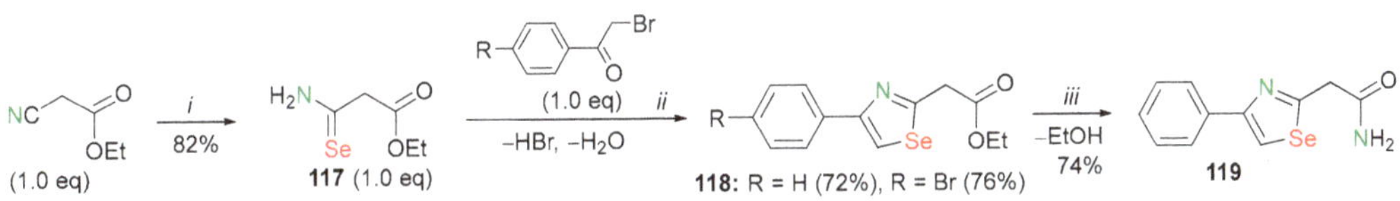

(i) H_2Se, Et_3N (cat), Ar, 0 °C to 5 °C, 3h; (ii) DMF, Ar, rt, 2 h; (iii) 25% NH_4OH aq, 1.0 eq), EtOH, reflux.

Scheme 44. Synthesis of trisubstituted 2-amino-1,3-selenazole derivatives **119**.

The multicomponent condensation of aromatic aldehydes, cyanoselenoacetamide, α-bromoketones, and dimedone was studied (Scheme 45) [88]. The reaction was carried out in DMF in the presence of morpholine at room temperature. This cascade transformation resulted in the formation of 2-amino-4-chromen-5(6*H*)-ones **123** in 77–79% yields. The reaction involved the formation of Knoevenagel alkenes **120** as intermediates. The Hantzsch-type reaction occurred to form vinylselenazoles **121**, which further reacted with dimedone to yield intermediate **122**. The latter underwent intramolecular cyclization into the final heterocyclic system **123** (Scheme 45) [88]. 3-Aminoselenoacrylamides are promising initial materials for the preparation of selenium heterocycles.

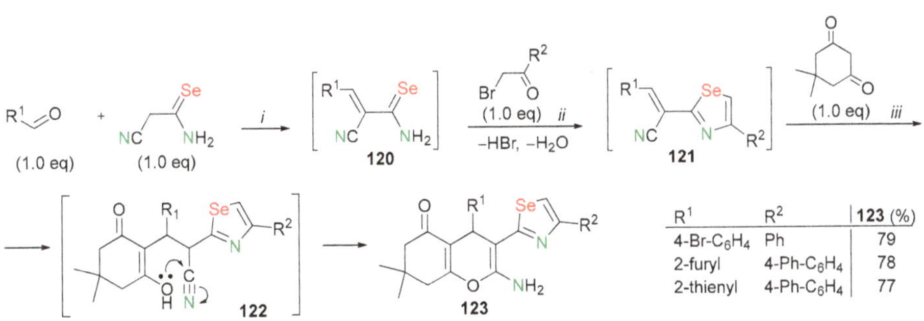

R^1	R^2	123 (%)
4-Br-C_6H_4	Ph	79
2-furyl	4-Ph-C_6H_4	78
2-thienyl	4-Ph-C_6H_4	77

(i) morpholine (cat), DMF, Ar, rt, 30 min; (ii) rt, 2 h; (iii) morpholine (1.0 eq), 25 h.

Scheme 45. Synthesis of 2-amino-4-chromen-5(6*H*)-ones **123**.

A convenient approach to producing 3-aminoselenoacrylamides **124** based on a three-component condensation reaction of triethylorthoformate, primary aromatic amines, and cyanoselenoacetamide was discovered (Scheme 46) [89].

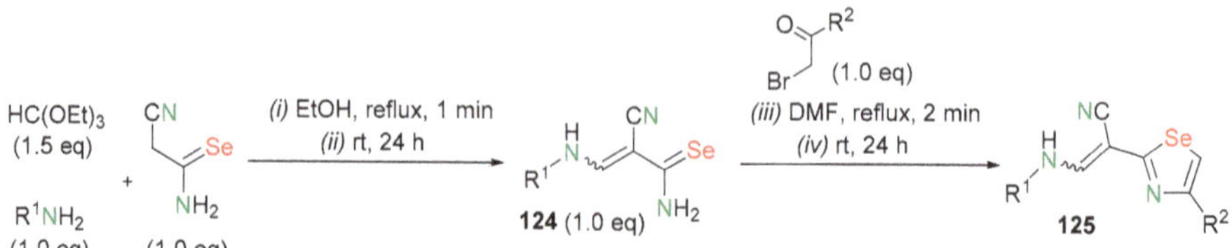

124 (%)	E/Z	R^1	R^2	125 (%)	E/Z
87	8/1	Ph	Ph	61	1/12
–	–	Ph	4-Me-C_6H_4	41	1/32
42	7/1	4-Me-C_6H_4	Ph	73	1/19

Scheme 46. Synthesis of 2,4-substituted 1,3-selenazole derivatives **125**.

The unsaturated selenoamides **124** were formed as a mixture of *Z*- and *E*-isomers by refluxing the reagents in EtOH. The selenoamides **124** readily took part in Hantzsch-type reactions with α-bromoketones to form 2,4-substituted 1,3-selenazole derivatives **125** predominantly as the *Z*-isomers (Scheme 46) [89].

The reactions of cyanoselenoacetamide, 2-furfurylidene acetoacetic ester, and phenacyl bromide in the presence of *N*-methylmorpholine at 20 °C under an argon atmosphere afforded functionalized selenazole **126** (68% yield), containing furan and dihydropyridine rings (Scheme 47) [90].

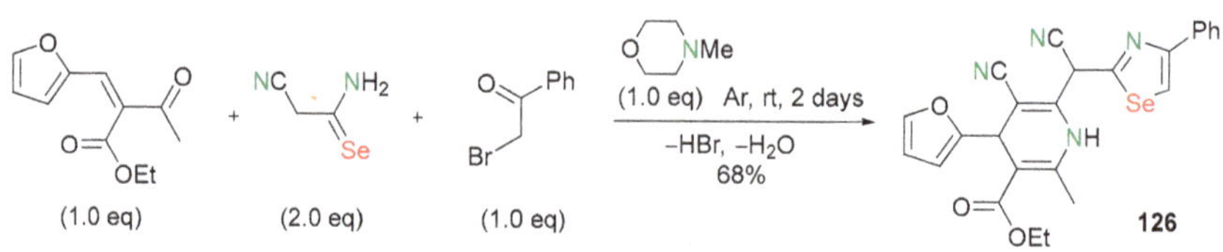

Scheme 47. Synthesis of compound **126**.

2.3. 2,5-Substituted 1,3-Selenazoles

An approach to 2,5-disubstituted 1,3-selenazoles based on the cycloisomerization of propargyl selenoamides, using readily available starting materials, was reported (Scheme 48) [91]. The intermediate selenoamides **128** were synthesized in situ from aromatic and aliphatic terminal propargyl amides **127** with neutral, electron-rich, and electron-deficient groups.

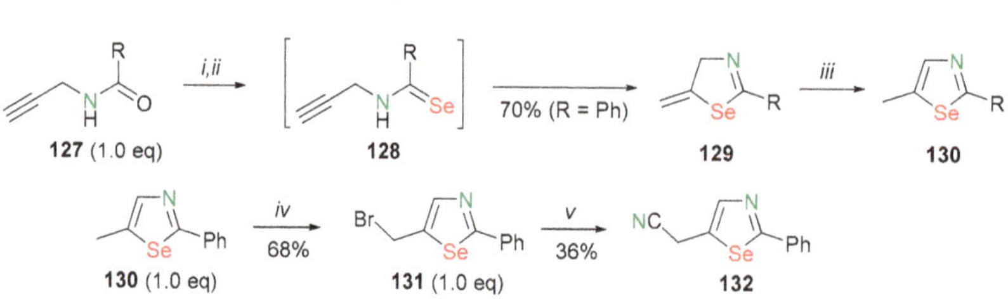

130 (%)	R
74	Ph
70	4-Br-C_6H_4
55	4-F-C_6H_4
42	3-Cl-C_6H_4
61	4-*i*-Pr-C_6H_4
88	4-MeO-C_6H_4
25	*n*-Pr
23	*i*-Pr
20	*t*-Bu
36	4-MeO-C_6H_4-CH_2

(i) PCl_5 (2.0 eq), DMF (0.4 eq), toluene, rt, 30 min; *(ii)* LiAlHSeH (1.4 eq), THF, rt, 1 h; *(iii)* (a) PipAcOH (1.0 eq), reflux, 2 h; (b) rt, 12 h; *(iv)* (a) NBS (1.1 eq), AIBN (0.1 eq), *hv*, CCl_4, rt, 1 h; (b) rt, 12 h; *(v)* KCN (1.2 eq), DMF, 0 °C to rt, 2 h.

Scheme 48. Synthesis of 2,5-disubstituted 1,3-selenazoles **130–132** with using Ishihara reagent.

This procedure used PCl_5 and a catalytic amount of DMF in toluene for the conversion of amides into iminochlorides, followed by the exchange of chlorine for selenium using the Ishihara reagent (LiAlHSeH), freshly prepared from elemental selenium and lithium alumohydride. The spontaneous *5-exo-dig* cyclization of intermediates **128** led to the formation of selenazolines **129**. Compound **129** (R = Ph) was isolated in 70% yield. The *exo*-double bond of compounds **129** was isomerized with piperidinium acetate, and 2,5-disubstituted 1,3-selenazoles **130** were prepared in up-to-88% yields. Methylselenazole **130** (R = Ph) was transformed into a bromomethyl derivative **131** (68% yield) using *N*-bromosuccinimide (NBS) in the presence of azobisisobutyronitrile (AIBN) in CCl_4 with irradiation and room temperature. The cyanide **132** was obtained from intermediate **131**, which can be considered a versatile building block for further derivatizations (Scheme 48) [91].

A red-shifted fluorescent indicator for magnesium was developed by the incorporation of selenium in the selenazole moiety of fura fluorophore. To introduce selenium into the structure of Mag-fura-2, 2-methyl-1,3-selenazole **133** was prepared by the reflux of freshly prepared selenoacetamide and ethyl 2-chloro-3-oxopropanoate in MeCN (Scheme 49) [92].

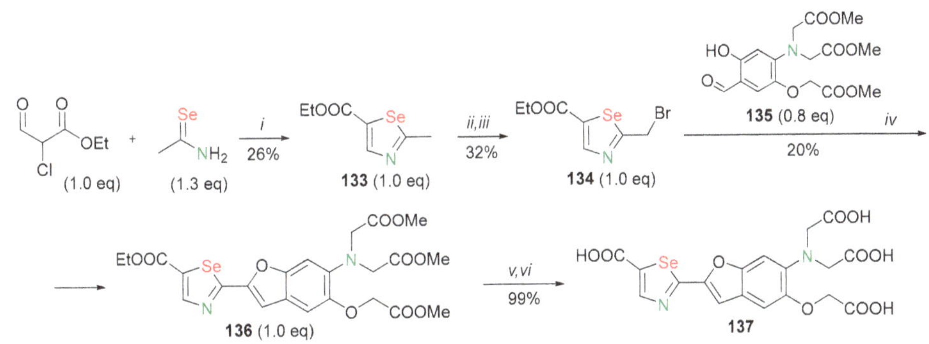

(i) (a) MeCN, rt, 15 min; (b) reflux, 6 h; *(ii)* 1,3-dibromo-5,5-dimethylhydantoin (2.0 eq), AIBN (0.3 eq), AcOH (cat), $PhCl/CCl_4$ (v/v=1/1), 75 °C, 3 h; *(iii)* $HPO(OEt)_2$ (1.2 eq), *i*-Pr_2EtN (1.2 eq), THF, Ar, 0 °C, 5 h;
(iv) K_2CO_3 (3.6 eq), DMF, Ar, 100 °C, 2 h; *(v)* KOH (12.0 eq), MeOH, H_2O, rt, 24 h; *(vi)* HCl or buffer (pH 7).

Scheme 49. Synthesis of Se-containing 'fura' fluorophore **137**.

Following nonselective radical bromination resulted in a mixture of 2-bromomethyl- and 2-dibromomethyl-1,3-selenazoles. The treatment of the mixture with Hünig's base and diethylphosphite made it possible to transform an undesirable dibromo by-product into pure 2-bromomethyl-1,3-selenazole **134** in a 32% yield. This precursor was used to finally produce the benzofuran fluorophore **136** (20% yield) via condensation with salicylaldehyde **135**, which contains the protected magnesium-binding group. Metal-responsive sensor **137** in the form of free acid may be generated in a quantitative yield via the base-catalyzed hydrolysis of the esters right before use. The ester form of the sensor is more stable during storage (Scheme 49) [92].

The replacement of one chalcogen atom in the acceptor oxazole fragment of the Mag-fura-2 indicator induces an increase in the Stokes shift and a large shift in emission wavelength without affecting, significantly, the basic chemical features of the compound [92].

A series of 2,5-disubstituted 1,3-selenazole derivatives, containing sulfonamide moiety, were synthesized (Scheme 50) [93,94]. 4-Cyanobenzenesulfonamide (**138**) was prepared in an 88% yield by the reaction of the corresponding sulfonyl chloride with an aqueous solution of ammonium hydroxide. Consistently, 4-sulfamoyl benzo selenoamide (**139**) was prepared in a 76% yield by the reaction of nitrile compound **138** with Na_2Se as the selenating reagent in refluxing ethanol. And lastly, the reaction of primary selenobenzamide **139** with various α-haloketones, including aliphatic or aromatic moieties, yielded various 2,5-disubstituted 1,3-selenazoles **140** in 55–83% yields (Scheme 50) [93,94]. Prepared selenazoles **140** demonstrated potent inhibitory effects on cell viability against breast (MDA-MB-

231) cancer cell lines and the human prostate (PC3), thus possessing effective antitumor activity. Also, compounds **140** showed antibacterial activity against *Helicobacter pylori* and *Burkholderia pseudomallei* [93,94].

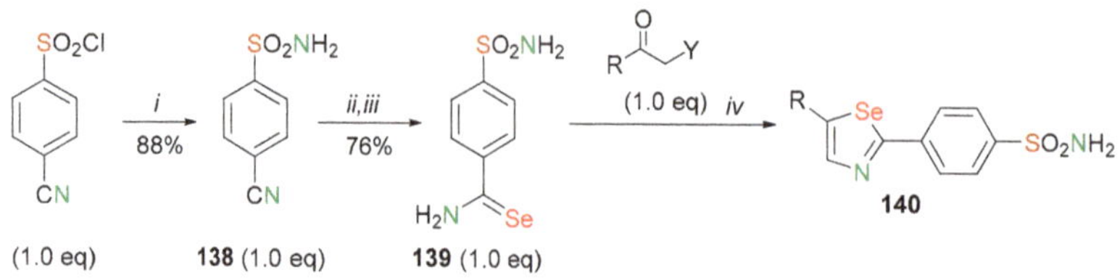

(*i*) 30% NH_4OH, THF, rt, 1 h; (*ii*) $NaBH_4$ (3.0 eq), Se (1.5 eq), EtOH, 0 °C, 30 min; (*iii*) reflux, 6 h; (*iv*) reflux, 20 min.

Y	R	**140** (%)
Cl	Me	61
Cl	HOOC	55
Cl	$ClCH_2$	70
Br	Ph	83
Br	4-F-C_6H_4	64
Br	4-Br-C_6H_4	60
Br	4-Cl-C_6H_4	61
Br	4-NO_2-C_6H_4	55
Br	3-MeO-C_6H_4	80

Scheme 50. Synthesis of 2,5-disubstituted 1,3-selenazole derivatives **140** containing sulfonamide moiety.

Primary alcohols were converted into 5-substituted 2-amino-1,3-selenazoles **141** under microwave irradiation (MW) using trichloroisocyanuric acid (TCCA) as the chlorine source and an oxidant, (2,2,6,6-tetramethylpiperidin-1-yl)oxyl (TEMPO), as a co-oxidant, and dimethyl selenourea (Scheme 51) [95]. Primary alcohols were added to a mixture of TCCA and TEMPO in CH_2Cl_2 and the mixture was stirred under microwave irradiation at 60 °C for 10 min. Upon the formation of the α-chlorocarbonyl compound, *t*-BuOH and dimethyl selenourea were added, and the mixture was additionally stirred under microwave irradiation at 100 °C for 15 min.

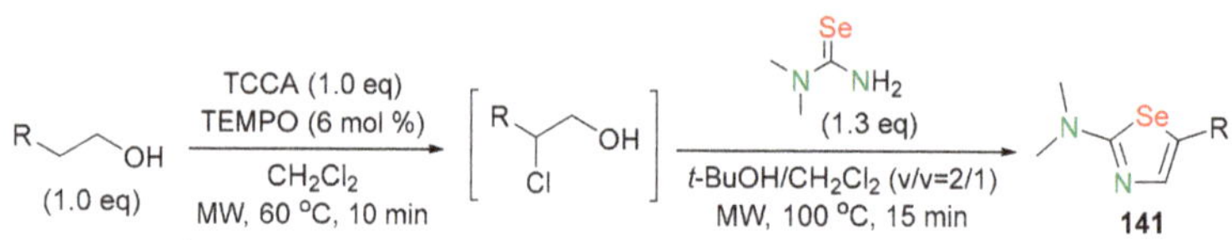

R	**141** (%)
n-Hex	52
n-$C_{10}H_{21}$	73
Ph	64

Scheme 51. Synthesis of 5-substituted 2-amino-1,3-thiazoles **141** using MW.

The reaction of selenazadienes **142** with α-haloketones **143** in the appropriate conditions yielded 5-acyl-2-amino-1,3-selenazoles **144** in high yields (64–99%) (Scheme 52) [96, 97].

The reaction of compounds **142** with 1,3-dichloro-2-propanone in the presence of Et_3N yielded bis [2-dialkylamino-5-(1,3-selenazolyl)] ketones **145** (82–91% yields) as major products together with 1,3-selenazoles **144** as the minor products (Scheme 52) [96,97].

According to the investigation results, bis[2-dimethylamino-5-(1,3-selenazolyl)] ketone showed the properties of effective and potentially useful superoxide anion-scavenger in vitro [97].

Five different selenazadienes (selenocarbamoylamidines) **147** were prepared in 75–98% yields by the reaction of *N*,*N*-dialkylselenoureas **146** and *N*,*N*-dimethylformamide dimethyl acetal in THF at room temperature.

The reaction of selenazadienes **147** with chloroacetyl chloride in THF yielded 1,3-selenazole-5-carboxylic acids **148** in moderate yields (40–80%). However, the expected acyl chloride derivative of 1,3-selenazole was not isolated.

The capture of the intermediate, obtained by the reaction of *N*,*N*-dimethyl-*N′*-(piperidinoselenocarbonyl)formamidine with chloroacetyl chloride, with various alcohols afforded 1,3-selenazole-5-carboxylates **149** in 50–58% yields (Scheme 53) [98,99].

Selenazadienes **150** were prepared by the condensation of *N*,*N*-unsubstituted selenoureas with *N*,*N*-dimethylformamide dimethylacetal at room temperature. Six kinds of 1,3-selenazole-5-carbonitriles **151** were obtained by reactions of selenazadienes **150** with chloroacetonitrile in the presence of Et_3N (Scheme 54) [100]. Chloroacetonitrile was added to a stirred solution of compound **150** in dry THF under an argon atmosphere at an ambient temperature. The reaction mixture was refluxed for 3 h. Then, Et_3N was added and the reaction mixture was refluxed. As a result, 2-organylamino-1,3-selenazole-5-carbonitrile **151** samples were prepared in 73–91% yields.

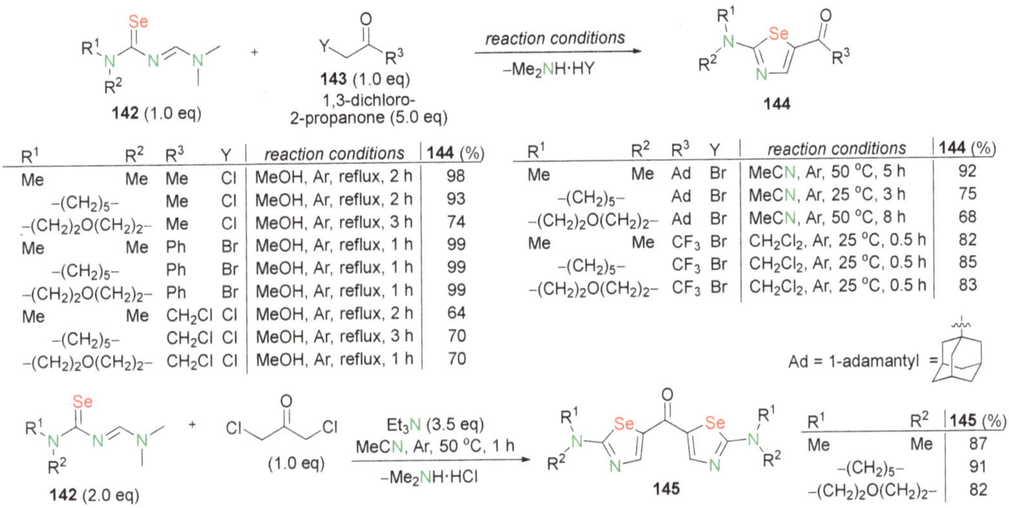

R^1	R^2	R^3	Y	reaction conditions	**144** (%)
Me	Me	Me	Cl	MeOH, Ar, reflux, 2 h	98
$-(CH_2)_5-$		Me	Cl	MeOH, Ar, reflux, 2 h	93
$-(CH_2)_2O(CH_2)_2-$		Me	Cl	MeOH, Ar, reflux, 3 h	74
Me	Me	Ph	Br	MeOH, Ar, reflux, 1 h	99
$-(CH_2)_5-$		Ph	Br	MeOH, Ar, reflux, 1 h	99
$-(CH_2)_2O(CH_2)_2-$		Ph	Br	MeOH, Ar, reflux, 1 h	99
Me	Me	CH_2Cl	Cl	MeOH, Ar, reflux, 2 h	64
$-(CH_2)_5-$		CH_2Cl	Cl	MeOH, Ar, reflux, 3 h	70
$-(CH_2)_2O(CH_2)_2-$		CH_2Cl	Cl	MeOH, Ar, reflux, 1 h	70
Me	Me	Ad	Br	MeCN, Ar, 50 °C, 5 h	92
$-(CH_2)_5-$		Ad	Br	MeCN, Ar, 25 °C, 3 h	75
$-(CH_2)_2O(CH_2)_2-$		Ad	Br	MeCN, Ar, 50 °C, 8 h	68
Me	Me	CF_3	Br	CH_2Cl_2, Ar, 25 °C, 0.5 h	82
$-(CH_2)_5-$		CF_3	Br	CH_2Cl_2, Ar, 25 °C, 0.5 h	85
$-(CH_2)_2O(CH_2)_2-$		CF_3	Br	CH_2Cl_2, Ar, 25 °C, 0.5 h	83

R^1	R^2	**145** (%)
Me	Me	87
$-(CH_2)_5-$		91
$-(CH_2)_2O(CH_2)_2-$		82

Scheme 52. Synthesis of 5-acyl-2-amino-1,3-selenazoles **144** and bis[2-dialkylamino-5-(1,3-selenazolyl)] ketones **145**.

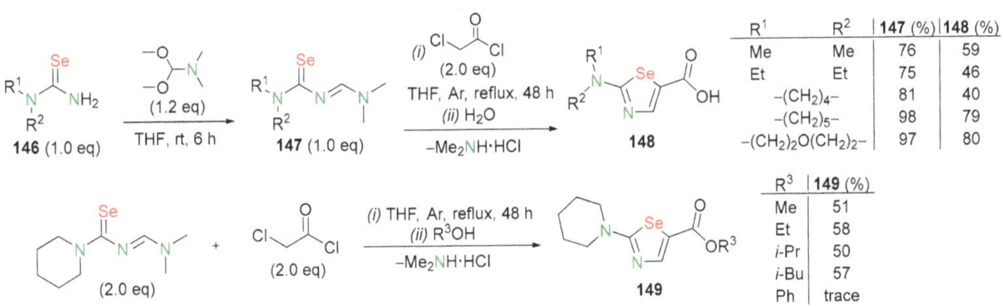

R^1	R^2	**147** (%)	**148** (%)
Me	Me	76	59
Et	Et	75	46
$-(CH_2)_4-$		81	40
$-(CH_2)_5-$		98	79
$-(CH_2)_2O(CH_2)_2-$		97	80

R^3	**149** (%)
Me	51
Et	58
i-Pr	50
i-Bu	57
Ph	trace

Scheme 53. Synthesis of 1,3-selenazole-5-carboxylic acids **147** and carboxylates **149**.

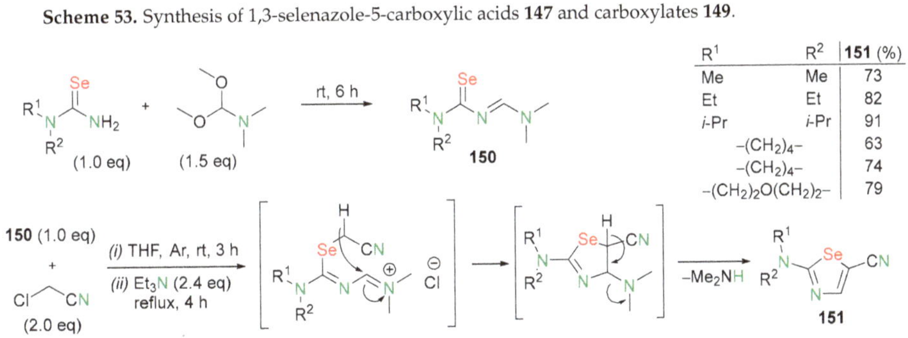

R^1	R^2	**151** (%)
Me	Me	73
Et	Et	82
i-Pr	*i*-Pr	91
$-(CH_2)_4-$		63
$-(CH_2)_4-$		74
$-(CH_2)_2O(CH_2)_2-$		79

Scheme 54. Synthesis of 2-organylamino-1,3-selenazole-5-carbonitrile **151**.

The reactions of *N*,*N*-dimethyl-*N*′-(piperidinoselenocarbonyl)-formamidine **152** with chloroacetyl chloride and then with amines yielded 1,3-selenazole-5-carboxamides **153**

(Scheme 55) [100]. The reaction of chloroacetyl chloride with selenoazadienes **152** was carried out in dry THF under an argon atmosphere at room temperature. After the addition of amines, the obtained mixture was refluxed for 5 min. After the working up of the reaction mixture and purification of the residue, 1,3-selenazole-5-carboxamides **153** were prepared in 40–82% yields (Scheme 55) [100].

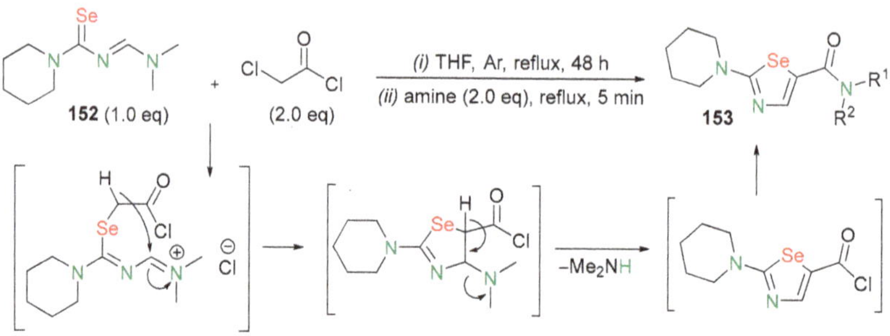

R^1	R^2	**153** (%)
n-Pr	H	80
n-Bu	H	82
Bn	H	75
$-(CH_2)_2O(CH_2)_2-$		40

Scheme 55. Synthesis of 1,3-selenazole-5-carboxamides **153**.

2-dialkylamino-1,3-selenazoles **154** were obtained in 26–93% yields by the reaction of α,β-unsaturated aldehydes with selenoureas in EtOH in the presence of $FeCl_3$ at an ambient temperature (Scheme 56) [101]. The reaction of selenoureas with α,β-unsaturated aldehydes was initiated by the nucleophilic addition of the nitrogen of the selenourea to the carbonyl carbon, resulting in the formation of 2-amino-1,3-selenazoles **154**. The reactions in MeOH, *i*-PrOH, and *t*-BuOH yielded 5-(alkoxymethyl)-2-amino-1,3-selenazole derivatives **155** in 51–79% yields (Scheme 56) [101].

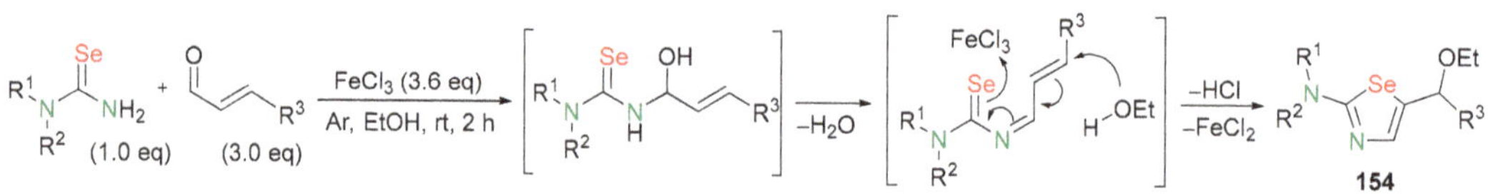

R^1	R^2	R^3	**154** (%)
Me	Me	H	68
Et	Et	H	74
$-(CH_2)_2O(CH_2)_2-$		H	63
$-(CH_2)_4-$		H	93
$-(CH_2)_4-$		Me	49
$-(CH_2)_4-$		Et	26

R^4	**155** (%)
Me	58
i-Pr	79
t-Bu	51

Scheme 56. Synthesis of 2-dialkylamino-1,3-selenazoles **154**,**155** in the presence of $FeCl_3$.

2.4. 2,4,5-Substituted 1,3-Selenazoles

The reaction of primary selenoamides with chloroacetyl chlorides afforded two products, 4-hydroxy-1,3-selenazoles **157**, which constitute the tautomeric enol form of compounds (**156**) (Scheme 57) [102].

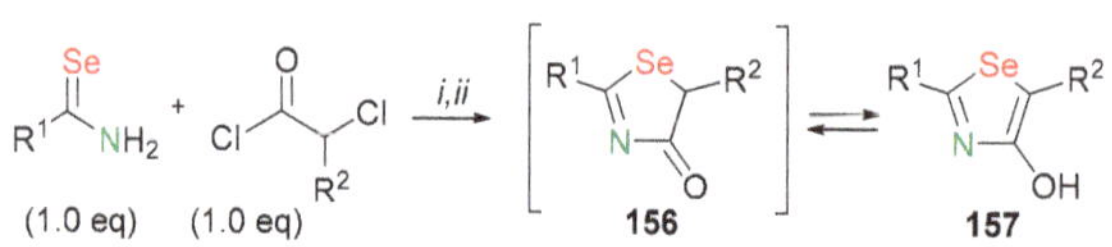

R^1	R^2	**157** (%)
4-Me-C_6H_4	Ph	63
$PhCH_2$	H	25

(*i*) CH_2Cl_2, Ar, 0 °C to rt, 1 h;
(*ii*) pyridine (2.0 eq), CH_2Cl_2, 0 °C, 2 h.

Scheme 57. Synthesis of 2,5-substituted 4-hydroxy-1,3-selenazoles **157**.

2-(4-Methoxyphenyl)-4-methyl-1,3-selenazole-5-ethyl carbonate **158** was synthesized starting from 4-methoxybenzaldehyde via intermediate 4-methoxybenzonitrile and 4-methoxybenzenecarboselenoamide (Scheme 58) [103,104]. 1,3-selenazole-5-ethyl carbonate **158** was obtained in an 82% yield by heating the mixture of 4-methoxybenzenecarboselenoamide and ethyl 2-chloroacetoacetate in EtOH under reflux. Selenazole **158** was used as a template for the synthesis of various 2,4,5-subtituted selenazoles modified by oxadiazole, pyrazole,1,2,4-triazole, tetrazole, 1,2,4-triazine, and succinic imide (Scheme 58) [103,104].

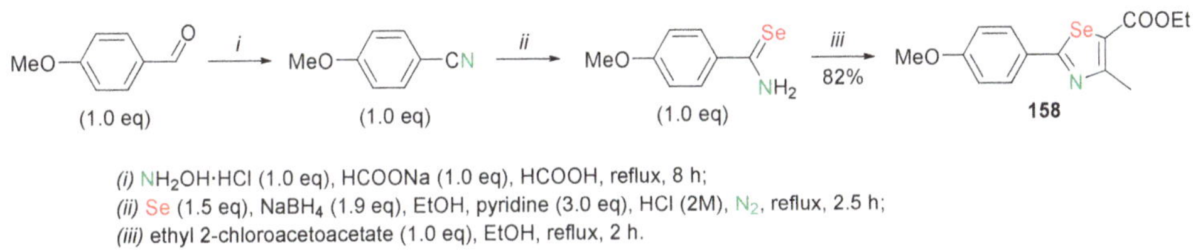

Scheme 58. Synthesis of 2-(4-methoxyphenyl)-4-methyl-1,3-selenazole-5-ethyl carbonate **158**.

Some of the modified 1,3-selenazoles have shown the inhibitory activities against cell division cycle 25B phosphatase (Cdc25B) and can be anticancer drugs leading compounds [103].

A number of new 2-phenyl-4-methyl-1,3-selenazole-5-carboxylic acid derivatives **166** and **167** were synthesized (Scheme 59) [105]. Commercially accessible 4-hydroxybenzonitrile as the starting material was transformed into selenocarboxamide **159** (55% yield) by reaction with NaHSe. The freshly prepared product **159** underwent cyclization with ethyl 2-chloroacetoacetate in refluxing ethanol to form the key intermediate **160** in 87% yield. The ortho-formylation of compound **160** with paraformaldehyde in the presence of anhydrous $MgCl_2$ and Et_3N was used for the synthesis of salicylaldehyde **161** with a high selectivity and isolated yield (63%). The reaction of salicylaldehyde **161** with $NH_2OH{\cdot}HCl$ using sodium formate as a base under reflux in formic acid provided compound **162** in a 59% yield. On the other hand, product **163** (72% yield) was prepared by the treatment of intermediate **160** with KNO_3 in concentrated H_2SO_4. Compounds **162** and **163** were alkylated with corresponding alkyl halide in DMF to provide 1,3-selenazoles **164** and **165,** which were subsequently hydrolyzed in the presence of NaOH followed by acidification to obtain the target compounds **166** and **167** (Scheme 59) [105].

Most of the products exhibited inhibitory activity against xanthine oxidase in vitro in the nanomolar range. Compound **166** (R = allyl) is considered the most potent xanthine oxidase inhibitor [105].

α-Tosyloxylketones, prepared by the treatment of ketones with HTIB, were involved in a reaction with primary selenoamides to provide a convenient method of synthesis of 1,3-selenazoles (Scheme 60) [52]. This method avoids the use of toxic and lachrymatory *α*-haloketones. 2,4,5-Substituted 1,3-selenazoles **168** were prepared in 54–81% yields from the corresponding ketones and selenoamides under reflux in a solvent mixture MeCN/MeOH.

Selenazole syntheses by the Sc-catalyzed cycloaddition reactions of 1,1-bisaryl-3-phenylselanylpropargyl alcohols with selenoamides was reported (Scheme 61) [106]. The reaction of selenamide with 1,1-bisarylpropargyl alcohols in $MeNO_2/H_2O$ resulted in 4-bisarylmethyl-1,3-selenazoles **169**. Initial 1,1-bisaryl-3-(phenylselanyl)prop-2-yn-1-ols were obtained using *n*-BuLi by the reaction of ethynyl phenyl selenide with diarylketones in THF under an argon atmosphere at room temperature (Scheme 61) [106].

The synthesis of 2-alkylthio-4,5-disubstituted-1,3-selenazoles **172–175** via the metallation of dimethyl *N*-(ethoxycarbonylmethyl)iminodithiocarbonate and further reaction with selenothioic acid *S*-ester was developed (Scheme 62) [107]. Using the LDA–THF system for the deprotonation of compound **170**, selenazoles **172** and **173** in 22% yields were isolated after chromatography on silica gel. In addition, two products, **174** and **175,** were isolated.

This reaction afforded four 2-alkylthio-substituted selenazoles **172–175**, which varied only by side chains attaching to the C-2 and C-4 functional groups (Scheme 62) [107].

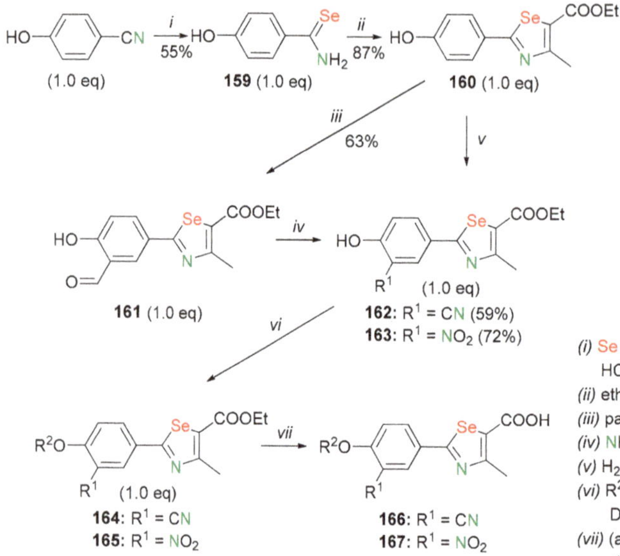

R^2	**164** (%)	**165** (%)	**166** (%)	**167** (%)
Et	88	87	81	88
n-Bu	82	78	86	85
i-Bu	72	73	84	83
i-Am	70	70	78	79
allyl	84	85	75	81
Bn	81	74	84	76
Me	85	–	91	–
n-Pr	76	–	83	–
i-Pr	72	–	89	–
c-Pr-CH_2	69	–	75	–
s-Bu	73	–	84	–
n-Am	72	–	79	–
propargyl	82	–	77	–

(*i*) Se (1.5 eq), $NaBH_4$ (1.9 eq), EtOH, pyridine (3.0 eq), HCl (2M), N_2, reflux, 2.5 h;
(*ii*) ethyl 2-chloroacetoacetate (1.0 eq), EtOH, reflux, 15 min;
(*iii*) paraformaldehyde, $MgCl_2$ (1.5 eq), Et_3N, MeCN, reflux, 8 h;
(*iv*) $NH_2OH{\cdot}HCl$ (1.0 eq), HCOONa (1.0 eq), HCOOH, reflux, 8 h;
(*v*) H_2SO_4 conc, KNO_3 (1.0 eq), 0 °C, 2 h;
(*vi*) R^2Hal (2.0 eq), K_2CO_3 (2.0 eq), KI (16.0 eq), DMF, 30 °C to 80 °C, 3 h;
(*vii*) (a) NaOH aq (1M), THF/EtOH (v/v=1/1), 60 °C, 1,5 h; (b) HCl aq (1M) (pH 2), rt, 20 min.

Scheme 59. Synthesis of 2-phenyl-4-methyl-1,3-selenazole-5-carboxylic acid derivatives **166,167**.

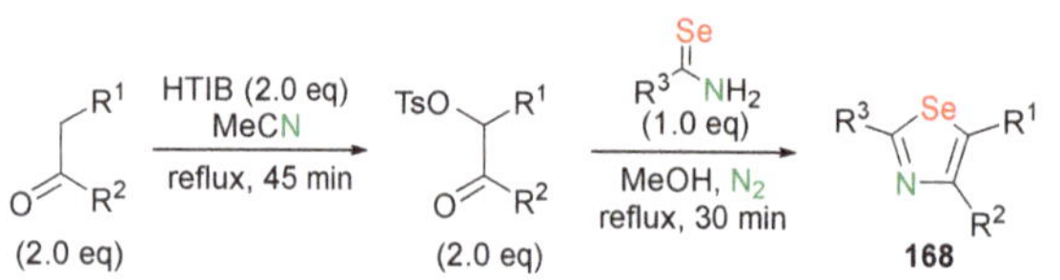

R^1	R^2	R^3	**168** (%)
PhCO	Me	Ph	65
–$(CH_2)_4$–		Ph	81
Me	Et	Ph	56
MeCO	Me	Ph	76
MeCO	Me	2,4-Cl_2-C_6H_3	75
PhCO	Me	4-NMe_2-C_6H_4	66
PhCO	Me	4-Me-C_6H_4	60
MeCO	Me	4-MeO-C_6H_4	54

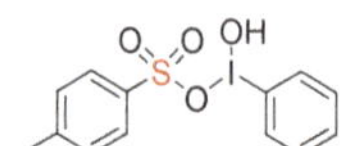

HTIB

Scheme 60. Synthesis of 2,4,5-Substituted 1,3-selenazoles **168** with using HTIB.

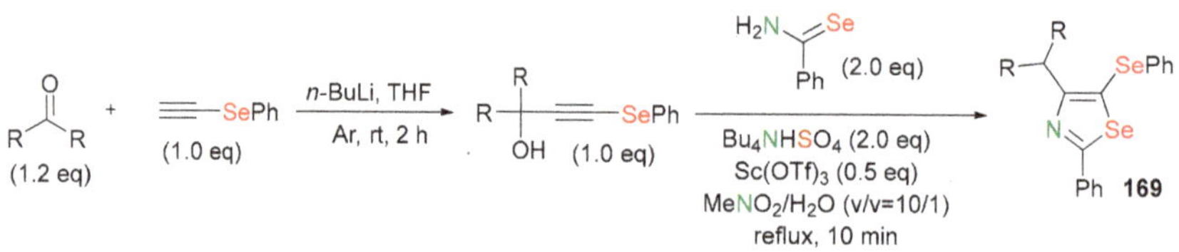

R	**169** (%)
Ph	55
2-thienyl	68

Scheme 61. Scandium-catalyzed synthesis of 4-bisarylmethyl-1,3-thiazoles **169**.

These selenazoles **172–175** are easily transformed into electron-rich diselenadiazafulvalenes and azino-diselenadiazafulvalenes with excellent reversible donor ability [107].

A one-pot synthesis of 2-amino-4-alkyl-5-(*R*-benzyl)selenazoles **178** and **179** was developed through the bromoarylation of methyl or cyclopropyl vinyl ketones and further cyclization with selenourea (Scheme 63) [108]. Anilines were first converted into arenediazonium bromides, which reacted with methyl vinyl ketone or cyclopropyl vinyl ketone in the presence of $CuBr_2$ to produce 4-aryl-3-bromobutan-2-ones **176** (40–71% yields) and 3-aryl-2-bromo-1-cyclopropylpropan-1-ones **177** (41–79% yields). These products reacted with selenourea without preliminary isolation. After nitrogen was no longer evolved, the reaction mixture was neutralized with $NaHCO_3$ and selenourea was added to the reaction mixture, which was heated at reflux temperature for 2 h. The 4-methyl-5-(*R*-benzyl)selenazol-2-amines **178** and 4-cyclopropyl-5-(*R*-benzyl)selenazol-2-amines **179** were prepared in 43–67% and 40–72% yields, respectively (Scheme 63) [108].

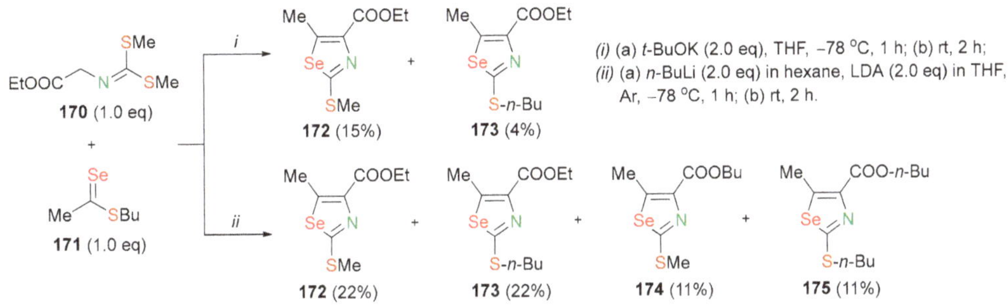

Scheme 62. Synthesis of 2-alkylthio-4,5-disubstituted-1,3-selenazoles **172–175**.

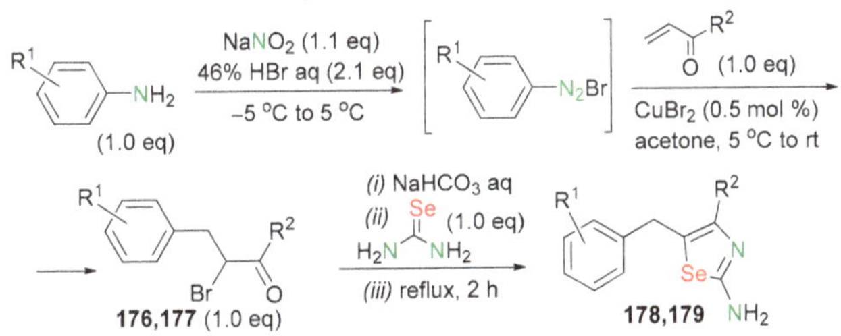

R^1	R^2 = Me: 176 (%)	R^2 = Me: 178 (%)	R^2 = c-Pr: 177 (%)	R^2 = c-Pr: 179 (%)
H	40	43	41	40
4-Me	43	43	42	45
4-CF_3O	71	67	79	78
2-F	51	57	72	72
4-F	48	54	51	56
2-CF_3	52	67	73	72
3-CF_3	69	66	60	62

Scheme 63. Synthesis of 2-amino-4-alkyl-5-(*R*-benzyl)selenazoles **178,179**.

An efficient and convenient protocol for the direct synthesis of 4-substituted 2-amino-5-nitro-1,3-selenazoles **181** from β-aminonitroalkenes and *N*-selenocyanatosaccharin **180** was developed (Scheme 64) [109]. *N*-selenocyanatosaccharin reagent was prepared from saccharin by the treatment with *tert*-butyl hypochlorite in MeOH and then with AgSeCN in CH_2Cl_2 at room temperature. Freshly prepared selenocyanatosaccharin **180** was added to β-nitroenamine in anhydrous THF at 0 °C and the reaction mixture was stirred for 30 min. Then, triethylamine was added, and the reaction mixture was stirred at room temperature for 3 h. After workup, 4-aryl-substituted 2-amino-5-nitro-1,3-selenazoles **181** were isolated in 40–76% yields. This metal-free method is convenient and features a short reaction time, mild reaction conditions, and a good compatibility of functional groups (Scheme 64) [109].

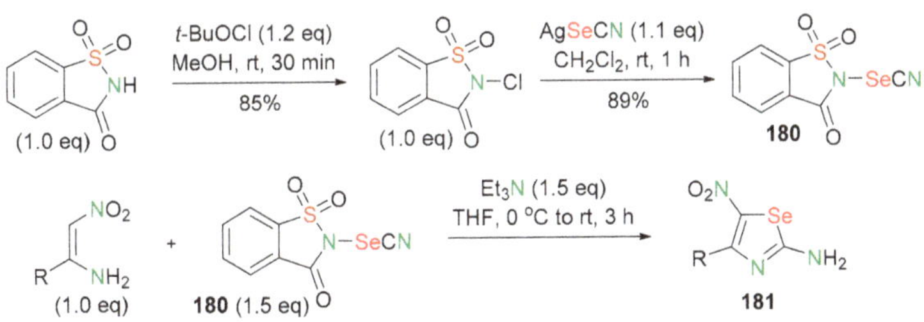

R	181 (%)
Ph	76
4-F-C_6H_4	47
3-Cl-C_6H_4	62
3,4-$(MeO)_2$-C_6H_3	52
4-Me-C_6H_4	52
2-Cl-C_6H_4	60
2-thienyl	45
3-Py	40

Scheme 64. Synthesis of 4-aryl-substituted 2-amino-5-nitro-1,3-selenazoles **181**.

An alternative one-pot method for the synthesis of the biologically promising 2-amino-1,3-selenazole skeleton under mild conditions was demonstrated (Scheme 65) [110]. The formation of a 2-amino-1,3-selenazole skeleton was made by the electrophilic selenocyanation of β-enaminones and β-enamino esters followed by intramolecular cyclization under basic conditions. The electrophilic compound, selenocyanogen, was obtained in situ from the reaction of $PhICl_2$ and KSeCN. Thus, 4,5-substituted 2-amino-1,3-selenazoles **182** were prepared in 62–90% yields. This method is based on the use of the easily accessible selenocynate salt as both the selenium and nitrogen sources for the formation of the 1,3-selenazole ring (Scheme 65) [110]. On the other hand, the classical Hantzsch approach involves the use of selenourea or selenoamide as the building block.

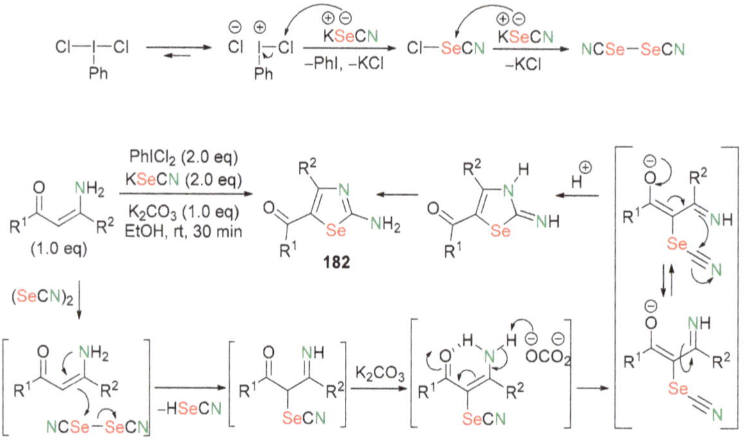

R¹	R²	182 (%)
MeO	Ph	90
MeO	4-F-C_6H_4	70
MeO	4-Cl-C_6H_4	67
MeO	4-Br-C_6H_4	62
MeO	4-Me-C_6H_4	85
MeO	4-Et-C_6H_4	82
MeO	4-MeO-C_6H_4	88
MeO	3-Br-C_6H_4	74
MeO	3-CF_3-C_6H_4	70
MeO	3-MeO-C_6H_4	85
MeO	2-I-C_6H_4	85
MeO	3,4-Cl_2-C_6H_3	68
MeO	3,4-$(MeO)_2$-C_6H_3	86
MeO	2-naphthyl	80
MeO	2-thienyl	84
MeO	Me	84
EtO	Ph	89
Ph	Ph	85
2-naphthyl	*n*-Pr	75
2-thienyl	4-Cl-C_6H_4	70
Me	Ph	90

Scheme 65. Synthesis of 4,5-substituted 2-amino-1,3-selenazoles **182** via selenocyanative cyclization of enamino esters and enaminones.

Cyclodextrins selectively catalyze reactions involving supramolecular catalysis. These reactions can effectively take place in water without the formation of any toxic wastes. In this way, 2-amino-4-alkyl- and 2-amino-4-arylselenazole-5-carboxylates **183** were synthesized in studies. The reactions were carried out by the formation of the β-cyclodextrin complex of the β-ketoester in water in situ, followed by the addition of NBS (*N*-bromosuccinimide) and selenourea. Stirring at the same temperature yielded selenazole derivatives **183** in 89–94% yields. Succinimide was obtained as a by-product, and this could be recycled to produce NBS, and β-cyclodextrin could be easily recovered and reused (Scheme 66) [111].

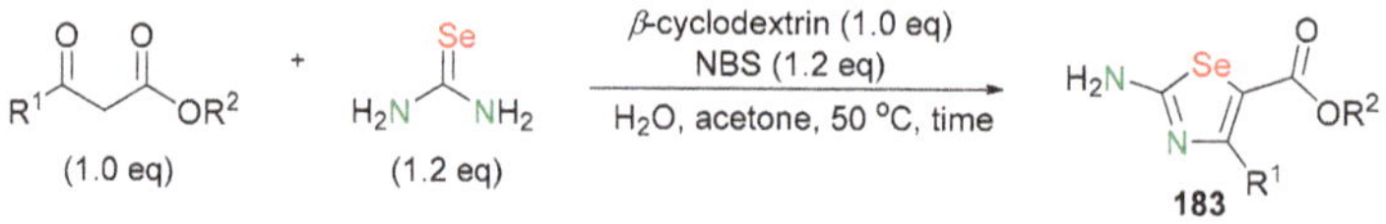

R¹	R²	time (h)	183 (%)
Me	Et	1.5	92
i-Pr	Et	2.0	90
i-Pr	Bn	2.0	93
t-Bu	Me	1.5	89
Ph	Et	1.3	94

Scheme 66. Synthesis of 2-amino-4-alkyl- and 2-amino-4-arylselenazole-5-carboxylates **183** in the presence of β-cyclodextrin.

An efficient protocol has been developed for the synthesis of selenazole-5-carboxylates **184** (94–96% yields) from β-ketoesters, selenourea, and *N*-bromosuccinamide (NBS) by using xanthan sulfuric acid (XSA) as an effective solid acid catalyst. The reaction process is very simple and the catalyst can be easily isolated from the reaction mixture and reused several times (Scheme 67) [112].

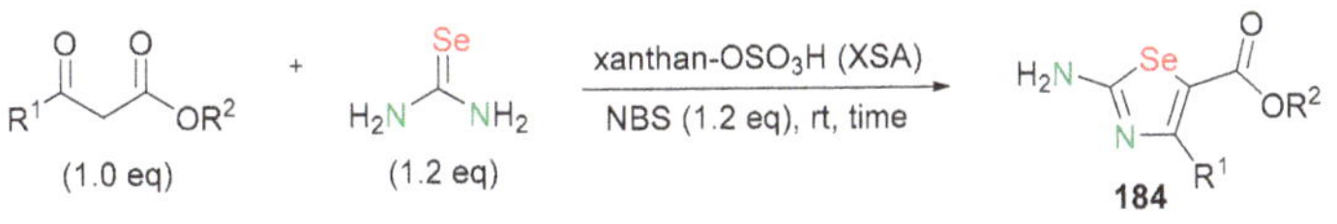

R^1	R^2	time (min)	**184** (%)
Me	Et	15	96
i-Pr	Et	15	97
i-Pr	Bn	15	95
t-Bu	Me	18	94
Ph	Et	15	97

Scheme 67. Synthesis of selenazole-5-carboxylates **184** by using xanthan sulfuric acid.

A series of methyl 2-amino-1,3-selenazole-5-carboxylates **187** bearing chiral azetidine-3-yl, pyrrolidin-2-yl, piperidin-2-yl, or piperidin-3-yl substituents at the C4 atom of the 1,3-selenazole ring were designed and synthesized (Scheme 68). The treatment of the carboxylic chiral heterocyclic amino acid with Meldrum's acid and 1-ethyl-3-(3-dimethylaminopropyl) carbodiimide (EDC) in the presence of 4-dimethylaminopyridine (DMAP), followed by heating the intermediate isolated from the reaction mixture in MeOH, afforded β-ketoesters **185**, which were then α-brominated with *N*-bromosuccinimide (NBS) in MeCN. The reaction of compounds **186** with selenourea at room temperature in MeOH yielded, to the target, methyl 2-amino-1,3-selenazole-5-carboxylates **187** in 17–42% yields (Scheme 68) [113,114].

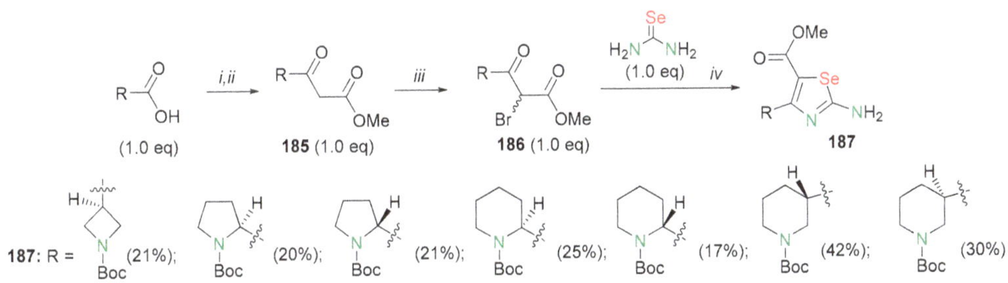

(i) Meldrum's acid (1.2 eq), DMAP (1.5 eq), EDC (1.2 eq), CH_2Cl_2, rt, 18 h; *(ii)* MeOH, reflux, 18 h; *(iii)* NBS (1.2 eq), MeCN, rt, 2 h; *(iv)* MeOH, rt, 3 h.

Scheme 68. Synthesis of 2-amino-1,3-selenazole-5-carboxylates **187**.

Prepared chiral amino acids **187** can probably be used as starting materials for the synthesis of more complex molecules with biological activity or be used as building blocks for the development of DNA-encoded chemical libraries [113,114].

The reaction of diethyl 2,4-dibromo-3-oxoglutarate **189** with selenourea afforded ethyl 2-(2-amino-5-ethoxycarbonyl-1,3-selenazol-4-yl)-2-bromoethanoate (**190**) (Scheme 69) [115]. At first, the dibromination of diethyl 3-oxoglutarate **188** with NBS in CCl_4 produced diethyl 2,4-dibromo-3-oxoglutarate **189** as a mixture of diastereomers, which were immediately used in the next reaction stage. The reaction of the dibrominated carbonyl compound **189** with selenourea was carried out under an argon atmosphere in EtOH at room temperature. After the working up of the reaction mixture and purifying the crude product, ethyl 2-(2-amino-5-ethoxycarbonyl-1,3-selenazol-4-yl)-2-bromoethanoate (**190**) was prepared in a 61% yield. Selenazole **190** is a convenient starting material for the synthesis of other 4,5-functionalized 2-amino-1,3-selenazoles. The treatment of selenazole **190** with acylating agents and various nucleophiles made it possible to obtain a series of new 4,5-disubstituted 2-amino-1,3-selenazoles **191–196** in 42–72% yields (Scheme 69) [115].

The cyclization of selenosemicarbazide (**198**) with desyl bromide resulted in 2-hydrazino-4,5-diphenyl-1,3-selenazole (**199**) in a 70% yield. Selenosemicarbazide (**198**) was obtained in a 42% yield through the intermediate cyclohexanone selenosemicarbazone (**197**) by the reflux of hydrazine hydrate with potassium selenocyanate and cyclohexanone in EtOH and further treatment of the compound **197** with *n*-BuOH, water, and EtOH (Scheme 70) [50,116]. Compound **199** is a useful building block for the synthesis of functionalized selenazoles such as 1,3-selenazoles substituted with pyrazole [50].

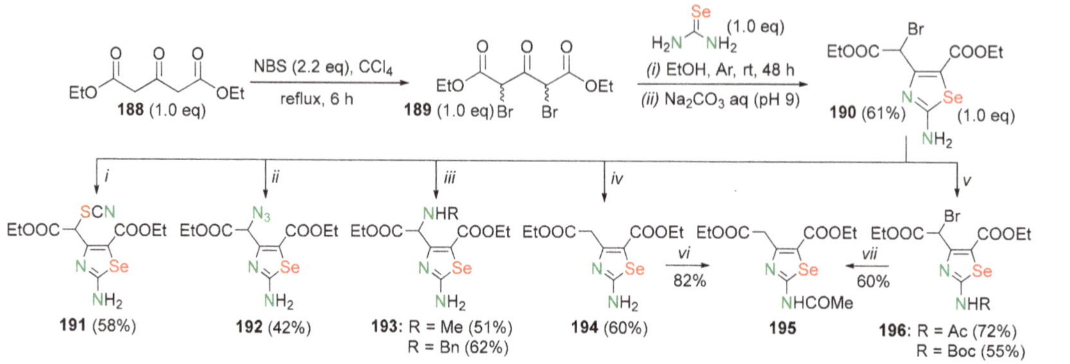

(i) KSCN (1.2 eq), acetone, Ar, rt, 48 h; *(ii)* NaN_3 (1.1 eq), DMSO, Ar, 60 °C, 6 h; *(iii)* $MeNH_2$·HCl (1.9 eq) or $BnNH_2$ (2.0 eq), K_2CO_3 (1.3 eq), Et_3N (2.6 eq), EtOH, Ar, rt, 5 days; *(iv)* EtSH (5.0 eq), Et_3N (1.9 eq), EtOH, Ar, rt, 5 days; *(v)* Ac_2O (1.9 eq) or Boc_2O (1.2 eq), DMAP (1.3 eq), CH_2Cl_2, Ar, rt, 6 h; *(vi)* Ac_2O (3.1 eq), DMAP (1.3 eq), CH_2Cl_2, Ar, rt, 6 h; *(vii)* EtSH (4.5 eq), Et_3N (1.7 eq), EtOH, Ar, rt, 5 days.

Scheme 69. Synthesis of ethyl 2-(2-amino-5-ethoxycarbonyl-1,3-selenazol-4-yl)-2-bromoethanoate (**190**) and its derivatives **191–196**.

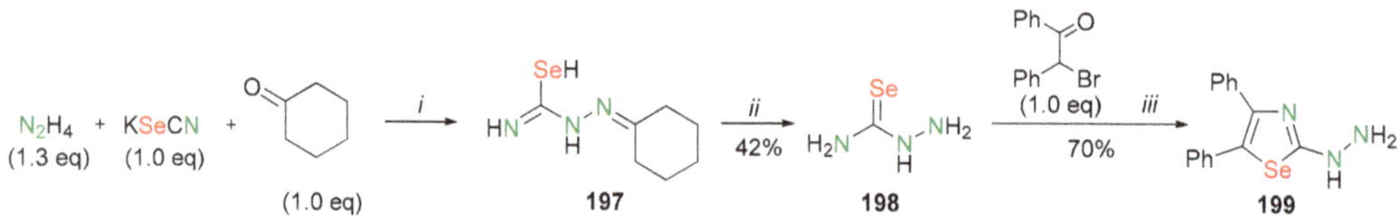

(i) EtOH, H_2O, HCl (pH 6), reflux 3 h; *(ii)* *n*-BuOH, H_2O, EtOH; *(iii)* Na (2.0 eq), EtOH, rt, 1.5 h.

Scheme 70. Synthesis of 2-hydrazino-4,5-diphenyl-1,3-selenazole (**199**).

A series of 2,4,5-trisubstituted selenazole compounds were designed and synthesized (Scheme 71) [117–119]. The starting carboxylic acids **200** were treated with $SOCl_2$ to yield acetyl chlorides **201**, which, without further purification, were reacted with aromatic compounds to yield products (**202**). Then, ketones **202** were treated with bromine to yield compounds (**203**). A suspension of products (**203**), KSeCN/SiO_2, and $AcONH_4/Al_2O_3$ was heated in toluene at 80 °C to afford 2,4,5-trisubstituted-1,3-selenazoles **204** with the amino group. Then, compounds (**204**) were acylated to yield 2,4,5-trisubstituted 1,3-selenazoles **205**. If compounds (**205**) had the methoxyl group, these compounds were treated with BBr_3 to provide phenolic compounds (**206**). Unfortunately, the yields of compounds **205** and **206** were not indicated (Scheme 71) [117–119].

Some examples of compounds (**206**) exhibited favorable PLTP (plasma phospholipid transfer protein)-inhibiting activity [119].

A consistent multicomponent one-pot reaction for the synthesis of 2-amino-1,3-selenazole **207** was developed (Scheme 72) [120]. *N,N*-Diethyl-benzamidine or acetamidine was added to a solution of isoselenocyanate in THF, and the resulting reaction mixture was stirred for 3 h at room temperature. To this reaction mixture, Et_3N was added, followed by the addition of halomethylene compound in THF, and the stirring continued until the completion of the reaction. Thus, amidino-selenourea synthesized from isoselenocyanate and *N,N*-diethyl-amidine reacted with various halomethylene compounds to furnish 2-amino-1,3-selenazole derivatives **207** in 65–89% yields (Scheme 72) [120].

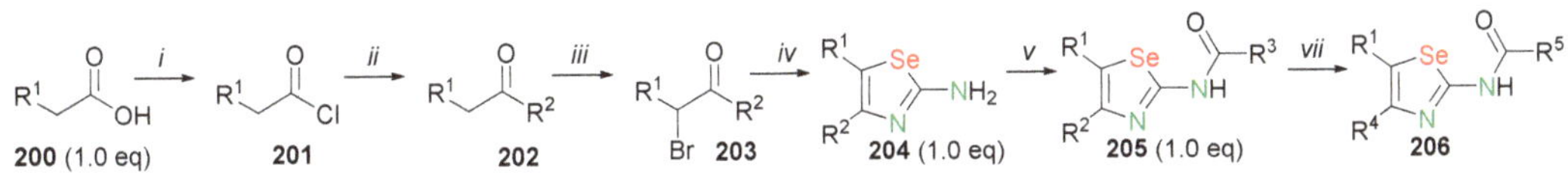

205: R^1	R^2	R^3
Et	4-MeO-C_6H_4	3,4-$(MeO)_2$-C_6H_3
Et	4-MeO-C_6H_4	3,4,5-$(MeO)_3$-C_6H_2
Et	4-MeO-C_6H_4	4-Cl-C_6H_4
Bn	4-MeO-C_6H_4	3,4-$(MeO)_2$-C_6H_3
Bn	4-MeO-C_6H_4	4-NO_2-C_6H_4
Bn	4-MeO-C_6H_4	3,4,5-$(MeO)_3$-C_6H_2
Ph	Ph	3,4,5-$(MeO)_3$-C_6H_2
Ph	Ph	3,4,5-F_3-C_6H_2
n-Bu	4-MeO-C_6H_4	3,4,5-$(MeO)_3$-C_6H_2
n-Bu	4-MeO-C_6H_4	3,4-$(MeO)_2$-C_6H_3

206: R^1	R^4	R^5
Et	4-OH-C_6H_4	3,4,5-$(OH)_3$-C_6H_2
Et	4-OH-C_6H_4	4-Cl-C_6H_4
Bn	4-OH-C_6H_4	3,4-$(OH)_2$-C_6H_3
Bn	4-OH-C_6H_4	3,4,5-$(OH)_3$-C_6H_2
Bn	4-OH-C_6H_4	Am
*Ph	Ph	3,4-$(OH)_2$-C_6H_3
Ph	Ph	3,4,5-$(OH)_3$-C_6H_2
**n*-Bu	4-OH-C_6H_4	3,4,5-$(OH)_3$-C_6H_2
n-Bu	4-OH-C_6H_4	Me
n-Bu	4-OH-C_6H_4	4-Cl-C_6H_4

(i) $SOCl_2$ (1.0 eq), CH_2Cl_2, reflux, 1 h; *(ii)* R^2H (1.0 eq), $AlCl_3$ (1.0 eq), CH_2Cl_2, rt, 2 h; *(iii)* Br_2 (1.0 eq), $AlCl_3$ (1.0 eq), CH_2Cl_2, 0 °C, 2 h; *(iv)* KSeCN/SiO_2 (5.0 eq), $AcONH_4/Al_2O_3$ (6.0 eq), toluene, 80 °C, 12 h; *(v)* R^3COCl (1.3 eq), NaOH (1.3 eq), H_2O, heating; *(vi)* BBr_3 (3.0 eq), CH_2Cl_2, 0 °C, 2.5 h.

*PLTP inhibiting activity

Scheme 71. Synthesis of 2,4,5-trisubstituted-1,3-selenazoles **205** and **206**.

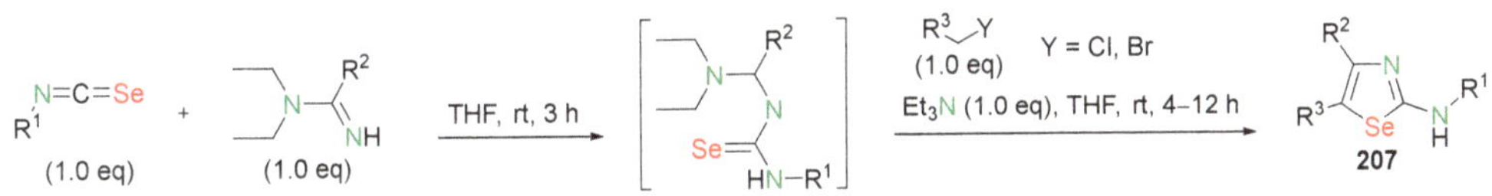

R^1	R^2	R^3	207 (%)
Bn	Ph	Ph-CO	89
Ph	Ph	4-Me-C_6H_4-CO	81
Ph	Ph	4-Cl-C_6H_4-CO	84
4-Et-C_6H_4	Me	Ph-CO	65
4-Et-C_6H_4	Me	4-Me-C_6H_4-CO	77
4-Et-C_6H_4	Me	4-Cl-C_6H_4-CO	66
4-Et-C_6H_4	Ph	Ph-CO	88
4-Et-C_6H_4	Ph	4-Me-C_6H_4-CO	78
4-Et-C_6H_4	Ph	4-Cl-C_6H_4-CO	70
4-Et-C_6H_4	Ph	2-Py	80
4-Et-C_6H_4	Ph	4-Py	88
4-Et-C_6H_4	Ph	2-quinolinyl	73
Ph	Ph	Ph-CO	89
Bn	Ph	4-Me-C_6H_4-CO	79
Bn	Ph	4-Cl-C_6H_4-CO	76
Ph	Ph	2-Py	74
4-Et-C_6H_4	Me	4-Py	87
Ph	Ph	2-quinolinyl	79
4-Cl-C_6H_4	Me	2-Py	78

Scheme 72. Synthesis of trisubstituted 2-amino-1,3-selenazole derivatives **207**.

Aryl-hydrazinyl-1,3-selenazoles **209** and aroyl-hydrazonyl-1,3-selenazoles **210** were obtained through the Hantzsch-type condensation reactions of α-halogenocarbonyl derivatives with selenosemicarbazides **208** under ordinary and microwave (MW) heating conditions (Scheme 73) [121]. Shorter reaction times and excellent yields were achieved under irradiation conditions. Starting arylidenselenosemicarbazones and 4-methoxybenzoylselenosemicarbazide were obtained by the reaction of methoxybenzoyl chloride or aromatic carbaldehydes with selenosemicarbazide.

Selenazoles **209, 210** showed moderate antiproliferative in vitro activity against two leukemia cell lines (HL60 and CCRF-CEM) and three carcinoma cell lines (HCT116, MDA-MB231, and U87MG) [121].

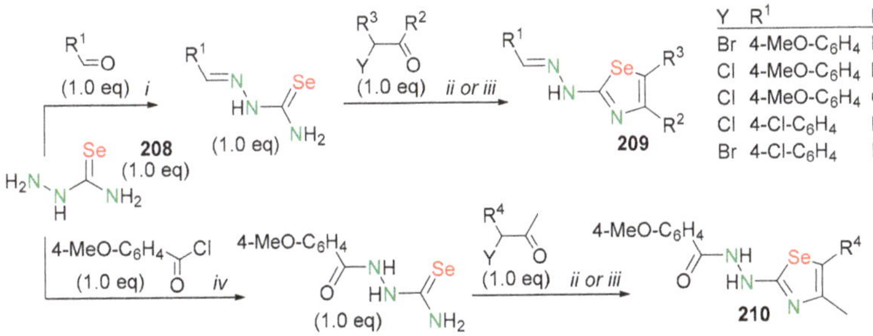

Y	R^1	R^2	R^3	209 (%) *(ii)*	209 (%) *(iii)*
Br	4-MeO-C_6H_4	EtOOC	H	52	93
Cl	4-MeO-C_6H_4	Me	MeCO	51	92
Cl	4-MeO-C_6H_4	$ClCH_2$	H	56	95
Cl	4-Cl-C_6H_4	Me	MeCO	56	94
Br	4-Cl-C_6H_4	EtOOC	H	60	91

Y	R^4	210 (%) *(ii)*	210 (%) *(iii)*
Br	EtOOC	57	94
Cl	MeOOC	63	93
Cl	H	67	87

(i) AcOH, EtOH, reflux, 4 h; *(ii)* DMF, acetone, rt, 24 h; *(iii)* DMF, acetone, MW, 60 °C, 1 h; *(iv)* 10% NaOH aq, rt, 1 h.

Scheme 73. Synthesis of aryl-hydrazinyl-1,3-selenazoles **209** and aroyl-hydrazonyl-1,3-selenazoles **210**.

Sugar derivatives of 2-amino-1,3-selenazole **215** were prepared as promising biologically active compounds (Scheme 74) [122]. The key intermediate for accessing such scaffold was *N*-benzoyl isoselenocyanate **211**, easily available by the treatment of potassium selenocyanate with benzoyl chloride. The coupling of compound **211** with hydrochloride **212** in CH_2Cl_2 in the presence of Et_3N afforded *N*-benzoyl selenourea **213** in a 98% yield. The treatment of selenourea **213** with phenacyl bromide in DMF at room temperature in the presence of diisopropylethylamine (DIEA) afforded phenacyl isoselenourea **214** in a 50% yield. The acidic treatment of isoselenourea **214**, using refluxing ethanolic AcOH, led to an intramolecular cyclodehydration to produce selenazole **215** in a 79% yield. The construction mechanism of compound **215** can be explained by the formation of appropriate intermediates (Scheme 74) [122].

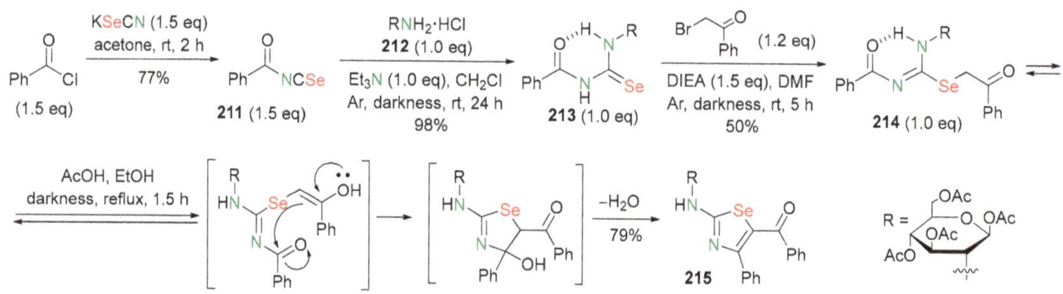

Scheme 74. Synthesis of sugar derivatives of 2-amino-1,3-selenazole **215**.

A convenient one-pot protocol for the synthesis of 2-amino selenazoles was developed that avoided using hazardous α-bromocarbonyl compounds (Scheme 75) [123]. The role of tribromoisocyanuric acid (TBCA) in the reaction was to produce α-bromo-β-ketoester in situ, which, in the presence of various selenoureas and 1,4-diazabicyclo[2.2.2]octane (DABCO), yielded the corresponding selenazoles. TBCA was added to a solution of β-ketoester in H_2O and the mixture was stirred at 70 °C for 20 min. Then, MeCN, the selenourea, and DABCO were added, and the mixture was stirred at 70 °C for 20 min. After the working up of the reaction mixture, three 4,5-substituted 2-aminoselenazoles **216** were prepared in 57–64% yields. Short reaction times, experimental simplicity, and easily accessible reagents make this protocol attractive for the synthesis of 1,3-selenazole-containing biologically active compounds (Scheme 75) [123].

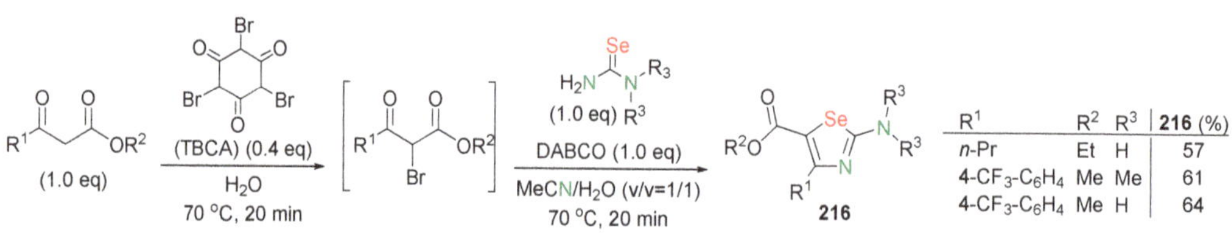

Scheme 75. Synthesis of 4,5-substituted 2-aminoselenazoles **216** using TBCA and DABCO.

A series of 2-amino-1,3-selenazole derivatives were synthesized using a transition-metal-free, multicomponent reaction through a base-promoted cascade cyclization process (Scheme 76) [124]. 2-amino-1,3-selenazoles **217** were obtained in 24–80% yields by the nucleophilic addition of amines to isoselenocyanate obtained in situ from elemental selenium and α,β-unsaturated isocyanides, followed by the Michael addition reaction and aromatization.

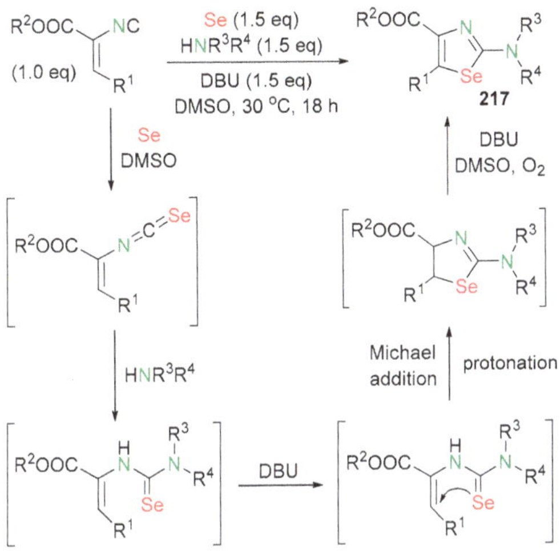

R^1	R^2	R^3	R^4	**217** (%)
Ph	Et	Et	Et	69
2-Me-C_6H_4	Et	Et	Et	74
3-Me-C_6H_4	Et	Et	Et	75
4-Me-C_6H_4	Et	Et	Et	67
4-MeO-C_6H_4	Et	Et	Et	76
Ph	Me	Et	Et	56
4-MeO-C_6H_4	Me	Et	Et	45
4-CF_3-C_6H_4	Et	Et	Et	35
2-CN-C_6H_4	Et	Et	Et	24
4-F-C_6H_4	Et	Et	Et	68
2-thienyl-C_6H_4	Et	Et	Et	57
2-Me-C_6H_4	Et	*n*-Pr	*n*-Pr	76
2-Me-C_6H_4	Et	*n*-Bu	*n*-Bu	78
2-Me-C_6H_4	Et	Me	HO$(CH_2)_2$	80
2-Me-C_6H_4	Et	Me	NC$(CH_2)_2$	25
Ph	Et	$-(CH_2)_4-$		42
Ph	Et	$-(CH_2)_5-$		48
Ph	Et	$-(CH_2)_2O(CH_2)_2-$		35
Ph	Et	$-(CH_2)_2$-1,2-benzo-CH_2-		50
2-Me-C_6H_4	Et	Me	4-CF_3-C_6H_4-O-C(Ph)-$(CH_2)_2$	53

Scheme 76. Synthesis of 2-amino-1,3-selenazole derivatives **217**.

5-acyl-1,3-selenazol-2-amines **220** were prepared from the reaction of 3,3-disubstituted 1-acylselenoureas and phenacyl bromides in the presence of Et_3N in refluxing ethanol for 15 min in 66–97% yields. At first, the reaction of acyl selenoureas with phenacyl bromide yielded salts **218**. The further Hantzsch-type cyclization of salt **218** was accompanied by the formation of intermediate heterocycle **219** with the elimination of HBr. The hydration of heterocycle **219** led to the formation of 1,3-selenazoles **220** (Scheme 77) [125].

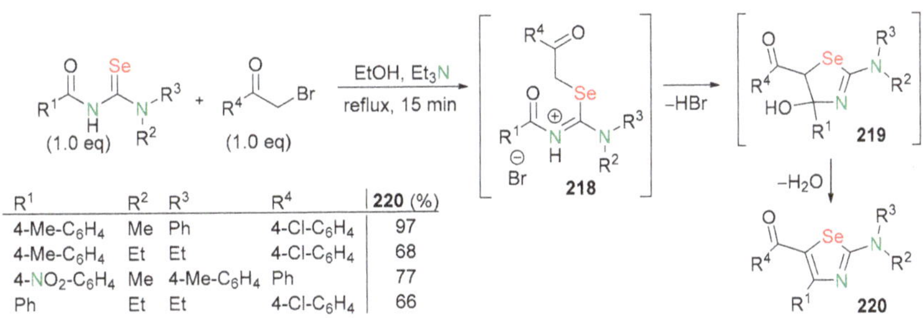

R^1	R^2	R^3	R^4	**220** (%)
4-Me-C_6H_4	Me	Ph	4-Cl-C_6H_4	97
4-Me-C_6H_4	Et	Et	4-Cl-C_6H_4	68
4-NO_2-C_6H_4	Me	4-Me-C_6H_4	Ph	77
Ph	Et	Et	4-Cl-C_6H_4	66

Scheme 77. Synthesis of 5-acyl-1,3-selenazol-2-amines **220**.

The synthetic methodology for the preparation of *N*,*N*-disubstituted 5-amino-1,3-selenazoles **222** was developed. The addition reaction of selenoamide dianions to thio- and selenoformamide in THF followed by treatment with iodine at 0 °C allowed researchers to produce 5-amino-2-selenazolines **221** in 44–91% yields. Further oxidation with iodine at room temperature afforded the 5-amino selenazoles **222** in 22–84% yields (Scheme 78) [126].

A very convenient procedure for the synthesis of 2,4,5-trisubstituted-1,3-selenazoles **227** was described (Scheme 79) [127,128]. The starting material for the preparation of selenazoles **227**, dimethyl cyanothioimidocarbonate **223**, was obtained in a good yield (71%) from cyanamide. Compound **223** was heated at 70 °C with secondary amines to form intermediates **224**. Freshly prepared Na_2Se was added to the reaction mixture under an argon atmosphere at 70 °C to form the intermediate selenolates **225**. The cyclization of the intermediate **226** was carried out in the presence of K_2CO_3. The compounds **227** were isolated as solids in 48–87% yields (Scheme 79) [127,128].

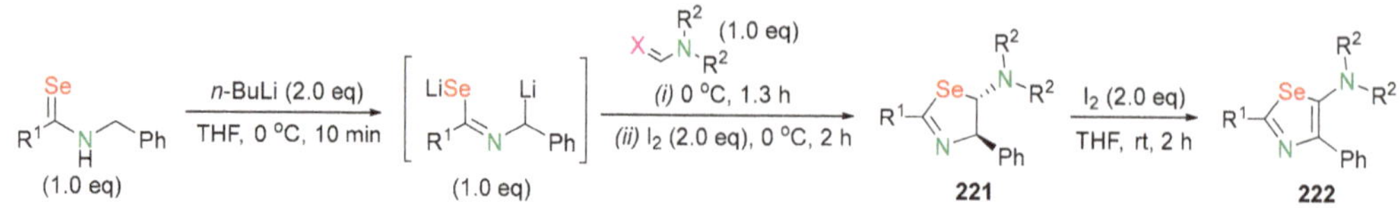

R^1	R^2	X	**221** (%)	**222** (%)
Ph	Me	S	91	46
4-Me-C_6H_4	Me	S	53	22
4-Cl-C_6H_4	Me	S	47	41

R^1	R^2	X	**221** (%)	**222** (%)
Ph	Ph	Se	47	84
2-Py	Ph	Se	78	46
4-Ph-C_6H_4	Ph	Se	44	32

Scheme 78. Synthesis of *N*,*N*-disubstituted 5-amino-1,3-selenazoles **222**.

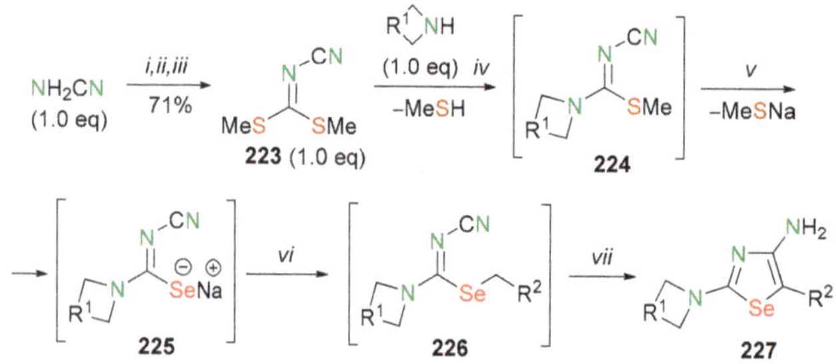

R^1	R^2	Y	halide (eq)	**227** (%)
$-CH_2OCH_2-$	CN	Cl	3.0	55
$-CH_2OCH_2-$	Ac	Cl	2.0	28
$-CH_2OCH_2-$	Ac	Cl	3.0	36
$-CH_2OCH_2-$	NO_2	Br	2.0	32
$-(CH_2)_2-$	CN	Cl	3.0	70
$-(CH_2)_2-$	COOEt	Br	2.0	52
$-(CH_2)_3-$	CN	Cl	3.0	67
$-(CH_2)_3-$	Bz	Cl	2.0	41
$-(CH_2)_3-$	4-Cl-C_6H_4-CO	Br	2.0	50

(i) KOH (2.0 eq), H_2O, rt, 45 min; *(ii)* CS_2 (1.1 eq), 40 °C, 2 h; *(iii)* MeI (2.0 eq), rt, 12 h; *(iv)* DMF, 70 °C, 1 h; *(v)* Na_2Se (1.0 eq), Ar, 70 °C, 20 min; *(vi)* YCH_2R^2 (2.0–3.0 eq), 70 °C (0 °C for $BrCH_2CN$), 2 h; *(vii)* (a) K_2CO_3 (1.0 eq), 70 °C, 1 h; (b) H_2O.

Scheme 79. Synthesis of 2,4,5-trisubstituted-1,3-selenazoles **227**.

2-dialkylamino-1,3-selenazoles **228** were produced by the reaction of ketones with *N*,*N*-unsubstituted selenoureas in the presence of $FeCl_3$. The reaction of ketones with *N*,*N*-unsubstituted selenoureas is activated by the nucleophilic addition of the nitrogen atom of the selenourea to the carbonyl carbon, resulting in the formation of 4,5-substituted 2-amino-1,3-selenazoles **228**. Selenazoles **228** were indicated to be potentially useful superoxide anion scavengers (Scheme 80) [129,130].

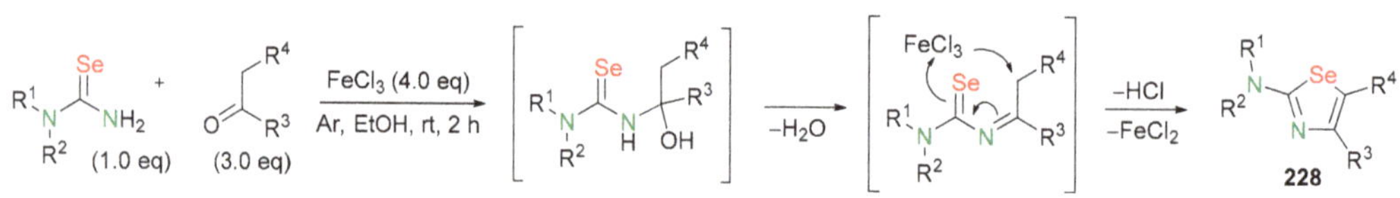

R^1	R^2	R^3	R^4	**228** (%)
Me	Me	$-(CH_2)_4-$		99
Et	Et	$-(CH_2)_4-$		87
$-(CH_2)_2O(CH_2)_2-$		$-(CH_2)_4-$		83
$-(CH_2)_4-$		$-(CH_2)_4-$		99

R^1	R^2	R^3	R^4	**228** (%)
$-(CH_2)_5-$		$-(CH_2)_4-$		97
$-(CH_2)_5-$		Et	Me	73
$-(CH_2)_5-$		Me	*n*-Pr	52
$-(CH_2)_5-$		Me	*i*-Pr	19

Scheme 80. Synthesis of 4,5-substituted 2-amino-1,3-selenazoles **228** in the presence of $FeCl_3$.

The reaction of selenoureas with methyl vinyl ketone yielded 2-amino-5-(1-ethoxymethyl)-1,3-selenazoles **229** (77–89% yields) as the only products (Scheme 81) [131]. The reaction of piperidine-1-carboselenoamide with (2*E*)-4-methylhex-2-ene yielded two types of selenazole derivatives, 2-piperidino-5-(1-ethoxyalkyl)-1,3-selenazoles **230** and 2-piperidino-1,3-selenazoles **231**, in moderate yields. The reaction with cyclohex-2-en-1-one proceeded in a similar manner to form products **232** and **233**.

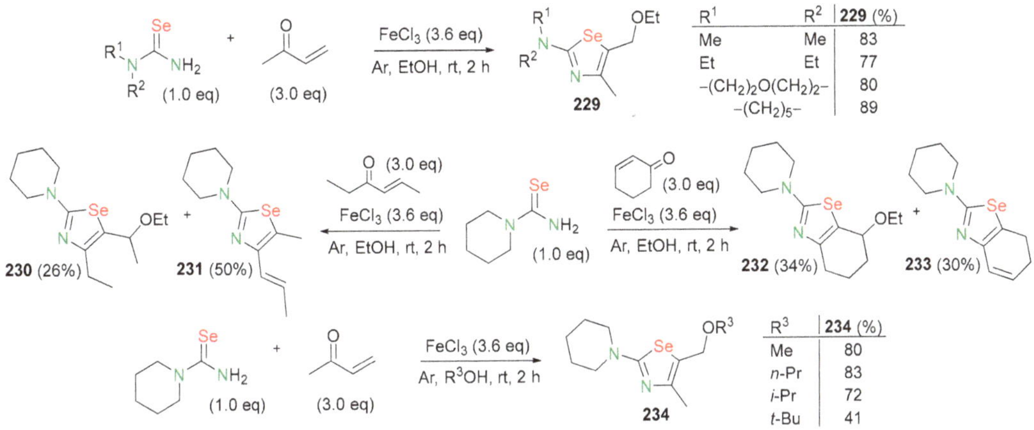

R^1	R^2	**229** (%)
Me	Me	83
Et	Et	77
$-(CH_2)_2O(CH_2)_2-$		80
$-(CH_2)_5-$		89

R^3	**234** (%)
Me	80
n-Pr	83
i-Pr	72
t-Bu	41

Scheme 81. Synthesis of 4,5-substituted 2-amino-1,3-selenazoles **229–234** from *α*,*β*-unsaturated ketones in the presence of $FeCl_3$.

Thus, selenoureas reacted with *α*,*β*-unsaturated ketones at both *α*-positions of carbonyl carbon to yield two types of products in a certain ratio. The reactions of piperidine-1-carboselenoamide with methyl vinyl ketone in MeOH, *n*-PrOH, *i*-PrOH, and *t*-BuOH led to the formation of 2-amino-5-(1-alkoxymethyl)-1,3-selenazole derivatives **234** in 41–83% yields (Scheme 81) [131].

A series of 1,3-selenazoles substituted with phthalazinone derivatives were prepared (Scheme 82) [132]. The treatment of phthalazin-1(2*H*)-one derivative with phenacyl bromide in refluxing acetone in the presence of K_2CO_3 afforded the ketones **235** in 60–77% yields.

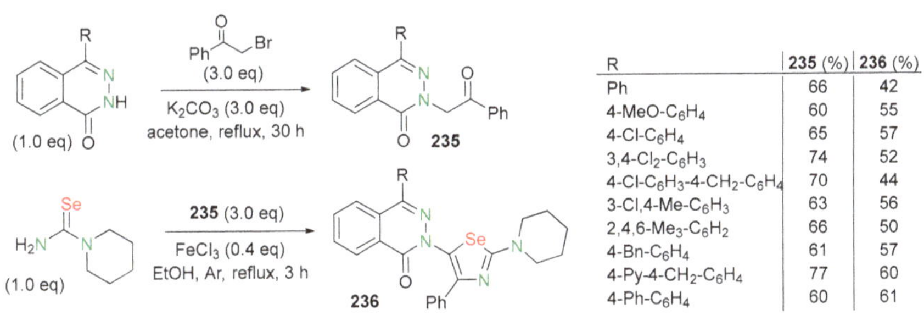

R	**235** (%)	**236** (%)
Ph	66	42
4-MeO-C_6H_4	60	55
4-Cl-C_6H_4	65	57
3,4-Cl_2-C_6H_3	74	52
4-Cl-C_6H_3-4-CH_2-C_6H_4	70	44
3-Cl,4-Me-C_6H_3	63	56
2,4,6-Me_3-C_6H_2	66	50
4-Bn-C_6H_4	61	57
4-Py-4-CH_2-C_6H_4	77	60
4-Ph-C_6H_4	60	61

Scheme 82. Synthesis of 1,3-selenazolyl phthalazinone derivatives **236**.

The reaction of 4-substituted 2-(2-phenylethyl)phthalazin-1(2*H*)-ones **235** with piperidine-containing selenoamide in EtOH under an argon atmosphere in the presence of $FeCl_3$ afforded 1,3-selenazolyl phthalazinone derivatives **236** in 42–61% yields (Scheme 82) [132].

An antimicrobial activity study showed that all synthesized 1,3-selenazoles **236** exhibited moderate-to-good antifungal and antibacterial activities against pathogenic strains [132].

4-aryl-1,3-selenazol-2-amines **241** with electron-withdrawing groups are conveniently accessible via the reaction of selenourea derivatives **240** with activated bromomethylene compounds (Scheme 83) [133].

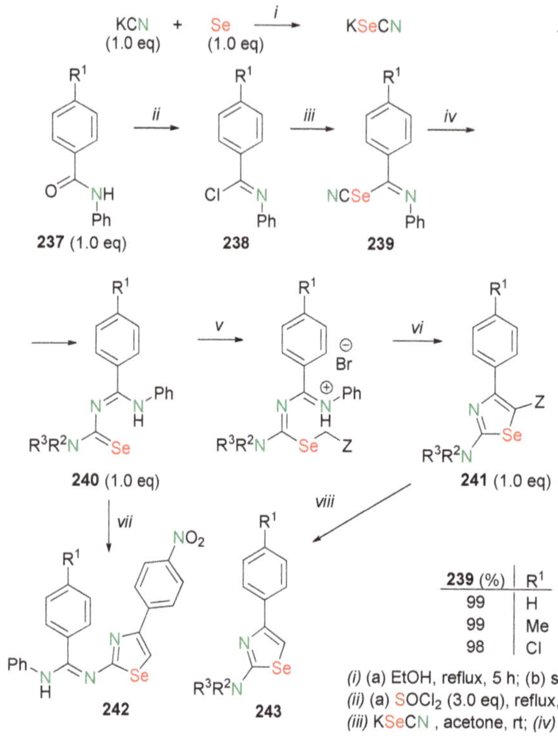

240 (%)	241 (%)	R^1	R^2	R^3	Z
92	90	H	$-(CH_2)_2O(CH_2)_2-$		4-NO_2-C_6H_4CO
–	88	H	$-(CH_2)_2O(CH_2)_2-$		EtOCO
74	97	H	$-(CH_2)_4-$		4-NO_2-C_6H_4CO
–	80	H	$-(CH_2)_4-$		EtOCO
80	98	H	$-(CH_2)_5-$		4-MeO-C_6H_4CO
65	–	H	H	H	–
84	95	H	Me	Ph	4-NO_2-C_6H_4CO
–	79	H	Me	Ph	EtOCO
–	85	H	Me	Ph	CN
85	96	Me	$-(CH_2)_2O(CH_2)_2-$		PhCO
–	95	Me	$-(CH_2)_2O(CH_2)_2-$		4-NO_2-C_6H_4CO
–	81	Me	$-(CH_2)_2O(CH_2)_2-$		H_2NCO
–	85	Me	$-(CH_2)_2O(CH_2)_2-$		EtOCO
85	90	Me	Me	Ph	CN
–	86	Me	Me	Ph	4-NC-C_6H_4
82	95	Cl	$-(CH_2)_2O(CH_2)_2-$		4-NO_2-C_6H_4CO
–	96	Cl	$-(CH_2)_2O(CH_2)_2-$		EtOCO
–	95	Cl	$-(CH_2)_2O(CH_2)_2-$		CN
97	92	Cl	*n*-Bu	*n*-Bu	4-NO_2-C_6H_4CO
–	94	Cl	*n*-Bu	*n*-Bu	EtOCO
75	–	Cl	H	H	–
92	99	Cl	Me	Ph	4-NO_2-C_6H_4CO
–	85	Cl	Me	Ph	EtOCO

239 (%)	R^1
99	H
99	Me
98	Cl

242 (%)	R^1
98	H
87	Cl

243 (%)	R^1	R^2	R^3
71	Me	$-(CH_2)_2O(CH_2)_2-$	
76	Cl	Me	Ph

(i) (a) EtOH, reflux, 5 h; (b) solvent evaporation and drying, 120 °C, 12 h; *(ii)* (a) $SOCl_2$ (3.0 eq), reflux, 3 h; (b) excess $SOCl_2$ evaporation and drying in vacuo 24 h; *(iii)* KSeCN , acetone, rt; *(iv)* NR^2R^3 (1.0 eq), rt, 30 min; then H_2O, a few h; *(v)* ZCH_2Br (1.0 eq), acetone, rt, a few min; *(vi)* (a) Et_3N or NaH (1.3 eq), acetone, rt, 30 min; (b) ice/H_2O, 1h; *(vii)* (a) 4-NO_2-C_6H_4CO-CH_2Br (1.0 eq), acetone, reflux, 30 min; (b) triturating in 5% NaOH aq, 1h; *(viii)* acetone, 5% NaOH aq (3.0 eq), reflux, 2 h.

Scheme 83. Synthesis of 1,3-selenazoles **237–243**.

The reaction of *N*-phenylbenzamides **237** with excess $SOCl_2$ under reflux yielded *N*-phenylbenzimidoyl chlorides **238**, which, upon treatment with KSeCN in acetone, produced imidoyl isoselenocyanates **239** in quantitative yields. They were converted into selenourea derivatives **240** by reaction with primary and secondary amines or NH_3. Compounds **(240)** reacted with activated bromomethylene compounds such as acetamides, acetonitriles, and 2-bromoacetates, as well as 4-cyanobenzyl bromide and phenacyl bromides, to yield 1,3-selenazol-2-amines **241** in 80–99% yields. The described synthesis of 1,3-selenazoles **241** is cheap and avoids using toxic compounds.

The reaction of selenourea derivatives **240** with an unsubstituted NH_2 group led to 1,3-selenazoles **242** in up to 98% yields. In this case, an alternative ring closure process occurred. On treatment with 5% aqueous NaOH solution, ethyl 1,3-selenazole-5-carboxylates **241** were saponified and decarboxylated to yield 5-unsubstituted 1,3-selenazoles **243** (Scheme 83) [133].

3. Conclusions

The analysis of literature data shows that the chemistry of 1,3-selenazoles, including the study of their properties and especially biological activity, is of increasing interest to scientists.

There are examples of reactions using catalytic systems with $FeCl_3$, $CuBr_2$, $Sc(OTf)_3$, PEG-400, benzyl triethylammonium chloride (BTEAC), NaF, $[Ir(cod)Cl]_2$, β-cyclodextrin, PCl_5, and xanthan sulfuric acid (XSA). The use of ultrasound or microwave irradiation significantly reduces the reaction time and increases the yields of the target products. The

Hantzsch reaction has continued its new development including various modifications in the synthesis of 1,3-selenazoles.

A large number of Se-containing compounds are used as selenation agents, including elemental Se, NaHSe (usually obtained from Se and $NaBH_4$), KSeCN, and Woollins' reagent.

Most of the developed methods relate to 2,4-substituted 1,3-selenazoles. Many reactions directed to 2,4-substituted 1,3-selenazole derivatives have used α-bromoketones, α-bromaldehydes, selenoureas, and selenoamides as starting reagents. Regarding the starting materials, α-bromoketones and α-bromaldehydes are available compounds. Selenourea and its derivatives are relatively expensive compounds and this circumstance can be considered a negative factor for the method, which uses the selenourea derivatives. The substituted selenoureas can be synthesized from secondary amines and organic isoselenocyanates. The latter compounds, in turn, serve as excellent starting materials for the synthesis of selenium/nitrogen heterocycles [134]. However, isoselenocyanates can be considered as not yet sufficiently accessible compounds. Although the classical method for obtaining organic isoselenocyanates is based on the reaction of isonitriles with elemental selenium, a number of other methods involve the use of highly toxic carbon diselenide [134].

In contrast, selenoamides are very accessible reagents. An effective and simple method that is commonly used to prepare selenoamides is the reaction of NaHSe with aryl or alkyl nitriles. Sodium hydroselenide can be easily prepared by the reaction of elemental selenium with sodium borohydride. Also, very convenient and reliable methods to obtain selenoamides are based on the reaction of nitriles with Woollins' reagent and P_2Se_5.

Thus, although both approaches based on selenourea and selenamide derivatives lead to the target heterocyclic products in high yields, selenoamides are more accessible starting compounds compared to the selenourea derivatives and the approach with the use of selenoamides has an advantage in terms of starting compound availability.

The synthesis of ferrocene-bearing 1,3-selenazoles **15** demonstrates the broad scope of this approach (Scheme 8). Selenazoles **15** exhibit high antibacterial activity against *Escherichia coli*, *Staphylococcus aureus*, and *Pseudomonas aeruginosa* [53]. Using this approach, it is possible to combine two cycles in one molecules and to obtain bridged ring systems. Compounds **(19)** containing two 1,3-selenazole rings, separated by carbon bridges of different lengths, were obtained from dinitriles, P_2Se_5, and phenacyl bromide (Scheme 9) [50]. The synthesis of compound **17**, bearing the oxazole and selenazole rings, was also developed.

Selenazoles **34** and **36**, which contain two chlorine atoms or two iodine atoms linked to the selenium atom, exhibit antiviral activity against HIV-1 (Scheme 14) [58].

The convenient synthesis of a chromenone derivative **45**, containing the 1,3-selenazol ring, was developed by two approaches (Scheme 17) [61].

Some approaches to 1,3-selenazoles use eco-friendly reaction conditions and catalyst-free protocols for the synthesis of target products and meet the requirements of green chemistry (for example, the synthesis of 2-amino-1,3-selenazoles **70** through Hantzsch-type condensation (Scheme 27)) [71].

Compounds **(72)** show relatively high activity and low toxicity against the fungal strains *C. albicans* and *C. neoformans* [72].

According to research results, the products with F, NO_2, CN, and MeO substituents and the cyclohexanone and 2-methylcyclohexanone derivatives **78** exhibit antifungal activity [73]. Compounds **(78)** with 3,4-dichlorophenyl and adamantanyl moieties show antibacterial activity [73].

The selenazoles **81** exhibit high antifungal activity against *Candida* spp. and antibacterial activity against Gram-positive bacteria. In addition, some compounds demonstrate significant anticonvulsant activity [74].

Compounds **(83)** were found to be active against Gram-positive bacteria, both pathogenic staphylococci and *Staphylococcus aureus* [75].

The selenazolyl-hydrazones **84** exhibit the properties of selective monoamine oxidase inhibitors with antioxidant and antiproliferative activities. These compounds are obtained by the reaction of α-bromocarbonyl derivatives and selenosemicarbazones [76].

Functionalized 1,3-selenazoles **86** were prepared based on selenosemicarbazide and 4-oxo-4*H*-chromene-3-carboxaldehyde. Compounds (**86**) show cytotoxic activity against Hep-G2, HCT-116, A549, and MCF-7 cancer cell lines [77].

Compounds **88** and **89** exhibit significant cytotoxic activities against two human cancer cell lines, hepatocarcinoma cells (Hep-G2) and prostate cancer cells (DU-145) [78].

The combination of the selenazole ring with sulfonamide moiety is favorable in respect to antitumor activity. Compounds (**140**) show potent inhibitory effects against breast cancer cell lines (MDA-MB-231) and human prostate cancer cells (PC3). In addition, these compounds show antibacterial activity [93,94].

Thus, significant progress has been made in the development and synthesis of 1,3-selenazole derivatives, and this area of research is currently being intensively developed. A number of promising compounds with high biological activity have been found that may find application in the future.

Author Contributions: Methodology and writing—original draft preparation, N.A.M.; editing and supervision, S.V.A.; formal analysis, A.S.F.; conceptualization and writing—review and editing, V.A.P.; data curation, M.V.M. All authors have read and agreed to the published version of the manuscript.

Funding: The financial support of the Russian Science Foundation (Grant No 22-13-00339) is gratefully acknowledged.

Institutional Review Board Statement: Not applicable.

Informed Consent Statement: Not applicable.

Data Availability Statement: Not applicable.

Acknowledgments: The authors thank Baikal Analytical Center SB RAS for providing the literature data.

Conflicts of Interest: The authors declare no conflicts of interest.

References

1. Longtin, R. A forgotten debate: Is selenocysteine the 21st amino acid? *J. Nat. Cancer Inst.* **2004**, *96*, 504–505. [CrossRef] [PubMed]
2. Iwaoka, M.; Arai, K. From sulfur to selenium. A new research arena in chemical biology and biological chemistry. *Curr. Chem. Biol.* **2013**, *7*, 2–24. [CrossRef]
3. Gandhil, U.H.; Nagaraja, T.P.; Prabhu, K.S. Selenoproteins and their role in oxidative stress and inflammation. *Curr. Chem. Biol.* **2013**, *7*, 65–73. [CrossRef]
4. Refaay, D.A.; Ahmed, D.M.; Mowafy, A.M.; Shaaban, S. Evaluation of novel multifunctional organoselenium compounds as potential cholinesterase inhibitors against Alzheimer's disease. *Med. Chem. Res.* **2022**, *31*, 894–904. [CrossRef]
5. Mamgain, R.; Kostic, M.; Singh, F.V. Synthesis and antioxidant properties of organoselenium compounds. *Curr. Med. Chem.* **2023**, *30*, 2421–2448. [CrossRef]
6. Obieziurska-Fabisiak, M.; Pacuła-Miszewska, A.J.; Laskowska, A.; Ścianowski, J. Organoselenium compounds as antioxidants. *Arkivoc* **2023**, *2023*, 69–92. [CrossRef]
7. Orlov, A.A.; Eletskaya, A.A.; Frolov, K.A.; Golinets, A.D.; Palyulin, V.A.; Krivokolysko, S.G.; Kozlovskaya, L.I.; Dotsenko, V.V.; Osolodkin, D.I. Probing chemical space of tick-borne encephalitis virus reproduction inhibitors with organoselenium compounds. *Arch. Pharm. Chem. Life Sci.* **2018**, *351*, e1700353. [CrossRef]
8. Qiao, J.; Zhao, C.; Liu, J.; Du, Y. Design and synthesis of selenazole-substituted ritonavir analogs. *Bioorg. Med. Chem. Lett.* **2018**, *28*, 2379–2381. [CrossRef]
9. Eibergen, N.R.; Im, I.; Patel, N.Y.; Hergenrother, P.J. Identification of a novel protein synthesis inhibitor active against gram-positive bacteria. *ChemBioChem* **2012**, *13*, 574–583. [CrossRef]
10. Marković, S.B.; Maciejewska, N.; Olszewski, M.; Višnjevac, A.; Puerta, A.; Padrón, J.M.; Novaković, I.; Kojić, S.; Fernandes, H.S.; Sousa, S.F.; et al. Study of the anticancer potential of Cd complexes of selenazoyl-hydrazones and their sulfur isosters. *Eur. J. Med. Chem.* **2022**, *238*, 114449. [CrossRef]
11. Wu, T.-Y.; Chen, X.-C.; Tang, G.-X.; Shao, W.; Li, Z.-C.; Chen, S.-B.; Huang, Z.-S.; Tan, J.-H. Development and characterization of benzoselenazole derivatives as potent and selective *c-MYC* transcription inhibitors. *J. Med. Chem.* **2023**, *66*, 5484–5499. [CrossRef] [PubMed]

12. Araškov, J.B.; Nikolić, M.; Armaković, S.; Armaković, S.; Rodić, M.; Višnjevac, A.; Padrón, J.M.; Todorović, T.R.; Filipović, N.R. Structural, antioxidant, antiproliferative and in–silico study of pyridine-based hydrazonyl–selenazoles and their sulphur isosteres. *J. Mol. Struct.* **2021**, *1240*, 130512. [CrossRef]
13. Al-Tamimi, A.-M.S.; Etxebeste-Mitxeltorena, M.; Sanmartín, C.; Jiménez-Ruiz, A.; Syrjänen, L.; Parkkila, S.; Selleri, S.; Carta, F.; Angeli, A.; Supuran, C.T. Discovery of new organoselenium compounds as antileishmanial agents. *Bioorg. Chem.* **2019**, *86*, 339–345. [CrossRef]
14. Wang, H.; Yue, Y.; Zhao, H.; Wu, H.; Jiang, K.; Li, S.; Zhao, M.; Lin, F. Neuroprotective effects of 2-substituted 1, 3-selenazole amide derivatives on amyloid-beta-induced toxicity in a transgenic caenorhabditis elegans model of Alzheimer's disease. *Neurotoxic. Res.* **2021**, *39*, 841–850. [CrossRef]
15. Šmelcerović, A.; Tomović, K.; Šmelcerović, Ž.; Petronijević, Ž.; Kocić, G.; Tomašič, T.; Jakopin, Ž.; Anderluh, M. Xanthine oxidase inhibitors beyond allopurinol and febuxostat; an overview and selection of potential leads based on in silico calculated physico-chemical properties, predicted pharmacokinetics and toxicity. *Eur. J. Med. Chem.* **2017**, *135*, 491–516. [CrossRef]
16. Jain, V.K.; Priyadarsini, K.I. Selenium compounds as promising antiviral agents. *New J. Chem.* **2024**, *48*, 6534–6552. [CrossRef]
17. Leyssen, P.; Van Lommel, A.; Drosten, C.; Schmitz, H.; De Clercq, E.; Neyts, J. A novel model for the study of the therapy of Flavivirus infections using the Modoc virus. *Virology* **2001**, *279*, 27–37. [CrossRef]
18. Paragas, J.; Whitehouse, C.A.; Endy, T.P.; Bray, M. A simple assay for determining antiviral activity against Crimean-Congo hemorrhagic fever virus. *Antivir. Res.* **2004**, *62*, 21–25. [CrossRef]
19. Chung, D.-H.; Sun, Y.; Parker, W.B.; Arterburn, J.B.; Bartolucci, A.; Jonsson, C.B. Ribavirin reveals a lethal threshold of allowable mutation frequency for Hantaan virus. *J. Virol.* **2007**, *81*, 11722–11729. [CrossRef]
20. Roth, J.P.; Li, J.K.-K.; Smee, D.F.; Morrey, J.D.; Barnard, D.L. A recombinant, infectious human parainfluenza virus type 3 expressing the enhanced green fluorescent protein for use in high-throughput antiviral assays. *Antivir. Res.* **2009**, *82*, 12–21. [CrossRef]
21. Tao, H.; Peng, L.; Zhang, Q. Synthesis of azole-enriched cyclic peptides by a clean solid-phase-based cyclization-cleavage strategy. *ACS Comb. Sci.* **2013**, *15*, 447–451. [CrossRef] [PubMed]
22. Tao, H.; Weng, Y.; Zhuo, R.; Chang, G.; Urbatsch, I.L.; Zhang, Q. Design and synthesis of selenazole-containing peptides for cocrystallization with P-glycoprotein. *ChemBioChem* **2011**, *12*, 868–873. [CrossRef] [PubMed]
23. Aller, S.G.; Yu, J.; Ward, A.; Weng, Y.; Chittaboina, S.; Zhuo, R.; Harrell, P.M.; Trinh, Y.T.; Zhang, Q.; Urbatsch, I.L.; et al. Structure of P-glycoprotein reveals a molecular basis for poly-specific drug binding. *Science* **2009**, *323*, 1718–1722. [CrossRef] [PubMed]
24. Singh, S.; Prasad, N.R.; Kapoor, K.; Chufan, E.E.; Patel, B.A.; Ambudkar, S.V.; Talele, T.T. Design, synthesis, and biological evaluation of (*S*)-valine thiazole-derived cyclic and noncyclic peptidomimetic oligomers as modulators of human P-glycoprotein (ABCB1). *ChemBioChem* **2014**, *15*, 157–169. [CrossRef]
25. Yu, L.-J.; Gai, W.; Yang, Q.-F.; Xiang, J.-F.; Sun, H.-X.; Li, Q.; Wang, L.-X.; Guan, A.-J.; Tang, Y.-L. Recognizing parallel-stranded G-quadruplex by cyanine dye dimer based on dual-site binding mode. *Chin. Chem. Lett.* **2015**, *26*, 705–708. [CrossRef]
26. Lukashov, S.S.; Makovenko, I.E.; Losytskyy, M.Y.; Slominskii, Y.L.; Yarmoluk, S.M. Interaction of cyanine dyes with nucleic acids. Meso-methylsubstituted trimethincyanines, as possible probes for fluorescent nucleic acid detection. *Biopolym. Cell.* **2001**, *17*, 448–454. [CrossRef]
27. Hong, S.Y.; Jun, H.; Yoon, S.S.; Kang, C.; Suh, M. New fluorescent stains for protein detection in sodium dodecyl sulfate–polyacrylamide gels. *Chem. Lett.* **2004**, *33*, 318–319. [CrossRef]
28. Volkova, K.D.; Kovalska, V.B.; Losytskyy, M.Y.; Bento, A.; Reis, L.V.; Santos, P.F.; Almeida, P.; Yarmoluk, S.M. Studies of benzothiazole and benzoselenazole squaraines as fluorescent probes for albumins detection. *J. Fluoresc.* **2008**, *18*, 877–882. [CrossRef]
29. Volkova, K.D.; Kovalska, V.B.; Losytskyy, M.Y.; Reis, L.V.; Santos, P.F.; Almeida, P.; Lynch, D.E.; Yarmoluk, S.M. Aza-substituted squaraines for the fluorescent detection of albumins. *Dyes Pigm.* **2011**, *90*, 41–47. [CrossRef]
30. Li, Z.-C.; Wu, T.-Y.; Zeng, S.-T.; Fang, L.; Mao, J.-X.; Chen, S.-B.; Huang, Z.-S.; Chen, X.-C.; Tan, J.-H. Benzoselenazolium-based hemicyanine dye for G-Quadruplex detection. *Bioorg. Med. Chem. Lett.* **2022**, *70*, 128801. [CrossRef]
31. Deligeorgiev, T.; Timcheva, I.; Maximova, V.; Gadjev, N.; Vassilev, A.; Jacobsen, J.-P.; Drexhage, K.-H. Homodimeric monomethine cyanine dyes SOSO-1 and TOTO-1-6C—Synthesis and fluorescence properties in the presence of nucleic acids. *Dyes Pigm.* **2004**, *61*, 79–84. [CrossRef]
32. Kurutos, A.; Ryzhova, O.; Trusova, V.; Gorbenko, G.; Gadjev, N.; Deligeorgiev, T.J. Symmetric *meso*-chloro-substituted pentamethine cyanine dyes containing benzothiazolyl/benzoselenazolyl chromophores novel synthetic approach and studies on photophysical properties upon interaction with bio-objects. *J. Fluoresc.* **2016**, *26*, 177–187. [CrossRef] [PubMed]
33. Jędrzejewska, B.; Bajorek, A.; Moraczewska, J. Interaction of carbocyanine dyes with DNA: Synthesis and spectroscopic studies. *Appl. Spectrosc.* **2013**, *67*, 672–680. [CrossRef] [PubMed]
34. Martynov, A.V.; Makhaeva, N.A.; Amosova, S.V. Synthesis of *E*,*E*-bis(chloromethylidene) derivatives of *N*-organylthiomorpholines and -selenomorpholines and their quaternary salts. *J. Sulfur Chem.* **2014**, *35*, 502–511. [CrossRef]
35. Potapov, V.A.; Ishigeev, R.S.; Amosova, S.V.; Borodina, T.N. Synthesis of a novel family of water-soluble 2*H*,3*H*-[1,3]thia- and -selenazolo[3,2-*a*]pyridin-4-ium heterocycles by annulation reactions. *Tetrahedron Lett.* **2019**, *60*, 475–479. [CrossRef]

36. Potapov, V.A.; Ishigeev, R.S.; Shkurchenko, I.V.; Zinchenko, S.V.; Amosova, S.V. Natural compounds and their structural analogs in regio- and stereoselective synthesis of new families of water-soluble 2*H*,3*H*-[1,3]thia- and -Selenazolo[3,2-*a*]pyridin-4-ium heterocycles by annulation reactions. *Molecules* **2020**, *25*, 376. [CrossRef]
37. Ishigeev, R.S.; Amosova, S.V.; Potapov, V.A. Regioselective synthesis of new [1,3]thiazolo- and [1,3]selenazolo[3,2-*a*]pyridinium derivatives. *Russ. J. Gen. Chem.* **2023**, *93*, S87–S93. [CrossRef]
38. Potapov, V.A.; Ishigeev, R.S.; Belovezhets, L.A.; Amosova, S.V. A novel family of selenazolo[3,2-a]pyridinium derivatives based on annulation reactions and comparative analysis of antimicrobial activity of the selenium and sulfur analogs of chalcogenazolo[3,2-*a*]pyridiniums. *Curr. Org. Chem.* **2025**, *29*. [CrossRef]
39. Koketsu, M.; Ishihara, H. Synthesis of 1,3-selenazine and 1,3-selenazole and their biological activities. *Curr. Org. Chem.* **2003**, *7*, 175–185. [CrossRef]
40. Ninomiya, M.; Garud, D.R.; Koketsu, M. Biologically significant selenium-containing heterocycles. *Coord. Chem. Rev.* **2011**, *255*, 2968–2990. [CrossRef]
41. Langer, P. Adventures in 1,3-selenazole chemistry. *Synlett* **2022**, *33*, 728–736. [CrossRef]
42. Elsherbini, M.; Hamama, W.S.; Zoorob, H.H. Recent advances in the chemistry of selenium-containing heterocycles: Five-membered ring systems. *Coord. Chem. Rev.* **2016**, *312*, 149–177. [CrossRef]
43. Banerjee, B.; Koketsu, M. Recent developments in the synthesis of biologically relevant selenium-containing scaffolds. *Coord. Chem. Rev.* **2017**, *339*, 104–127. [CrossRef]
44. Ruberte, A.C.; Sanmartin, C.; Aydillo, C.; Sharma, A.K.; Plano, D. Development and therapeutic potential of selenazo compounds. *J. Med. Chem.* **2020**, *63*, 1473–1489. [CrossRef] [PubMed]
45. Below, H.; Pfeiffer, W.-D.; Geisler, K.; Lalk, M.; Langer, P. 1,3-Selenazole. *Eur. J. Org. Chem.* **2005**, *2005*, 3637–3639. [CrossRef]
46. Below, H.; Pfeiffer, W.-D.; Geisler, K.; Saghyan, A.S.; Fischere, C.; Langer, P. Synthesis of 2-unsubstituted 1,3-selenazoles by cyclization of selenoformamide with α-bromocarbonyl compounds. *J. Heterocycl. Chem.* **2015**, *52*, 592–596. [CrossRef]
47. Geisler, K.; Jacobs, A.; Künzler, A.; Mathes, M.; Girrleit, I.; Zimmermann, B.; Bulka, E.; Pfeiffer, W.-D.; Langer, P. Efficient synthesis of primary selenocarboxylic amides by reaction of nitriles with phosphorous (V) selenide. *Synlett* **2002**, *2002*, 1983–1986. [CrossRef]
48. Geisler, K.; Künzler, A.; Below, H.; Bulka, E.; Pfeiffer, W.-D.; Langer, P. Efficient synthesis of 2-unsubstituted 1,3-selenazoles. *Synlett* **2003**, *2003*, 1195–1197. [CrossRef]
49. Geisler, K.; Künzler, A.; Below, H.; Bulka, E.; Pfeiffer, W.-D.; Langer, P. Synthesis and reactivity of 2-acyl-1,3-selenazoles. *Synthesis* **2003**, *2004*, 97–105. [CrossRef]
50. Geisler, K.; Pfeiffer, W.-D.; Künzler, A.; Below, H.; Bulka, E.; Langer, P. Synthesis of 1,3-selenazoles and bis(selenazoles) from primary selenocarboxylic amides and selenourea. *Synthesis* **2004**, *35*, 875–884. [CrossRef]
51. Hua, G.; Du, J.; Slawin, A.M.Z.; Woollins, J.D. A synthetic and structural study of arylselenoamides and 2,4-diaryl-1,3- selenazoles. *Synlett* **2014**, *25*, 2189–2195. [CrossRef]
52. Zhang, P.-F.; Chen, Z.-C. Hypervalent iodine in synthesis 53: Synthesis of 2,4-disubstituted and 2,4,5-trisubstituted 1,3-selenazoles. *Synthesis* **2000**, *2000*, 1219–1222. [CrossRef]
53. Al-Rubaie, A.Z.; Al-Jadaan, S.A.S.; Muslim, S.K.; Saeed, E.A.; Ali, E.T.; Al-Hasani, A.K.J.; Al-Salman, H.N.K.; Al-Fadal, S.A.M. Synthesis, characterization and antibacterial activity of some new ferrocenyl selenazoles and 3,5-diferrocenyl-1,2,4-selenadiazole. *J. Organomet. Chem.* **2014**, *774*, 43–47. [CrossRef]
54. Koketsu, M.; Takenaka, Y.; Ishihara, H. Syntheses of cyanoselenoamides and diselenoamides: Conversion into selenazoles and selenazines. *Heteroat. Chem.* **2003**, *14*, 106–110. [CrossRef]
55. Yoshimatsu, M.; Yamamoto, T.; Sawa, A.; Kato, T.; Tanabe, G.; Muraoka, O. α-Sulfanyl and α-selanyl propadienyl cations: Regioselective generations and cycloadditions with thioamides and selemides controlled by $MeNO_2$-H_2O system. *Org. Lett.* **2009**, *11*, 2952–2955. [CrossRef]
56. Hua, G.; Du, J.; Slawin, A.M.Z.; Woollins, J.D. 2,4-Diaryl-1,3-chalcogen azoles bearing pentafluorosulfanyl SF5 groups: A synthetic and structural study. *J. Org. Chem.* **2014**, *79*, 3876–3886. [CrossRef]
57. Zhang, P.-F.; Chen, Z.-C. Hypervalent iodine in synthesis 50: A novel method of synthesis of selenazoles by cyclocondensation of selenoamides and alkynyl(phenyl)iodonium salts. *J. Heterocycl. Chem.* **2001**, *38*, 503–505. [CrossRef]
58. Al-Rubaie, A.Z.; Al-Masoudi, W.A.; Hameed, A.J.; Yousif, L.Z.; Graia, M. Synthesis, reaction and antiviral activity of 2,4-diaryl-1,3-selenazoles. *J. Korean Chem. Soc.* **2008**, *52*, 36–46. [CrossRef]
59. Majnooni, S.; Duffield, J.; Price, J.; Khosropour, A.R.; Zali-Boeini, H.; Beyzavi, M.H. Aryliodoazide synthons: A different approach for diversified synthesis of 2-aminothiazole, 1,3-thiazole, and 1,3-selenazole scaffolds. *ACS Comb. Sci.* **2019**, *21*, 516–521. [CrossRef]
60. Zhao, H.-C.; Shi, Y.-P.; Liu, Y.-M.; Li, C.-W.; Xuan, L.-N.; Wang, P.; Zhang, K.; Chen, B.-Q. Synthesis and antitumor-evaluation of 1,3-selenazole-containing 1,3,4-thiadiazole derivatives. *Bioorg. Med. Chem. Lett.* **2013**, *23*, 6577–6579. [CrossRef]
61. Dyachenko, V.D.; Pugach, Y.Y. Synthesis of 2-(4′-morpholin-4″-yl-5*H*-chromeno-[2,3-*d*]pyrimidin-2′-yl)phenol from salicylaldehyde and substituted acrylonitriles. *Russ. J. Gen. Chem.* **2012**, *82*, 921–926. [CrossRef]
62. Dyachenko, I.V.; Dyachenko, V.D.; Abakarov, G.M.; Nenajdenko, V.G. New derivatives of (carbamoselenoyl)acetic acid for the synthesis of functionally substituted selenazoles. *Russ. J. Org. Chem.* **2021**, *57*, 1188–1190. [CrossRef]

63. Battula, S.R.K.; Putta, V.P.R.K.; Subbareddy, G.V.; Chakravarthy, I.E.; Saravanan, V. A divergent and metal free synthesis of sulfoximine tethered imidazoles, imidazopyridines, imidazothiazoles, imidazobenzothiazines, thiazoles and selenazoles. *Org. Biomol. Chem.* **2017**, *15*, 3742–3755. [CrossRef] [PubMed]
64. Qiao, J.; Liu, Y.; Du, Y. Method to build 2,4-substituted selenazole from β-azido diselenide and carboxylic acid: A formal synthesis of selenazofurin. *Tetrahedron* **2018**, *74*, 3061–3068. [CrossRef]
65. Banothu, J.; Vaarla, K.; Bavantula, R.; Crooks, P.A. Sodium fluoride as an efficient catalyst for the synthesis of 2,4-disubstituted-1,3-thiazoles and selenazoles at ambient temperature. *Chin. Chem. Lett.* **2014**, *25*, 172–175. [CrossRef]
66. Ramesh, G.; Janardhan, B.; Rajitha, B. Green approach: An efficient synthesis of 2,4-disubstituted-1,3-thiazoles and selenazoles in aqueous medium under ultrasonic irradiation. *Res. Chem. Intermed.* **2015**, *41*, 8099–8109. [CrossRef]
67. Narender, M.; Reddy, M.S.; Kumar, V.P.; Reddy, V.P.; Nageswar, Y.V.D.; Rao, K.R. Supramolecular synthesis of selenazoles using selenourea in water in the presence of β-cyclodextrin under atmospheric pressure. *J. Org. Chem.* **2007**, *72*, 1849–1851. [CrossRef]
68. Madhav, J.V.; Kuarm, B.S.; Rajitha, B. Solid-state synthesis of 1,3-selenazoles employing $CuPy_2Cl_2$ as a Lewis acid catalyst. *Synth. Commun.* **2008**, *38*, 3514–3522. [CrossRef]
69. Hassell-Hart, S.; Speranzini, E.; Srikwanjai, S.; Hossack, E.; Roe, S.M.; Fearon, D.; Akinbosede, D.; Hare, S.; Spencer, J. Synthesis of a thiazole library via an iridium-catalyzed sulfur ylide insertion reaction. *Org. Lett.* **2022**, *24*, 7924–7927. [CrossRef]
70. Lalithamba, H.S.; Narendra, N.; Naik, S.A.; Sureshbabu, V.V. Ultrasound mediated synthesis of 2-amino-1,3-selenazoles derived from Fmoc/Boc/Z-α-amino acids. *Arkivoc* **2010**, *2010*, 77–90. [CrossRef]
71. Facchinetti, V.; Avellar, M.M.; Nery, A.C.S.; Gomes, C.R.B.; Vasconcelos, T.R.A.; de Souza, M.V.N. An eco-friendly, Hantzsch-based, solvent-free approach to 2-aminothiazoles and 2-aminoselenazoles. *Synthesis* **2016**, *48*, 437–440. [CrossRef]
72. Kuchar, J.; Reinhold, K.; Rösgen, V.; Nöthling, N.; Lehmann, C.W.; Mohr, F. Synthesis, reactivity and antimicrobial activity of a series of 2-arylamino-1,3-selenazoles. *Molecules* **2021**, *26*, 7695. [CrossRef] [PubMed]
73. Łączkowski, K.Z.; Misiura, K.; Biernasiuk, A.; Malm, A. Discovery and evaluation of efficient selenazoles with high antifungal activity against *Candida* spp. *Med. Chem.* **2015**, *11*, 118–127. [CrossRef] [PubMed]
74. Łączkowski, K.Z.; Biernasiuk, A.; Baranowska-Łączkowska, A.; Zielińska, S.; Sałat, K.; Furgała, A.; Misiura, K.; Malm, A. Synthesis, antimicrobial and anticonvulsant screening of small library of tetrahydro-2*H*-thiopyran-4-yl based thiazoles and selenazoles. *J. Enzyme Inhib. Med. Chem.* **2016**, *31*, 24–39. [CrossRef]
75. Łączkowski, K.Z.; Motylewska, K.; Baranowska-Łączkowska, A.; Biernasiuk, A.; Misiura, K.; Malm, A.; Fernández, B. Synthesis, antimicrobial evaluation and theoretical prediction of NMR chemical shifts of thiazole and selenazole derivatives with high antifungal activity against *Candida* spp. *J. Mol. Struct.* **2016**, *1108*, 427–437. [CrossRef]
76. Elshaflu, H.; Todorović, T.R.; Nikolić, M.; Lolić, A.; Višnjevac, A.; Hagenow, S.; Padrón, J.M.; García-Sosa, A.T.; Djordjević, I.S.; Grubišić, S.; et al. Selenazolyl-hydrazones as novel selective MAO inhibitors with antiproliferative and antioxidant activities: Experimental and in-silico studies. *Front. Chem.* **2018**, *6*, 247. [CrossRef]
77. Ali, T.E.; Assiri, M.A.; Aboelwafa, H.R. Synthesis and cytotoxicity properties of some novel functionalized 2-{2-[(4-oxo-4*H*-chromen-3-yl)methylene]-hydrazinyl}-1,3-selenazoles. *Heterocycles* **2021**, *102*, 920–929. [CrossRef]
78. Zaharia, V.; Ignat, A.; Ngameni, B.; Kuete, V.; Moungang, M.L.; Fokunang, C.N.; Vasilescu, M.; Palibroda, N.; Cristea, C.; Silaghi-Dumitrescu, L.; et al. Heterocycles 23: Synthesis, characterization and anticancer activity of new hydrazinoselenazole derivatives. *Med. Chem. Res.* **2013**, *22*, 5670–5679. [CrossRef]
79. Kanoh, K.; Ishihara, H.; Koketsu, M. Preparation of 2-dialkylamino-1,3-selenazoles by reaction of *N,N*-unsubstituted selenoureas with α,α-diketones. *Heterocycles* **2007**, *74*, 1009–1014. [CrossRef]
80. Hua, G.; Du, J.; Slawin, A.M.Z.; Woollins, J.D. Synthesis and structural study of novel selenation derivatives of *N,N*-dialkylcyanamides. *ChemistrySelect* **2016**, *1*, 6810–6817. [CrossRef]
81. Geisler, K.; Pfeiffer, W.-D.; Müller, C.; Nobst, E.; Bulka, E.; Langer, P. Synthesis and functionalization of 4-halomethyl-1,3-selenazoles. *Synthesis* **2003**, *2003*, 1215–1220. [CrossRef]
82. Madhav, B.; Murthy, S.N.; Anil Kumar, B.S.P.; Ramesh, K.; Nageswar, Y.V.D. A tandem one-pot aqueous phase synthesis of thiazoles/selenazoles. *Tetrahedron Lett.* **2012**, *53*, 3835–3838. [CrossRef]
83. Potewar, T.M.; Ingale, S.A.; Srinivasan, K.V. An efficient and eco-friendly synthesis of 2-amino-1,3-selenazoles in an ionic liquid/water system under ambient conditions. *Arkivoc* **2008**, *2008*, 117–125. [CrossRef]
84. Maradolla, M.; Chandramouli, G.V.P. A convenient synthesis of 2-amino-1,3-selenazoles using ionic liquid and microwave irradiation. *Phosphorus Sulfur Silicon Relat. Elem.* **2011**, *186*, 1650–1654. [CrossRef]
85. Hua, G.; Du, J.; Slawin, A.M.Z.; Woollins, J.D. Synthesis and single crystal structures of substituted-1,3-selenazol-2-amines. *Molecules* **2017**, *22*, 46. [CrossRef]
86. Madhu, C.; Panguluri, N.R.; Narendra, N.; Panduranga, V.; Sureshbabu, V.V. One-pot synthesis of orthogonally protected dipeptide selenazoles employing N a -amino selenocarboxamides and a -bromomethyl ketones. *Tetrahedron Lett.* **2014**, *55*, 6831–6835. [CrossRef]
87. Dyachenko, I.V.; Dyachenko, V.D. Ethyl 3-amino-3-selenoxopropanoate as a new reagent for the synthesis of selenium-containing heterocycles. *Russ. J. Gen. Chem.* **2015**, *85*, 1673–1676. [CrossRef]
88. Dyachenko, I.V.; Dyachenko, V.D.; Dorovatovskii, P.V.; Khrustalev, V.N.; Nenaidenko, V.G. Multicomponent synthesis of thiazole, selenazole, pyrane, and pyridine derivatives, initiated by the Knoevenagel reaction. *Russ. J. Org. Chem.* **2019**, *55*, 215–226. [CrossRef]

89. Frolov, K.A.; Dotsenko, V.V.; Krivokolysko, S.G. Synthesis of 3-aminoselenoacrylamides and 3-amino-2-(selenazol-2-yl)acrylonitriles. *Chem. Heterocycl. Compd.* **2013**, *48*, 1886–1888. [CrossRef]
90. Dyachenko, V.D.; Roman, S.V. Unexpected formation of ethyl 5-cyano-6-[cyano(4′-phenyl-1,3′-selenazol-2-yl)methyl]-4-(furan-2′-yl)-2-methyl-1,4-dihydropyridine-3-carboxylate. *Russ. J. Gen. Chem.* **2011**, *81*, 442–443. [CrossRef]
91. Pizzo, C.; Mahler, S.G. Synthesis of selenazoles by *in situ* cycloisomerization of propargyl selenoamides using oxygen−selenium exchange reaction. *J. Org. Chem.* **2014**, *79*, 1856–1860. [CrossRef] [PubMed]
92. Afzal, M.S.; Pitteloud, J.-P.; Buccella, D. Enhanced ratiometric fluorescent indicators for magnesium based on azoles of the heavier chalcogens. *Chem. Commun.* **2014**, *50*, 11358–11361. [CrossRef] [PubMed]
93. Angeli, A.; Trallori, E.; Ferraroni, M.; Di Cesare Mannelli, L.; Ghelardini, C.; Supuran, C.T. Discovery of new 2, 5-disubstituted 1,3-selenazoles as selective human carbonic anhydrase IX inhibitors with potent anti-tumor activity. *Eur. J. Med. Chem.* **2018**, *157*, 1214–1222. [CrossRef] [PubMed]
94. Angeli, A.; Pinteala, M.; Maier, S.S.; Del Prete, S.; Capasso, C.; Simionescu, B.C.; Supuran, C.T. Inhibition of bacterial α-, β- and γ-class carbonic anhydrases with selenazoles incorporating benzenesulfonamide moieties. *J. Enzyme Inhib. Med. Chem.* **2019**, *34*, 244–249. [CrossRef] [PubMed]
95. Macias-Benitez, P.; Sierra-Padilla, A.; Guerra, F.M.; Moreno-Dorado, F.J. Microwave-assisted one-pot telescoped synthesis of 2-amino-1,3-thiazoles, selenazoles, imidazo[1,2-a]pyridines, and other heterocycles from alcohols. *J. Org. Chem.* **2024**, *89*, 4628–4646. [CrossRef]
96. Koketsu, M.; Kogami, M.; Ando, H.; Ishihara, H. Preparation of 5-acyl-2-amino-1,3-selenazoles by the reaction of selenazadienes with α-haloketone. *Synthesis* **2006**, *2006*, 31–36. [CrossRef]
97. Sekiguchi, A.; Nishina, A.; Kimura, H.; Fukumoto, R.-h.; Kogami, M.; Ishihara, H.; Koketsu, M. Bis-(2-amino-5-selenazoyl) Ketone as a Superoxide Anion-Scavenger. *Biol. Pharm. Bull.* **2006**, *29*, 1404–1407. [CrossRef]
98. Koketsu, M.; Nada, F.; Mio, T.; Ishihara, H. Syntheses of 4-selenazolones. Hetero Diels-Alder reaction of the selenazadienes with DMAD. *Heterocycles* **2003**, *60*, 1211–1218. [CrossRef]
99. Koketsu, M.; Mio, T.; Ishihara, H. Facile preparation of 1,3-selenazole-5-carboxylic acids and the carboxylates by reaction of selenazadienes with chloroacetyl chloride preparation of 1,3-selenazole-5-carboxylates. *Synthesis* **2004**, *2004*, 233–236. [CrossRef]
100. Koketsu, M.; Imagawa, M.; Mio, T.; Ishihara, H. Preparation of 1,3-selenazoles using selenazadienes. *J. Heterocycl. Chem.* **2005**, *42*, 831–834. [CrossRef]
101. Koketsu, M.; Kanoh, K.; Ishihara, H. Synthesis of 2-amino-1,3-selenazoles by reaction of *N*,*N*-unsubstituted selenoureas with α,β-unsaturated aldehydes. *Heterocycles* **2006**, *68*, 2647–2652. [CrossRef]
102. Koketsu, M.; Takenaka, Y.; Ishihara, H. Preparation of 1,3-selenazol-4-one derivatives from primary selenoamides and haloacyl halides. *Synthesis* **2001**, *2001*, 731–734. [CrossRef]
103. Zhang, C.; Sun, X.; Pu, Y.; Li, C.; Sun, L.; Wang, J.; Li, Y. Design, synthesis and activities of multiheterocyclic modified novel molecules using 1,3-selenazole as template. *Chin. J. Org. Chem.* **2017**, *37*, 440–454. [CrossRef]
104. Zhang, C.; Li, Y.; Li, J.; Wang, J.; Wang, H.; Gong, R. Synthesis of two novel 1,3-selenazole carbonyl hydrazones and their specific fluorescent recognition toward acetate ions. *Chin. J. Appl. Chem.* **2018**, *35*, 197–205. [CrossRef]
105. Guan, Q.; Cheng, Z.; Ma, X.; Wang, L.; Feng, D.; Cui, Y.; Bao, K.; Wu, L.; Zhang, W. Synthesis and bioevaluation of 2-phenyl-4-methyl-1,3-selenazole-5-carboxylic acids as potent xanthine oxidase inhibitors. *Eur. J. Med. Chem.* **2014**, *85*, 508–516. [CrossRef]
106. Yoshimatsu, M.; Matsui, M.; Yamamoto, T.; Sawa, A. Convenient preparation of 4-arylmethyl- and 4-hetarylmethyl thiazoles by regioselective cycloaddition reactions of 3-sulfanyl- and selanylpropargyl alcohols. *Tetrahedron* **2010**, *66*, 7975–7987. [CrossRef]
107. Časar, Z.; Majcen-Le Maréchal, A.; Lorcy, D. A novel approach to a substituted 1,3-selenazole core as a precursor of electron-rich olefins: Diselenadiazafulvalene and azino-diselenadiazafulvalene. *New J. Chem.* **2003**, *27*, 1622–1626. [CrossRef]
108. Ostapiuk, Y.V.; Ostapiuk, M.Y.; Barabash, O.V.; Kravets, M.; Herzberger, C.; Namyslo, J.C.; Obushak, M.D.; Schmidt, A. One-pot syntheses of substituted 2-aminothiazoles and 2-amino-selenazoles via Meerwein arylation of alkyl vinyl ketones. *Synthesis* **2022**, *54*, 3658–3666. [CrossRef]
109. Yang, H.; Chen, Y.; Xu, X.; Li, Z. Metal-free synthesis of thiocyanated aminonitroalkenes and 2-aminothiazoles/selenazoles from β-aminonitroalkenes and *N*-thio/selenocyanatosaccharin. *Synlett* **2023**, *34*, 176–182. [CrossRef]
110. Tao, S.; Xu, L.; Yang, K.; Zhang, J.; Du, Y. Construction of the 2-amino-1,3-selenazole skeleton via $PhICl_2$/KSeCN-mediated selenocyanation/cyclization. *Org. Lett.* **2022**, *24*, 4187–4191. [CrossRef]
111. Narender, M.; Reddy, M.S.; Kumar, V.P.; Srinivas, B.; Sridhar, R.; Venkata, Y.; Nageswar, D.; Rao, K.R. Aqueous-phase one-pot synthesis of 2-aminothiazole- or 2-aminoselenazole-5-carboxylates from β-keto esters, thiourea or selenourea, and *N*-bromosuccinimide under supramolecular catalysis synthesis of 2-aminothiazole- and 2-aminoselenazole-5-carboxylates. *Synthesis* **2007**, *2007*, 3469–3472. [CrossRef]
112. Kuarm, B.S.; Madhav, J.V.; Rajitha, B. Xanthan sulfuric acid: An efficient bio-supported and recyclable solid acid catalyst for the synthesis of 2-aminothiazole-5-carboxylates and 2-aminoselenazole-5-carboxylates. *Lett. Org. Chem.* **2011**, *8*, 549–553. [CrossRef]
113. Malinauskienė, V.; Kveselytė, A.; Dzedulionytė, K.; Bieliauskas, A.; Burinskas, S.; Sløk, F.A.; Šačkus, A. *L*-Proline and related chiral heterocyclic amino acids as scaffolds for the synthesis of functionalized 2-amino-1,3-selenazole-5-carboxylates. *Chem. Heterocycl. Compd.* **2018**, *54*, 469–473. [CrossRef]
114. Dzedulionytė, K.; Voznikaitė, P.; Bieliauskas, A.; Malinauskienė, V.; Sløk, F.A.; Šačkus, A. Methyl 2-amino-4-[1-(*tert*-butoxycarbonyl)azetidin-3-yl]-1,3-selenazole-5-carboxylate. *Molbank* **2021**, *2021*, M1207. [CrossRef]

115. Kanapickaitė, V.; Martynaitis, V.; Šačkus, A. Facile synthesis of novel functionalized 1,3-selenazoles. *Arkivoc* **2009**, *2009*, 268–276. [CrossRef]
116. Fry, D.J.; Keogh, P.J. Production of selenium compounds. Patent GB 1326379 A, 8 August 1973.
117. Kodomari, M.; Aoyama, T.; Suzuki, Y. One-pot synthesis of 2-aminothiazoles using supported reagents. *Tetrahedron Lett.* **2002**, *43*, 1717–1720. [CrossRef]
118. Poirier, D.; Mérand, Y.; Labrie, F. Synthesis of 17β-estradiol derivatives with *N*-butyl, *N*-methyl alkylamides side chain at position 15. *Tetrahedron* **1991**, *47*, 7751–7766. [CrossRef]
119. Ling, C.; Zheng, Z.; Jiang, X.C.; Zhong, W.; Li, S. Synthesis of a series of novel 2,4,5-trisubstituted selenazole compounds as potential PLTP inhibitors. *Bioorg. Med. Chem. Lett.* **2010**, *20*, 5123–5125. [CrossRef]
120. Ranjan, A.; Yerande, R.; Jadhav, M.; Yerande, S.G.; Dethem, D.H. One-pot synthesis of 2-amino-1,3-selenazole via an intermediary amidinoselenourea. *Eur. J. Org. Chem.* **2015**, *2015*, 3230–3234. [CrossRef]
121. Ignat (Grozav), A.; Gaina, L.; Kuete, V.; Silaghi-Dumitrescu, L.; Efferth, T.; Zaharia, V. Microwave-assisted synthesis of new selenazole derivatives with antiproliferative activity. *Molecules* **2013**, *18*, 4679–4688. [CrossRef]
122. López, Ó.; Maza, S.; Ulgar, V.; Maya, I.; Fernández-Bolaños, J.G. Synthesis of sugar-derived isoselenocyanates, selenoureas, and selenazoles. *Tetrahedron* **2009**, *65*, 2556–2566. [CrossRef]
123. Pereira, F.A.C.; de Andrade, V.S.C.; Souza, E.A.; de Mattos, M.C.S.; Oliveira, D.F. 2-Aminoselenazoles and 2-aminothiazoles: One-pot synthesis and control of the fungus Colletotrichum lindemuthianum in common beans. *Pest Manag. Sci.* **2022**, *78*, 1665–1676. [CrossRef] [PubMed]
124. Liu, H.-W.; Fang, Y.; Wang, S.-Y.; Ji, S.-J. Base-promoted multicomponent reactions: A synthesis of 2-amino-1,3-selenazole derivatives. *J. Org. Chem.* **2020**, *85*, 3508–3516. [CrossRef] [PubMed]
125. Mohr, F. Old selenium heterocycles revisited: Synthesis, spectroscopic, and structural characterization of *N*-Acyl-1,3-selenazol-2(3*H*)-imines and 5-acyl-1,3-selenazol-2-amines from acylselenourea derivatives. *J. Heterocycl. Chem.* **2014**, *51*, 1435–1441. [CrossRef]
126. Murai, T.; Yamaguchi, K.; Hori, F.; Maruyama, T. Reaction of selenoamide dianions with thio- and selenoformamides leading to the formation of 5-aminoselenazoles: Photophysical and electrochemical properties. *J. Org. Chem.* **2014**, *79*, 4930–4939. [CrossRef]
127. Thomae, D.; Perspicace, E.; Hesse, S.; Kirsch, G.; Seck, P. Synthesis of substituted [1,3]thiazolo[4,5-*b*]pyridines and [1,3]thiazolo-[4,5-*d*][1,2,3]triazines. *Tetrahedron* **2008**, *64*, 9309–9314. [CrossRef]
128. Thomae, D.; Perspicace, E.; Xu, Z.; Henryon, D.; Schneider, S.; Hesse, S.; Kirsch, G.; Seck, P. One-pot synthesis of new 2,4,5-trisubstituted 1,3-thiazoles and 1,3-selenazoles. *Tetrahedron* **2009**, *65*, 2982–2988. [CrossRef]
129. Sekiguchi, A.; Nishina, A.; Kimura, H.; Fukumoto, R.-h.; Kanoh, K.; Ishihara, H.; Koketsu, M. Superoxide anion-scavenging effect of 2-amino-1,3-selenazoles. *Chem. Pharm. Bull.* **2005**, *53*, 1439–1442. [CrossRef]
130. Koketsu, M.; Kanoh, K.; Ando, H.; Ishihara, H. A facile synthesis of 2-amino-1,3-selenazole by reaction of *N,N*-unsubstituted selenourea with ketone. *Heteroat. Chem.* **2006**, *17*, 88–92. [CrossRef]
131. Koketsu, M.; Kanoh, K.; Ishihara, H. Preparation of 2-amino-1,3-selenazoles by reaction of *N,N*-unsubstituted selenoureas with α,β-unsaturated ketones in alcohol. *Heterocycles* **2006**, *68*, 2145–2152. [CrossRef]
132. El-Shamy, I.E.; Abdel-Mohsen, A.M.; Al-Shehri, M.M.; El-Hashash, M.A.; Al-Shamrani, K.M. Selenium containing heterocycles: Synthesis and antimicrobial evaluation of some new 4-substituted-2-(4-phenyl-2-(piperidin-1-yl)-1, 3-selenazol-5-yl) phthalazin-1(2*H*)-ones. *Life Sci. J.* **2014**, *11*, 385–391. [CrossRef]
133. Zhou, Y.; Linden, A.; Heimgartner, H. Selenium-containing heterocycles from isoselenocyanates: Synthesis of 1,3-selenazoles from *N*-phenylimidoyl isoselenocyanates. *Helv. Chim. Acta* **2000**, *83*, 1576–1598. [CrossRef]
134. Garud, D.R.; Koketsu, M.; Ishihara, H. Isoselenocyanates: A powerful tool for the synthesis of selenium-containing heterocycles. *Molecules* **2007**, *12*, 504–535. [CrossRef] [PubMed]

biomolecules

MDPI

Review

Unlocking the Potential of Bioactive Compounds in Pancreatic Cancer Therapy: A Promising Frontier

Silvia Brugiapaglia [1,†], Ferdinando Spagnolo [2,3,†] and Claudia Curcio [1,*]

1 Department of Molecular Biotechnology and Health Sciences, Piazza Nizza 44b, 10126 Turin, Italy; silvia.brugiapaglia@unito.it
2 School of Advanced Defence Studies, Defence Research & Analysis Institute, Piazza della Rovere 83, 00165 Rome, Italy; ferdinando.spagno57@edu.unito.it
3 Defense Institute for Biomedical Sciences, Via Santo Stefano Rotondo 4, 00184 Rome, Italy
* Correspondence: claudia.curcio@unito.it
† These authors contributed equally to this work.

Abstract: Pancreatic ductal adenocarcinoma (PDA) is a highly challenging malignancy to treat, with a high mortality rate and limited therapeutic options. Despite advances in cancer research, the prognosis for patients diagnosed with PDA is often poor due to late-stage detection and resistance to conventional therapies. Consequently, there is growing interest in the potential of bioactive compounds as alternative or adjuvant treatments, given their ability to target multiple aspects of cancer biology, offering a more holistic approach to treatment. In the context of PDA, certain bioactive compounds, such as polyphenols (found in fruits, vegetables, and tea), flavonoids, carotenoids and compounds in cruciferous vegetables, have shown potential in inhibiting cancer cell growth, reducing inflammation, and promoting cancer cell apoptosis. This review aims to elucidate the mechanisms, by which these bioactive compounds exert their effects, modulating the oxidative stress, influencing inflammatory pathways and regulating cell survival and death. It also highlights current clinical trials that are paving the way toward incorporating these natural agents into mainstream treatment strategies, with the goal of boosting the efficacy of conventional therapies for PDA.

Keywords: pancreatic cancer; bioactive compounds; flavonoids

1. Introduction

Cancer universally represents one of the largest public health concerns, substantially contributing to global disease burden and mortality, with complex etiological factors including genetic predisposition, environmental exposures, and lifestyle choices contributing to its onset and progression [1,2]. Among lifestyle factors, diet has emerged as a crucial modifiable risk factor that can influence cancer development and progression [2].

Bioactive compounds are extra nutritional constituents that typically occur in small quantities in foods [3], such as vegetables, whole grains, and spices—has garnered significant attention for their potential protective and therapeutic effects against various types of cancer [4]. Bioactive compounds, including polyphenols, flavonoids, carotenoids, and glucosinolates, have been shown to possess a range of anticancer properties, such as antioxidant activity, anti-inflammatory effects, and the ability to modulate key signaling pathways involved in tumorigenesis [4–7]. These compounds can influence mechanisms such as oxidative stress, cell proliferation, apoptosis, and metastasis, all of which are critical in the development and spread of cancer. Moreover, emerging evidence suggests that bioactive compounds may also enhance the efficacy of conventional cancer treatments [8–12] even against the aggressive pancreatic ductal adenocarcinoma (PDA).

PDA is a devastating disease with a five-year survival rate of about 13% [13]. Resistance to conventional treatment options and the toxicity of current chemotherapy agents, such as gemcitabine, makes PDA a vital target for the development of novel therapeutic agents [14].

This review explores the potential of dietary bioactive compounds in the aggressive context of PDA prevention and therapy, with a focus on their mechanisms of action, their role in modulating cancer-related pathways, and the emerging clinical trials that examine their impact on cancer outcomes.

2. Bioactive Agents Against PDA

In the following sections, we will discuss some of the most notable bioactive compounds investigated for their potential roles in preventing and managing pancreatic ductal adenocarcinoma. By examining their mechanisms of action and their contributions to mainstream treatment approaches, we aim to provide a comprehensive overview of how these natural agents might enhance therapeutic strategies against this highly aggressive malignancy.

As illustrated in Figure 1, the following paragraph will describe the main bioactive compounds and their biological mechanisms.

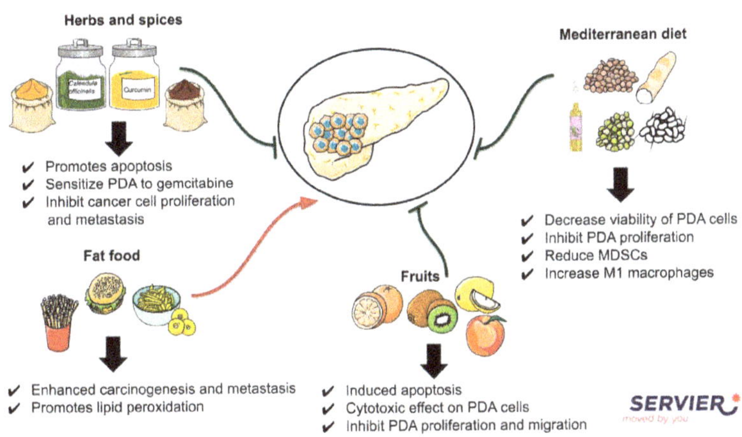

Figure 1. Mechanisms of action of PDA-targeting bioactive compounds. The figure summarizes the main biological effects of the bioactive compounds, highlighting key molecular targets and signaling pathways involved.

2.1. Olive Oil

The mediterranean diet is primarily a plant-based dietary pattern, consisting of a high intake of fruit, vegetables, legumes, nuts and seeds, whole grains, spices, herbs, and olive oil [15]. Owing to its food composition, the mediterranean diet is a dietary pattern rich in protective nutrients and bioactive compounds able to prevent several diseases, including obesity and cancer [16]. One of the major differences between mediterranean diets and other healthy diets is the high intake of olives and olive oil; the annual intake of olive oil in Mediterranean countries can range from 15.3 to 23 kg per capita [17,18]. Many of the health benefits associated with consuming olive oil have been attributed to its high concentration of biophenols [18]. Adherence to a Mediterranean diet is associated with a reduced risk for heart disease and most cancers, including PDA [19,20].

A recent meta-analysis pooling data from one case–control and three cohort studies found that mediterranean diet was not statistically significantly associated with PDA risk [21]. However, findings from two more recent prospective studies observed reductions for high mediterranean diet adherence of between 18% and 43% in the risk of PDA [22]. Pooled data from observational studies support that high olive oil consumption may protect against upper aerodigestive (composed of oral cavity, pharynx, and larynx) and total gastrointestinal and esophageal subtype cancer risk, but not against colorectal and gastric cancers risk [15,23]. An Italian case–control study showed an inverse relationship between olive oil and PDA [24]. Additionally, oleocanthal-rich extra virgin olive oils can induce lysosomal membrane permeabilization in breast and prostate cancer cells, leading to cellular toxicity [25]. In mouse models engineered to develop pancreatic neuroendocrine tumors, oleocanthal treatment reduced tumor burden and extended lifespan [25].

Studies have shown that olive biophenols, such as oleuropein and hydroxytyrosol (HT), can reduce the viability of PDA cells in vitro [26,27]. These compounds induced morphological changes and caused G2/M cell cycle arrest in PDA cells, suggesting potential therapeutic effects [27]. HT, a compound derived from olive oil, has been shown to have various health benefits, though its therapeutic effect on PDA remains debated. HT inhibited the proliferation of Panc02 cells through the STAT3/Cyclin D1 signaling pathway and in HT-treated tumor-bearing mice, orthotopic PDA tumors were suppressed, with a reduction in myeloid-derived suppressor cells (MDSCs) and an increase in M1 macrophages [28]. Additionally, HT suppressed the expression of immunosuppressive molecules in bone marrow-derived MDSCs, down-regulating C/EBPβ and the phosphorylation of STAT3 [28]. Furthermore, HT enhanced the anti-tumor effects of an anti-CD47 antibody in vivo.

These findings collectively suggest that olive oil and its bioactive compounds may offer potential benefits in the context of PDA. However, further clinical trials are necessary to fully understand their efficacy and therapeutic potential.

2.2. *Vitamin E Tocotrienols*

One of the most compelling groups of anti-tumor bioactive compounds in cereal grains are vitamin E tocotrienols [29]. Tocotrienols are unsaturated, naturally occurring vitamin E compounds, which exist as four isoforms: α-, β-, δ-, and γ-tocotrienol [30]. In PDA, vitamin E δ-tocotrienol (VEDT) is the most potent anticancer agent among the four isomers, both in vitro and in vivo [31]. It was shown that oral administration of 100 mg/kg/day of VEDT to mice resulted in satisfactory bioavailability in mouse pancreas tissue with no significant toxicity [32]. Of note, VEDT administered for almost 1 year, prolonged the survival and delayed pancreatic intraepithelial neoplasia lesions in the LSL-KRASG12D/PDX-1-Cre genetic mouse model of PDA [33].

2.3. *Marigold Supercritical Fluid Extract*

The use of supercritical fluids in green technology, with a particular focus on supercritical CO_2 in the extraction of compounds with low polarity, can be enhanced by the incorporation of distinct co-solvents to optimize extraction performance.

The antitumoral properties and mechanism of action of a supercritical CO_2 extract from *Calendula officinalis*, more commonly known as marigold, in the context of PDA, have been investigated [34]. It was observed that marigold supercritical fluid extract diminished the PDA cell viability in a dose dependent manner inducing apoptotic cell death, increasing the percentage of necrotic cells, inhibiting the anchorage-independent cell growth, and synergizing with the chemotherapeutic drug 5-fluorouracil, used in clinics [35–37]. Additionally, marigold supercritical fluid extract stimulates the expression of BMP8B, causing an energetic imbalance that ultimately results in autophagy-induced cell death [34]. Given that metabolic reprogramming is a well-known hallmark of cancer, the direct influence of marigold extract on PDA cell metabolism warrants further investigation into its potential as co-adjuvant in patient therapy.

Another study highlights that marigold extracts may also protect pancreatic β-cells from oxidative damage induced by arsenic exposure [38]. This protective effect is mediated through the activation of the Nrf2 signaling pathway, highlighting the potential of marigold extracts in preserving pancreatic cell function [38].

While preclinical studies have demonstrated that marigold extracts possess antitumoral properties against PDA cells [34–37], there is a lack of clinical trials directly evaluating marigold extract as a treatment for PDA patients.

2.4. *Polyunsaturate Fatty Acids*

Epidemiological studies suggest a correlation between dietary fat intake and carcinogenesis, with several authors proposing that certain polyunsaturated fatty acids (PUFAs) may have modulatory effects [39–43]. Although n-3 fatty acids (like eicosapentanoic acid (EPA) and docosahexaenoic acid (DHA)) are supposed to suppress tumor proliferation, n-6 PUFA (e.g., linoleic acid) enhanced carcinogenesis and metastasis in various trials [44]. While the beneficial effects of PUFAs remain a topic of debate in the literature, several studies highlight their antioxidant properties, suggesting a role in enhancing cellular defense mechanisms and inhibiting the arachidonic acid cascade [45]. Conversely, other studies propose that the tumor-suppressive effects of n-3 PUFAs may result from their preferential accumulation in tumor cells. This selective enrichment may render tumor cells more susceptible to oxidative stress and lipid peroxidation, thereby inducing cytotoxic effects and contributing to the inhibition of tumor growth [46–48]. In addition, Gregor and colleagues demonstrated that lipid peroxidation was increased in tumor-free pancreas tissue of a standard high fat diet rich in n-6 PUFA-10 mg N-nitrosobis-2-oxypropylamine, compared to the high fat diet rich in n-6 PUFA, the diet rich in n-3 PUFA—and a diet containing a mixture of n-3, n-6 and n-9 PUFA, which might be explained by the fact that n-6 PUFA are prone to lipid peroxidation initiated by reactive oxygen species and may promote carcinogenesis and metastasis [45]. In contrast to previous studies [46,49,50], the level of lipid peroxidation was decreased in intratumoral tissue compared to tumor-free pancreas [45].

2.5. *Citrus Fruits*

The role of fruits and vegetables consumption on pancreatic cancer risk has not been established yet, as most of the studies on the topic reported non-significant results [51].

Citrus, the genus *Citrus L.* of the family *Rutaceae*, subfamily *Aurantioideae* [52], is one of the most important fruit crops, including pomelo, sweet orange, sour, lemon, lime, citron, grapefruit, kumquat, and hybrids are rich in phytonutrients, offering various health benefits [53–55]. They are a major source of flavonoids, such as hesperidin, narirutin, and rutin, which have antioxidant, anti-inflammatory, and anticancer properties [56]. Citrus fruits are also high in carotenoids like β-carotene, lutein, and lycopene, which may reduce the risk of cardiovascular disease, macular degeneration, and cancer [57]. Additionally, citrus limonoids, including limonin and nomilin, are known for their antioxidant, anticancer, and hypocholesterolemic effects [58].

A plethora of epidemiological studies, encompassing case–control and cohort designs, have been conducted to investigate the association between the consumption of citrus fruits and the risk of developing PDA [59–69]. A number of these studies have indicated an inverse relationship between the intake of citrus fruits and the likelihood of contracting the aforementioned disease [59,61,62,66–69]. Of note, when total fruit intake was divided into citrus fruit or yellow-orange fruit, the intake of yellow-orange fruit was positively associated with pancreatic cancer risk among normal-weight participants [64]. Larsson et al. emphasize the potential overstatement of relationships observed in case–control studies, due to the assessment of dietary intake occurring subsequent to the diagnosis of pancreatic cancer, a process which may result in the occurrence of recall bias. Furthermore, selection bias is a problem in situations of low participation rates among controls because those who participate are likely to be more health conscious and, therefore, are likely to consume more fruits and vegetables than nonrespondents. The potential for selection bias may be introduced if the case series is restricted to cases in which subjects are still alive at the time of interview. Case–control studies of pancreatic cancer are particularly vulnerable to bias, owing to the elevated and expeditious fatality rates associated with the condition [60]. In addition, Silverman et al. demonstrate, a statistically significant correlation between obesity and a 50–60% increased risk of PDA, which remains consistent across both sex and race demographics [63]. No important associations were observed with citrus fruit and juice consumption in a prospective study on United States adults with no reported history of cancer [65].

Citrus bioactive compounds have the ability to inhibit multiple stages of breast [70], colon [71], prostate [72], lung [73] and PDA [74,75]. Lime juice extracts inhibited cancer cell growth in a dose-dependent manner, with the methanol extract showing the highest activity [74]. Protein-level analysis for p53, Bax, Bcl-2, and caspase-3 indicated that the extracts promoted apoptosis in the cancer cells [76]. Additionally, limonoids—such as limonin glucoside, limonexic acid, isolimonexic acid, and limonin—extracted from lime seeds also inhibited PDA cells through apoptosis [75].

2.6. *Flavonoids*

Flavonoids are bioactive compounds found in various sources, such as citrus fruit, apples, green tea, berries, and grapes [77,78]. Consumption of flavonoid-rich foods has been linked to a reduced risk of several diseases, including obesity, cancer, and heart disease [79]. Flavonoids are the most common of the plant polyphenolics [80] and are thought to have chemoprotective properties [81].

Based on the degree of substitution, flavonoids are further subdivided into chalcones, flavanones, flavones, flavonols, flavanols, isoflavones, and anthocyanins [82]. Moreover, studies reveal that flavonoids, such as isoliquiritigenin [83], apigenin [84], quercetin [85], among others exert significant anticancer effects in various cancers [86].

Flavones have shown inhibition of PDA cell growth in vitro [87], and quercetin (a flavonol) has demonstrated inhibition of PDA growth and prevention of metastasis in vitro and in vivo [88].

Epidemiological studies on flavonoids and PDA suggest an inverse association with intake of specific flavonoids, but results are inconsistent and based on few cases [89–91].

2.6.1. Broussoflavonol B

Broussoflavonol B (Bf-B), a flavonoid compound identified in the roots of Daphne giraldii Nitsche, has been extensively investigated for its potential anti-inflammatory, antioxidant, and anticancer properties [92]. The results demonstrated that Bf-B with diisopentenyl has potent cytotoxic effects on PANC-1 cancer cells. AURKA, PLK1, and MET might serve as key targets for Bf-B inhibition of disease progression in PDA patients [92]. The results demonstrated that Bf-B inhibits the proliferation and migration of PANC-1 and BXPC-3 cells and induces cell cycle S-phase arrest, apoptosis, and DNA damage [92].

2.6.2. Isorhamnetin

Isorhamnetin (ISO), 3′-methylquercetin, is a dietary flavonoid found in numerous plants such as red onion, broccoli, gingko biloba leaves, sea buckthorns, apples, pears, green grapes [93]. ISO showed anti-proliferative effect on several types of cancer cells such as skin, colon, breast, and PDA cells by inducing apoptosis, inhibiting proliferation, modulating signaling pathways, and exerting antioxidant effect [94–97]. Recently, it was also shown that the cytostatic effect of ISO on human cancer-associated-fibroblast (CAFs) impacts the tumor growth and development of chemoresistance. In particular, Ganbold and colleagues, demonstrated that in PDA-derived CAFs, ISO induce cell cycle arrest at G2/M phase associated with activation of p21, impaired mitochondrial homeostasis, and inhibition of inflammatory mediators gene expression [93]. In addition, it was shown in a PDA xenograft mouse model, that the combined administration of gemcitabine and flavopiridol demonstrated a significant reduction in tumor volume and induction of apoptosis [98].

2.6.3. Apiin, Rhoifolin and Vitexin

Apiin, rhoifolin, and vitexin are flavonoid glycosides found in various plants, each with distinct bioactive properties that contribute to their potential health benefits. Apiin, primarily found in celery and parsley, has shown anti-inflammatory, antioxidant, and anticancer activities, and may also help in regulating blood pressure and supporting cardiovascular health [99]. Rhoifolin, typically present in citrus fruits, has demonstrated anticancer, anti-inflammatory, and antioxidant effects, with studies suggesting it inhibits tumor cell proliferation and metastasis while modulating key signaling pathways like NF-κB and MAPK [100]. Vitexin, found in plants such as passionflower and hawthorn, is known for its anti-inflammatory, antioxidant, and neuroprotective properties [101].

Cell viability assay revealed that apiin, rhoifolin, and vitexin could inhibit proliferation of PDA cell lines, with rhoifolin showing the maximum inhibitory effect [102]. Rhoifolin inhibited cell proliferation and promoted apoptosis of PDA cells, which was associated with up-regulated JNK and p-JNK as well as down-regulated p-AKT [102]. Rhoifolin also inhibited cell migration and invasion and increased the antioxidant capacity in PANC-1 and ASPC-1. In addition, AKT activator or JNK inhibitor effectively reversed the anticancer effects of rhoifolin in PDA [102].

2.6.4. Hispidulin

Hispidulin (4′, 5, 7-trihydroxy-6-methoxyflavone) is one of the most studied flavonoids, primarily present in plants of the *Asteraceae* [103–105] and *Lamiaceae families* [106]. Hispidulin has a wide range of biological activities, including anti-inflammatory, antifungal, antiplatelet, anticonvulsant, antiosteoporotic, and notably anticancer activities [107]. Moreover, hispidulin exhibits synergistic anti-tumor effects when combined with some common clinical anticancer drugs. Indeed, hispidulin enhances the chemosensitivity of bladder cancer cells to gemcitabine and 5-Fluorouracil by suppressing the HIF-1α/P-gp signaling cascade [108], sensitizes renal cell carcinoma cells to sunitinib-induced growth suppression, G0/G1 arrest, and apoptosis by regulating the Stat3 pathway [109], enhances the anti-tumor activity of temozolomide by promoting ROS generation and regulating the AMPK/mTOR signaling pathway in glioblastoma [110] and sensitizes SKOV3 cells (human ovarian cancer cells) to TRAIL-induced apoptosis and converts TRAIL-resistant cells to TRAIL-sensitive cells [111]. The combination of hispidulin and chemotherapeutic drugs reduces the efflux of chemotherapeutic drugs, enhances the chemosensitivity of cancer cells, and reverses drug resistance [112]. In a human PDA mouse xenograft model, oral administration of hispidulin has been shown to suppress tumor growth and angiogenesis, without significant toxicity [112]. Similarly, in vitro results indicate that endothelial cells are more sensitive to hispidulin compared to PDA cells, and hispidulin inhibits VEGF-induced cell migration and tubular formation in endothelial cells [112].

2.6.5. Isoorientin

Isoorientin is a 6-C-glycosylflavone, present in many plant species, such as corn (*Zea mays*) silks and pollens, kudzu (*Pueraria tuberosa*), *Patrinia villosa* [113]. Isoorientin exhibits antioxidant, antiviral, analgesic, antitumor, and anti-inflammatory activities [114–116]. In PDA, isoorientin significantly inhibited cell survival, induced apoptosis, and reduced malignancy by reversing epithelial–mesenchymal transition, matrix metalloproteinase expression, and decreasing vascular endothelial growth factor levels [117]. Furthermore, the AMP-activated protein kinase (AMPK) signaling pathway was strongly activated by isoorientin treatment [117]. However, in PDA cells transfected with a lentivirus to interfere with the expression of the PRKAA1 (protein kinase AMP-activated catalytic subunit alpha 1) gene, there were no significant differences in apoptosis rates or malignancy biomarker expression between the isoorientin-treated and untreated groups [117].

2.6.6. Naringenin

Naringenin ((2S)-5,7-dihydroxy-2-(4-hydroxyphenyl)-2,3-dihydrochromen-4-one) is a flavanone, a type of flavonoid, and is colorless and odorless [118,119]. Naringenin is the most abundant in grapefruit, yuzu, pummelo, orange, tangerine and lime [120]. Naringenin inhibited PDA by suppressing the TGF-β signaling pathway, a key regulator of epithelial–mesenchymal transition (EMT). It also reduced cell migration through caspase-3 cleavage, elevated reactive oxygen species levels, and induced cell death via apoptosis signal-regulating kinase (ASK)-1. By inhibiting the TGF-β/Smad-3 pathway, naringenin decreased the expression of EMT markers [121]. Naringenin augmented the sensitivity of PANC-1 cells to gemcitabine [121]. Finally, naringenin increased ROS levels in PDA SNU-213 cells and induced ASK-1-mediated cell death [122]. A reduction in the expression of p38, JNK, p58 and peroxiredoxin-1, a regulator of oxidative stress and cell homeostasis, was observed when SNU-213 cells were treated with naringenin [122].

2.6.7. Kaempferol

Kaempferol (KAE), a natural flavonoid widely present in a variety of plant-based foods (i.e., Leafy greens, fruits, cruciferous vegetables, herbs, tea, legumes), with significant anti-tumor and anti-inflammatory properties. Recent studies have explored its ability to sensitize PDA cells and mouse models to Erlotinib [123]. In vitro, the combination of KAE and Erlotinib markedly inhibited cell proliferation and promoted apoptosis, compared to Erlotinib alone. Network pharmacology analysis suggested that KAE enhances Erlotinib's effect in PDA, potentially through the PI3K/AKT signaling pathway and EGFR TKI resistance mechanisms [123]. Notably, survival analysis revealed that PDA patients with high EGFR expression had lower survival rates. In vivo, the combined treatment of KAE and Erlotinib significantly reduced the volume and weight of subcutaneously grafted tumors [123].

2.6.8. Puerarin

Puerarin is a natural flavonoid extracted from the roots of the kudzu plant or the kudzu vine [124]. Puerarin has various pharmacological effects, such as enhancing the circulatory system function, reducing myocardial oxygen consumption, decreasing blood sugar, and preventing hypertension and arteriosclerosis [125]. Puerarin induced mitochondrial-dependent apoptosis in PDA cell lines by disrupting the balance between Bcl-2 and Bax [124]. It also inhibited PDA cell migration and invasion by counteracting epithelial–mesenchymal transition [124]. In a nude mouse model, puerarin administration reduced

PDA growth and metastasis [124]. Mechanistically, puerarin exerted its therapeutic effects by suppressing the Akt/mTOR signaling pathway. Notably, puerarin is bound to the kinase domain of the mTOR protein, altering the activity of surrounding amino acid residues associated with the ATP-Mg2+ complex [124]. Additionally, puerarin impaired glucose uptake and metabolism by decreasing the oxygen consumption rate and extracellular acidification rate, both of which were dependent on HIF-1α and the glucose transporter GLUT1 [124].

2.6.9. Fisetin

Fisetin is a flavonoid that occurs naturally in a variety of plant species and has a wide range of functionalities, including anti-inflammatory, antioxidant and anticancer properties [126,127]. Fisetin treatment was shown to inhibit the growth of chemoresistant PDA cells [128]. It induced apoptosis and suppressed the invasion of AsPC-1 PDA cells by inhibiting DR3-mediated NF-κB activation. cDNA array analysis revealed that fisetin altered the expression of over twenty genes, with the most significant decrease observed in DR3 expression, and a parallel increase in IκBα, the NF-κB inhibitor. Down-regulation of DR3 led to reduced activation of NF-κB/p65, MMP-9, and XIAP, all of which are associated with chemoresistance in PDA cells [128]. Additionally, transient knockdown of DR3 using RNA interference, along with blocking the DR3 receptor with an extracellular domain antibody, significantly enhanced fisetin-induced effects on cell proliferation, invasion, and apoptosis, accompanied by decreased MMP-9, XIAP, and NF-κB DNA binding activity [128].

2.6.10. Wogonin

Wogonin is a flavonoid compound extracted from the root of *Scutellaria baicalensis* [129]. It has antioxidant activity, and antiinflammatory, anti-tumor, immunomodulatory, neuroprotective effects [130]. Wogonin also acts as a chemosensitizer, reducing drugresistance in cancer therapy. When wogonin is used in combination with anticancer drugs such as etoposide, doxorubicin, 5-FU, and cisplatin [131], it can induce tumor cell apoptosis [132] and protect normal cells from side effects. In addition, Xing et al., reported that wogonin enhanced the sensitivity of ovarian cancer cells to gemcitabine by inhibiting the PI3K/Akt signaling pathway [133], while bioinformatics results predicted that wogonin promoted PDA cell apoptosis by inhibiting protein kinase B (Akt) signaling, thereby enhancing the sensitivity of gemcitabine to PDA [134].

It is reported that FV-429, a derivative of the natural flavonoid wogonin, inhibited the invasion and metastasis of PDA cells by modulating Epithelial–mesenchymal transition-related proteins [135]. In addition, FV-429 inhibits migration, invasion, and metastasis of human PDA cells by affecting the Hippo/YAP1 pathway both in vivo and in vitro [135].

2.6.11. Isoliquiritigenin

Isoliquiritigenin (ISL) is a bioactive flavonoid isolated from licorice, the ground root of *Glycyrrhiza glabra* [136]. ISL exhibits numerous pharmacological properties, such as anti-inflammatory, anti-microbial, antioxidative, anticancer as well as immunoregulatory effects [137]. Zhang and colleagues showed that ISL inhibited PDA cell growth and induced apoptosis, both in vitro and in vivo [136]. ISL caused accumulation of autophagosome through blockade of late stage autophagic flux [136]. Of note, ISL synergistically sensitized the cytotoxic effect of gemcitabine and 5-fluorouracil on PDA cells as both drugs induced autophagy. Molecular docking analysis has indicated that ISL acted by direct targeting of p38 MAPK, which was confirmed by ISL-induced phosphorylation of p38. The autophagy flux induced by p38 inhibitor SB203580 was blocked by ISL, with further increasing toxicity of ISL in PDA cells [136].

2.6.12. Luteolin

Luteolin (Lut), is a flavonoid, specifically a flavone, found in celery, green pepper, parsley, and perilla leaf. Lut suppressed pancreatocarcinogenesis and reduced the expression of dihydropyrimidine dehydrogenase (DPYD), an enzyme that degrades pyrimidines such as 5-fluorouracil, in PDA [138]. Lut exhibits strong antioxidant activity and shows antiinflammatory and antitumor effects against different tumor among which PDA [138–143]. Recently, it was observed the therapeutic effects of the combined treatment with 5-FU and Lut in PDA resulted in remarkable therapeutic effects both in vitro and in vivo, whereas 5-FU or Lut alone showed no significant effects [138].

2.6.13. Anthocyanins

Anthocyanins, found in various pigmented plants as secondary metabolites, represent a class of dietary polyphenols known for their bioactive properties, demonstrating health-promoting effects against several chronic diseases [144]. Cyanidin-3-O-glucoside (C3G) is a major anthocyanin found in various fruits, particularly berries, and is known for its antioxidant and anti-inflammatory properties. Upon ingestion, C3G undergoes metabolism in the human body, where it is hydrolyzed by intestinal enzymes into its aglycone form, cyanidin, which is further absorbed and distributed [145,146]. The bioavailability of C3G is relatively low, but its metabolites may still exert beneficial effects on various cellular processes, including reducing oxidative stress and modulating inflammatory pathways. Notably, studies have shown that C3G can induce apoptosis in colon cancer and glioblastoma cells by modulating oxidative stress pathways [147]. Additionally, research indicates that C3G exerts protective effects on pancreatic beta cells by alleviating palmitic acid-induced dysfunction through the regulation of endoplasmic reticulum stress pathways [147].

Kuntz et al., showed that 60 min after a single anthocyanins dose, plasma extracts from volunteers inhibited migration of the PDA cell line PANC-1 in vitro [148]. In a later study, the same group showed that plasma extracts had different effects on cancer cell migration in vitro depending on the cancer cell line [149]. In PANC-1, extracted-plasma metabolites after the administration of anthocyanins-rich juice reduced cell migration significantly in comparison to plasma extracts after placebo. By contrast, no reduction was observed for the migration of AsPC-1. The observed discrepancy was attributed to the diminished migration and expression of cell adhesion molecules in PANC-1 cancer cells in vitro, as evidenced by the activation of FAK- and NF-kB-pathways, along with the reduction in ROS [149].

Anthocyanins exert antitumor effects through multiple mechanisms. In the early stages, they inhibit inflammation and prevent normal cell transformation by regulating antioxidant enzymes. During carcinogenesis, they target key signaling pathways like MAPK and AP-1, inhibiting RTK activity and causing cell cycle arrest and DNA repair [150]. In later stages, anthocyanins promote cancer cell apoptosis by activating caspases and reduce metastasis by targeting VEGF signaling. Additionally, they help overcome multidrug resistance, improving chemotherapy sensitivity. These actions are mediated through several molecular pathways, including Ras-MAPK, PI3K/Akt, and NF-κB [150].

While anthocyanins has demonstrated anticancer properties in preclinical studies [149–151], there is a notable absence of clinical trials specifically investigating its efficacy in treating PDA. Further research is necessary to fully understand the therapeutic potential and mechanisms of anthocyanins in PDA prevention and treatment.

2.6.14. Xanthohumol, Resveratrol, Phenethyl Isothiocyanate, Indole-3-Carbinol

Xanthohumol, resveratrol, phenethyl isothiocyanate (PEITC), and indole-3-carbinol are bioactive compounds with promising health benefits, particularly in cancer prevention and treatment.

Xanthohumol (XN) is a prenylated flavonoid compound primarily found in hops (*Humulus lupulus*), the key ingredient used in beer production. Xanthohumol exhibits antioxidant, anti-inflammatory, and anticancer properties by modulating various cellular pathways involved in tumor growth and metastasis [152,153]. The regulation by xanthohumol of the Nrf2/NF-kB/mTOR/AKT pathways induce a strong antioxidant and anti-inflammatory effect, among others the acceleration of autophagy through increased synthesis of Bcl-2 proteins, inhibition of the synthesis of VEGF responsible for angiogenesis and phosphorylation of Hexokinase II [152].

Resveratrol, a polyphenol found in grapes, berries, and red wine, is known for its antioxidant, anti-inflammatory, and anticancer effects, including its ability to inhibit tumor cell proliferation and induce apoptosis [154,155]. Notably, it was observed that resveratrol might be involved in regulating EMT in the PDA microenvironment [156].

PEITC, derived from cruciferous vegetables like watercress, is a potent anticancer agent, known for its ability to induce cancer cell death and inhibit metastasis through various molecular mechanisms [157,158]. PEITC targets crucial cellular signaling pathways involved in cancer progression, notably the NF-κB, Akt, and MAPK pathways [159]. Regarding the studies with animal models, Stan et al. demonstrated that oral administration of PEITC reduced pancreatic cancer cell growth in a MIAPaca2 xenograft animal model of 6 weeks old [160].

Indole-3-carbinol, also found in cruciferous vegetables such as broccoli and cabbage, has demonstrated anticancer effects by regulating estrogen metabolism, modulating signaling pathways involved in cell growth, and promoting apoptosis in cancer cells [157]. Interestingly, the mixture of xanthohumol and PEITC was found to be the most potent modulator of the Nrf2 pathway in human PDA cell line [161].

Together, these compounds contribute to cancer prevention and therapy, with ongoing research exploring their combined therapeutic potential.

2.7. Chinese Herbs

Chinese herbs have long been utilized in traditional medicine for their potential therapeutic properties, particularly in the treatment of cancer. Several studies have highlighted the bioactive compounds derived from Chinese herbs that show promise in PDA treatment. For instance, compounds such as curcumin, found in *Curcuma longa* (turmeric), and berberine, isolated from *Coptis chinensis*, have been shown to exhibit anticancer properties by modulating various cellular pathways involved in cancer cell proliferation, apoptosis, and metastasis [162–165]. Additionally, the active ingredients in *Glycyrrhiza uralensis* (licorice) have demonstrated anti-inflammatory and anti-tumor activities, which may aid in reducing the aggressiveness of PDA [166]. Recent research underscores the potential of these herbs as adjuncts to conventional therapies, aiming to enhance the effectiveness of treatment while minimizing side effects [167]. However, more clinical trials are necessary to validate these findings and determine their safety and efficacy in PDA treatment.

2.7.1. Curcumin

Curcumin, a component of turmeric (*Curcuma longa*), is one such agent that has been shown to suppress the transcription factor nuclear factor-κB (NF-κB), which is implicated in proliferation, survival, angiogenesis, and chemoresistance [168]. It was shown that curcumin can sensitize PDA to gemcitabine in vitro and in vivo. In vitro, curcumin inhibited the proliferation of various PDA cell lines, potentiated the apoptosis induced by gemcitabine, and inhibited constitutive NF-κB activation in the cells [168]. In vivo, tumors from nude mice injected with PDA cells and treated with a combination of curcumin and gemcitabine showed significant reductions in volume, Ki-67 proliferation index, NF-κB activation, and expression of NF-κB–regulated gene products (cyclin D1, c-myc, Bcl-2, Bcl-xL, cellular inhibitor of apoptosis protein-1, cyclooxygenase-2, matrix metalloproteinase, and vascular endothelial growth factor) compared with tumors from control mice treated with olive oil only [168]. The combination treatment was also highly effective in suppressing angiogenesis [168].

2.7.2. Thymoquinone (From *Nigella sativa* Seeds)

Thymoquinone is the principal bioactive compound obtained from the seeds of *Nigella sativa*, a plant commonly referred to as black cumin. Traditional remedies have utilized these seeds for centuries, and modern research has validated thymoquinone's anti-inflammatory, antioxidant, and proapoptotic properties in a range of malignancies. In PDA, it has been shown to suppress cell growth and survival by modulating pivotal pathways such as NF-κB and PI3K/Akt, both of which play essential roles in tumor proliferation and angiogenesis [169,170]. Notably, this compound also appears to heighten the efficacy of conventional chemotherapeutics, including gemcitabine and oxaliplatin, possibly by enhancing cancer cell sensitivity to apoptosis [171–173]. Researchers suggest that its capacity to overcome drug resistance stems from the downregulation of multiple survival pathways [174]. While most findings to date are based on in vitro and in vivo preclinical models, the evidence strongly supports the potential of thymoquinone as part of a combination strategy against PDA [175]. Further investigation into its clinical application and synergistic effects with standard therapies is warranted to define optimal treatment protocols.

2.7.3. Alpinumisoflavone

Alpinumisoflavone (AIF) is a prenylated isoflavone originated in *Cudrania tricuspidate* with versatile bioactive properties, including anticancer activity. As it is widely spread in East Asia and known for its versatile bioactive properties, it was utilized for a medical herb in traditional Chinese medicine [176]. Different studies showed that AIF suppressed cell proliferation, migration, and invasion capacity of tumoral cells [177], favor apoptosis [178] and suppress the tumor growth and metastatization [177,179,180]. As a result, AIF has garnered significant attention as a potential chemotherapeutic adjuvant [181]. Recent in silico studies have suggested AIF's capability to target receptors associated with the angiogenesis pathway [182]. In addition, different studies have demonstrated that combining AIF with standard anticancer drugs enhances its therapeutic efficacy against several cancers including PDA [176,183]. In fact, AIF: (i) suppressed PDA cell viability, (ii) disrupted the normal formation of PANC-1 and MIAPaca2 spheroids in vitro model, (iii) weakened cell migratory ability by downregulating mesenchymal proteins, (iv) strengthened apoptosis induction, oxidative stress, mitochondrial calcium dysregulation, depolarization, and (v) OXPHOS impairment in PDA cells [176]. Of note, the combination of AIF and gemcitabine synergistically induced mitochondrial dysfunction in PDA cells [176].

2.7.4. Piperlongumine

Piperlongumine is an alkaloid extracted from long pepper, a staple in various traditional Asian medical practices. One of its distinguishing features is the selective increase in oxidative stress in cancer cells, which in turn triggers apoptosis without major harm to normal cells [184]. Studies focused on PDA indicate that piperlongumine can curtail tumor growth and migration by elevating intracellular reactive oxygen species and suppressing pro-survival pathways [185]. In addition, it disrupts NF-κB signaling, reducing the expression of genes essential for tumor progression [186]. Several preclinical investigations point to its potential synergy with gemcitabine, as enhanced oxidative stress can sensitize PDA cells to chemotherapy [187]. Transcriptome analyses further reveal that the compound engages both oxidative and endoplasmic reticulum stress mechanisms, providing a multifaceted assault on tumor cells [187]. Although clinical data are still lacking, these promising laboratory findings suggest that piperlongumine may hold significant therapeutic value, especially in combination with existing standard-of-care treatments.

2.7.5. Honokiol

Honokiol is a polyphenol derived from the bark of Magnolia officinalis, traditionally prized in Chinese medicine for its broad-ranging benefits, including antioxidant and anti-inflammatory effects. Research on PDA has shown that honokiol interferes with oncogenic signaling networks—namely STAT3 and NF-κB—thereby inhibiting tumor cell proliferation and enhancing apoptosis [188]. It also appears to impede the complex interplay between tumor and stromal cells, a critical factor in PDA progression and metastasis [189]. Notably, combining honokiol with conventional chemotherapy, such as gemcitabine, has led to improved anticancer outcomes in preclinical models, partly due to its capacity to block multiple pathways linked to resistance [188]. Additional findings suggest that honokiol exerts an antiangiogenic effect, limiting the formation of new blood vessels, and thus restraining tumor growth [190]. Although these data remain largely at the laboratory stage, the compound's ability to target diverse cancer-promoting mechanisms underscores its potential for integration into future therapeutic regimens for PDA malignancies.

3. Active Clinical Trials with Bioactive Compounds for PDA

The favorable outcomes observed in both in vitro and in vivo experiments employing bioactive compounds in the treatment of PDA have prompted clinicians and oncologists to investigate the potential of these dietary supplements to enhance postoperative recovery (Table 1). The aim of these study is to examine the effect of dietary prescription with and without nutrition supplementation in PDA patients. In fact, an accelerated recovery may improve outcomes after surgery following complex abdominal operations resulting in a shorter length of stay in PDA patients. It may also help patients to mobilize more quickly and return to the home setting, decrease hospital-acquired infectious complications, and increase potential cost savings. In Table 2, the effectiveness of conventional chemotherapy regimens and dietary supplement is shown. Given the aggressive nature and poor prognosis of PDA, investigating the combined effectiveness of conventional chemotherapy regimens and dietary supplements offers a promising strategy to enhance therapeutic outcomes, overcome chemoresistance, and improve patient quality of life.

Table 1. Clinical trial with dietary supplement in PDA treatment.

NCT	Therapy	Sponsor
NCT02681601	Omega rich fish oil supplement	Jonsson Comprehensive Cancer Center
NCT02517268	Liquid or solid post operative diet	Sidney Kimmel Cancer Center at Thomas Jefferson University [191]
NCT03187028	Diet alone vs. diet+exercise	University of Alabama at Birmingham
NCT06833658	Aromatherapy with essential oil intervention	Peking University First Hospital
NCT06852014	Peptamen 1.6 supplement	Fundación Pública Andaluza para la Investigación de Málaga en Biomedicina y Salud
NCT03244683	Oral Nutritional Supplementation combined with resistance training	Ohio State University

Table 1. *Cont.*

NCT	Therapy	Sponsor
NCT02940067	Components of the mediterranean diet and exercise training	Royal Surrey County Hospital NHS Foundation Trust
NCT04306874	High-protein nutritional supplementation	Thomas Jefferson University
NCT03167814	No long-chain triglycerides	Helsinki University Central Hospital
NCT06069297	Exercise training, nutritional therapy and anxiety reducing techniques	IRCCS San Raffaele

Table 2. Clinical trial with dietary supplement associated with conventional chemotherapy PDA therapy.

NCT	Therapy	Sponsor
NCT06090916	Conventional vs. Dietary and physical activity using MyFitnessPal smartphone app	Jonsson Comprehensive Cancer Center
NCT06050395	Anti-inflammatory and pro-inflammatory dietary patterns	H. Lee Moffitt Cancer Center and Research Institute
NCT06595160	Plant-based diet	Emory University
NCT06149546	High protein, high energy diet, Fish oil supplement, Pancreatic Enzymes	Cancer Trials Ireland
NCT05420259	Exercise and Dietary Intervention	Hospital Beatriz Ângelo
NCT04837118	Dietary intervention	M.D. Anderson Cancer Center
NCT04188990	Dietary advice, Oral Nutritional Supplementation, Enteral Feeding or Parenteral Nutrition	Hospital Galdakao-Usansolo
NCT02336087	Curcumin, vitamin D, vitamin K2, vitamin K1, B-6, high selenium broccoli sprouts, epigallocatechin gallate, L-carnitine, garlic extract, genistein, zinc amino chelate, mixed toxopherols, ascorbic acid, D-limonene	City of Hope Medical Center
NCT03958019	Supervised and self-managed exercise, dietary counseling, and education sessions	University of Dublin, Trinity College [192]
NCT06412510	30 gm protein supplement (high calorie, high protein supplement or low fat/low sugar, high protein supplement) and exercise intervention	Case Comprehensive Cancer Center
NCT02607826	Short-term Starvation	University Hospital Tuebingen

Clinical studies have demonstrated that both diet and exercise significantly impact recovery and the effectiveness of chemotherapy in PDA patients. Rosebrock and colleagues found that exercise is feasible and safe for these patients, leading to improvements in quality of life, reduction in cancer-related fatigue, and increased muscle strength [193]. Similarly, research indicates that regular physical activity can help manage side effects of chemotherapy, such as fatigue, and may enhance the efficacy of treatment. Moreover, a balanced diet rich in fruits, vegetables, lean proteins, and whole grains supports nutritional status, reduces the risk of malnutrition, and may alleviate chemotherapy side effects like nausea and fatigue [194].

Incorporating exercise into daily routines has also been associated with improved mental health, quality of life, and reduced anxiety and depression among cancer patients. Additionally, studies suggest that regular exercise and a balanced diet can positively influence survival rates, underscoring the importance of these interventions during cancer treatment [195].

Integrating appropriate diet and exercise regimens into the care plan of PDA patients undergoing chemotherapy can lead to improved recovery, enhanced treatment efficacy, and better overall outcomes. However, it is essential to tailor these interventions to individual patient needs and consult healthcare professionals before making significant changes to diet or physical activity.

4. Discussion

PDA is one of the most aggressive cancers because it often remains undetected until advanced stages, shows a dense fibrotic tissue around the tumor cells (desmoplasia), and develops resistance to common treatments. In recent years, many studies have explored how natural bioactive compounds, found in fruits, vegetables, spices, and medicinal plants, might improve treatment results for this type of cancer. These substances include polyphenols (for example, curcumin or resveratrol), flavonoids (such as quercetin, apigenin, and hispidulin), tocotrienols (especially δ-tocotrienol), and certain isothiocyanates. Research shows that they can interfere with multiple signaling pathways in cancer cells, like NF-κB and STAT3 (which promote inflammation and cell survival), PI3K/Akt (involved in cell growth), and processes that lead to tumor invasion (like the epithelial–mesenchymal transition). By working on several targets at once, these molecules might prevent the tumor from quickly becoming resistant, a frequent problem when a therapy only blocks a single pathway.

A key advantage of these bioactive compounds is how they sometimes boost the effectiveness of chemotherapy. For instance, curcumin can strengthen gemcitabine's impact by lowering the levels of proteins that help cancer cells avoid death (like Bcl-2) and reducing NF-κB activity [196]. Flavonoids like quercetin and kaempferol may also help gemcitabine work better, mainly by increasing the oxidative stress in cancer cells and blocking proteins that make them drug-resistant [197]. Moreover, some of these substances seem to modify the tumor's microenvironment, reducing the presence of immunosuppressive cells (e.g., myeloid-derived suppressor cells) and breaking the harmful exchange of signals between the tumor and the surrounding tissue [198,199].

Even with these encouraging discoveries, it is not simple to bring these findings into the clinic. Many of these compounds have poor oral bioavailability, meaning only a small portion is absorbed when taken by mouth. This has led to experiments with new techniques, like nano-encapsulation, to deliver them more effectively to the tumor [200–202]. Additionally, the different extraction methods from natural sources can cause variability in the final purity of these products, making it difficult to compare studies or reproduce results [203,204]. Although some clinical trials focus on combining nutritional supplements (like fish oil high in n-3 fatty acids or specific high-protein formulas) with standard anticancer treatments, more robust investigations are needed to see if these truly improve survival and quality of life [205].

Finally, recent attention has turned to combining these natural compounds with healthy diets and exercise programs. Physical activity and balanced nutrition are known to reduce overall inflammation and enhance general health, possibly adding to the beneficial effects of bioactive agents. In the future, using these strategies together with chemotherapy, dietary supplements, and lifestyle changes, could improve outcomes for patients with this highly resilient cancer. To reach this goal, well-designed clinical trials will be essential for defining the best doses, the most effective combinations, and the ideal schedules for administration, ensuring safer and more successful therapies.

5. Conclusions

Despite the limited therapeutic options and poor prognosis associated with PDA, the integration of bioactive compounds into current treatment strategies represents a promising area of research. Preclinical studies have demonstrated that various natural compounds—such as polyphenols, isothiocyanates, and omega-3 fatty acids—can modulate key molecular pathways involved in tumor growth, inflammation, and chemoresistance. While these findings provide a strong rationale for the adjunctive use of bioactive compounds alongside conventional chemotherapy, clinical evidence remains limited and often inconclusive. Rigorous, well-designed clinical trials are urgently needed to validate their efficacy, determine optimal dosing regimens, and assess potential interactions with standard treatments. Advancing this line of research may pave the way for more effective, integrative therapeutic approaches in pancreatic cancer care.

Author Contributions: Conceptualization, S.B. and C.C.; methodology, F.S.; data curation, C.C.; writing—original draft preparation, S.B. and F.S.; writing—review and editing, C.C. All authors have read and agreed to the published version of the manuscript.

Funding: This research received no external funding.

Institutional Review Board Statement: Not applicable.

Informed Consent Statement: Not applicable.

Data Availability Statement: No new data were created or analyzed in this study.

Conflicts of Interest: The authors declare no conflicts of interest.

References

1. Wu, Z.; Xia, F.; Lin, R. Global Burden of Cancer and Associated Risk Factors in 204 Countries and Territories, 1980–2021: A Systematic Analysis for the GBD 2021. *J. Hematol. Oncol.* **2024**, *17*, 119. [CrossRef]
2. Britten, O.; Tosi, S. The Role of Diet in Cancer: The Potential of Shaping Public Policy and Clinical Outcomes in the UK. *Genes Nutr.* **2024**, *19*, 15. [CrossRef]
3. Kris-Etherton, P.M.; Hecker, K.D.; Bonanome, A.; Coval, S.M.; Binkoski, A.E.; Hilpert, K.F.; Griel, A.E.; Etherton, T.D. Bioactive Compounds in Foods: Their Role in the Prevention of Cardiovascular Disease and Cancer. *Am. J. Med.* **2002**, *113*, 71–88. [CrossRef]
4. Delgado-Gonzalez, P.; Garza-Treviño, E.N.; de la Garza Kalife, D.A.; Quiroz Reyes, A.; Hernández-Tobías, E.A. Bioactive Compounds of Dietary Origin and Their Influence on Colorectal Cancer as Chemoprevention. *Life* **2023**, *13*, 1977. [CrossRef]
5. Ağagündüz, D.; Şahin, T.Ö.; Yılmaz, B.; Ekenci, K.D.; Duyar Özer, Ş.; Capasso, R. Cruciferous Vegetables and Their Bioactive Metabolites: From Prevention to Novel Therapies of Colorectal Cancer. *Evid. Based Complement. Altern. Med.* **2022**, *2022*, 1534083. [CrossRef]
6. Akhtar, M.F.; Saleem, A.; Rasul, A.; Faran Ashraf Baig, M.M.; Bin-Jumah, M.; Abdel Daim, M.M. Anticancer Natural Medicines: An Overview of Cell Signaling and Other Targets of Anticancer Phytochemicals. *Eur. J. Pharmacol.* **2020**, *888*, 173488. [CrossRef]
7. Paul, J.K.; Azmal, M.; Haque, A.S.N.B.; Talukder, O.F.; Meem, M.; Ghosh, A. Phytochemical-Mediated Modulation of Signaling Pathways: A Promising Avenue for Drug Discovery. *Adv. Redox Res.* **2024**, *13*, 100113. [CrossRef]
8. Sznarkowska, A.; Kostecka, A.; Meller, K.; Bielawski, K.P. Inhibition of Cancer Antioxidant Defense by Natural Compounds. *Oncotarget* **2016**, *8*, 15996–16016. [CrossRef]
9. Mileo, A.M.; Miccadei, S. Polyphenols as Modulator of Oxidative Stress in Cancer Disease: New Therapeutic Strategies. *Oxid. Med. Cell. Longev.* **2016**, *2016*, 6475624. [CrossRef]
10. Rizeq, B.; Gupta, I.; Ilesanmi, J.; AlSafran, M.; Rahman, M.M.; Ouhtit, A. The Power of Phytochemicals Combination in Cancer Chemoprevention. *J. Cancer* **2020**, *11*, 4521–4533. [CrossRef]
11. Nisar, S.; Masoodi, T.; Prabhu, K.S.; Kuttikrishnan, S.; Zarif, L.; Khatoon, S.; Ali, S.; Uddin, S.; Akil, A.A.-S.; Singh, M.; et al. Natural Products as Chemo-Radiation Therapy Sensitizers in Cancers. *Biomed. Pharmacother.* **2022**, *154*, 113610. [CrossRef]
12. Salami, C.; Mbakidi, J.-P.; Audonnet, S.; Brassart-Pasco, S.; Bouquillon, S. Extraction of Curcuminoids and Carvacrol with Biobased Ionic Liquids—Evaluation of Anti-Cancer Properties of Curcuminoid Extracts. *Molecules* **2025**, *30*, 1180. [CrossRef]
13. Stoop, T.F.; Javed, A.A.; Oba, A.; Koerkamp, B.G.; Seufferlein, T.; Wilmink, J.W.; Besselink, M.G. Pancreatic Cancer. *Lancet* **2025**, *405*, 1182–1202. [CrossRef]
14. Vincent, A.; Herman, J.; Schulick, R.; Hruban, R.H.; Goggins, M. Pancreatic Cancer. *Lancet* **2011**, *378*, 607–620. [CrossRef]
15. Almanza-Aguilera, E.; Cano, A.; Gil-Lespinard, M.; Burguera, N.; Zamora-Ros, R.; Agudo, A.; Farràs, M. Mediterranean Diet and Olive Oil, Microbiota, and Obesity-Related Cancers. From Mechanisms to Prevention. *Semin. Cancer Biol.* **2023**, *95*, 103–119. [CrossRef] [PubMed]
16. Mentella, M.C.; Scaldaferri, F.; Ricci, C.; Gasbarrini, A.; Miggiano, G.A.D. Cancer and Mediterranean Diet: A Review. *Nutrients* **2019**, *11*, 2059. [CrossRef] [PubMed]

17. Australian Olive Industry Research, Development and Extension Plan 2010–2015. Available online: https://agrifutures.com.au/product/australian-olive-industry-research-development-and-extension-plan-2010-2015/ (accessed on 13 April 2025).
18. Serra-Majem, L.; Ngo de la Cruz, J.; Ribas, L.; Tur, J.A. Olive Oil and the Mediterranean Diet: Beyond the Rhetoric. *Eur. J. Clin. Nutr.* **2003**, *57*, S2–S7. [CrossRef] [PubMed]
19. Trichopoulou, A.; Lagiou, P.; Kuper, H.; Trichopoulos, D. Cancer and Mediterranean Dietary Traditions. *Cancer Epidemiol. Biomark. Prev.* **2000**, *9*, 869–873.
20. Kapiszewska, M.; Sołtys, E.; Visioli, F.; Cierniak, A.; Zajac, G. The Protective Ability of the Mediterranean Plant Extracts against the Oxidative DNA Damage. The Role of the Radical Oxygen Species and the Polyphenol Content. *J. Physiol. Pharmacol.* **2005**, *56* (Suppl. 1), 183–197.
21. Schwingshackl, L.; Schwedhelm, C.; Galbete, C.; Hoffmann, G. Adherence to Mediterranean Diet and Risk of Cancer: An Updated Systematic Review and Meta-Analysis. *Nutrients* **2017**, *9*, 1063. [CrossRef]
22. Luu, H.N.; Paragomi, P.; Jin, A.; Wang, R.; Neelakantan, N.; van Dam, R.M.; Brand, R.E.; Koh, W.-P.; Yuan, J.-M. Quality Diet Index and Risk of Pancreatic Cancer: Findings from the Singapore Chinese Health Study. *Cancer Epidemiol. Biomark. Prev.* **2021**, *30*, 2068–2078. [CrossRef]
23. Markellos, C.; Ourailidou, M.-E.; Gavriatopoulou, M.; Halvatsiotis, P.; Sergentanis, T.N.; Psaltopoulou, T. Olive Oil Intake and Cancer Risk: A Systematic Review and Meta-Analysis. *PLoS ONE* **2022**, *17*, e0261649. [CrossRef]
24. Soler, M.; Chatenoud, L.; La Vecchia, C.; Franceschi, S.; Negri, E. Diet, Alcohol, Coffee and Pancreatic Cancer: Final Results from an Italian Study. *Eur. J. Cancer Prev.* **1998**, *7*, 455. [CrossRef]
25. Goren, L.; Zhang, G.; Kaushik, S.; Breslin, P.A.S.; Du, Y.-C.N.; Foster, D.A. (-)-Oleocanthal and (-)-Oleocanthal-Rich Olive Oils Induce Lysosomal Membrane Permeabilization in Cancer Cells. *PLoS ONE* **2019**, *14*, e0216024. [CrossRef] [PubMed]
26. Goldsmith, C.D.; Vuong, Q.V.; Sadeqzadeh, E.; Stathopoulos, C.E.; Roach, P.D.; Scarlett, C.J. Phytochemical Properties and Anti-Proliferative Activity of Olea Europaea L. Leaf Extracts against Pancreatic Cancer Cells. *Molecules* **2015**, *20*, 12992–13004. [CrossRef]
27. Goldsmith, C.D.; Bond, D.R.; Jankowski, H.; Weidenhofer, J.; Stathopoulos, C.E.; Roach, P.D.; Scarlett, C.J. The Olive Biophenols Oleuropein and Hydroxytyrosol Selectively Reduce Proliferation, Influence the Cell Cycle, and Induce Apoptosis in Pancreatic Cancer Cells. *Int. J. Mol. Sci.* **2018**, *19*, 1937. [CrossRef]
28. Wang, B.; Yang, L.; Liu, T.; Xun, J.; Zhuo, Y.; Zhang, L.; Zhang, Q.; Wang, X. Hydroxytyrosol Inhibits MDSCs and Promotes M1 Macrophages in Mice with Orthotopic Pancreatic Tumor. *Front. Pharmacol.* **2021**, *12*, 759172. [CrossRef]
29. Aggarwal, B.; Nesaretnam, K. Vitamin E Tocotrienols: Life beyond Tocopherols. *Genes Nutr.* **2012**, *7*, 1. [CrossRef]
30. Aggarwal, B.B.; Sundaram, C.; Prasad, S.; Kannappan, R. Tocotrienols, the Vitamin E of the 21st Century: Its Potential against Cancer and Other Chronic Diseases. *Biochem. Pharmacol.* **2010**, *80*, 1613–1631. [CrossRef]
31. Husain, K.; Francois, R.A.; Yamauchi, T.; Perez, M.; Sebti, S.M.; Malafa, M.P. Vitamin E δ-Tocotrienol Augments the Antitumor Activity of Gemcitabine and Suppresses Constitutive NF-κB Activation in Pancreatic Cancer. *Mol. Cancer Ther.* **2011**, *10*, 2363–2372. [CrossRef] [PubMed]
32. Husain, K.; Francois, R.A.; Hutchinson, S.Z.; Neuger, A.M.; Lush, R.; Coppola, D.; Sebti, S.; Malafa, M.P. Vitamin E Delta-Tocotrienol Levels in Tumor and Pancreatic Tissue of Mice after Oral Administration. *Pharmacology* **2009**, *83*, 157–163. [CrossRef] [PubMed]
33. Husain, K.; Centeno, B.A.; Chen, D.-T.; Fulp, W.J.; Perez, M.; Zhang Lee, G.; Luetteke, N.; Hingorani, S.R.; Sebti, S.M.; Malafa, M.P. Prolonged Survival and Delayed Progression of Pancreatic Intraepithelial Neoplasia in LSL-KrasG12D/+;Pdx-1-Cre Mice by Vitamin E δ-Tocotrienol. *Carcinogenesis* **2013**, *34*, 858–863. [CrossRef] [PubMed]
34. Gómez de Cedrón, M.; Mouhid, L.; García-Carrascosa, E.; Fornari, T.; Reglero, G.; Ramírez de Molina, A. Marigold Supercritical Extract as Potential Co-Adjuvant in Pancreatic Cancer: The Energetic Catastrophe Induced via BMP8B Ends Up with Autophagy-Induced Cell Death. *Front. Bioeng. Biotechnol.* **2019**, *7*, 455. [CrossRef]
35. Martin, D.; Navarro Del Hierro, J.; Villanueva Bermejo, D.; Fernández-Ruiz, R.; Fornari, T.; Reglero, G. Bioaccessibility and Antioxidant Activity of Calendula Officinalis Supercritical Extract as Affected by in Vitro Codigestion with Olive Oil. *J. Agric. Food Chem.* **2016**, *64*, 8828–8837. [CrossRef]
36. García-Risco, M.R.; Mouhid, L.; Salas-Pérez, L.; López-Padilla, A.; Santoyo, S.; Jaime, L.; Ramírez de Molina, A.; Reglero, G.; Fornari, T. Biological Activities of Asteraceae (Achillea Millefolium and Calendula Officinalis) and Lamiaceae (Melissa Officinalis and Origanum Majorana) Plant Extracts. *Plant Foods Hum. Nutr.* **2017**, *72*, 96–102. [CrossRef]
37. Mouhid, L.; Gómez de Cedrón, M.; Vargas, T.; García-Carrascosa, E.; Herranz, N.; García-Risco, M.; Reglero, G.; Fornari, T.; Ramírez de Molina, A. Identification of Antitumoral Agents against Human Pancreatic Cancer Cells from Asteraceae and Lamiaceae Plant Extracts. *BMC Complement. Altern. Med.* **2018**, *18*, 254. [CrossRef]
38. Mei, Z.; Dai, J.; Liu, G.; He, Z.; Gu, S. Roles of Aqueous Extract of Marigold on Arsenic-Induced Oxidative Damage in Pancreatic Islet β-Cells. *J. Biosci. Med.* **2024**, *12*, 19–34. [CrossRef]

39. Mahjourian, M.; Anjom-Shoae, J.; Mohammadi, M.A.; Feinle-Bisset, C.; Sadeghi, O. Associations of Dietary Fat Types (MUFA, PUFA, SFA) and Sources (Animal, Plant) with Colorectal Cancer Risk: A Comprehensive Systematic Review and Dose-Response Meta-Analysis of Prospective Cohort Studies. *Cancer Epidemiol.* **2025**, *95*, 102768. [CrossRef]
40. Žuža Praštalo, M.; Pokimica, B.; Arsić, A.; Ilich, J.Z.; Vučić, V. Current Evidence on the Impact of Diet, Food, and Supplement Intake on Breast Cancer Health Outcomes in Patients Undergoing Endocrine Therapy. *Nutrients* **2025**, *17*, 456. [CrossRef]
41. Shrivastava, R.; Gupta, A.; Mehta, N.; Das, D.; Goyal, A. Dietary Patterns and Risk of Oral and Oropharyngeal Cancers: A Systematic Review and Meta-Analysis. *Cancer Epidemiol.* **2024**, *93*, 102650. [CrossRef]
42. Li, X.; Zhang, H.; Yang, H.; Song, Y.; Zhang, F.; Wang, A. Modifiable Risk Factors for Breast Cancer: Insights From Systematic Reviews. *Public Health Nurs.* **2025**, *42*, 1060–1071. [CrossRef] [PubMed]
43. Conigliaro, T.; Boyce, L.M.; Lopez, C.A.; Tonorezos, E.S. Food Intake During Cancer Therapy: A Systematic Review. *Am. J. Clin. Oncol.* **2020**, *43*, 813–819. [CrossRef] [PubMed]
44. Appel, M.J.; van Garderen-Hoetmer, A.; Woutersen, R.A. Effects of Dietary Linoleic Acid on Pancreatic Carcinogenesis in Rats and Hamsters. *Cancer Res.* **1994**, *54*, 2113–2120.
45. Gregor, J.I.; Heukamp, I.; Kilian, M.; Kiewert, C.; Schimke, I.; Kristiansen, G.; Walz, M.K.; Jacobi, C.A.; Wenger, F.A. Does Enteral Nutrition of Dietary Polyunsaturated Fatty Acids Promote Oxidative Stress and Tumour Growth in Ductal Pancreatic Cancer?: Experimental Trial in Syrian Hamster. *Prostaglandins Leukot. Essent. Fat. Acids* **2006**, *74*, 67–74. [CrossRef] [PubMed]
46. Schønberg, S.A.; Rudra, P.K.; Nøding, R.; Skorpen, F.; Bjerve, K.S.; Krokan, H.E. Evidence That Changes in Se-Glutathione Peroxidase Levels Affect the Sensitivity of Human Tumour Cell Lines to n-3 Fatty Acids. *Carcinogenesis* **1997**, *18*, 1897–1904. [CrossRef]
47. Devery, R.; Miller, A.; Stanton, C. Conjugated Linoleic Acid and Oxidative Behaviour in Cancer Cells. *Biochem. Soc. Trans.* **2001**, *29*, 341–344. [CrossRef]
48. Maehle, L.; Lystad, E.; Eilertsen, E.; Einarsdottír, E.; Høstmark, A.T.; Haugen, A. Growth of Human Lung Adenocarcinoma in Nude Mice Is Influenced by Various Types of Dietary Fat and Vitamin E. *Anticancer. Res.* **1999**, *19*, 1649–1655.
49. Roebuck, B.D. Dietary Fat and the Development of Pancreatic Cancer. *Lipids* **1992**, *27*, 804–806. [CrossRef]
50. Wenger, F.A.; Jacobi, C.A.; Kilian, M.; Zieren, J.; Zieren, H.U.; Müller, J.M. Does Dietary Alpha-Linolenic Acid Promote Liver Metastases in Pancreatic Carcinoma Initiated by BOP in Syrian Hamster? *Ann. Nutr. Metab.* **1999**, *43*, 121–126. [CrossRef]
51. American Institute for Cancer Research, World Cancer Research Fund (Ed.) *Food, Nutrition, Physical Activity and the Prevention of Cancer: A Global Perspective: A Project of World Cancer Research Fund International*; American Institute for Cancer Research: Washington, DC, USA, 2007; ISBN 978-0-9722522-2-5.
52. Wu, G.A.; Terol, J.; Ibanez, V.; López-García, A.; Pérez-Román, E.; Borredá, C.; Domingo, C.; Tadeo, F.R.; Carbonell-Caballero, J.; Alonso, R.; et al. Genomics of the Origin and Evolution of Citrus. *Nature* **2018**, *554*, 311–316. [CrossRef] [PubMed]
53. Lu, X.; Zhao, C.; Shi, H.; Liao, Y.; Xu, F.; Du, H.; Xiao, H.; Zheng, J. Nutrients and Bioactives in Citrus Fruits: Different Citrus Varieties, Fruit Parts, and Growth Stages. *Crit. Rev. Food Sci. Nutr.* **2023**, *63*, 2018–2041. [CrossRef]
54. Zhang, M.; Zhu, S.; Yang, W.; Huang, Q.; Ho, C.-T. The Biological Fate and Bioefficacy of Citrus Flavonoids: Bioavailability, Biotransformation, and Delivery Systems. *Food Funct.* **2021**, *12*, 3307–3323. [CrossRef] [PubMed]
55. Wang, Y.-C.; Chuang, Y.-C.; Ku, Y.-H. Quantitation of Bioactive Compounds in Citrus Fruits Cultivated in Taiwan. *Food Chem.* **2007**, *102*, 1163–1171. [CrossRef]
56. Buer, C.S.; Imin, N.; Djordjevic, M.A. Flavonoids: New Roles for Old Molecules. *J. Integr. Plant Biol.* **2010**, *52*, 98–111. [CrossRef]
57. Stewart, I.; Wheaton, T.A. Conversion of β-Citraurin to Reticulataxanthin and β-Apo-8′-Carotenal to Citranaxanthin during the Isolation of Carotenoids from *Citrus*. *Phytochemistry* **1973**, *12*, 2947–2951. [CrossRef]
58. Manners, G.D. Citrus Limonoids: Analysis, Bioactivity, and Biomedical Prospects. *J. Agric. Food Chem.* **2007**, *55*, 8285–8294. [CrossRef]
59. Polesel, J.; Talamini, R.; Negri, E.; Bosetti, C.; Boz, G.; Lucenteforte, E.; Franceschi, S.; Serraino, D.; La Vecchia, C. Dietary Habits and Risk of Pancreatic Cancer: An Italian Case-Control Study. *Cancer Causes Control* **2010**, *21*, 493–500. [CrossRef]
60. Larsson, S.C.; Håkansson, N.; Näslund, I.; Bergkvist, L.; Wolk, A. Fruit and Vegetable Consumption in Relation to Pancreatic Cancer Risk: A Prospective Study. *Cancer Epidemiol. Biomark. Prev.* **2006**, *15*, 301–305. [CrossRef]
61. Chan, J.M.; Wang, F.; Holly, E.A. Vegetable and Fruit Intake and Pancreatic Cancer in a Population-Based Case-Control Study in the San Francisco Bay Area. *Cancer Epidemiol. Biomark. Prev.* **2005**, *14*, 2093–2097. [CrossRef]
62. Olsen, G.W.; Mandel, J.S.; Gibson, R.W.; Wattenberg, L.W.; Schuman, L.M. Nutrients and Pancreatic Cancer: A Population-Based Case-Control Study. *Cancer Causes Control* **1991**, *2*, 291–297. [CrossRef] [PubMed]
63. Silverman, D.T.; Swanson, C.A.; Gridley, G.; Wacholder, S.; Greenberg, R.S.; Brown, L.M.; Hayes, R.B.; Swanson, G.M.; Schoenberg, J.B.; Pottern, L.M.; et al. Dietary and Nutritional Factors and Pancreatic Cancer: A Case-Control Study Based on Direct Interviews. *J. Natl. Cancer Inst.* **1998**, *90*, 1710–1719. [CrossRef] [PubMed]

64. Nöthlings, U.; Murphy, S.P.; Wilkens, L.R.; Henderson, B.E.; Kolonel, L.N. Dietary Glycemic Load, Added Sugars, and Carbohydrates as Risk Factors for Pancreatic Cancer: The Multiethnic Cohort Study. *Am. J. Clin. Nutr.* **2007**, *86*, 1495–1501. [CrossRef]
65. Coughlin, S.S.; Calle, E.E.; Patel, A.V.; Thun, M.J. Predictors of Pancreatic Cancer Mortality among a Large Cohort of United States Adults. *Cancer Causes Control* **2000**, *11*, 915–923. [CrossRef] [PubMed]
66. Bueno de Mesquita, H.B.; Maisonneuve, P.; Runia, S.; Moerman, C.J. Intake of Foods and Nutrients and Cancer of the Exocrine Pancreas: A Population-Based Case-Control Study in The Netherlands. *Int. J. Cancer* **1991**, *48*, 540–549. [CrossRef]
67. Stolzenberg-Solomon, R.Z.; Pietinen, P.; Taylor, P.R.; Virtamo, J.; Albanes, D. Prospective Study of Diet and Pancreatic Cancer in Male Smokers. *Am. J. Epidemiol.* **2002**, *155*, 783–792. [CrossRef]
68. Ji, B.T.; Chow, W.H.; Gridley, G.; Mclaughlin, J.K.; Dai, Q.; Wacholder, S.; Hatch, M.C.; Gao, Y.T.; Fraumeni, J.F. Dietary Factors and the Risk of Pancreatic Cancer: A Case-Control Study in Shanghai China. *Cancer Epidemiol. Biomark. Prev.* **1995**, *4*, 885–893.
69. Lin, Y.; Kikuchi, S.; Tamakoshi, A.; Yagyu, K.; Obata, Y.; Inaba, Y.; Kurosawa, M.; Kawamura, T.; Motohashi, Y.; Ishibashi, T. Dietary Habits and Pancreatic Cancer Risk in a Cohort of Middle-Aged and Elderly Japanese. *Nutr. Cancer* **2006**, *56*, 40–49. [CrossRef]
70. Sergeev, I.N.; Ho, C.-T.; Li, S.; Colby, J.; Dushenkov, S. Apoptosis-Inducing Activity of Hydroxylated Polymethoxyflavones and Polymethoxyflavones from Orange Peel in Human Breast Cancer Cells. *Mol. Nutr. Food Res.* **2007**, *51*, 1478–1484. [CrossRef]
71. Jayaprakasha, G.K.; Mandadi, K.K.; Poulose, S.M.; Jadegoud, Y.; Nagana Gowda, G.A.; Patil, B.S. Novel Triterpenoid from Citrus Aurantium L. Possesses Chemopreventive Properties against Human Colon Cancer Cells. *Bioorg. Med. Chem.* **2008**, *16*, 5939–5951. [CrossRef]
72. Tang, M.; Ogawa, K.; Asamoto, M.; Hokaiwado, N.; Seeni, A.; Suzuki, S.; Takahashi, S.; Tanaka, T.; Ichikawa, K.; Shirai, T. Protective Effects of Citrus Nobiletin and Auraptene in Transgenic Rats Developing Adenocarcinoma of the Prostate (TRAP) and Human Prostate Carcinoma Cells. *Cancer Sci.* **2007**, *98*, 471–477. [CrossRef] [PubMed]
73. Hung, J.-Y.; Hsu, Y.-L.; Ko, Y.-C.; Tsai, Y.-M.; Yang, C.-J.; Huang, M.-S.; Kuo, P.-L. Didymin, a Dietary Flavonoid Glycoside from Citrus Fruits, Induces Fas-Mediated Apoptotic Pathway in Human Non-Small-Cell Lung Cancer Cells in Vitro and in Vivo. *Lung Cancer* **2010**, *68*, 366–374. [CrossRef]
74. Patil, J.R.; Chidambara Murthy, K.N.; Jayaprakasha, G.K.; Chetti, M.B.; Patil, B.S. Bioactive Compounds from Mexican Lime (*Citrus aurantifolia*) Juice Induce Apoptosis in Human Pancreatic Cells. *J. Agric. Food Chem.* **2009**, *57*, 10933–10942. [CrossRef]
75. Patil, J.R.; Jayaprakasha, G.K.; Murthy, K.N.C.; Chetti, M.B.; Patil, B.S. Characterization of *Citrus aurantifolia* Bioactive Compounds and Their Inhibition of Human Pancreatic Cancer Cells through Apoptosis. *Microchem. J.* **2010**, *94*, 108–117. [CrossRef]
76. Murthy, K.N.C.; Jayaprakasha, G.K.; Safe, S.; Patil, B.S. Citrus Limonoids Induce Apoptosis and Inhibit the Proliferation of Pancreatic Cancer Cells. *Food Funct.* **2021**, *12*, 1111–1120. [CrossRef]
77. Modanwal, S.; Maurya, A.K.; Mulpuru, V.; Mishra, N. Exploring Flavonoid Derivatives as Potential Pancreatic Lipase Inhibitors for Obesity Management: An in Silico and in Vitro Study. *Mol. Divers.* 2024; *Epub ahead of print*.
78. Bondonno, N.P.; Dalgaard, F.; Kyrø, C.; Murray, K.; Bondonno, C.P.; Lewis, J.R.; Croft, K.D.; Gislason, G.; Scalbert, A.; Cassidy, A.; et al. Flavonoid Intake Is Associated with Lower Mortality in the Danish Diet Cancer and Health Cohort. *Nat. Commun.* **2019**, *10*, 3651. [CrossRef] [PubMed]
79. Modanwal, S.; Maurya, A.K.; Mishra, S.K.; Mishra, N. Development of QSAR Model Using Machine Learning and Molecular Docking Study of Polyphenol Derivatives against Obesity as Pancreatic Lipase Inhibitor. *J. Biomol. Struct. Dyn.* **2023**, *41*, 6569–6580. [CrossRef]
80. Chun, O.K.; Chung, S.J.; Song, W.O. Estimated Dietary Flavonoid Intake and Major Food Sources of U.S. Adults. *J. Nutr.* **2007**, *137*, 1244–1252. [CrossRef]
81. Ramos, S. Cancer Chemoprevention and Chemotherapy: Dietary Polyphenols and Signalling Pathways. *Mol. Nutr. Food Res.* **2008**, *52*, 507–526. [CrossRef]
82. Redondo-Blanco, S.; Fernández, J.; Gutiérrez-Del-Río, I.; Villar, C.J.; Lombó, F. New Insights toward Colorectal Cancer Chemotherapy Using Natural Bioactive Compounds. *Front. Pharmacol.* **2017**, *8*, 109. [CrossRef]
83. Zhao, T.-T.; Xu, Y.-Q.; Hu, H.-M.; Gong, H.-B.; Zhu, H.-L. Isoliquiritigenin (ISL) and Its Formulations: Potential Antitumor Agents. *Curr. Med. Chem.* **2019**, *26*, 6786–6796. [CrossRef] [PubMed]
84. Perrott, K.M.; Wiley, C.D.; Desprez, P.-Y.; Campisi, J. Apigenin Suppresses the Senescence-Associated Secretory Phenotype and Paracrine Effects on Breast Cancer Cells. *Geroscience* **2017**, *39*, 161–173. [CrossRef] [PubMed]
85. Xavier, C.P.R.; Lima, C.F.; Rohde, M.; Pereira-Wilson, C. Quercetin Enhances 5-Fluorouracil-Induced Apoptosis in MSI Colorectal Cancer Cells through P53 Modulation. *Cancer Chemother. Pharmacol.* **2011**, *68*, 1449–1457. [CrossRef]
86. George, V.C.; Dellaire, G.; Rupasinghe, H.P.V. Plant Flavonoids in Cancer Chemoprevention: Role in Genome Stability. *J. Nutr. Biochem.* **2017**, *45*, 1–14. [CrossRef]

87. Ujiki, M.B.; Ding, X.-Z.; Salabat, M.R.; Bentrem, D.J.; Golkar, L.; Milam, B.; Talamonti, M.S.; Bell, R.H.; Iwamura, T.; Adrian, T.E. Apigenin Inhibits Pancreatic Cancer Cell Proliferation through G2/M Cell Cycle Arrest. *Mol. Cancer* **2006**, *5*, 76. [CrossRef] [PubMed]
88. Mouria, M.; Gukovskaya, A.S.; Jung, Y.; Buechler, P.; Hines, O.J.; Reber, H.A.; Pandol, S.J. Food-Derived Polyphenols Inhibit Pancreatic Cancer Growth through Mitochondrial Cytochrome C Release and Apoptosis. *Int. J. Cancer* **2002**, *98*, 761–769. [CrossRef]
89. Arts, I.C.W.; Jacobs, D.R.; Gross, M.; Harnack, L.J.; Folsom, A.R. Dietary Catechins and Cancer Incidence among Postmenopausal Women: The Iowa Women's Health Study (United States). *Cancer Causes Control* **2002**, *13*, 373–382. [CrossRef]
90. Nöthlings, U.; Murphy, S.P.; Wilkens, L.R.; Henderson, B.E.; Kolonel, L.N. Flavonols and Pancreatic Cancer Risk: The Multiethnic Cohort Study. *Am. J. Epidemiol.* **2007**, *166*, 924–931. [CrossRef]
91. Bobe, G.; Weinstein, S.J.; Albanes, D.; Hirvonen, T.; Ashby, J.; Taylor, P.R.; Virtamo, J.; Stolzenberg-Solomon, R.Z. Flavonoid Intake and Risk of Pancreatic Cancer in Male Smokers (Finland). *Cancer Epidemiol. Biomark. Prev.* **2008**, *17*, 553–562. [CrossRef]
92. Choe, H.; Wang, Z.; Huang, J.; Yang, Y.; Zhao, Z.; Jo, H.; Pak, H.; Ali, T.; Ding, K.; Ma, J.; et al. Broussoflavonol B Induces S-Phase Arrest and Apoptosis in Pancreatic Cancer Cells by Modulating the Cell Cycle Checkpoint through Inhibition of the AURKA/PLK1 Pathway. *Cancer Cell Int.* **2025**, *25*, 100. [CrossRef]
93. Ganbold, M.; Louphrasitthiphol, P.; Miyamoto, T.; Miyazaki, Y.; Oda, T.; Tominaga, K.; Isoda, H. Isorhamnetin Exerts Anti-Proliferative Effect on Cancer-Associated Fibroblasts by Inducing Cell Cycle Arrest. *Biomed. Pharmacother.* **2025**, *185*, 117954. [CrossRef]
94. Kim, J.-E.; Lee, D.-E.; Lee, K.W.; Son, J.E.; Seo, S.K.; Li, J.; Jung, S.K.; Heo, Y.-S.; Mottamal, M.; Bode, A.M.; et al. Isorhamnetin Suppresses Skin Cancer through Direct Inhibition of MEK1 and PI3-K. *Cancer Prev. Res.* **2011**, *4*, 582–591. [CrossRef] [PubMed]
95. Li, C.; Yang, X.; Chen, C.; Cai, S.; Hu, J. Isorhamnetin Suppresses Colon Cancer Cell Growth through the PI3K-Akt-mTOR Pathway. *Mol. Med. Rep.* **2014**, *9*, 935–940. [CrossRef] [PubMed]
96. Wang, J.-L.; Quan, Q.; Ji, R.; Guo, X.-Y.; Zhang, J.-M.; Li, X.; Liu, Y.-G. Isorhamnetin Suppresses PANC-1 Pancreatic Cancer Cell Proliferation through S Phase Arrest. *Biomed. Pharmacother.* **2018**, *108*, 925–933. [CrossRef] [PubMed]
97. Hu, S.; Huang, L.; Meng, L.; Sun, H.; Zhang, W.; Xu, Y. Isorhamnetin Inhibits Cell Proliferation and Induces Apoptosis in Breast Cancer via Akt and Mitogen-activated Protein Kinase Kinase Signaling Pathways. *Mol. Med. Rep.* **2015**, *12*, 6745–6751. [CrossRef]
98. Chung, K.H.; Cho, I.R.; Paik, W.H.; Kim, Y.-T.; Lee, S.H.; Ryu, J.K. Enhanced Anti-Tumor Effect of Flavopiridol in Combination with Gemcitabine in Pancreatic Cancer. *Anticancer. Res.* **2024**, *44*, 1097–1108. [CrossRef]
99. Allemailem, K.S.; Almatroudi, A.; Alharbi, H.O.A.; AlSuhaymi, N.; Alsugoor, M.H.; Aldakheel, F.M.; Khan, A.A.; Rahmani, A.H. Apigenin: A Bioflavonoid with a Promising Role in Disease Prevention and Treatment. *Biomedicines* **2024**, *12*, 1353. [CrossRef]
100. Kılıç, C.S.; Kışla, M.M.; Amasya, G.; Şengel-Türk, C.T.; Alagöz, Z.A.; Özkan, A.M.G.; Ateş, İ.; Gümüşok, S.; Herrera-Bravo, J.; Sharifi-Rad, J.; et al. Rhoifolin: A Promising Flavonoid with Potent Cytotoxic and Anticancer Properties: Molecular Mechanisms and Therapeutic Potential. *EXCLI J.* **2025**, *24*, 289–320. [CrossRef]
101. Babaei, F.; Moafizad, A.; Darvishvand, Z.; Mirzababaei, M.; Hosseinzadeh, H.; Nassiri-Asl, M. Review of the Effects of Vitexin in Oxidative Stress-Related Diseases. *Food Sci. Nutr.* **2020**, *8*, 2569–2580. [CrossRef]
102. Zheng, B.; Zheng, Y.; Zhang, N.; Zhang, Y.; Zheng, B. Rhoifolin from Plumula Nelumbinis Exhibits Anti-Cancer Effects in Pancreatic Cancer via AKT/JNK Signaling Pathways. *Sci. Rep.* **2022**, *12*, 5654. [CrossRef]
103. Cui, B.; Lee, Y.H.; Chai, H.; Tucker, J.C.; Fairchild, C.R.; Raventos-Suarez, C.; Long, B.; Lane, K.E.; Menendez, A.T.; Beecher, C.W.; et al. Cytotoxic Sesquiterpenoids from Ratibida Columnifera. *J. Nat. Prod.* **1999**, *62*, 1545–1550. [CrossRef] [PubMed]
104. Flamini, G.; Antognoli, E.; Morelli, I. Two Flavonoids and Other Compounds from the Aerial Parts of *Centaurea bracteata* from Italy. *Phytochemistry* **2001**, *57*, 559–564. [CrossRef] [PubMed]
105. Fullas, F.; Hussain, R.A.; Chai, H.B.; Pezzuto, J.M.; Soejarto, D.D.; Kinghorn, A.D. Cytotoxic Constituents of Baccharis Gaudichaudiana. *J. Nat. Prod.* **1994**, *57*, 801–807. [CrossRef] [PubMed]
106. Kavvadias, D.; Monschein, V.; Sand, P.; Riederer, P.; Schreier, P. Constituents of Sage (Salvia Officinalis) with in Vitro Affinity to Human Brain Benzodiazepine Receptor. *Planta Med.* **2003**, *69*, 113–117. [CrossRef]
107. Chao, S.-W.; Su, M.-Y.; Chiou, L.-C.; Chen, L.-C.; Chang, C.-I.; Huang, W.-J. Total Synthesis of Hispidulin and the Structural Basis for Its Inhibition of Proto-Oncogene Kinase Pim-1. *J. Nat. Prod.* **2015**, *78*, 1969–1976. [CrossRef]
108. Gao, H.; Xie, J.; Peng, J.; Han, Y.; Jiang, Q.; Han, M.; Wang, C. Hispidulin Inhibits Proliferation and Enhances Chemosensitivity of Gallbladder Cancer Cells by Targeting HIF-1α. *Exp. Cell Res.* **2015**, *332*, 236–246. [CrossRef]
109. Gao, H.; Jiang, Q.; Han, Y.; Peng, J.; Wang, C. Hispidulin Potentiates the Antitumor Effect of Sunitinib against Human Renal Cell Carcinoma in Laboratory Models. *Cell Biochem. Biophys.* **2015**, *71*, 757–764. [CrossRef]
110. Wang, Y.; Liu, W.; He, X.; Fei, Z. Hispidulin Enhances the Anti-Tumor Effects of Temozolomide in Glioblastoma by Activating AMPK. *Cell Biochem. Biophys.* **2015**, *71*, 701–706. [CrossRef]

111. Yang, J.-M.; Hung, C.-M.; Fu, C.-N.; Lee, J.-C.; Huang, C.-H.; Yang, M.-H.; Lin, C.-L.; Kao, J.-Y.; Way, T.-D. Hispidulin Sensitizes Human Ovarian Cancer Cells to TRAIL-Induced Apoptosis by AMPK Activation Leading to Mcl-1 Block in Translation. *J. Agric. Food Chem.* **2010**, *58*, 10020–10026. [CrossRef]
112. Liu, K.; Zhao, F.; Yan, J.; Xia, Z.; Jiang, D.; Ma, P. Hispidulin: A Promising Flavonoid with Diverse Anti-Cancer Properties. *Life Sci.* **2020**, *259*, 118395. [CrossRef]
113. Li, Y.; Zhao, Y.; Tan, X.; Liu, J.; Zhi, Y.; Yi, L.; Bai, S.; Du, Q.; Li, Q.X.; Dong, Y. Isoorientin Inhibits Inflammation in Macrophages and Endotoxemia Mice by Regulating Glycogen Synthase Kinase 3β. *Mediat. Inflamm.* **2020**, *2020*, 8704146. [CrossRef] [PubMed]
114. Tunalier, Z.; Koşar, M.; Küpeli, E.; Çaliş, İ.; Başer, K.H.C. Antioxidant, Anti-Inflammatory, Anti-Nociceptive Activities and Composition of *Lythrum salicaria* L. Extracts. *J. Ethnopharmacol.* **2007**, *110*, 539–547. [CrossRef] [PubMed]
115. Küpeli, E.; Aslan, M.; Gürbüz, İ.; Yesilada, E. Evaluation of in Vivo Biological Activity Profile of Isoorientin. *Z. für Naturforschung C* **2004**, *59*, 787–790. [CrossRef]
116. Yuan, L.; Wang, J.; Xiao, H.; Xiao, C.; Wang, Y.; Liu, X. Isoorientin Induces Apoptosis through Mitochondrial Dysfunction and Inhibition of PI3K/Akt Signaling Pathway in HepG2 Cancer Cells. *Toxicol. Appl. Pharmacol.* **2012**, *265*, 83–92. [CrossRef]
117. Ye, T.; Su, J.; Huang, C.; Yu, D.; Dai, S.; Huang, X.; Chen, B.; Zhou, M. Isoorientin Induces Apoptosis, Decreases Invasiveness, and Downregulates VEGF Secretion by Activating AMPK Signaling in Pancreatic Cancer Cells. *OTT* **2016**, *9*, 7481–7492. [CrossRef] [PubMed]
118. Shin, W.; Kim, S.; Chun, K.S. Structure of (R,S)-Hesperetin Monohydrate. *Acta Crystallogr. Sect. C* **1987**, *43*, 1946–1949. [CrossRef]
119. Esaki, S.; Nishiyama, K.; Sugiyama, N.; Nakajima, R.; Takao, Y.; Kamiya, S. Preparation and Taste of Certain Glycosides of Flavanones and of Dihydrochalcones. *Biosci. Biotechnol. Biochem.* **1994**, *58*, 1479–1485. [CrossRef]
120. Shin, J.H.; Shin, S.H. A Comprehensive Review of Naringenin, a Promising Phytochemical with Therapeutic Potential. *J. Microbiol. Biotechnol.* **2024**, *34*, 2425–2438. [CrossRef]
121. Lou, C.; Zhang, F.; Yang, M.; Zhao, J.; Zeng, W.; Fang, X.; Zhang, Y.; Zhang, C.; Liang, W. Naringenin Decreases Invasiveness and Metastasis by Inhibiting TGF-β-Induced Epithelial to Mesenchymal Transition in Pancreatic Cancer Cells. *PLoS ONE* **2012**, *7*, e50956. [CrossRef]
122. Park, H.J.; Choi, Y.J.; Lee, J.H.; Nam, M.J. Naringenin Causes ASK1-Induced Apoptosis via Reactive Oxygen Species in Human Pancreatic Cancer Cells. *Food Chem. Toxicol.* **2017**, *99*, 1–8. [CrossRef]
123. Zhang, Z.; Guo, Y.; Chen, M.; Chen, F.; Liu, B.; Shen, C. Kaempferol Potentiates the Sensitivity of Pancreatic Cancer Cells to Erlotinib via Inhibition of the PI3K/AKT Signaling Pathway and Epidermal Growth Factor Receptor. *Inflammopharmacology* **2021**, *29*, 1587–1601. [CrossRef] [PubMed]
124. Zhu, H.; Xiao, Y.; Guo, H.; Guo, Y.; Huang, Y.; Shan, Y.; Bai, Y.; Lin, X.; Lu, H. The Isoflavone Puerarin Exerts Anti-Tumor Activity in Pancreatic Ductal Adenocarcinoma by Suppressing mTOR-Mediated Glucose Metabolism. *Aging* **2021**, *13*, 25089–25105. [CrossRef] [PubMed]
125. Ahmad, B.; Khan, S.; Liu, Y.; Xue, M.; Nabi, G.; Kumar, S.; Alshwmi, M.; Qarluq, A.W. Molecular Mechanisms of Anticancer Activities of Puerarin. *CMAR* **2020**, *12*, 79–90. [CrossRef]
126. Arai, Y.; Watanabe, S.; Kimira, M.; Shimoi, K.; Mochizuki, R.; Kinae, N. Dietary Intakes of Flavonols, Flavones and Isoflavones by Japanese Women and the Inverse Correlation between Quercetin Intake and Plasma LDL Cholesterol Concentration. *J. Nutr.* **2000**, *130*, 2243–2250. [CrossRef]
127. Ravichandran, N.; Suresh, G.; Ramesh, B.; Vijaiyan Siva, G. Fisetin, a Novel Flavonol Attenuates Benzo(a)Pyrene-Induced Lung Carcinogenesis in Swiss Albino Mice. *Food Chem. Toxicol.* **2011**, *49*, 1141–1147. [CrossRef]
128. Murtaza, I.; Adhami, V.M.; Hafeez, B.B.; Saleem, M.; Mukhtar, H. Fisetin, a Natural Flavonoid, Targets Chemoresistant Human Pancreatic Cancer AsPC-1 Cells through DR3-Mediated Inhibition of NF-κB. *Int. J. Cancer* **2009**, *125*, 2465–2473. [CrossRef]
129. Banik, K.; Khatoon, E.; Harsha, C.; Rana, V.; Parama, D.; Thakur, K.K.; Bishayee, A.; Kunnumakkara, A.B. Wogonin and Its Analogs for the Prevention and Treatment of Cancer: A Systematic Review. *Phytother. Res.* **2022**, *36*, 1854–1883. [CrossRef] [PubMed]
130. Huynh, D.L.; Ngau, T.H.; Nguyen, N.H.; Tran, G.-B.; Nguyen, C.T. Potential Therapeutic and Pharmacological Effects of Wogonin: An Updated Review. *Mol. Biol. Rep.* **2020**, *47*, 9779–9789. [CrossRef]
131. Huynh, D.L.; Sharma, N.; Kumar Singh, A.; Singh Sodhi, S.; Zhang, J.-J.; Mongre, R.K.; Ghosh, M.; Kim, N.; Ho Park, Y.; Kee Jeong, D. Anti-Tumor Activity of Wogonin, an Extract from Scutellaria Baicalensis, through Regulating Different Signaling Pathways. *Chin. J. Nat. Med.* **2017**, *15*, 15–40. [CrossRef]
132. Wu, K.; Teng, M.; Zhou, W.; Lu, F.; Zhou, Y.; Zeng, J.; Yang, J.; Liu, X.; Zhang, Y.; Ding, Y.; et al. Wogonin Induces Cell Cycle Arrest and Apoptosis of Hepatocellular Carcinoma Cells by Activating Hippo Signaling. *Anticancer Agents Med. Chem.* **2022**, *22*, 1551–1560. [CrossRef]
133. Xing, F.; Sun, C.; Luo, N.; He, Y.; Chen, M.; Ding, S.; Liu, C.; Feng, L.; Cheng, Z. Wogonin Increases Cisplatin Sensitivity in Ovarian Cancer Cells Through Inhibition of the Phosphatidylinositol 3-Kinase (PI3K)/Akt Pathway. *Med. Sci. Monit.* **2019**, *25*, 6007–6014. [CrossRef] [PubMed]

134. Zhang, T.; Liu, M.; Liu, Q.; Xiao, G.G. Wogonin Increases Gemcitabine Sensitivity in Pancreatic Cancer by Inhibiting Akt Pathway. *Front. Pharmacol.* **2022**, *13*, 1068855. [CrossRef] [PubMed]
135. Wang, Z.; Pan, X.; Ma, X.; Zhang, Y.; Gao, Y.; Guo, Y.; Zhou, Y. FV-429 Suppresses Cancer Cell Migration and Invasion by EMT via the Hippo/YAP1 Pathway in Pancreatic Cancer Cells. *Anti-Cancer Drugs*, 2025; *Epub ahead of print.*
136. Zhang, Z.; Chen, W.; Zhang, S.; Bai, J.; Liu, B.; Yung, K.K.-L.; Ko, J.K.-S. Isoliquiritigenin Inhibits Pancreatic Cancer Progression through Blockade of P38 MAPK-Regulated Autophagy. *Phytomedicine* **2022**, *106*, 154406. [CrossRef] [PubMed]
137. Peng, F.; Du, Q.; Peng, C.; Wang, N.; Tang, H.; Xie, X.; Shen, J.; Chen, J. A Review: The Pharmacology of Isoliquiritigenin. *Phytother. Res.* **2015**, *29*, 969–977. [CrossRef]
138. Kato, H.; Sato, M.; Naiki-Ito, A.; Inaguma, S.; Sano, M.; Komura, M.; Nagayasu, Y.; Xiaochen, K.; Kato, A.; Matsuo, Y.; et al. The Role of DPYD and the Effects of DPYD Suppressor Luteolin Combined with 5-FU in Pancreatic Cancer. *Cancer Med.* **2024**, *13*, e70124. [CrossRef]
139. Johnson, J.L.; Dia, V.P.; Wallig, M.; Gonzalez de Mejia, E. Luteolin and Gemcitabine Protect Against Pancreatic Cancer in an Orthotopic Mouse Model. *Pancreas* **2015**, *44*, 144–151. [CrossRef]
140. Jiang, Z.-Q.; Li, M.-H.; Qin, Y.-M.; Jiang, H.-Y.; Zhang, X.; Wu, M.-H. Luteolin Inhibits Tumorigenesis and Induces Apoptosis of Non-Small Cell Lung Cancer Cells via Regulation of MicroRNA-34a-5p. *Int. J. Mol. Sci.* **2018**, *19*, 447. [CrossRef]
141. Pandurangan, A.K.; Esa, N.M. Luteolin, a Bioflavonoid Inhibits Colorectal Cancer through Modulation of Multiple Signaling Pathways: A Review. *Asian Pac. J. Cancer Prev.* **2014**, *15*, 5501–5508. [CrossRef]
142. Sagawa, H.; Naiki-Ito, A.; Kato, H.; Naiki, T.; Yamashita, Y.; Suzuki, S.; Sato, S.; Shiomi, K.; Kato, A.; Kuno, T.; et al. Connexin 32 and Luteolin Play Protective Roles in Non-Alcoholic Steatohepatitis Development and Its Related Hepatocarcinogenesis in Rats. *Carcinogenesis* **2015**, *36*, 1539–1549. [CrossRef]
143. Naiki-Ito, A.; Naiki, T.; Kato, H.; Iida, K.; Etani, T.; Nagayasu, Y.; Suzuki, S.; Yamashita, Y.; Inaguma, S.; Onishi, M.; et al. Recruitment of miR-8080 by Luteolin Inhibits Androgen Receptor Splice Variant 7 Expression in Castration-Resistant Prostate Cancer. *Carcinogenesis* **2020**, *41*, 1145–1157. [CrossRef]
144. Oumeddour, D.Z.; Al-Dalali, S.; Zhao, L.; Zhao, L.; Wang, C. Recent Advances on Cyanidin-3-O-Glucoside in Preventing Obesity-Related Metabolic Disorders: A Comprehensive Review. *Biochem. Biophys. Res. Commun.* **2024**, *729*, 150344. [CrossRef] [PubMed]
145. Czank, C.; Cassidy, A.; Zhang, Q.; Morrison, D.J.; Preston, T.; Kroon, P.A.; Botting, N.P.; Kay, C.D. Human Metabolism and Elimination of the Anthocyanin, Cyanidin-3-Glucoside: A (13)C-Tracer Study. *Am. J. Clin. Nutr.* **2013**, *97*, 995–1003. [CrossRef] [PubMed]
146. Tan, J.; Li, Y.; Hou, D.-X.; Wu, S. The Effects and Mechanisms of Cyanidin-3-Glucoside and Its Phenolic Metabolites in Maintaining Intestinal Integrity. *Antioxidants* **2019**, *8*, 479. [CrossRef]
147. Sun, W.; Zhang, N.-D.; Zhang, T.; Li, Y.-N.; Xue, H.; Cao, J.-L.; Hou, W.-S.; Liu, J.; Wang, Y.; Jin, C.-H. Cyanidin-3-O-Glucoside Induces the Apoptosis of Human Gastric Cancer MKN-45 Cells through ROS-Mediated Signaling Pathways. *Molecules* **2023**, *28*, 652. [CrossRef]
148. Kuntz, S.; Kunz, C.; Rudloff, S. Inhibition of Pancreatic Cancer Cell Migration by Plasma Anthocyanins Isolated from Healthy Volunteers Receiving an Anthocyanin-Rich Berry Juice. *Eur. J. Nutr.* **2017**, *56*, 203–214. [CrossRef]
149. Mostafa, H.; Behrendt, I.; Meroño, T.; González-Domínguez, R.; Fasshauer, M.; Rudloff, S.; Andres-Lacueva, C.; Kuntz, S. Plasma Anthocyanins and Their Metabolites Reduce in Vitro Migration of Pancreatic Cancer Cells, PANC-1, in a FAK- and NF-kB Dependent Manner: Results from the ATTACH-Study a Randomized, Controlled, Crossover Trial in Healthy Subjects. *Biomed. Pharmacother.* **2023**, *158*, 114076. [CrossRef] [PubMed]
150. Lin, B.-W.; Gong, C.-C.; Song, H.-F.; Cui, Y.-Y. Effects of Anthocyanins on the Prevention and Treatment of Cancer. *Br. J. Pharmacol.* **2017**, *174*, 1226–1243. [CrossRef]
151. Posadino, A.M.; Giordo, R.; Ramli, I.; Zayed, H.; Nasrallah, G.K.; Wehbe, Z.; Eid, A.H.; Gürer, E.S.; Kennedy, J.F.; Aldahish, A.A.; et al. An Updated Overview of Cyanidins for Chemoprevention and Cancer Therapy. *Biomed. Pharmacother.* **2023**, *163*, 114783. [CrossRef]
152. Piekara, J.; Piasecka-Kwiatkowska, D. Antioxidant Potential of Xanthohumol in Disease Prevention: Evidence from Human and Animal Studies. *Antioxidants* **2024**, *13*, 1559. [CrossRef]
153. Tuli, H.S.; Aggarwal, V.; Parashar, G.; Aggarwal, D.; Parashar, N.C.; Tuorkey, M.J.; Varol, M.; Sak, K.; Kumar, M.; Buttar, H.S. Xanthohumol: A Metabolite with Promising Anti-Neoplastic Potential. *Anticancer. Agents Med. Chem.* **2022**, *22*, 418–432. [CrossRef]
154. Krajka-Kuźniak, V.; Szaefer, H.; Stefański, T.; Sobiak, S.; Cichocki, M.; Baer-Dubowska, W. The Effect of Resveratrol and Its Methylthio-Derivatives on the Nrf2-ARE Pathway in Mouse Epidermis and HaCaT Keratinocytes. *Cell. Mol. Biol. Lett.* **2014**, *19*, 500–516. [CrossRef]
155. Berman, A.Y.; Motechin, R.A.; Wiesenfeld, M.Y.; Holz, M.K. The Therapeutic Potential of Resveratrol: A Review of Clinical Trials. *npj Precis. Onc.* **2017**, *1*, 35. [CrossRef] [PubMed]

156. Sukocheva, O.A. Resveratrol and Pancreatic Cancers: Questions and Future Perspectives. *World J. Gastrointest. Oncol.* **2025**, *17*, 100342. [CrossRef]
157. Krajka-Kuźniak, V.; Paluszczak, J.; Szaefer, H.; Baer-Dubowska, W. The Activation of the Nrf2/ARE Pathway in HepG2 Hepatoma Cells by Phytochemicals and Subsequent Modulation of Phase II and Antioxidant Enzyme Expression. *J. Physiol. Biochem.* **2015**, *71*, 227–238. [CrossRef] [PubMed]
158. Coscueta, E.R.; Sousa, A.S.; Reis, C.A.; Pintado, M.M. Phenylethyl Isothiocyanate: A Bioactive Agent for Gastrointestinal Health. *Molecules* **2022**, *27*, 794. [CrossRef] [PubMed]
159. M Ezzat, S.; M Merghany, R.; M Abdel Baki, P.; Ali Abdelrahim, N.; M Osman, S.; A Salem, M.; Peña-Corona, S.I.; Cortés, H.; Kiyekbayeva, L.; Leyva-Gómez, G.; et al. Nutritional Sources and Anticancer Potential of Phenethyl Isothiocyanate: Molecular Mechanisms and Therapeutic Insights. *Mol. Nutr. Food Res.* **2024**, *68*, e2400063. [CrossRef]
160. Stan, S.D.; Singh, S.V.; Whitcomb, D.C.; Brand, R.E. Phenethyl Isothiocyanate Inhibits Proliferation and Induces Apoptosis in Pancreatic Cancer Cells in Vitro and in a MIAPaca2 Xenograft Animal Model. *Nutr. Cancer* **2014**, *66*, 747–755. [CrossRef]
161. Cykowiak, M.; Krajka-Kuźniak, V.; Baer-Dubowska, W. Combinations of Phytochemicals More Efficiently than Single Components Activate Nrf2 and Induce the Expression of Antioxidant Enzymes in Pancreatic Cancer Cells. *Nutr. Cancer* **2022**, *74*, 996–1011. [CrossRef]
162. Wan, Q.; Ren, Q.; Qiao, S.; Lyu, A.; He, X.; Li, F. Therapeutic Potential of Flavonoids from Traditional Chinese Medicine in Pancreatic Cancer Treatment. *Front. Nutr.* **2024**, *11*, 1477140. [CrossRef]
163. Coppinger, C.; Pomales, B.; Movahed, M.R.; Marefat, M.; Hashemzadeh, M. Berberine: A Multi-Target Natural PCSK9 Inhibitor with the Potential to Treat Diabetes, Alzheimer's, Cancer and Cardiovascular Disease. *Curr. Rev. Clin. Exp. Pharmacol.* **2024**, *19*, 312–326. [CrossRef]
164. Vlavcheski, F.; O'Neill, E.J.; Gagacev, F.; Tsiani, E. Effects of Berberine against Pancreatitis and Pancreatic Cancer. *Molecules* **2022**, *27*, 8630. [CrossRef] [PubMed]
165. Davoodvandi, A.; Sadeghi, S.; Alavi, S.M.A.; Alavi, S.S.; Jafari, A.; Khan, H.; Aschner, M.; Mirzaei, H.; Sharifi, M.; Asemi, Z. The Therapeutic Effects of Berberine for Gastrointestinal Cancers. *Asia Pac. J. Clin. Oncol.* **2024**, *20*, 152–167. [CrossRef]
166. Yang, J.; Xu, T.; Wang, H.; Wang, L.; Cheng, Y. Mechanisms of Berberine in Anti-Pancreatic Ductal Adenocarcinoma Revealed by Integrated Multi-Omics Profiling. *Sci. Rep.* **2024**, *14*, 22929. [CrossRef]
167. Schwarz, R.E.; Donohue, C.A.; Sadava, D.; Kane, S.E. Pancreatic Cancer in Vitro Toxicity Mediated by Chinese Herbs SPES and PC-SPES: Implications for Monotherapy and Combination Treatment. *Cancer Lett.* **2003**, *189*, 59–68. [CrossRef] [PubMed]
168. Kunnumakkara, A.B.; Guha, S.; Krishnan, S.; Diagaradjane, P.; Gelovani, J.; Aggarwal, B.B. Curcumin Potentiates Antitumor Activity of Gemcitabine in an Orthotopic Model of Pancreatic Cancer through Suppression of Proliferation, Angiogenesis, and Inhibition of Nuclear Factor-kappaB-Regulated Gene Products. *Cancer Res.* **2007**, *67*, 3853–3861. [CrossRef] [PubMed]
169. Chehl, N.; Chipitsyna, G.; Gong, Q.; Yeo, C.J.; Arafat, H.A. Anti-Inflammatory Effects of the Nigella Sativa Seed Extract, Thymoquinone, in Pancreatic Cancer Cells. *HPB* **2009**, *11*, 373–381. [CrossRef]
170. El-Far, A.H. Thymoquinone Anticancer Discovery: Possible Mechanisms. *Curr. Drug Discov. Technol.* **2015**, *12*, 80–89. [CrossRef]
171. Woo, C.C.; Kumar, A.P.; Sethi, G.; Tan, K.H.B. Thymoquinone: Potential Cure for Inflammatory Disorders and Cancer. *Biochem. Pharmacol.* **2012**, *83*, 443–451. [CrossRef]
172. Banerjee, S.; Azmi, A.S.; Padhye, S.; Singh, M.W.; Baruah, J.B.; Philip, P.A.; Sarkar, F.H.; Mohammad, R.M. Structure-Activity Studies on Therapeutic Potential of Thymoquinone Analogs in Pancreatic Cancer. *Pharm. Res.* **2010**, *27*, 1146–1158. [CrossRef]
173. Mu, G.-G.; Zhang, L.-L.; Li, H.-Y.; Liao, Y.; Yu, H.-G. Thymoquinone Pretreatment Overcomes the Insensitivity and Potentiates the Antitumor Effect of Gemcitabine Through Abrogation of Notch1, PI3K/Akt/mTOR Regulated Signaling Pathways in Pancreatic Cancer. *Dig. Dis. Sci.* **2015**, *60*, 1067–1080. [CrossRef]
174. Yusufi, M.; Banerjee, S.; Mohammad, M.; Khatal, S.; Venkateswara Swamy, K.; Khan, E.M.; Aboukameel, A.; Sarkar, F.H.; Padhye, S. Synthesis, Characterization and Anti-Tumor Activity of Novel Thymoquinone Analogs against Pancreatic Cancer. *Bioorg. Med. Chem. Lett.* **2013**, *23*, 3101–3104. [CrossRef]
175. Azimi, H.; Khakshur, A.A.; Abdollahi, M.; Rahimi, R. Potential New Pharmacological Agents Derived From Medicinal Plants for the Treatment of Pancreatic Cancer. *Pancreas* **2015**, *44*, 11–15. [CrossRef]
176. Lee, W.; Song, G.; Bae, H. In Vitro and in Silico Study of the Synergistic Anticancer Effect of Alpinumisoflavone with Gemcitabine on Pancreatic Ductal Adenocarcinoma through Suppression of Ribonucleotide Reductase Subunit-M1. *Eur. J. Pharm. Sci.* **2025**, *204*, 106969. [CrossRef] [PubMed]
177. Zhang, Y.; Yang, H.; Sun, M.; He, T.; Liu, Y.; Yang, X.; Shi, X.; Liu, X. Alpinumisoflavone Suppresses Hepatocellular Carcinoma Cell Growth and Metastasis via NLRP3 Inflammasome-Mediated Pyroptosis. *Pharmacol. Rep.* **2020**, *72*, 1370–1382. [CrossRef]
178. Han, Y.; Yang, X.; Zhao, N.; Peng, J.; Gao, H.; Qiu, X. Alpinumisoflavone Induces Apoptosis in Esophageal Squamous Cell Carcinoma by Modulating miR-370/PIM1 Signaling. *Am. J. Cancer Res.* **2016**, *6*, 2755–2771. [PubMed]

179. Namkoong, S.; Kim, T.-J.; Jang, I.-S.; Kang, K.-W.; Oh, W.-K.; Park, J. Alpinumisoflavone Induces Apoptosis and Suppresses Extracellular Signal-Regulated Kinases/Mitogen Activated Protein Kinase and Nuclear Factor-κB Pathways in Lung Tumor Cells. *Biol. Pharm. Bull.* **2011**, *34*, 203–208. [CrossRef] [PubMed]
180. Wang, T.; Jiang, Y.; Chu, L.; Wu, T.; You, J. Alpinumisoflavone Suppresses Tumour Growth and Metastasis of Clear-Cell Renal Cell Carcinoma. *Am. J. Cancer Res.* **2017**, *7*, 999–1015.
181. Zhang, B.-X.; Hou, Z.-L.; Yan, W.; Zhao, Q.-L.; Zhan, K.-T. Multi-Dimensional Flexible Reduced Graphene Oxide/Polymer Sponges for Multiple Forms of Strain Sensors. *Carbon.* **2017**, *125*, 199–206. [CrossRef]
182. Alos, H.C.; Billones, J.B.; Castillo, A.L.; Vasquez, R.D. Alpinumisoflavone against Cancer Pro-Angiogenic Targets: In Silico, In Vitro, and In Ovo Evaluation. *DARU J. Pharm. Sci.* **2022**, *30*, 273–288. [CrossRef]
183. Hong, T.; Ham, J.; Song, G.; Lim, W. Alpinumisoflavone Disrupts Endoplasmic Reticulum and Mitochondria Leading to Apoptosis in Human Ovarian Cancer. *Pharmaceutics* **2022**, *14*, 564. [CrossRef]
184. Dhillon, H.; Chikara, S.; Reindl, K.M. Piperlongumine Induces Pancreatic Cancer Cell Death by Enhancing Reactive Oxygen Species and DNA Damage. *Toxicol. Rep.* **2014**, *1*, 309–318. [CrossRef] [PubMed]
185. Yamaguchi, Y.; Kasukabe, T.; Kumakura, S. Piperlongumine Rapidly Induces the Death of Human Pancreatic Cancer Cells Mainly through the Induction of Ferroptosis. *Int. J. Oncol.* **2018**, *52*, 1011–1022. [CrossRef]
186. Wang, Y.; Wu, X.; Zhou, Y.; Jiang, H.; Pan, S.; Sun, B. Piperlongumine Suppresses Growth and Sensitizes Pancreatic Tumors to Gemcitabine in a Xenograft Mouse Model by Modulating the NF-Kappa B Pathway. *Cancer Prev. Res.* **2016**, *9*, 234–244. [CrossRef]
187. Dhillon, H.; Mamidi, S.; McClean, P.; Reindl, K.M. Transcriptome Analysis of Piperlongumine-Treated Human Pancreatic Cancer Cells Reveals Involvement of Oxidative Stress and Endoplasmic Reticulum Stress Pathways. *J. Med. Food* **2016**, *19*, 578–585. [CrossRef] [PubMed]
188. Arora, S.; Bhardwaj, A.; Srivastava, S.K.; Singh, S.; McClellan, S.; Wang, B.; Singh, A.P. Honokiol Arrests Cell Cycle, Induces Apoptosis, and Potentiates the Cytotoxic Effect of Gemcitabine in Human Pancreatic Cancer Cells. *PLoS ONE* **2011**, *6*, e21573. [CrossRef]
189. Averett, C.; Bhardwaj, A.; Arora, S.; Srivastava, S.K.; Khan, M.A.; Ahmad, A.; Singh, S.; Carter, J.E.; Khushman, M.; Singh, A.P. Honokiol Suppresses Pancreatic Tumor Growth, Metastasis and Desmoplasia by Interfering with Tumor-Stromal Cross-Talk. *Carcinogenesis* **2016**, *37*, 1052–1061. [CrossRef]
190. Qin, T.; Li, J.; Xiao, Y.; Wang, X.; Gong, M.; Wang, Q.; Zhu, Z.; Zhang, S.; Zhang, W.; Cao, F.; et al. Honokiol Suppresses Perineural Invasion of Pancreatic Cancer by Inhibiting SMAD2/3 Signaling. *Front. Oncol.* **2021**, *11*, 728583. [CrossRef] [PubMed]
191. Lavu, H.; McCall, N.S.; Winter, J.M.; Burkhart, R.A.; Pucci, M.; Leiby, B.E.; Yeo, T.P.; Cannaday, S.; Yeo, C.J. Enhancing Patient Outcomes While Containing Costs after Complex Abdominal Operation: A Randomized Controlled Trial of the Whipple Accelerated Recovery Pathway. *J. Am. Coll. Surg.* **2019**, *228*, 415–424. [CrossRef]
192. O'Neill, L.; Murphy, F.; Reidy, D.; Poisson, C.; Hussey, J.; Guinan, E. Development and Initial Implementation of a Clinical Monitoring Strategy in a Non-Regulated Trial: A Research Note from the ReStOre II Trial. *HRB Open Res.* **2023**, *6*, 46. [CrossRef]
193. Rosebrock, K.; Sinn, M.; Uzunoglu, F.G.; Bokemeyer, C.; Jensen, W.; Salchow, J. Effects of Exercise Training on Patient-Specific Outcomes in Pancreatic Cancer Patients: A Scoping Review. *Cancers* **2023**, *15*, 5899. [CrossRef] [PubMed]
194. McLaughlin, M.; Christie, A.; Campbell, A. Case Report of Exercise to Attenuate Side Effects of Treatment for Pancreatic Cancer. *Case Rep. Oncol.* **2019**, *12*, 845–854. [CrossRef]
195. Mehra, K.; Berkowitz, A.; Sanft, T. Diet, Physical Activity, and Body Weight in Cancer Survivorship. *Med. Clin. North Am.* **2017**, *101*, 1151–1165. [CrossRef]
196. Zoi, V.; Galani, V.; Lianos, G.D.; Voulgaris, S.; Kyritsis, A.P.; Alexiou, G.A. The Role of Curcumin in Cancer Treatment. *Biomedicines* **2021**, *9*, 1086. [CrossRef]
197. Lotfi, N.; Yousefi, Z.; Golabi, M.; Khalilian, P.; Ghezelbash, B.; Montazeri, M.; Shams, M.H.; Baghbadorani, P.Z.; Eskandari, N. The Potential Anti-Cancer Effects of Quercetin on Blood, Prostate and Lung Cancers: An Update. *Front. Immunol.* **2023**, *14*, 1077531. [CrossRef]
198. Pistollato, F.; Giampieri, F.; Battino, M. The Use of Plant-Derived Bioactive Compounds to Target Cancer Stem Cells and Modulate Tumor Microenvironment. *Food Chem. Toxicol.* **2015**, *75*, 58–70. [CrossRef] [PubMed]
199. Chen, Y.; Fan, W.; Zhao, Y.; Liu, M.; Hu, L.; Zhang, W. Progress in the Regulation of Immune Cells in the Tumor Microenvironment by Bioactive Compounds of Traditional Chinese Medicine. *Molecules* **2024**, *29*, 2374. [CrossRef]
200. Bazana, M.T.; Codevilla, C.F.; de Menezes, C.R. Nanoencapsulation of Bioactive Compounds: Challenges and Perspectives. *Curr. Opin. Food Sci.* **2019**, *26*, 47–56. [CrossRef]
201. Guía-García, J.L.; Charles-Rodríguez, A.V.; Reyes-Valdés, M.H.; Ramírez-Godina, F.; Robledo-Olivo, A.; García-Osuna, H.T.; Cerqueira, M.A.; Flores-López, M.L. Micro and Nanoencapsulation of Bioactive Compounds for Agri-Food Applications: A Review. *Ind. Crops Prod.* **2022**, *186*, 115198. [CrossRef]
202. Chowdhury, S.; Kar, K.; Mazumder, R. Exploration of Different Strategies of Nanoencapsulation of Bioactive Compounds and Their Ensuing Approaches. *Future J. Pharm. Sci.* **2024**, *10*, 72. [CrossRef]

203. Usman, I.; Hussain, M.; Imran, A.; Afzaal, M.; Saeed, F.; Javed, M.; Afzal, A.; Ashfaq, I.; Al Jbawi, E.; Saewan, S.A. Traditional and Innovative Approaches for the Extraction of Bioactive Compounds. *Int. J. Food Prop.* **2022**, *25*, 1215–1233. [CrossRef]
204. Azmir, J.; Zaidul, I.S.M.; Rahman, M.M.; Sharif, K.M.; Mohamed, A.; Sahena, F.; Jahurul, M.H.A.; Ghafoor, K.; Norulaini, N.A.N.; Omar, A.K.M. Techniques for Extraction of Bioactive Compounds from Plant Materials: A Review. *J. Food Eng.* **2013**, *117*, 426–436. [CrossRef]
205. Corradetti, B.; Vaiasicca, S.; Mantovani, M.; Virgili, E.; Bonucci, M.; Hammarberg Ferri, I. Bioactive Immunomodulatory Compounds: A Novel Combinatorial Strategy for Integrated Medicine in Oncology? BAIC Exposure in Cancer Cells. *Integr. Cancer Ther.* **2019**, *18*, 1534735419866908. [CrossRef] [PubMed]

biomolecules

MDPI

Review

Pharmacological Potential and Mechanisms of Bisbenzylisoquinoline Alkaloids from Lotus Seed Embryos

Yan Liu [1], Cong Wang [1], Qiong Liao [2,*] and Canwei Du [1,*]

[1] Hunan Engineering Research Center of Lotus Deep Processing and Nutritional Health Sciences, School of Life and Health Sciences, Hunan University of Science and Technology, Xiangtan 411201, China; liuyan@mail.hnust.edu.cn (Y.L.); 2409010227@mail.hnust.edu.cn (C.W.)

[2] Hunan Provincial Key Laboratory of Animal Intestinal Function and Regulation, College of Life Sciences, Hunan Normal University, Changsha 410006, China

* Correspondence: liaoqiong2023@hunnu.edu.cn (Q.L.); ducw2023@hnust.edu.cn (C.D.)

Abstract: Lotus seed embryos, a key component in traditional Chinese medicine, have attracted growing scientific interest due to their wide-ranging therapeutic potential. Among the bioactive compounds found in lotus seed embryos, three bisbenzylisoquinoline alkaloids—liensinine, isoliensinine, and neferine—stand out for their diverse pharmacological activities. These alkaloids are known to exhibit significant antitumor, anti-inflammatory, antihypertensive, neuroprotective, and antifibrotic effects, which make them promising candidates for the treatment of various chronic and acute diseases. Recent studies have highlighted their ability to modulate key signaling pathways involved in cancer progression, inflammation, fibrosis, and neurodegeneration. The precise mechanisms underlying their actions include modulation of oxidative stress, inhibition of pro-inflammatory cytokines, regulation of apoptosis, and modulation of cellular metabolism. This review aims to provide an in-depth overview of the pharmacological relevance of these alkaloids, focusing on their mechanisms of action and their therapeutic potential across different disease models. By synthesizing current evidence from preclinical studies, this review also lays a solid scientific foundation for future research, supporting the rational design and development of lotus-derived compounds for clinical application.

Keywords: lotus seed embryos; liensinine; isoliensinine; neferine; pharmacology; mechanism

1. Introduction

Nelumbo nucifera Gaertn, commonly referred to as lotus, holy lotus, Indian lotus, or Chinese water lily [1], is a perennial aquatic herb of the family Nelumbonaceae. Characterized by its rhizomatous underground stems, peltate floating leaves, and distinctive nut-like fruits, the lotus exhibits exceptional ecological adaptability and multifunctionality. All anatomical structures of the plant—roots, stems, leaves, flowers, and seeds—possess significant economic, nutritional, and medicinal value. Based on morphological and agricultural characteristics, cultivated lotus varieties are classified into three major groups: root-stem lotus, cultivated for its thickened, edible rhizomes; seed lotus, selected for prolific seed yield and floral integrity; and flower lotus, prized for its ornamental diversity, with extensive variation in petal number, shape, and pigmentation [2]. Beyond its agricultural applications, the lotus has long been revered in Chinese traditional medicine for its wide-ranging pharmacological effects. Multiple studies have substantiated its efficacy in mitigating obesity, oxidative stress, inflammation, metabolic disorders, hepatic dysfunction, cardiovascular pathologies, malignancies, and neurodegenerative conditions, with notable antiviral, hypoglycemic, hypolipidemic, and memory-enhancing effects. Various clinical preparations—both single-herb and multi-component formulations—utilizing lotus extracts have been successfully implemented in integrative medicine [3].

Lotus seeds (Figure 1A), the edible and medicinally valuable reproductive structures of the plant, consist of four key anatomical components: the external seed coat, nutrient-rich cotyledons, internal seed coat, and the plumule (embryonic axis). Archaeological evidence indicates the use of lotus seeds for over 7000 years, reflecting their longstanding role as both food and therapeutic agents. Phytochemical investigations have revealed a complex matrix of bioactive constituents in lotus seeds, including flavonoids, glycosides, phenolic acids, and alkaloids, supporting their classification as functional foods with substantial pharmacological utility [4]. The lotus plumule (Figure 1B), embedded between the cotyledons, was first recorded in the *Compendium of Materia Medica*, noted for its bitter flavor and effects in clearing internal heat, alleviating psychological distress, calming mind, and promoting urination [5]. Chemically, the plumule contains a diverse array of chemical compounds, including alkaloids, flavonoids, organic acids, sterols, volatile oils, monosaccharides, water-soluble polysaccharides, and trace elements [6].

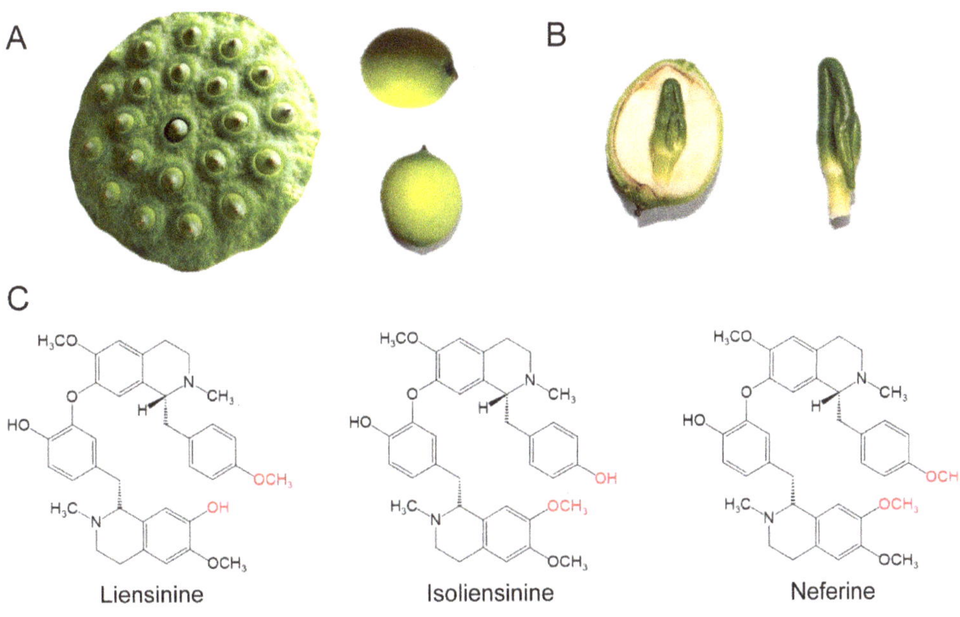

Figure 1. Morphology, anatomy, and chemical structures of lotus seed embryos and their alkaloids. (**A**) Morphology of lotus seeds and seed pods, showing the external structure of lotus and the embedded seeds. (**B**) Anatomical structure of lotus plumule, showing the internal germ and its distinct green plumule part. (**C**) Chemical structures of liensinine, isoliensinine, and neferine. They share a bisbenzylisoquinoline skeleton but differ in the substitution of different groups marked in red in the figure.

Among these chemical compounds, bisbenzylisoquinoline alkaloids are the most pharmacologically prominent, with approximately 20 distinct types identified to date that are widely distributed in roots, seeds, flowers and leaves. Liensinine, isoliensinine, and neferine are the most abundant alkaloids in the lotus seed embryo, each differing in concentration but sharing a closely related chemical structure [7] (Figure 1C). The crude extract of germinated lotus seed embryo was extracted and purified by ionic liquids pH-zone-refining countercurrent chromatography, with the gain of 37.3 mg liensinine,

57.7 mg isoliensinine and 179.9 mg neferine from 1.00 g germinated lotus seed embryo [8]. These three alkaloids have demonstrated therapeutic efficacy across a broad spectrum of pathological conditions, with significant potential in cancer treatment. In addition to their antitumor activity, these alkaloids exhibit diverse pharmacological properties, including antioxidant, astringent, emollient, diuretic, antidiabetic, antihyperlipidemic, antiaging, anti-ischemic, antiviral, anti-inflammatory, antiallergic, and hepatoprotective effects [9]. Their multifaceted actions across multiple physiological systems underscore their relevance as candidates for drug development.

This review aims to systematically summarize the latest research progress regarding these three major bisbenzylisoquinoline alkaloids in various diseases, with particular emphasis on molecular mechanism–based evidence. By offering a more targeted and practical reference, this work provided valuable insights for researchers and contributes to advancing the exploration of these natural compounds in drug development and clinical applications.

2. Antitumor Potential

Cancer arises from the disruption of normal regulatory mechanisms, leading to excessive proliferation and metastatic dissemination via the lymphatic and circulatory systems [10]. Conventional therapeutic modalities, including chemotherapy, radiotherapy, and targeted therapy, remain the primary interventions. However, these approaches are frequently associated with substantial financial burden and debilitating adverse effects, limiting their long-term tolerability and effectiveness [11]. Recent advances in biomedical research have facilitated the development of more refined anticancer strategies, including the incorporation of bioactive compounds derived from Chinese traditional medicine.

Bisbenzylisoquinoline alkaloids derived from the lotus plumule, including liensinine, isoliensinine, and neferine, have gained increasing attention for their potent anticancer activity (Figure 2). Increasing in vitro studies have demonstrated the antitumor mechanisms of liensinine, isoliensinine, and neferine across a wide spectrum of cancer cell lines, including colon (SW480), lung (PC9, A549), liver (HepG2, Huh-7), bladder (T24), gall bladder (GBC-SD, NOZ), prostate (LNCaP, DU-145, PC3), breast (MDA-MB-231, MCF-7), cervical (HeLa, SiHa), and esophageal (KYSE30, KYSE150, KYSE510) cancers, as well as non-tumorigenic mammary epithelial cells (MCF-10A), with multiple experimental approaches employed to assess cellular responses. For example, MTT and clonogenic survival assays have been used to quantify proliferation dynamics. Flow cytometric analysis has been performed to assess cell cycle distribution and apoptosis, while Hoechst 33342 staining has been applied to examine nuclear morphology. Mechanistic investigations have included Western blot analysis to measure the expression of regulatory proteins associated with cell cycle progression and apoptotic signaling pathways [12].

Liensinine has been shown to significantly suppress the proliferation of gastric cancer cells (BGC-823 and SGC-7901) when exposed to concentrations ranging from 20 to 120 μM. This suppression occurs primarily through the generation of reactive oxygen species (ROS) and the inhibition of PI3K/AKT signaling, leading to apoptosis and cell cycle arrest [13]. Liensinine at 80 μM disrupted redox homeostasis by increasing intracellular ROS, thereby impairing antioxidant defenses and inhibiting activation of the JAK2/STAT3 signaling pathway, promoting cell cycle arrest and apoptosis [14]. In gallbladder cancer models (GBC-SD and NOZ), liensinine induces G2/M arrest and apoptosis through inhibition of the zinc finger protein X-linked (ZFX)-induced PI3K/AKT pathway [12]. In non-small cell lung cancer (NSCLC), liensinine modulates autophagy by inducing autophagosome accumulation while simultaneously blocking autophagic flux by impairing lysosomal degradation capacity [15]. In hepatocellular carcinoma (HCC), target-based virtual screening revealed that liensinine interacted with the Kv10.1 potassium channel, inhibiting its currents in a concentration-dependent manner [16]. Further investigations in HUH7 and Hep1-6 cell lines demonstrated that liensinine inhibits cell viability, migration, and proliferation while promoting apoptosis [17]. In intrahepatic cholangiocarcinoma (ICC), in vitro assays confirmed its ability to suppress cell proliferation and inhibit epithelial–mesenchymal transition (EMT) through modulation of the HIF-1α-mediated TGF-β1/p-Smad3 signaling pathway [18]. In bladder cancer T24 cells, liensinine decreases the expression levels of p-AKT, CDK2, and CDK4 while increasing the expression of γH2AX, a marker of DNA damage and senescence [19]. In breast cancer, liensinine at 60 μM exhibits significant antitumor effects in MDA-MB-231 and MCF-7 cancer cells by increasing the Bax/Bcl-2 ratio, activating caspase-3, and inducing PARP cleavage, culminating in apoptosis, cell cycle arrest, and inhibition of migration and invasion [20]. Furthermore, liensinine at 20 μM binds to the S1 G-quadruplex structure in the fibroblast growth factor receptor 2 (FGFR2) promoter with a high affinity and selectivity, inhibiting FGFR2 expression at both the transcriptional and translational levels, providing a novel approach for breast cancer treatment [21]. In colorectal cancer, liensinine acts synergistically with oxaliplatin by inhibiting autophagy via suppressing HIF-1α/eNOS signaling, a key regulatory axis in oxaliplatin resistance [22].

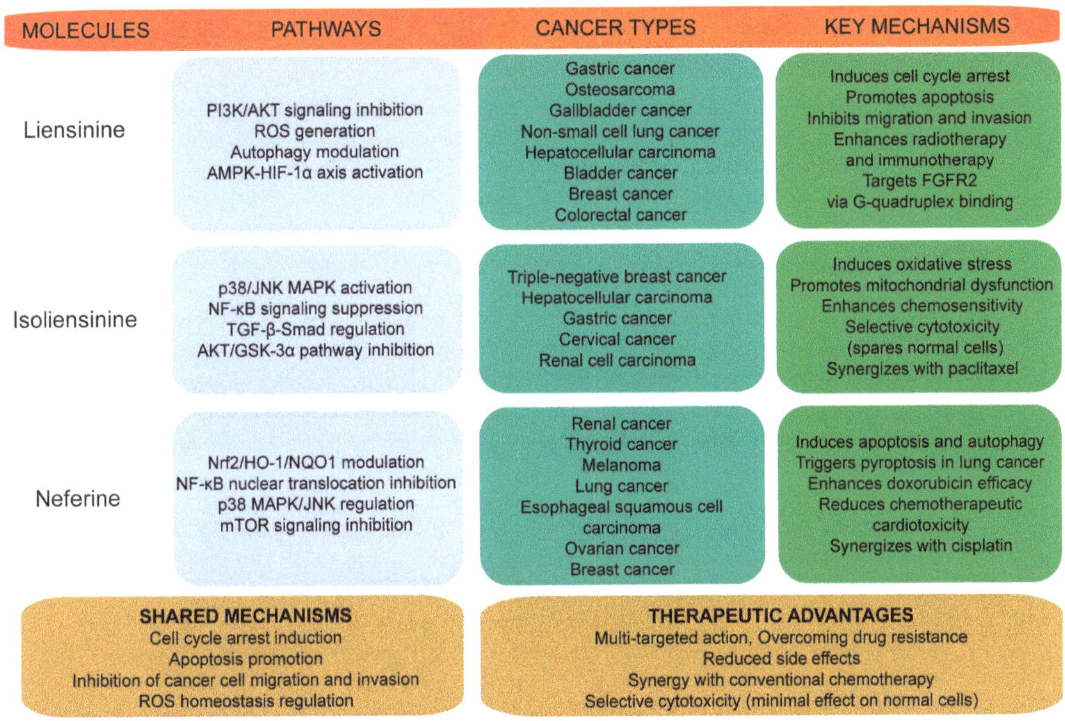

Figure 2. Anticancer mechanisms of liensinine, isoliensinine, and neferine across various tumor types. This figure illustrates how liensinine, isoliensinine, and neferine induce apoptosis, autophagy, and regulate the cell cycle across different cancer types. It highlights their key mechanisms of action and common pathways, along with their potential therapeutic value in cancer treatment.

Isoliensinine exhibits potent anticancer activity across various cancer types, with notable efficacy against triple-negative breast cancer. It triggers apoptotic cell death by inducing oxidative stress and activating both the p38 and JNK MAPK signaling pathways, showing selective cytotoxicity with minimal effects on normal human mammary epithelial cells (MCF-10A) [23]. In hepatocellular carcinoma, isoliensinine induces apoptosis in HepG2, Huh-7, and H22 cell lines by suppressing NF-κB signaling through promoting the dephosphorylation of the P65 subunit at Ser536 [24,25]. Additional studies demonstrate that 10 μM isoliensinine targets TGFBR1 and regulates the TGF-β-Smad signaling cascade to suppress the proliferation and migration of gastric cancer cell lines HGC27 and AGS [26]. Functional assays also identify isoliensinine as a novel AKT-binding ligand capable of suppressing the AKT/GSK-3α pathway and inducing cell cycle arrest and apoptosis in cervical cancer models [27]. Isoliensinine increases intracellular and mitochondrial ROS levels, disrupts mitochondrial membrane potential, impairs mitochondrial function, and suppresses cell viability in a dose-dependent manner at concentrations ranging from 0 to 40 μM [28]. Moreover, isoliensinine enhances the efficacy of conventional chemotherapeutics; co-administration with paclitaxel significantly amplifies cytotoxicity in multidrug-resistant HCT-15 colorectal cancer cells primarily through potentiation of mitochondria-mediated apoptosis [29].

Neferine exhibits broad-spectrum anticancer activity across multiple tumor types through diverse molecular mechanisms. In renal cancer models, neferine treatment induces apoptotic cell death through suppression of the NF-κB signaling pathway, mediated by caspase-dependent cleavage of the p65 (RelA) subunit [30]. In thyroid cancer, neferine demonstrates significant antitumor activity by inhibiting cell proliferation and promoting apoptosis via nuclear factor E2-associated factor 2 (Nrf2)/HO-1/NQO1 signaling modulation with 5 and 10 μM neferine [31]. It regulates p38 MAPK/JNK1/2 pathways to modulate melanoma proliferation, apoptosis, and oxidative stress [32], and inhibits TGF-β signaling to induce MST1/ROS-mediated pyroptosis in lung cancer cells [33]. In esophageal squamous cell carcinoma (ESCC), neferine exerts antiproliferative and pro-apoptotic effects by activating the ROS-dependent JNK signaling cascade, inducing G2/M phase arrest and apoptotic cell death [34]. In ovarian cancer cells, neferine selectively induces autophagy by activating the p38 MAPK/JNK pathway and inhibiting mTOR signaling, with minimal cytotoxicity in non-malignant oviduct epithelial cells [35]. In human prostate cancer cell lines (DU145 and LNCaP), neferine induces cell death by dual modulation of autophagic flux and JNK signaling pathway activation [36]. Subsequent investigations revealed its ability to induce ROS-mediated autophagy and apoptosis in androgen receptor (AR)-positive prostate cancer cells [37]. In cervical cancer cell lines (HeLa and SiHa), neferine upregulates key apoptotic proteins while downregulating antiapoptotic molecules such as Bcl-2, procaspase-3, procaspase-9, and translationally controlled tumor protein (TCTP) [38]. In MDA-MB-231 breast cancer cells, neferine modulates miR-374a and FGFR-2 expression, influencing the PI3K/AKT and MEK/ERK signaling pathways to inhibit proliferation [39]. Recent miRNA multi-omics analyses of endometrial cancer cells revealed that neferine activates mitochondrial apoptotic signaling through Ca^{2+}-mediated endoplasmic reticulum stress and PI3K/AKT modulation, thereby promoting apoptosis [40]. Neferine has also been shown to potentiate the efficacy of doxorubicin (DOX) in human lung adenocarcinoma A549 cells through ROS-mediated apoptosis and MAPK activation, as well as potently inhibiting NF-κB nuclear translocation [41,42]. Additional studies on A549 cells have shown that neferine downregulates NF-κB and Bcl-2, upregulates Bax and cytochrome c, and activates caspase cascades, leading to apoptosis through DNA fragmentation. It also induces excessive ROS production, MAPK activation, lipid peroxidation, depletion of antioxidant reserves, mitochondrial membrane potential loss, intracellular calcium accumulation, and G1 phase cell cycle arrest in a dose-dependent manner [43]. It also potentiates cisplatin efficacy, lowering required doses and reducing toxicity [44].

Evidence from various in vivo animal models strongly supports the antitumor potential of these alkaloids, demonstrating their efficacy in suppressing tumor growth and metastasis with manageable toxicity. Liensinine has shown significant antitumor activity in several cancer models. In prostate cancer models, these alkaloids suppress 5-α-reductase activity and downregulate androgen receptor expression through modulation of the PI3K/AKT signaling pathway [45]. In hepatocellular carcinoma, it suppressed tumor growth in mouse models with efficacy comparable to cisplatin and oxaliplatin, while inducing fewer adverse effects [16]. Liensinine also enhanced the efficacy of radiotherapy and immunotherapy in subcutaneous xenograft and orthotopic liver cancer models. Similarly, in intrahepatic cholangiocarcinoma (ICC) models, liensinine significantly inhibited tumor growth in vivo [18]. Neferine has been extensively evaluated in a range of in vivo models. It has demonstrated antitumor activity in breast cancer models by inhibiting tumor growth and inducing apoptosis [35,46]. In pulmonary carcinogenesis models in Wistar rats, neferine effectively suppressed tumor development, restoring the expression of key proteins like p53, Bax, and caspase-3 [47]. Neferine has also been shown to enhance the efficacy of conventional chemotherapies. It potentiates the anticancer effects of cisplatin while reducing required dosages, thereby mitigating chemoresistance and toxicity [44].

In vitro evidence indicates that liensinine, isoliensinine and neferine inhibit multiple cancer cell lines by inducing ROS, cell-cycle arrest and mitochondrial/apoptotic pathways, and by modulating signaling cascades such as PI3K/AKT, MAPK/JNK and NF-κB. Distinctively, liensinine targets Kv10.1 channels and FGFR2 G-quadruplexes, isoliensinine directly binds AKT and modulates TGFBR1, while neferine uniquely reverses chemoresistance and induces pyroptosis. In vivo studies provide preliminary efficacy data and suggest synergy with radiotherapy, chemotherapy and immunotherapy. However, a translational gap remains—many in vitro effective concentrations are in the micromolar range while achievable systemic exposures following oral administration are likely lower. To support clinical translation, future work should prioritize comprehensive in vivo pharmacokinetics, standardized formulation development, and rigorous efficacy testing across multiple tumor models.

3. Neuroprotective Potential

Alzheimer's disease (AD) is a multifactorial neurodegenerative disorder, with cholinergic and amyloid hypotheses [48]. Bisbenzylisoquinoline alkaloids from lotus seed embryos exhibit notable neuroprotective effects through modulation of oxidative stress, apoptosis, autophagy, and inflammatory pathways. Liensinine, isoliensinine, and neferine suppress tau hyperphosphorylation by inhibiting the Ca^{2+}-CaM/CaMKII pathway, thereby reducing tau accumulation and protecting $A\beta_{25-35}$-damaged PC12 cells [49]. In vitro studies with APP695swe SH-SY5Y cells and in vivo transgenic *C. elegans* confirm that liensinine and neferine enhance cell viability, inhibit Aβ accumulation, reduce ROS, and activate autophagy to mitigate Aβ/tau toxicity [50]. Moreover, Tianjin Anti-cancer Special Extract of Nelumbo nucifera (TASENN), a formulation enriched in these alkaloids, reduces tau phosphorylation and neurofibrillary tangle (NFT) formation in APP/PS1 mice and protects PC12 cells from Aβ damage [51].

Liensinine, neferine, and isoliensinine exhibit strong antineuroinflammatory effects in lipopolysaccharide (LPS)-activated microglia by reducing nitric oxide (NO) and pro-inflammatory cytokines, including tumor necrosis factor-alpha (TNF-α), interleukin-1 beta (IL-1β), and interleukin-6 (IL-6). They inhibit IκBα phosphorylation/degradation through suppression of ROS generation and free radical scavenging [52]. Lotus plumule alkaloids have also demonstrated neuroprotective effects across various other neurological disease models, including Parkinson's disease (PD), ischemic brain injury, hypoxic–ischemic encephalopathy, epilepsy, and depression. In PD models involving LPS-activated microglial cultures and MPTP-treated mice, neferine has been shown to markedly suppress neuroinflammation by inhibiting the NF-κB signaling pathway. This effect is mediated through the prevention of IκBα phosphorylation and p65 nuclear translocation, resulting in attenuated pro-inflammatory cytokine production [53].

Liensinine has also been investigated in models of ischemia–reperfusion (I/R) injury. Pretreatment with liensinine or rapamycin (RAPA) in an external I/R neuronal model has been shown to increase cell viability, reduce cellular damage and apoptosis, and inhibit autophagy. Liensinine exerts its neuroprotective efficacy through positive regulation of the PI3K/Akt signaling pathway [54].

In rats with permanent brain occlusion, neferine reduces neurological deficits, infarct volume, oxidative stress, and apoptosis, accompanied by lowering 4-hydroxynonenal (4-HNE), NO, neuronal nitric oxide (nNOS), calcium, Bax, caspase-3 and increased Hsp70, Bcl-2 expression [55]. Neferine also protects against hypoxic–ischemic encephalopathy (HIE) by suppressing neuronal pyroptosis. It downregulates caspase-1, ASC, and GS-DMD, reduces IL-18 and IL-1β, and alleviates neuroinflammation and oxidative stress. In $CoCl_2$-induced HIE models, neferine further inhibits nod-like receptor family pyrin domain (NLRP3) inflammasome–mediated pyroptosis, highlighting its therapeutic potential against neonatal brain injury [56]. In the context of excitotoxicity, neferine has been shown to inhibit glutamate release induced by the potassium channel blocker 4-aminopyridine in a dose-dependent manner. In addition, neferine activates 5-Hydroxytryptamine (5-HT1A) receptors in cortical synaptosomes, thereby reducing calcium influx and glutamate release via Gi/o protein activation and inhibition of the adenylate cyclase (AC)/cAMP/protein kinase A (PKA) cascade, thereby contributing to neuroprotection [57]. In kainic acid–induced seizures, neferine lowers NLRP3, caspase-1, IL-18, IL-1β, IL-6, TNF-α, while enhancing glutamate and synaptic protein expression, reducing glial activation and neuronal injury, indicating its potential as a treatment for epilepsy [58]. In ischemic stroke models, it protects PC12 cells from oxidative stress, preserves mitochondrial function, boosts adenosine triphosphate (ATP), and activates Nrf2 signaling; in vivo, it improves neural scores, infarct volume, cerebral blood flow, and antioxidant enzyme function [59]. Additionally, in chronic stress–induced depression, neferine enhances monoamine neurotransmitter secretion and modulates gut microbiota, particularly *Lactobacillus*, suggesting antidepressant potential [60].

Preclinical in vitro and in vivo studies indicate that liensinine, isoliensinine and neferine exert neuroprotective effects by suppressing neuroinflammation, mitigating oxidative stress, reducing tau hyperphosphorylation and modulating autophagy/mitochondrial homeostasis in models of Alzheimer's disease, Parkinson's disease and ischemic injury (Figure 3). While mechanistic data are promising, further dose–response, long-term safety, behavioral outcome and multi-species in vivo studies are needed to define translational potential.

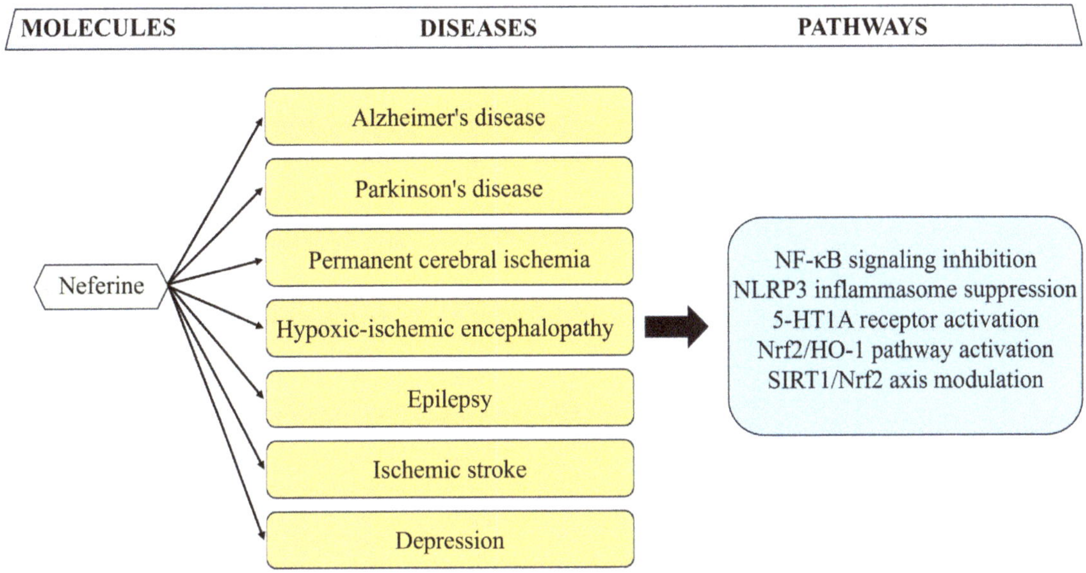

Figure 3. Neuroprotective potential of neferine in AD, PD, and other neurogenic diseases. The figure shows the neuroprotective potential of neferine in neurodegenerative diseases by modulating oxidative stress and neuroinflammation.

4. Anti-Inflammatory Potential

Bisbenzylisoquinoline alkaloids have demonstrated significant anti-inflammatory activity across a range of models (Table 1). Liensinine modulates macrophage ferroptosis and ameliorates acute liver injury by regulating iron metabolism and autophagy, specifically inhibiting ferritinophagy-mediated iron release through blocking autophagosome-lysosome fusion [61]. In LPS-induced acute lung injury (ALI), it acts through PI3K/AKT/mTOR signaling to reduce inflammatory responses [62]. Additionally, liensinine alleviates oxidative stress and inhibits NF-κB phosphorylation and NLRP3 inflammasome synthesis, reducing pro-inflammatory cytokine release in sepsis models [63]. Liensinine also suppresses IL-1β-induced chondrocyte inflammation via NF-κB inhibition and improves cartilage damage in osteoarthritis models [64], indicating its potential utility in osteoarthritis treatment.

Neferine has also been shown to exert anti-inflammatory effects. In vitro experiments using TNF-α/IFN-γ-stimulated human keratinocytes (HaCaT) have demonstrated that neferine suppresses phosphorylation of p38, JNK, and ERK. In vivo, neferine treatment substantially alleviates 2,4-dinitrochlorobenzene (DNCB)-induced atopic dermatitis-like symptoms in mice, supporting its medical potential for the treatment of inflammatory skin conditions [65]. Neferine has also been shown to attenuate vascular inflammation by downregulating the mRNA and protein expression levels of intercellular adhesion molecule 1 (ICAM1) and vascular cell adhesion molecule 1 (VCAM1) via the inhibition of NF-κB signaling [66]. In lysophosphatidylcholine (LPC)-stimulated human umbilical vein endothelial cells (HUVECs), neferine exhibits endothelial-protective effects byregulating the dimethylarginine dimethylaminohydrolase (DDAH)-asymmetric dimethylarginine (ADMA)-NO pathway. Mechanistic studies have reported that neferine preserves DDAH enzymatic activity, reduces pathological ADMA accumulation, and lowers oxidative stress markers such as ROS and malondialdehyde (MDA), thereby restoring NO bioavailability and attenuating LPC-induced endothelial dysfunction [67].

Table 1. Mechanism of anti-inflammatory action of liensinine, isoliensinine, and neferine in the treatment of different diseases.

Alkaloid	Disease	Pathway
Liensinine	Acute liver injury	Inhibiting ferritinophagy and autophagosome-lysosome fusion
	Acute lung injury	Regulating PI3K/AKT/mTOR pathway, reducing LPS-induced acute lung injury and inflammatory factors
	Intestinal injury	Inhibiting NF-κB phosphorylation and NLRP3 inflammasome synthesis
	Chondrocyte inflammatory	Inhibiting NF-κB signaling pathway
Isoliensinine	Osteoarthritis	Inhibiting MAPK/NF-κB pathway activation, reducing extracellular matrix degradation, NLRP3, MMP 3 protein expression, and inflammation
Neferine	Atopic dermatitis-like skin Inflammation	Decreasing phosphorylation of p38, JNK, and ERK proteins
	Vascular inflammatory	Inhibiting NF-κB signaling, reducing mRNA and protein expression of ICAM1 and VCAM1
	Human lysophosphatidylcholine-stimulated human umbilical vein endothelial cells	Regulating DDAH-ADM pathway by restoring DDAH activity, increasing NO concentration, and decreasing ADMA, ROS, and MDA levels

Isoliensinine demonstrates anti-inflammatory and chondroprotective effects in osteoarthritis by reducing extracellular matrix degradation, NLRP3 activation, and Matrix metalloproteinase 3 (MMP-3) expression while inhibiting MAPK/NF-κB signaling and chondrocyte pyroptosis [68].

Both in vitro and in vivo studies demonstrate that liensinine, isoliensinine and neferine attenuate inflammatory responses by inhibiting NF-κB, MAPK and NLRP3 pathways and by modulating ferritinophagy/autophagy, leading to reduced pro-inflammatory cytokine production and tissue protection. Heterogeneity among models and dosing regimens highlights the need for standardized comparative studies to define optimal indications and dosing.

5. Anti-Hypertensive Potential

Bisbenzylisoquinoline alkaloids have demonstrated significant blood pressure-lowering effects (Table 2).

Table 2. Mechanism of Antihypertensive action of liensinine, isoliensinine, and neferine in the treatment of different diseases.

Alkaloid	Disease	Pathway
Liensinine	Ang II-induced hypertension	Inhibits MAPK and TGF-β1/Smad2/3 signaling, reduces collagen deposition, and attenuates aortic wall thickening
Isoliensinine	Ang II- induced proliferation	Antagonizes Ang II effects, downregulates PDGF-β, bFGF, proto-oncogenes and c-Fos/c-Myc, and inhibits VSMC proliferation
Neferine	Pulmonary hypertension	Reduces RVSP, inhibits maladaptive right ventricular hypertrophy, restores extracellular matrix homeostasis, and suppresses proliferative signaling
	Hypertensive cardiomyocyte apoptosis	Inhibits activation of PI3K/AKT and TGF-β1/Smad2/3 signaling pathways
	Hypertensive vascular remodeling	Reduces cardiomyocyte apoptosis, restores mitochondrial membrane potential, and decreases ROS accumulation

Neferine has been shown to effectively improve vascular remodeling in spontaneously hypertensive rats by inhibiting abnormal activation of the PI3K/AKT and TGF-β1/Smad2/3 signaling pathways [69]. In hypoxia induced H9c2 cells, neferine treatment significantly reduces cardiomyocyte apoptosis, restores mitochondrial membrane potential, and decreases ROS accumulation. Similar outcomes have been observed in spontaneously hypertensive rats treated with neferine at various concentrations in vivo, confirming its ability to effectively mitigate apoptosis in hypertensive cardiomyocytes [70].

Isoliensinine dose-dependently inhibits angiotensin II (Ang II)-induced proliferation of porcine coronary artery smooth muscle cells (CASMC) by suppressing platelet-derived growth factor-β (PDGF-β), basic fibroblast growth factor (bFGF), and proto-oncogenes (c-Fos, c-Myc), and modulating heat shock protein 70 (HSP70) expression, indicating its multi-target action against Ang II-activated vascular remodeling [71]. Liensinine exhibits hypotensive effects in Ang II-induced hypertension by reducing abdominal aortic wall thickening, limiting collagen deposition in ventral aortic vessels, attenuating proliferative markers (PCNA) expression, and suppressing the reduction of α-smooth muscle actin (α-SMA). It also blocks activation of MAPK/TGF-β1/Smad 2/3 signaling, thereby suppressing Ang II-induced vascular remodeling in hypertensive models [72].

These compounds have shown therapeutic potential in pulmonary hypertension, as evidenced by network pharmacology and experimental studies. Notably, studies have reported therapeutic effects on pulmonary hemodynamics and cardiac remodeling, with the three alkaloids significantly reducing right ventricular systolic pressure (RVSP), inhibiting maladaptive right ventricular hypertrophy, restoring extracellular matrix homeostasis, and suppressing proliferative signaling in pulmonary hypertension [73]. Preclinical data suggest that liensinine, isoliensinine and neferine ameliorate hypertension and vascular remodeling by inhibiting VSMC proliferation and modulating MAPK/TGF-β/PI3K-AKT signaling, thereby improving hemodynamics and structural remodeling in animal models. Further long-term in vivo studies with hemodynamic endpoints and safety profiling are required to determine clinical relevance.

6. Anti-Fibrotic Potential

Multiple studies have demonstrated the significant antifibrotic potential of liensinine, isoliensinine, and neferine, highlighting their ability to modulate key fibrotic pathways and reduce fibrosis in various tissues. These alkaloids have been shown to effectively inhibit the progression of fibrosis in experimental models, suggesting their promising therapeutic value for fibrotic diseases (Figure 4).

In lung fibroblasts, all three alkaloids have been shown to inhibit TGF-β-induced proliferation in both normal cells and fibroblasts derived from pulmonary fibrosis, significantly downregulating α-SMA protein expression and suppressing fibroblast migration capacity in both cell types [74].

Neferine demonstrates broad antifibrotic activity. In pulmonary fibrosis models, it restores superoxide dismutase (SOD) activity, reduces MDA and myeloperoxidase (MPO) levels, and decreases pro-inflammatory cytokines (TNF-α, IL-6, endothelin-1). It also suppresses NF-κB activation and TGF-β1 upregulation in macrophages [75]. Additionally, neferine alleviates doxorubicin-induced cardiotoxicity by enhancing Sirtuin 1 expression and activity, modulating TGF-β1 and p53, thereby reducing fibrosis, hypertrophy, and apoptosis in cardiomyocytes [76]. In atrial fibrosis, it reverses mitochondrial depolarization, lowers ROS, and boosts GSH and SOD activity—inhibiting atrial fibrillation and fibrosis via Nrf2/HO-1 activation and TGF-β/p-Smad2/3 suppression [77]. Neferine also attenuates endometrial fibrosis by inhibiting fibronectin, collagen I, CTGF, and α-SMA expression through TGF-β/ERK modulation [78]. In diabetic cardiomyopathy, it reduces collagen overproduction and ECM deposition in cardiac fibroblasts via coordinated inhibition of TGF-β1/Smad, ERK, and p38 MAPK pathways [79].

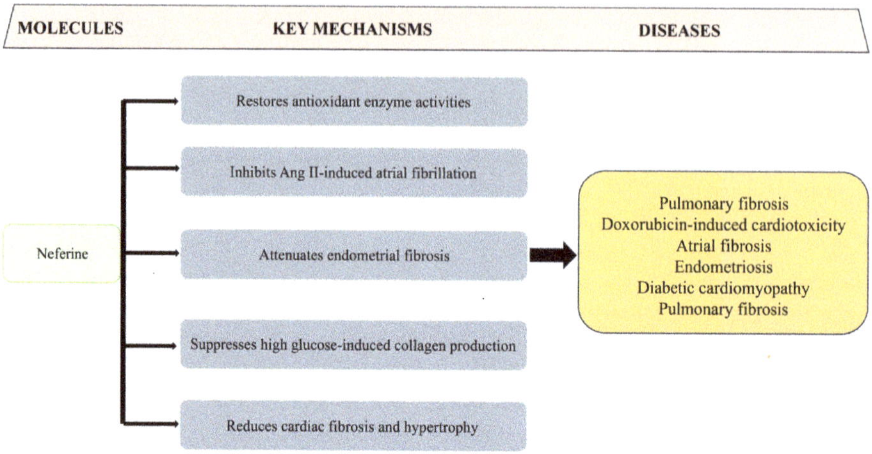

Figure 4. Proposed mechanisms underlying the antifibrotic actions of neferine. Highlights the antifibrotic activities of the alkaloid in different fibrosis models.

Isoliensinine also exerts antifibrotic effects. Notably, in spontaneously hypertensive rats, isoliensinine significantly alleviates renal injury and collagen accumulation by inhibiting TGF-β1/Smad2/3 signaling, leading to improved renal structure and function [80].

Liensinine, isoliensinine and neferine demonstrate antifibrotic activity across pulmonary, cardiac, endometrial and diabetic cardiomyopathy models, primarily via suppression of TGF-β/Smad, ERK/MAPK signaling and oxidative stress. Additional organ-specific dosing and long-term efficacy studies are warranted to validate therapeutic potential.

7. Antiarrhythmic Potential

Extensive preclinical studies have demonstrated that the three principal alkaloids of lotus plumule exhibit potent antiarrhythmic activity.

Neferine exerts concentration-dependent electrophysiological effects on cardiac tissue. In guinea pig ventricular myocardium, it prolongs action potential duration (APD), reduces the maximum upstroke velocity (Vmax), and diminishes the amplitude of slow response action potentials. It also suppresses ouabain-induced afterdepolarizations and decreases contractility. These effects result from concurrent inhibition of cardiac INa, IK, and ICa currents, which collectively stabilize myocardial electrical activity [81]. Neferine further blocks Kv4.3 channels in both open and inactivated states, contributing to its antiarrhythmatic properties in rabbit hearts [82]. In human embryonic kidney 293 (HEK 293) cells expressing Nav1.5 channels, neferine delays activation time to peak, prolongs inactivation, and slows recovery from inactivation, confirming Nav1.5 inhibition as a key antiarrhythmic mechanism [83]. It also enhances the pharmacokinetics and efficacy of amiodarone against both supraventricular and ventricular arrhythmias [84]. Additionally, neferine inhibits human ether-à-go-go-related gene (HERG) K^+ channels by altering their gating kinetics, further supporting its antiarrhythmic role [85].

Liensinine demonstrates concentration-dependent (1–100 μM) antiarrhythmic effects, reducing systolic force in isolated left atrium and decreasing spontaneous beating rate in right atria [86]. Both liensinine and neferine suppress ventricular arrhythmias, though neferine is more potent at concentrations below 10 μM. This difference is attributed to neferine's ability to bind both open and inactivated states of HERG channels, whereas liensinine only binds the open state [87].

Isoliensinine also exhibits antiarrhythmic activity by effectively suppressing early (EADs) and delayed afterdepolarizations (DADs) in rabbit ventricular myocytes. It achieves this through inhibition of INaL and ICaL currents, thereby stabilizing ventricular electrical activity [88].

Electrophysiological study indicated that liensinine, isoliensinine and neferine modulate multiple cardiac ion channels, prolong action potential duration and suppress EAD/DAD, suggesting antiarrhythmic potential. Given the dose-dependent channel effects and proarrhythmic risk, comprehensive safety and drug-interaction studies in larger animal models are necessary.

8. Antioxidant Potential

Lotus plumule alkaloids demonstrate clinically relevant antioxidant properties (Figure 5).

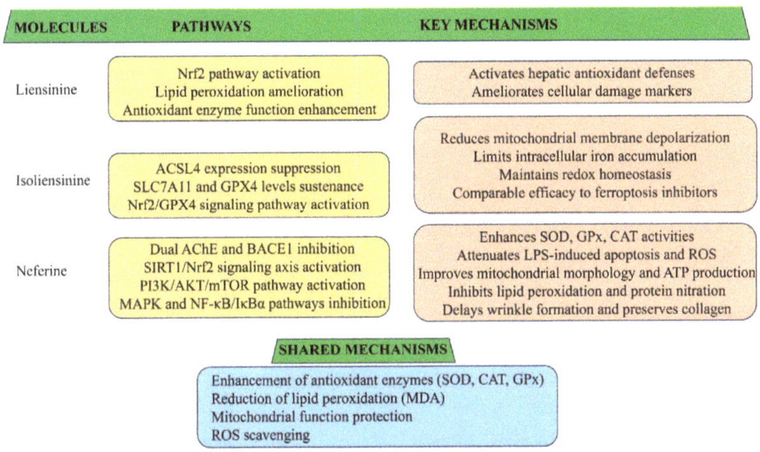

Figure 5. Proposed mechanisms underlying the antioxidant actions of liensinine, isoliensinine, and neferine, showing that how the alkaloids reduce oxidative stress and protect cells in various models.

Neferine reduces oxidative stress and inflammation in CCl_4-induced liver fibrosis by inhibiting MAPK and NF-κB/IκBα pathways. It enhances antioxidant enzyme such as SOD, GSH peroxidase (GPx), and catalase (CAT), while reducing MDA levels [89]. In LPS-induced septic myocardial injury, neferine improves cell viability and mitochondrial function in H9c2 cardiomyocytes via PI3K/AKT/mTOR activation, reducing apoptosis and ROS production [90]. Under hypoxia/reoxygenation stress, it activates SIRT1/Nrf2 signaling, upregulates heme oxygenase-1, and alleviates mitochondrial dysfunction [91]. Neferine also demonstrates neuroprotection through NF-κB suppression, inhibition of lipid peroxidation and protein nitration, modulation of NO homeostasis, and dual inhibition of acetylcholinesterase (AChE) and β-site amyloid precursor protein cleaving enzyme 1 (BACE1) [92]. In aluminum chloride-induced AD mice, oral administration of neferine has been shown to inhibit ROS production, increase antioxidant SOD and CAT expression, and reduce GSH levels, collectively demonstrating its significant antioxidant pharmacotherapeutic potential [93]. In UV-induced skin photoaging, neferine effectively delays wrinkle formation by enhancing SOD and GPx activities, reducing oxidative stress, inhibiting epidermal hypertrophy, and preserving collagen integrity [94].

Isoliensinine protects HT-22 hippocampal neurons from glutamate-induced ferroptosis via Nrf2/GPX4 activation. It reduces mitochondrial depolarization, iron overload, and accumulation of MDA and ROS, while maintaining GPx, SOD, solute carrier family 7 member 11 (SLC7A11), and GPx4 levels and suppressing acyl-CoA synthetase long-chain family member 4 (ACSL4). Its radical-scavenging and iron-chelating capacities are comparable to those of classic ferroptosis inhibitors [95].

Liensinine demonstrates potent hepatoprotective activity against septic liver injury by activating the Nrf2 pathway, ameliorating lipid peroxidation, and enhancing antioxidant enzyme function. It significantly mitigates markers of cellular damage and strengthens hepatic antioxidant defense mechanisms, supporting its potential as a therapeutic candidate for sepsis-associated liver dysfunction [96].

All three alkaloids show significant inhibitory effects on t-BHP-induced oxidative stress and cytotoxicity in HepG2 cells. Neferine exhibits the strongest protective effect by inhibiting ROS and thiobarbituric acid reactive substance (TBARS) formation, reducing lactate dehydrogenase (LDH) release, and increasing GSH levels. Overall, these compounds provide substantial protection against oxidative damage in vitro [97]. Liensinine, isoliensinine and neferine enhance endogenous antioxidant defenses and suppress oxidative markers across hepatic, cardiac, and neural models, indicating broad cytoprotective effects. Translational studies should define dosing and long-term outcomes in chronic disease models.

9. Antidiabetic Potential

Neferine enhances glucose metabolism in L6 myoblasts by promoting glucose transporter 4 (GLUT4) expression and translocation to the plasma membrane, thereby increasing membrane fusion and glucose uptake. These effects are mediated through activation of the G protein-PLC-PKC and adenosine monophosphate (AMP)-activated protein kinase (AMPK) pathways, which upregulate GLUT4 and facilitate its membrane fusion. Neferine also elevates intracellular Ca^{2+} via the G protein-PLC-IP3-IP3R axis, further stimulating glucose uptake [98]. In models of type 2 diabetes, neferine regulates chemokine signaling in the superior cervical ganglion (SCG), downregulating both chemokine ligand 5 (CCL5) and its receptor (CCR5) at the transcriptional level. This suppresses aberrant neuronal signaling within the SCG and ameliorates cardiovascular autonomic neuropathy, suggesting a therapeutic role in diabetic neuroinflammation [99]. Additionally, neferine reduces ROS, normalizes SOD and MDA levels, and inhibits PI3K/Akt and NF-κB pathways in vascular models, thereby protecting endothelial cells from apoptosis and suggesting benefits in diabetes-induced vascular dysfunction [100].

Isoliensinine also demonstrates antidiabetic and antidyslipidemic activity in both in vitro and in vivo models. In L6 cells, isoliensinine administration has been shown to increase GLUT4 translocation by 2.5-fold. In KK-Ay diabetic mice, treatment exerts significant positive effects on serum insulin levels, fasting blood glucose, and body weight, as well as on GLUT4 protein and phosphorylated AMPK levels [101].

Preclinical data indicate that neferine and isoliensinine improve glucose metabolism by promoting GLUT4 translocation, activating AMPK and modulating autonomic/chemokine signaling (Figure 6). Chronic diabetic animal studies are needed to confirm metabolic outcomes, safety and interactions with standard antidiabetic agents.

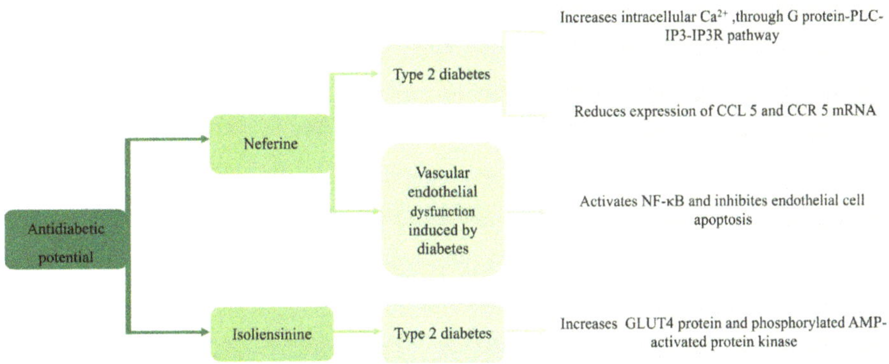

Figure 6. Proposed mechanisms underlying the antidiabetic actions of isoliensinine and neferine in the treatment of different diseases, showing the potential of isoliensinine and neferine on glucose uptake and vascular dysfunction in diabetic models.

10. Other Potential

Liensinine demonstrates therapeutic potential in non-alcoholic fatty liver disease (NAFLD) by improving metabolic disorders, insulin resistance, and dyslipidemia in high-fat diet (HFD)-fed mice. These benefits are mediated through TAK1/AMPK signaling activation, which reduces lipid accumulation in palmitic acid-treated cells and suppresses excessive ROS generation [102]. Additionally, liensinine attenuates ischemic myocardial injury and DNA damage by inhibiting Wnt/β-catenin signaling [103]. It also acts as a novel inhibitor of myostatin (MSTN), disrupting its binding to activin receptor IIB (ActRIIB) and downregulating atrophy-related proteins muscle RING-finger protein-1 (MuRF-1) and muscle atrophy F-box (MAFbx)/Atrogin-1, thereby promoting skeletal muscle regeneration [104].

Isoliensinine effectively inhibits osteoclast differentiation and bone resorption by competitively blocking receptor activator of nuclear factor κB ligand (RANKL)-RANK interaction, leading to suppression of MAPK (p38, JNK, ERK) and NF-κB pathways. It also reduces ROS accumulation and stabilizes Ca^{2+} flux in osteoclast precursors [105]. In ovariectomized (OVX) mice, isoliensinine alleviates bone loss by inhibiting NF-κB signaling and improving bone microarchitectural parameters such as trabecular thickness, number, and bone volume fraction [106].

Neferine exhibits broad protective effects across multiple disease models. In diabetic nephropathy, it reduces blood glucose, creatinine, urea, and miR-17-5p levels, activates Nrf2, enhances antioxidant activities (SOD, GPx), and ameliorates renal fibrosis [107]. In ovalbumin (OVA)-induced asthma models, neferine exerts immunomodulatory effects at doses of 20 mg/kg and 40 mg/kg by reducing systemic and pulmonary inflammation. It decreases Th2 cytokines, IgE, and pro-inflammatory mediators, improves airway resistance, and suppresses pulmonary inflammation and collagen deposition via MAPK inhibition [108]. Neferine also exhibits potent antiplatelet and antithrombotic activities. In murine models, it dose-dependently inhibits platelet activation, adhesion, and aggregation induced by collagen, thrombin, U46619 (washed platelets), and ADP (platelet-rich plasma). At 6 mg/kg, neferine significantly reduces thrombus formation in vivo and uniquely promotes disintegration of preformed platelet aggregates, a property not typically observed with conventional antiplatelet agents [109]. During adipogenesis in 3T3-L1 cells, neferine has been shown to inhibit lipid accumulation in a dose-dependent manner. Neferine increases AMPK activity, reduces fat production, promotes lipid metabolism, and enhances acetyl-CoA carboxylase (ACC) phosphorylation, thereby promoting fatty acid oxidation [110]. Furthermore, it mitigates aging-related liver dysfunction by inducing autophagy via DAPK1 and JNK signaling, restoring mitochondrial function, and reducing senescence markers [111].

Beyond the primary pharmacological activities, the alkaloids show effects on bone remodeling, platelet function, anti-aging and renal protection, indicating multi-target potential. Most evidence is from single short-term models, systematic and reproducible in vivo studies are required to evaluate feasibility and risk-benefit.

11. Discussion

Lotus plumule, a key botanical in traditional Chinese medicine, exhibits considerable therapeutic potential due to its rich alkaloid content. These bioactive alkaloids display a broad spectrum of pharmacological activities, encompassing anticancer, neuroprotective, anti-inflammatory, antifibrotic, antihypertensive, antiarrhythmic, antioxidant, and antidiabetic effects. Their low toxicity and multifaceted efficacy establish lotus plumule as a promising candidate in integrative medicine, offering a natural therapeutic alternative with reduced adverse effects compared to synthetic agents.

Mechanistically, these alkaloids act on target cells by modulating key signaling pathways such as PI3K/AKT, NF-κB, TGF-β, and MAPK/JNK. They also induce autophagy, including mitophagy and ferritinophagy, suppressing the proliferation of pathological cells. These bioactive compounds regulate programmed cell death pathways, control cell cycle checkpoints, and dynamically influence intracellular signaling, collectively contributing to their therapeutic effects.

Growing global interest in preventive healthcare has led to increased recognition of the therapeutic potential of natural botanicals. Among these, lotus plumule has emerged as one of the prominent medicinal agents for pharmacological investigation. Preclinical evidence from in vivo and in vitro studies supports the therapeutic promise of its bioactive alkaloids. However, challenges remain regarding clinical translation. While the pre-clinical data summarized herein demonstrates the remarkable pharmacological potential of lotus plumule alkaloids, a critical translational challenge must be acknowledged. The effective concentrations (IC_{50}/EC_{50}) in most in vitro studies are in the micromolar range, whereas the achievable plasma concentrations following oral ingestion of a traditional lotus plumule tea infusion are expected to be orders of magnitude lower, likely in the nanomolar range, due to limited solubility and poor bioavailability inherent to many bisbenzylisoquinoline alkaloids [112,113]. Pharmacokinetic studies on neferine support this notion, showing low oral bioavailability and rapid metabolism [114]. Consequently, while occasional consumption of lotus plumule tea may offer mild health benefits, potentially attributable to synergistic effects of its various constituents, it is unlikely to produce the robust therapeutic effects observed in laboratory settings. At the same time lotus-plumule products also contain higenamine, a plant-derived β2-agonist that is prohibited by the World Anti-Doping Agency (WADA) at all times. In a human study using commercially available lotus-plumule products, repeated intake equivalent to ~680 μg higenamine per dose led all participants to exceed the WADA reporting threshold of 10 ng/mL for free higenamine in urine, underscoring a tangible risk of adverse analytical findings (AAFs) in doping control. Given the batch-to-batch variability reported for plant-derived products and the possibility of undeclared higenamine, not only should athletes and individuals subject to anti-doping rules avoid lotus-plumule–containing products, but the general population should also use such products prudently and only under appropriate guidance [115,116]. To harness the full potential of these compounds for modern therapeutics, future research must prioritize the development of standardized, concentrated extracts and innovative formulations.

Beyond pharmacokinetic limitations, current research on lotus plumule alkaloids is constrained by an incomplete mechanistic understanding. Studies have predominantly focused on the PI3K/AKT axis, while other potentially relevant signaling pathways remain underexplored. This narrow scope hinders comprehensive insight into the therapeutic actions of the compounds. Furthermore, the pharmacological activity of lotus plumule alkaloids has primarily been evaluated in cellular systems and small animal models, with limited progression to clinical studies [117]. A substantial evidence gap remains regarding their efficacy and mechanisms in human subjects. Continued research, particularly in vivo and clinical investigations, is essential to elucidate the pharmacodynamics of individual alkaloids and establish their applicability in disease prevention and treatment.

In addition to the pharmacological insights, the therapeutic development of bisbenzylisoquinoline alkaloids from lotus seed embryos requires a structured translational framework. Firstly, mechanistic validation across multiple disease models is essential to confirm the consistency of signaling pathways such as PI3K/AKT, MAPK, NF-κB, and TGF-β/Smad. Secondly, dose–response and pharmacokinetic/pharmacodynamic profiling should be performed to determine therapeutic windows and optimize efficacy–toxicity balance. Thirdly, formulation strategies aimed at enhancing solubility, stability, and oral bioavailability are critical to address the limitations of these natural alkaloids. Fourthly, systematic safety and toxicity studies in animal models will provide indispensable data for translational evaluation. Finally, early-phase clinical trials guided by well-defined biomarkers of efficacy and safety represent the most appropriate step toward establishing therapeutic regimens. This staged approach highlights both the translational promise and the rigorous validation required before clinical application.

Author Contributions: Conceptualization, Y.L. and C.D.; methodology, Y.L. and C.W.; software, Y.L. and C.W.; formal analysis, Y.L. and C.D.; investigation, Y.L.; data curation, Y.L. and C.W.; writing—original draft preparation, Y.L. and C.D.; writing—review and editing, Y.L., C.W., Q.L. and C.D.; visualization, Q.L. and C.D.; supervision, Q.L. and C.D.; project administration, C.D.; funding acquisition, Q.L. and C.D. All authors have read and agreed to the published version of the manuscript.

Funding: This research was supported by National Natural Science Foundation of China (82304448), and Natural Science Foundation of Hunan Province (2024JJ6225, 2024JJ8216).

Institutional Review Board Statement: Not applicable.

Informed Consent Statement: Not applicable.

Data Availability Statement: No data.

Conflicts of Interest: The authors declare no conflicts of interest.

References

1. Bishayee, A.; Patel, P.A.; Sharma, P.; Thoutireddy, S.; Das, N. Lotus (*Nelumbo nucifera* Gaertn.) and Its Bioactive Phytocompounds: A Tribute to Cancer Prevention and Intervention. *Cancers* **2022**, *14*, 529. [CrossRef]
2. Yang, M.; Zhu, L.; Pan, C.; Xu, L.; Liu, Y.; Ke, W.; Yang, P. Transcriptomic Analysis of the Regulation of Rhizome Formation in Temperate and Tropical Lotus (*Nelumbo nucifera*). *Sci. Rep.* **2015**, *5*, 13059. [CrossRef]
3. Chen, G.; Zhu, M.; Guo, M. Research advances in traditional and modern use of *Nelumbo nucifera*: Phytochemicals, health promoting activities and beyond. *Crit. Rev. Food Sci. Nutr.* **2019**, *59* (Suppl. S1), S189–S209. [CrossRef]
4. Arooj, M.; Imran, S.; Inam-Ur-Raheem, M.; Rajoka, M.S.R.; Sameen, A.; Siddique, R.; Sahar, A.; Tariq, S.; Riaz, A.; Hussain, A.; et al. Lotus seeds (*Nelumbinis semen*) as an emerging therapeutic seed: A comprehensive review. *Food Sci. Nutr.* **2021**, *9*, 3971–3987. [CrossRef]
5. Pharmacopoeia Commission of the Ministry of Public Health of the People's Republic of China. *Pharmacopoeia of the People's Republic of China*, 1990 ed.; People's Medical Publishing House: Beijing, China, 1992.

6. Zhao, X.; Zhao, R.; Yang, X.; Sun, L.; Bao, Y.; Shuai Liu, Y.; Blennow, A.; Liu, X. Recent advances on bioactive compounds, biosynthesis mechanism, and physiological functions of *Nelumbo nucifera*. *Food Chem.* **2023**, *412*, 135581. [CrossRef]
7. Yang, M.; Zhu, L.; Li, L.; Li, J.; Xu, L.; Feng, J.; Liu, Y. Digital Gene Expression Analysis Provides Insight into the Transcript Profile of the Genes Involved in Aporphine Alkaloid Biosynthesis in Lotus (*Nelumbo nucifera*). *Front. Plant Sci.* **2017**, *8*, 80. [CrossRef] [PubMed]
8. Fang, Y.; Li, Q.; Shao, Q.; Wang, B.; Wei, Y. A general ionic liquid pH-zone-refining countercurrent chromatography method for separation of alkaloids from *Nelumbo nucifera* Gaertn. *J. Chromatogr. A* **2017**, *1507*, 63–71. [CrossRef]
9. Mukherjee, P.K.; Mukherjee, D.; Maji, A.K.; Rai, S.; Heinrich, M. The sacred lotus (*Nelumbo nucifera*)—Phytochemical and therapeutic profile. *J. Pharm. Pharmacol.* **2009**, *61*, 407–422. [CrossRef]
10. Bray, F.; Laversanne, M.; Sung, H.; Ferlay, J.; Siegel, R.L.; Soerjomataram, I.; Jemal, A. Global cancer statistics 2022: GLOBOCAN estimates of incidence and mortality worldwide for 36 cancers in 185 countries. *CA Cancer J. Clin.* **2024**, *74*, 229–263. [CrossRef] [PubMed]
11. Tang, X.; Zheng, Y.; Jiao, D.; Chen, J.; Liu, X.; Xiong, S.; Chen, Q. Anlotinib Inhibits Cell Proliferation, Migration and Invasion via Suppression of c-Met Pathway and Activation of ERK1/2 Pathway in H446 Cells. *Anticancer. Agents Med. Chem.* **2021**, *21*, 747–755. [CrossRef] [PubMed]
12. Shen, Y.; Bian, R.; Li, Y.; Gao, Y.; Liu, Y.; Xu, Y.; Song, X.; Zhang, Y. Liensinine induces gallbladder cancer apoptosis and G2/M arrest by inhibiting ZFX-induced PI3K/AKT pathway. *Acta Biochim. Biophys. Sin.* **2019**, *51*, 607–614. [CrossRef]
13. Yang, J.H.; Yu, K.; Si, X.K.; Li, S.; Cao, Y.J.; Li, W.; Zhang, J.X. Liensinine inhibited gastric cancer cell growth through ROS generation and the PI3K/AKT pathway. *J. Cancer* **2019**, *10*, 6431–6438. [CrossRef]
14. Jia, F.; Liu, Y.; Dou, X.; Du, C.; Mao, T.; Liu, X. Liensinine Inhibits Osteosarcoma Growth by ROS-Mediated Suppression of the JAK2/STAT3 Signaling Pathway. *Oxid. Med. Cell Longev.* **2022**, *2022*, 8245614. [CrossRef] [PubMed]
15. Chang, M.; Ding, S.; Dong, X.; Shang, X.; Li, Y.; Xie, L.; Song, X.; Song, X. Liensinine Inhibits Cell Growth and Blocks Autophagic Flux in Nonsmall-Cell Lung Cancer. *J. Oncol.* **2022**, *2022*, 1533779. [CrossRef]
16. Ma, B.; Shi, S.; Guo, W.; Zhang, H.; Zhao, Z.; An, H. Liensinine, a Novel and Food-Derived Compound, Exerts Potent Antihepatoma Efficacy via Inhibiting the Kv10.1 Channel. *J. Agric. Food Chem.* **2024**, *72*, 4689–4702. [CrossRef]
17. Liu, J.; Zhang, X.; Fan, X.; Liu, P.; Mi, Z.; Tan, H.; Rong, P. Liensinine reshapes the immune microenvironment and enhances immunotherapy by reprogramming metabolism through the AMPK-HIF-1α axis in hepatocellular carcinoma. *J. Exp. Clin. Cancer Res.* **2025**, *44*, 208. [CrossRef]
18. Zhu, X.; Bao, W.; Xie, X.; Chen, B.; Li, R.; Zhao, J.; Wu, L.; Yu, Z.; Li, S.; Zhu, Q.; et al. Liensinine inhibits progression of intrahepatic cholangiocarcinoma by regulating TGF-β1/P-smad3 signaling through HIF-1a. *Mol. Carcinog.* **2024**, *63*, 772–784. [CrossRef] [PubMed]
19. Jiang, H.; Zhu, S.; Wu, B.; Su, Y.; Wang, Q.; Lei, Y.; Shao, Q.; Gao, Y.; Gao, K.; Wu, G. CDK2 and CDK4 targeted liensinine inhibits the growth of bladder cancer T24 cells. *Chem. Biol. Interact.* **2023**, *382*, 110624. [CrossRef] [PubMed]
20. Kang, E.J.; Lee, S.K.; Park, K.K.; Son, S.H.; Kim, K.R.; Chung, W.Y. Liensinine and Nuciferine, Bioactive Components of *Nelumbo nucifera*, Inhibit the Growth of Breast Cancer Cells and Breast Cancer-Associated Bone Loss. *Evid. Based Complement. Altern. Med.* **2017**, *2017*, 1583185. [CrossRef]
21. Zhang, L.; Tan, W.; Zhou, J.; Xu, M.; Yuan, G. Investigation of G-quadruplex formation in the FGFR2 promoter region and its transcriptional regulation by liensinine. *Biochim. Biophys. Acta Gen. Subj.* **2017**, *1861*, 884–891. [CrossRef]
22. Feng, Z.; Zhang, S.; Han, Q.; Chu, T.; Wang, H.; Yu, L.; Zhang, W.; Liu, J.; Liang, W.; Xue, J.; et al. Liensinine sensitizes colorectal cancer cells to oxaliplatin by targeting HIF-1α to inhibit autophagy. *Phytomedicine* **2024**, *129*, 155647. [CrossRef] [PubMed]
23. Zhang, X.; Wang, X.; Wu, T.; Li, B.; Liu, T.; Wang, R.; Liu, Q.; Liu, Z.; Gong, Y.; Shao, C. Isoliensinine induces apoptosis in triple-negative human breast cancer cells through ROS generation and p38 MAPK/JNK activation. *Sci. Rep.* **2015**, *5*, 12579. [CrossRef]
24. Shu, G.; Yue, L.; Zhao, W.; Xu, C.; Yang, J.; Wang, S.; Yang, X. Isoliensinine, a Bioactive Alkaloid Derived from Embryos of *Nelumbo nucifera*, Induces Hepatocellular Carcinoma Cell Apoptosis through Suppression of NF-κB Signaling. *J. Agric. Food Chem.* **2015**, *63*, 8793–8803. [CrossRef] [PubMed]
25. Shu, G.; Zhang, L.; Jiang, S.; Cheng, Z.; Wang, G.; Huang, X.; Yang, X. Isoliensinine induces dephosphorylation of NF-kB p65 subunit at Ser536 via a PP2A-dependent mechanism in hepatocellular carcinoma cells: Roles of impairing PP2A/I2PP2A interaction. *Oncotarget* **2016**, *7*, 40285–40296. [CrossRef]
26. Hu, J.; Dai, S.; Yuan, M.; Li, F.; Xu, S.; Gao, L. Isoliensinine suppressed gastric cancer cell proliferation and migration by targeting TGFBR1 to regulate TGF-β-smad signaling pathways. *Front. Pharmacol.* **2024**, *15*, 1438161. [CrossRef]
27. Li, H.L.; Cheng, Y.; Zhou, Z.W.; Long, H.Z.; Luo, H.Y.; Wen, D.D.; Cheng, L.; Gao, L.C. Isoliensinine induces cervical cancer cell cycle arrest and apoptosis by inhibiting the AKT/GSK3α pathway. *Oncol. Lett.* **2022**, *23*, 8. [CrossRef]
28. Wu, M.J.; Chang, Y.T.; Chuang, T.Y.; Ko, W.S.; Lu, C.C.; Shieh, J.J. Targeting Mitophagy Using Isoliensinine as a Therapeutic Strategy for Renal Cell Carcinoma Treatment. *Free Radic. Biol. Med.* **2025**, *233*, 132–147. [CrossRef]

29. Manogaran, P.; Anandan, A.; Vijaya Padma, V. Isoliensinine augments the therapeutic potential of paclitaxel in multidrug-resistant colon cancer stem cells and induced mitochondria-mediated cell death. *J. Biochem. Mol. Toxicol.* **2023**, *37*, e23395. [CrossRef]
30. Kim, E.A.; Sung, E.G.; Song, I.H.; Kim, J.Y.; Sung, H.J.; Sohn, H.Y.; Park, J.Y.; Lee, T.J. Neferine-induced apoptosis is dependent on the suppression of Bcl-2 expression via downregulation of p65 in renal cancer cells. *Acta Biochim. Biophys. Sin.* **2019**, *51*, 734–742. [CrossRef]
31. Li, S.; Zhang, Y.; Zhang, J.; Yu, B.; Wang, W.; Jia, B.; Chang, J.; Liu, J. Neferine Exerts Ferroptosis-Inducing Effect and Antitumor Effect on Thyroid Cancer through Nrf2/HO-1/NQO1 Inhibition. *J. Oncol.* **2022**, *2022*, 7933775. [CrossRef] [PubMed]
32. Xie, J.; Chen, M.H.; Ying, C.P.; Chen, M.Y. Neferine induces p38 MAPK/JNK1/2 activation to modulate melanoma proliferation, apoptosis, and oxidative stress. *Ann. Transl. Med.* **2020**, *8*, 1643. [CrossRef]
33. Zhong, P.C.; Liu, Z.W.; Xing, Q.C.; Chen, J.; Yang, R.P. Neferine inhibits the development of lung cancer cells by downregulating TGF-β to regulate MST1/ROS-induced pyroptosis. *Kaohsiung, J. Med. Sci.* **2023**, *39*, 1106–1118. [CrossRef]
34. An, K.; Zhang, Y.; Liu, Y.; Yan, S.; Hou, Z.; Cao, M.; Liu, G.; Dong, C.; Gao, J.; Liu, G. Neferine induces apoptosis by modulating the ROS-mediated JNK pathway in esophageal squamous cell carcinoma. *Oncol. Rep.* **2020**, *44*, 1116–1126. [CrossRef]
35. Xu, L.; Zhang, X.; Li, Y.; Lu, S.; Lu, S.; Li, J.; Wang, Y.; Tian, X.; Wei, J.J.; Shao, C.; et al. Neferine induces autophagy of human ovarian cancer cells via p38 MAPK/JNK activation. *Tumour Biol.* **2016**, *37*, 8721–8729. [CrossRef]
36. Nazim, U.M.; Yin, H.; Park, S.Y. Neferine treatment enhances the TRAIL-induced apoptosis of human prostate cancer cells via autophagic flux and the JNK pathway. *Int. J. Oncol.* **2020**, *56*, 1152–1161. [CrossRef] [PubMed]
37. Dasari, S.; Pathak, N.; Thomas, A.; Bitla, S.; Kumar, R.; Munirathinam, G. Neferine Targets the Oncogenic Characteristics of Androgen-Dependent Prostate Cancer Cells via Inducing Reactive Oxygen Species. *Int. J. Mol. Sci.* **2023**, *24*, 14242. [CrossRef] [PubMed]
38. Dasari, S.; Bakthavachalam, V.; Chinnapaka, S.; Venkatesan, R.; Samy, A.; Munirathinam, G. Neferine, an alkaloid from lotus seed embryo targets HeLa and SiHa cervical cancer cells via pro-oxidant anticancer mechanism. *Phytother. Res.* **2020**, *34*, 2366–2384. [CrossRef] [PubMed]
39. Liu, Z.; Zhang, S.; Wang, T.; Shao, H.; Gao, J.; Wang, Y.; Ge, Y. Neferine inhibits MDA-MB-231 cells growth and metastasis by regulating miR-374a/FGFR-2. *Chem. Biol. Interact.* **2019**, *309*, 108716. [CrossRef]
40. Ma, F.F.; Ma, R.H.; Thakur, K.; Zhang, J.G.; Cao, H.; Wei, Z.J.; Simal-Gandara, J. miRNA omics reveal neferine induces apoptosis through Ca(2+)mediated endoplasmic reticulum stress pathway in human endometrial cancer. *Phytomedicine* **2024**, *134*, 155988. [CrossRef]
41. Awasthi, S.; Sharma, R.; Singhal, S.S.; Zimniak, P.; Awasthi, Y.C. RLIP76, a novel transporter catalyzing ATP-dependent efflux of xenobiotics. *Drug Metab. Dispos.* **2002**, *30*, 1300–1310. [CrossRef]
42. Poornima, P.; Kumar, V.B.; Weng, C.F.; Padma, V.V. Doxorubicin induced apoptosis was potentiated by neferine in human lung adenocarcima, A549 cells. *Food Chem. Toxicol.* **2014**, *68*, 87–98. [CrossRef]
43. Poornima, P.; Weng, C.F.; Padma, V.V. Neferine, an alkaloid from lotus seed embryo, inhibits human lung cancer cell growth by MAPK activation and cell cycle arrest. *Biofactors* **2014**, *40*, 121–131. [CrossRef]
44. Sivalingam, K.S.; Paramasivan, P.; Weng, C.F.; Viswanadha, V.P. Neferine Potentiates the Antitumor Effect of Cisplatin in Human Lung Adenocarcinoma Cells Via a Mitochondria-Mediated Apoptosis Pathway. *J. Cell Biochem.* **2017**, *118*, 2865–2876. [CrossRef]
45. Liu, C.M.; Wu, Z.; Pan, B.; An, L.; Zhu, C.; Zhou, J.; Jiang, Y. The antiandrogenic effect of neferine, liensinine, and isoliensinine by inhibiting 5-α-reductase and androgen receptor expression via PI3K/AKT signaling pathway in prostate cancer. *Pharmazie* **2021**, *76*, 225–231. [CrossRef]
46. Yang, L.; Xiong, Y.; Sun, Z.; Lin, X.; Ni, H. Neferine Inhibits 7,12-Dimethylbenz(a)anthracene-Induced Mammary Tumorigenesis by Suppression of Cell Proliferation and Induction of Apoptosis via Modulation of the PI3K/AKT/NF-κB Signaling Pathway. *J. Environ. Pathol. Toxicol. Oncol.* **2021**, *40*, 51–61. [CrossRef]
47. Sivalingam, K.; Amirthalingam, V.; Ganasan, K.; Huang, C.Y.; Viswanadha, V.P. Neferine suppresses diethylnitrosamine-induced lung carcinogenesis in Wistar rats. *Food Chem. Toxicol.* **2019**, *123*, 385–398. [CrossRef]
48. Breijyeh, Z.; Karaman, R. Comprehensive Review on Alzheimer's Disease: Causes and Treatment. *Molecules* **2020**, *25*, 5789. [CrossRef]
49. Meng, X.L.; Liu, S.Y.; Xue, J.S.; Gou, J.M.; Wang, D.; Liu, H.S.; Chen, C.L.; Xu, C.B. Protective effects of Liensinine, Isoliensinine, and Neferine on PC12 cells injured by amyloid-β. *J. Food Biochem.* **2022**, *46*, e14303. [CrossRef] [PubMed]
50. Wu, M.C.; Gao, Y.H.; Zhang, C.; Ma, B.T.; Lin, H.R.; Jiang, J.Y.; Xue, M.F.; Li, S.; Wang, H.B. Liensinine and neferine exert neuroprotective effects via the autophagy pathway in transgenic Caenorhabditis elegans. *BMC Complement. Med. Ther.* **2023**, *23*, 386. [CrossRef] [PubMed]
51. Meng, X.L.; Xue, J.S.; Su, S.J.; Gou, J.M.; Lu, J.; Chen, C.L.; Xu, C.B. Total alkaloids from the seed embryo of *Nelumbo nucifera* Gaertn. improve cognitive impairment in APP/PS1 mice and protect Aβ-damaged PC12 cells. *Nutr. Neurosci.* **2023**, *26*, 1243–1257. [CrossRef] [PubMed]

52. Meng, X.-L.; Zheng, L.-C.; Liu, J.; Gao, C.-C.; Qiu, M.-C.; Liu, Y.-Y.; Lu, J.; Wang, D.; Chen, C.-L. Inhibitory effects of three bisbenzylisoquinoline alkaloids on lipopolysaccharide-induced microglial activation. *RSC Adv.* **2017**, *7*, 18347–18357. [CrossRef]
53. Li, T.; Zhai, Y.X.; Zheng, T.; Xu, B. Neferine exerts anti-inflammatory activity in BV-2 microglial cells and protects mice with MPTP-induced Parkinson's disease by inhibiting NF-κB activation. *Mol. Med. Rep.* **2023**, *28*, 235. [CrossRef]
54. Qiao, W.; Zang, Z.; Li, D.; Shao, S.; Li, Q.; Liu, Z. Liensinine ameliorates ischemia-reperfusion-induced brain injury by inhibiting autophagy via PI3K/AKT signaling. *Funct. Integr. Genom.* **2023**, *23*, 140. [CrossRef]
55. Sengking, J.; Oka, C.; Yawoot, N.; Tocharus, J.; Chaichompoo, W.; Suksamrarn, A.; Tocharus, C. Protective Effect of Neferine in Permanent Cerebral Ischemic Rats via Anti-Oxidative and Anti-Apoptotic Mechanisms. *Neurotox. Res.* **2022**, *40*, 1348–1359. [CrossRef]
56. Zhu, J.J.; Yu, B.Y.; Huang, X.K.; He, M.Z.; Chen, B.W.; Chen, T.T.; Fang, H.Y.; Chen, S.Q.; Fu, X.Q.; Li, P.J.; et al. Neferine Protects against Hypoxic-Ischemic Brain Damage in Neonatal Rats by Suppressing NLRP3-Mediated Inflammasome Activation. *Oxid. Med. Cell Longev.* **2021**, *2021*, 6654954. [CrossRef]
57. Yeh, K.C.; Hung, C.F.; Lin, Y.F.; Chang, C.; Pai, M.S.; Wang, S.J. Neferine, a bisbenzylisoquinoline alkaloid of *Nelumbo nucifera*, inhibits glutamate release in rat cerebrocortical nerve terminals through 5-HT(1A) receptors. *Eur. J. Pharmacol.* **2020**, *889*, 173589. [CrossRef]
58. Lin, T.Y.; Hung, C.Y.; Chiu, K.M.; Lee, M.Y.; Lu, C.W.; Wang, S.J. Neferine, an Alkaloid from Lotus Seed Embryos, Exerts Antiseizure and Neuroprotective Effects in a Kainic Acid-Induced Seizure Model in Rats. *Int. J. Mol. Sci.* **2022**, *23*, 4130. [CrossRef] [PubMed]
59. Wu, C.; Chen, J.; Yang, R.; Duan, F.; Li, S.; Chen, X. Mitochondrial protective effect of neferine through the modulation of nuclear factor erythroid 2-related factor 2 signalling in ischaemic stroke. *Br. J. Pharmacol.* **2019**, *176*, 400–415. [CrossRef] [PubMed]
60. Dong, Z.; Xie, Q.; Xu, F.; Shen, X.; Hao, Y.; Li, J.; Xu, H.; Peng, Q.; Kuang, W. Neferine alleviates chronic stress-induced depression by regulating monoamine neurotransmitter secretion and gut microbiota structure. *Front. Pharmacol.* **2022**, *13*, 974949. [CrossRef] [PubMed]
61. Li, J.; Huang, Q.; Lv, M.; Ma, W.; Sun, J.; Zhong, X.; Hu, R.; Ma, M.; Han, Z.; Zhang, W.; et al. Role of liensinine in sensitivity of activated macrophages to ferroptosis and in acute liver injury. *Cell Death Discov.* **2023**, *9*, 189. [CrossRef]
62. Wang, C.; Zou, K.; Diao, Y.; Zhou, C.; Zhou, J.; Yang, Y.; Zeng, Z. Liensinine alleviates LPS-induced acute lung injury by blocking autophagic flux via PI3K/AKT/mTOR signaling pathway. *Biomed. Pharmacother.* **2023**, *168*, 115813. [CrossRef] [PubMed]
63. Wang, G.; Ma, F.; Xie, K.; Li, X.; Tan, X.; Xia, Y.; Wang, Y.; Dong, J. Liensinine alleviates mouse intestinal injury induced by sepsis through inhibition of oxidative stress, inflammation, and cell apoptosis. *Int. Immunopharmacol.* **2024**, *127*, 111335. [CrossRef]
64. Wang, L.; Shao, T.; Liu, C.; Han, Z.; Zhang, S.; Dong, Y.; Han, T.; Cheng, B.; Ren, W. Liensinine inhibits IL-1β-stimulated inflammatory response in chondrocytes and attenuates papain-induced osteoarthritis in rats. *Int. Immunopharmacol.* **2024**, *138*, 112601. [CrossRef]
65. Yang, C.C.; Hung, Y.L.; Ko, W.C.; Tsai, Y.J.; Chang, J.F.; Liang, C.W.; Chang, D.C.; Hung, C.F. Effect of Neferine on DNCB-Induced Atopic Dermatitis in HaCaT Cells and BALB/c Mice. *Int. J. Mol. Sci.* **2021**, *22*, 8237. [CrossRef]
66. Zhong, Y.; He, S.; Huang, K.; Liang, M. Neferine suppresses vascular endothelial inflammation by inhibiting the NF-κB signaling pathway. *Arch. Biochem. Biophys.* **2020**, *696*, 108595. [CrossRef]
67. Peng, Z.Y.; Zhang, S.D.; Liu, S.; He, B.M. Protective effect of neferine on endothelial cell nitric oxide production induced by lysophosphatidylcholine: The role of the DDAH-ADMA pathway. *Can. J. Physiol. Pharmacol.* **2011**, *89*, 289–294. [CrossRef]
68. Zhang, Z.; Ma, J.; Yi, Y.; Wang, S.; He, Y.; Liu, Y.; Meng, K.; Wang, Y.; Ma, W. Isoliensinine suppresses chondrocytes pyroptosis against osteoarthritis via the MAPK/NF-κB signaling pathway. *Int. Immunopharmacol.* **2024**, *143 Pt 3*, 113589. [CrossRef]
69. Zeng, W.; Zhang, X.; Lu, Y.; Wen, Y.; Xie, Q.; Yang, X.; He, S.; Guo, Z.; Li, J.; Shen, A.; et al. Neferine ameliorates hypertensive vascular remodeling modulating multiple signaling pathways in spontaneously hypertensive rats. *Biomed. Pharmacother.* **2023**, *158*, 114203. [CrossRef]
70. Guo, Z.; Wu, M.; Chen, L.; Chen, H.; Wu, J.; Xie, Q.; Lin, G.; Lian, D.; Peng, J.; Shen, A. Neferine attenuates hypertensive cardiomyocyte apoptosis and modulates key signaling pathways: An in vivo and in vitro study. *Eur. J. Pharmacol.* **2025**, *994*, 177393. [CrossRef] [PubMed]
71. Xiao, J.H.; Zhang, Y.L.; Feng, X.L.; Wang, J.L.; Qian, J.Q. Effects of isoliensinine on angiotensin II-induced proliferation of porcine coronary arterial smooth muscle cells. *J. Asian Nat. Prod. Res.* **2006**, *8*, 209–216. [CrossRef] [PubMed]
72. Jia, P.; Chen, D.; Zhu, Y.; Wang, M.; Zeng, J.; Zhang, L.; Cai, Q.; Lian, D.; Zhao, C.; Xu, Y.; et al. Liensinine improves AngII-induced vascular remodeling via MAPK/TGF-β1/Smad2/3 signaling. *J. Ethnopharmacol.* **2023**, *317*, 116768. [CrossRef]
73. Xiao, X.; Luo, F.; Fu, M.; Jiang, Y.; Liu, S.; Liu, B. Evaluating the therapeutic role of selected active compounds in *Plumula Nelumbinis* on pulmonary hypertension via network pharmacology and experimental analysis. *Front. Pharmacol.* **2022**, *13*, 977921. [CrossRef]
74. Ren, H.L.; Zhang, J.H.; Xiao, J.H. Benzylisoquinoline alkaloids inhibit lung fibroblast activation mainly via inhibiting TGF-β1/Smads and ERK1/2 pathway proteins. *Heliyon* **2023**, *9*, e16849. [CrossRef]

75. Zhao, L.; Wang, X.; Chang, Q.; Xu, J.; Huang, Y.; Guo, Q.; Zhang, S.; Wang, W.; Chen, X.; Wang, J. Neferine, a bisbenzylisoquinline alkaloid attenuates bleomycin-induced pulmonary fibrosis. *Eur. J. Pharmacol.* **2010**, *627*, 304–312. [CrossRef]
76. Lohanathan, B.P.; Rathinasamy, B.; Huang, C.Y.; Viswanadha, V.P. Neferine attenuates doxorubicin-induced fibrosis and hypertrophy in H9c2 cells. *J. Biochem. Mol. Toxicol.* **2022**, *36*, e23054. [CrossRef] [PubMed]
77. Jiang, X.X.; Zhang, R.; Wang, H.S. Neferine mitigates angiotensin II-induced atrial fibrillation and fibrosis via upregulation of Nrf2/HO-1 and inhibition of TGF-β/p-Smad2/3 pathways. *Aging* **2024**, *16*, 8630–8644. [CrossRef]
78. Xia, Y.; Guo, Y.; Zhou, J.; Fan, L.; Xie, J.; Wang, Y.; Du, H.; Ni, X. Neferine mediated TGF-β/ERK signaling to inhibit fibrosis in endometriosis. *Am. J. Transl. Res.* **2023**, *15*, 3240–3253.
79. Liu, X.; Song, X.; Lu, J.; Chen, X.; Liang, E.; Liu, X.; Zhang, M.; Zhang, Y.; Du, Z.; Zhao, Y. Neferine inhibits proliferation and collagen synthesis induced by high glucose in cardiac fibroblasts and reduces cardiac fibrosis in diabetic mice. *Oncotarget* **2016**, *7*, 61703–61715. [CrossRef] [PubMed]
80. Yao, M.; Lian, D.; Wu, M.; Zhou, Y.; Fang, Y.; Zhang, S.; Zhang, W.; Yang, Y.; Li, R.; Chen, H.; et al. Isoliensinine Attenuates Renal Fibrosis and Inhibits TGF-β1/Smad2/3 Signaling Pathway in Spontaneously Hypertensive Rats. *Drug Des. Devel Ther.* **2023**, *17*, 2749–2762. [CrossRef]
81. Li, G.R.; Li, X.G.; Lü, F.H. Effects of neferine on transmembrane potentials of guinea pig myocardium. *Acta Pharmacol. Sin.* **1989**, *10*, 406–410.
82. Wang, C.; Chen, Y.F.; Quan, X.Q.; Wang, H.; Zhang, R.; Xiao, J.H.; Wang, J.L.; Zhang, C.T.; Xiang, J.Z.; Tang, Q. Effects of neferine on Kv4.3 channels expressed in HEK293 cells and ex vivo electrophysiology of rabbit hearts. *Acta Pharmacol. Sin.* **2015**, *36*, 1451–1461. [CrossRef] [PubMed]
83. Wang, C.; Wang, H.; Xiao, J.H.; Wang, J.L.; Xiang, J.Z.; Tang, Q. Inhibitory effects of neferine on Nav1.5 channels expressed in HEK293 cells. *J. Huazhong Univ. Sci. Technol. Med. Sci.* **2016**, *36*, 487–493. [CrossRef]
84. Wan, J.; Zhao, L.; Xu, C.; Zhang, S.; Zhang, Z.; Zeng, C.; Chang, M.; Xiao, J.; Wang, J. Effects of neferine on the pharmacokinetics of amiodarone in rats. *Biomed. Chromatogr.* **2011**, *25*, 858–866. [CrossRef]
85. Gu, D.F.; Li, X.L.; Qi, Z.P.; Shi, S.S.; Hu, M.Q.; Liu, D.M.; She, C.B.; Lv, Y.J.; Li, B.X.; Yang, B.F. Blockade of HERG K+ channel by isoquinoline alkaloid neferine in the stable transfected HEK293 cells. *Naunyn Schmiedebergs Arch. Pharmacol.* **2009**, *380*, 143–151. [CrossRef]
86. Wang, J.L.; Nong, Y.; Jing, M.X. Effects of liensinine on haemodynamics in rats and the physiologic properties of isolated rabbit atria. *Acta Pharm. Sinica* **1992**, *27*, 881–885.
87. Dong, Z.X.; Zhao, X.; Gu, D.F.; Shi, Y.Q.; Zhang, J.; Hu, X.X.; Hu, M.Q.; Yang, B.F.; Li, B.X. Comparative effects of liensinine and neferine on the human ether-a-go-go-related gene potassium channel and pharmacological activity analysis. *Cell Physiol. Biochem.* **2012**, *29*, 431–442. [CrossRef] [PubMed]
88. Liu, Z.; Hu, L.; Zhang, Z.; Song, L.; Zhang, P.; Cao, Z.; Ma, J. Isoliensinine Eliminates Afterdepolarizations Through Inhibiting Late Sodium Current and L-Type Calcium Current. *Cardiovasc. Toxicol.* **2021**, *21*, 67–78. [CrossRef]
89. Wang, Y.; Wang, S.; Wang, R.; Li, S.; Yuan, Y. Neferine Exerts Antioxidant and Anti-Inflammatory Effects on Carbon Tetrachloride-Induced Liver Fibrosis by Inhibiting the MAPK and NF-κB/IκBα Pathways. *Evid. Based Complement. Altern. Med.* **2021**, *2021*, 4136019. [CrossRef]
90. Qi, Z.; Wang, R.; Liao, R.; Xue, S.; Wang, Y. Neferine Ameliorates Sepsis-Induced Myocardial Dysfunction Through Anti-Apoptotic and Antioxidative Effects by Regulating the PI3K/AKT/mTOR Signaling Pathway. *Front. Pharmacol.* **2021**, *12*, 706251. [CrossRef]
91. Lu, C.; Jiang, B.; Xu, J.; Zhang, X.; Jiang, N. Neferine protected cardiomyocytes against hypoxia/oxygenation injury through SIRT1/Nrf2/HO-1 signaling. *J. Biochem. Mol. Toxicol.* **2023**, *37*, e23398. [CrossRef] [PubMed]
92. Jung, H.A.; Jin, S.E.; Choi, R.J.; Kim, D.H.; Kim, Y.S.; Ryu, J.H.; Kim, D.W.; Son, Y.K.; Park, J.J.; Choi, J.S. Anti-amnesic activity of neferine with antioxidant and anti-inflammatory capacities, as well as inhibition of ChEs and BACE1. *Life Sci.* **2010**, *87*, 420–430. [CrossRef]
93. Yin, S.; Ran, Q.; Yang, J.; Zhao, Y.; Li, C. Nootropic effect of neferine on aluminium chloride-induced Alzheimer's disease in experimental models. *J. Biochem. Mol. Toxicol.* **2020**, *34*, e22429. [CrossRef]
94. Khan, A.; Bai, H.; Khan, A.; Bai, Z. Neferine prevents ultraviolet radiation-induced skin photoaging. *Exp. Ther. Med.* **2020**, *19*, 3189–3196. [CrossRef]
95. Long, H.Z.; Li, F.J.; Gao, L.C.; Zhou, Z.W.; Luo, H.Y.; Xu, S.G.; Dai, S.M.; Hu, J.D. Isoliensinine activated the Nrf2/GPX4 pathway to inhibit glutamate-induced ferroptosis in HT-22 cells. *J. Biochem. Mol. Toxicol.* **2024**, *38*, e23794. [CrossRef] [PubMed]
96. Zhang, X.; Yuan, S.; Fan, H.; Zhang, W.; Zhang, H. Liensinine alleviates sepsis-induced acute liver injury by inhibiting the NF-κB and MAPK pathways in an Nrf2-dependent manner. *Chem. Biol. Interact.* **2024**, *396*, 111030. [CrossRef] [PubMed]
97. Xie, Y.; Zhang, Y.; Zhang, L.T.; Zeng, S.X.; Guo, Z.B.; Zheng, B.D. Protective effects of alkaloid compounds from Nelumbinis Plumula on tert-butyl hydroperoxide-induced oxidative stress. *Molecules* **2013**, *18*, 10285–10300. [CrossRef]
98. Zhao, P.; Tian, D.; Song, G.; Ming, Q.; Liu, J.; Shen, J.; Liu, Q.H.; Yang, X. Neferine Promotes GLUT4 Expression and Fusion With the Plasma Membrane to Induce Glucose Uptake in L6 Cells. *Front. Pharmacol.* **2019**, *10*, 999. [CrossRef]

99. Li, G.; Xu, H.; Zhu, S.; Xu, W.; Qin, S.; Liu, S.; Tu, G.; Peng, H.; Qiu, S.; Yu, S.; et al. Effects of neferine on CCL5 and CCR5 expression in SCG of type 2 diabetic rats. *Brain Res. Bull.* **2013**, *90*, 79–87. [CrossRef]
100. Guan, G.; Han, H.; Yang, Y.; Jin, Y.; Wang, X.; Liu, X. Neferine prevented hyperglycemia-induced endothelial cell apoptosis through suppressing ROS/Akt/NF-κB signal. *Endocrine* **2014**, *47*, 764–771. [CrossRef]
101. Yang, X.; Huang, M.; Yang, J.; Wang, J.; Zheng, S.; Ma, X.; Cai, J.; Deng, S.; Shu, G.; Yang, G. Activity of Isoliensinine in Improving the Symptoms of Type 2 Diabetic Mice via Activation of AMP-Activated Kinase and Regulation of PPARγ. *J. Agric. Food Chem.* **2017**, *65*, 7168–7178. [CrossRef] [PubMed]
102. Liang, L.; Ye, S.; Jiang, R.; Zhou, X.; Zhou, J.; Meng, S. Liensinine alleviates high fat diet (HFD)-induced non-alcoholic fatty liver disease (NAFLD) through suppressing oxidative stress and inflammation via regulating TAK1/AMPK signaling. *Int. Immunopharmacol.* **2022**, *104*, 108306. [CrossRef] [PubMed]
103. Shen, F.; Wu, C.; Zhong, X.; Ma, E.; Peng, J.; Zhu, W.; Wo, D.; Ren, D.N. Liensinine prevents ischemic injury following myocardial infarction via inhibition of Wnt/β-catenin signaling activation. *Biomed. Pharmacother.* **2023**, *162*, 114675. [CrossRef]
104. Zheng, Y.; Yu, Y.; Feng, J.; Ling, M.; Wang, X. Unveiling the Potential of *Nelumbo nucifera*-Derived Liensinine to Target The Myostatin Protein and to Counteract Muscle Atrophy. *J. Agric. Food Chem.* **2024**, *72*, 2240–2249. [CrossRef]
105. Deng, W.; Li, H.; Zhang, Y.; Lin, Y.; Chen, C.; Chen, J.; Huang, Y.; Zhou, Y.; Tang, Y.; Ding, J.; et al. Isoliensinine suppresses bone loss by targeted inhibition of RANKL-RANK binding. *Biochem. Pharmacol.* **2023**, *210*, 115463. [CrossRef]
106. Liu, H.; Gu, R.; Huang, Q.; Liu, Y.; Liu, C.; Liao, S.; Feng, W.; Xie, T.; Zhao, J.; Xu, J.; et al. Isoliensinine Suppresses Osteoclast Formation Through NF-κB Signaling Pathways and Relieves Ovariectomy-Induced Bone Loss. *Front. Pharmacol.* **2022**, *13*, 870553. [CrossRef]
107. Hongmei, H.; Maojun, Y.; Ting, L.I.; Dandan, W.; Ying, L.I.; Xiaochi, T.; Lu, Y.; Shi, G.U.; Yong, X.U. Neferine inhibits the progression of diabetic nephropathy by modulating the miR-17-5p/nuclear factor E2-related factor 2 axis. *J. Tradit. Chin. Med.* **2024**, *44*, 44–53. [CrossRef] [PubMed]
108. Zhu, T.; Xiao, X.; Dong, Y.; Yuan, C. Neferine alleviates ovalbumin-induced asthma via MAPK signaling pathways in mice. *Allergol. Immunopathol.* **2023**, *51*, 135–142. [CrossRef]
109. Zhou, Y.J.; Xiang, J.Z.; Yuan, H.; Liu, H.; Tang, Q.; Hao, H.Z.; Yin, Z.; Wang, J.; Ming, Z.Y. Neferine exerts its antithrombotic effect by inhibiting platelet aggregation and promoting dissociation of platelet aggregates. *Thromb. Res.* **2013**, *132*, 202–210. [CrossRef]
110. Park, M.; Han, J.; Lee, H.J. Anti-Adipogenic Effect of Neferine in 3T3-L1 Cells and Primary White Adipocytes. *Nutrients* **2020**, *12*, 1858. [CrossRef]
111. Kageyama, S.; Maejima, Y.; Morioka, Y.; Escareal, Z.; Sato, Y.; Namba, T. Neferine Attenuates Aging-Related Liver Dysfunction by Suppressing Cellular Aging via Mitochondrial Reactivation. *Biol. Pharm. Bull.* **2024**, *47*, 1953–1960. [CrossRef] [PubMed]
112. Cheng, Y.; Li, H.L.; Zhou, Z.W.; Long, H.Z.; Luo, H.Y.; Wen, D.D.; Cheng, L.; Gao, L.C. Isoliensinine: A Natural Compound with "Drug-Like" Potential. *Front. Pharmacol.* **2021**, *12*, 630385. [CrossRef] [PubMed]
113. Hu, L.; Wang, Y.; Shu, C.; Yu, J.; Chen, Y.; Li, Y.; Tao, R.; Yang, H.; Dou, L. Pharmacokinetics, bioavailability and metabolism of neferine in rat by LC-MS/MS and LC-HRMS. *Biomed. Chromatogr.* **2021**, *35*, e5193. [CrossRef]
114. Wei, X.; Zhang, M.; Yang, M.; Ogutu, C.; Li, J.; Deng, X. Lotus (*Nelumbo nucifera*) benzylisoquinoline alkaloids: Advances in chemical profiling, extraction methods, pharmacological activities, and biosynthetic elucidation. *Veg. Res.* **2024**, *4*, e005. [CrossRef]
115. Yen, C.C.; Tung, C.W.; Chang, C.W.; Tsai, C.C.; Hsu, M.C.; Wu, Y.T. Potential Risk of Higenamine Misuse in Sports: Evaluation of Lotus Plumule Extract Products and a Human Study. *Nutrients* **2020**, *12*, 285. [CrossRef]
116. Rangelov Kozhuharov, V.; Ivanov, K.; Ivanova, S. Higenamine in Plants as a Source of Unintentional Doping. *Plants* **2022**, *11*, 354. [CrossRef] [PubMed]
117. Bharathi Priya, L.; Huang, C.Y.; Hu, R.M.; Balasubramanian, B.; Baskaran, R. An updated review on pharmacological properties of neferine-A bisbenzylisoquinoline alkaloid from *Nelumbo nucifera*. *J. Food Biochem.* **2021**, *45*, e13986. [CrossRef] [PubMed]

biomolecules

MDPI

Article

Pharmacological Modulation of the Unfolded Protein Response as a Therapeutic Approach in Cutaneous T-Cell Lymphoma

Nadia St. Thomas [1], Benjamin N. Christopher [1], Leticia Reyes [1], Reeder M. Robinson [1], Lena Golick [1], Xiaoyi Zhu [2], Eli Chapman [2] and Nathan G. Dolloff [1,3,4,*]

1 Department of Pharmacology and Immunology, Medical University of South Carolina, 173 Ashley Ave., MSC509, Charleston, SC 29425, USA; stthoman@musc.edu (N.S.T.); christob@musc.edu (B.N.C.); reyesl@musc.edu (L.R.); robinree@musc.edu (R.M.R.); golick@musc.edu (L.G.)
2 Department of Pharmacology and Therapeutics, Center for Inflammation Science and Systems Medicine, University of Florida Scripps Institute for Biomedical Innovation and Technology, Jupiter, FL 33458, USA; xiaoyizhu@ufl.edu (X.Z.); chapmaneli@ufl.edu (E.C.)
3 Hollings Cancer Center, Medical University of South Carolina, Charleston, SC 29425, USA
4 Zucker Institute for Innovation Commercialization, Charleston, SC 29425, USA
* Correspondence: dolloffn@musc.edu; Tel.: +1-843-876-2204

Abstract: Cutaneous T-cell lymphoma (CTCL) is a rare T-cell malignancy characterized by inflamed and painful rash-like skin lesions that may affect large portions of the body's surface. Patients experience recurrent infections due to a compromised skin barrier and generalized immunodeficiency resulting from a dominant Th2 immune phenotype of CTCL cells. Given the role of the unfolded protein response (UPR) in normal and malignant T-cell development, we investigated the impact of UPR-inducing drugs on the viability, transcriptional networks, and Th2 phenotype of CTCL. We found that CTCL cells were >5-fold more sensitive to the proteasome inhibitor bortezomib (Btz) and exhibited a distinct signaling and transcriptional response compared to normal CD4+ cells. The CTCL response was dominated by the induction of the HSP70 family member *HSPA6* (HSP70B') and, to a lesser extent, *HSPA5* (BiP/GRP78). To understand the significance of these two factors, we used a novel isoform selective small-molecule inhibitor of HSPA5/6 (JG-023). JG-023 induced pro-apoptotic UPR signaling and enhanced the cytotoxic effects of proteasome inhibitors and other UPR-inducing drugs in CTCL but not normal T cells. Interestingly, JG-023 also selectively suppressed the production of Th2 cytokines in CTCL and normal CD4+ T cells. Conditioned media (CM) from CTCL were immunosuppressive to normal T cells through an IL-10-dependent mechanism. This immunosuppression could be reversed by JG-023, other HSP70 inhibitors, Btz, and combinations of these UPR-targeted drugs. Our study points to the importance of the UPR in the pathology of CTCL and demonstrates the potential of proteasome and targeted HSPA5/6 inhibitors for therapy.

Keywords: cutaneous T-cell lymphoma; ER stress; unfolded protein response; bortezomib; HSP70; HSPA5; HSPA6

1. Introduction

Cutaneous T-cell lymphoma (CTCL) is characterized by the infiltration and expansion of malignant clonal T lymphocytes in the skin. Skin patches and plaques are pruritic and painful and may progress to confluent erythema that can ulcerate and affect >80% of the body surface area [1]. CTCL is a heterogenous group of malignancies, with Mycoses fungoides (MF), Sézary Syndrome (SS), and primary cutaneous CD30+ anaplastic large cell lymphoma (pcALCL) representing 80–85% of cases [2]. MF remains localized to the skin,

https://doi.org/10.3390/biom15010076

whereas the more aggressive leukemic form, SS, is found in the peripheral blood and lymph nodes and carries a significantly worse prognosis. Early-stage MF is a treatable and indolent disease, whereas advanced MF and SS are incurable, debilitating, and more aggressive, with a median survival estimated at less than 1.5 years and poor health-related quality of life [3,4]. For these patients, topical therapies are ineffective, and the standard of care includes the retinoid analog, bexarotene, interferon-α, HDAC inhibitors (vorinostat/ZOLINZA and romidepsin/ISTODAX), and, more recently, the development of monoclonal antibodies, brentuximab/ADCETRIS and mogamulizumab/POTELIGEO, which target CD30 and CCR4, respectively. These newer agents have shown promising response rates. However, they are short-lived, and a large international study showed that survival rates have not increased in the U.S. despite the emergence of these new drugs [5]. Therefore, there is a need for innovative new treatments for CTCL, which necessitates the identification of novel therapeutic targets through a more complete understanding of the fundamental cellular and molecular disease underpinnings.

Genetic analysis has identified mutational events that are candidate drivers of MF and SS, including loss of tumor suppressors and epigenetic regulators and dysregulation of the NF-κB, MAPK, and JAK signaling pathways [6–8]. These studies shed light on the intrinsic events that influence cell-autonomous cell growth, survival, and proliferation. However, much less is known about the extrinsic factors in the tumor microenvironment (TME) that spur disease onset, progression, and therapeutic resistance. The CTCL TME, like that of other tumor types, is complex and unique to the organ site of the primary tumor and the repertoire of cell types that mediate bilateral communication between malignant cells and normal bystanders. The architecture of the CTCL TME is dictated by the anatomy of the dermis which supports skin-resident keratinocytes, Langerhans cells, melanocytes, and fibroblasts, along with immune cell populations, including non-malignant T and B cells, dendritic cells, neutrophils, NK cells, and others [9–12]. The immunophenotype of MF and SS is predominantly CD4+, with MF showing characteristics of skin-resident CD4+ effector memory T cells and SS more closely resembling circulating central memory CD4+ T cells [13]. Malignant T cells from CTCL lesions show a strong type 2 T helper (Th2) bias characterized by the production of Th2 cytokines (Il-4, IL-5, IL-10, IL-13) and the low production of Th1 cytokines (IFNγ, TNFα, and IL-2) [9,14]. Immunologic abnormalities observed in patients that are characteristic of a Th2-dominant imbalance include T-cell and NK-cell suppression, increased eosinophilia, and increased serum IgE and IgA [15,16], M2 macrophage polarization, suppression of proinflammatory cytokine secretion, and T-cell anergy [17–19]. Patients with advanced disease exhibit a generalized state of systemic immunosuppression and are highly susceptible to cutaneous infections, bacteremia, and pneumonia which are leading causes of death [20–23]. Little is known about the disease-specific regulators that govern the Th2 phenotype of CTCL or how this affects disease progression and clinical outcomes. Investigation in this area could offer a more complete understanding of the disease and lead to new therapeutic strategies that restore immune balance and resolve CTCL lesions.

The unfolded protein response pathway (UPR) is a multifaceted cellular response induced by an increased cellular demand for protein production, folding, and modification [24]. When challenged by stress signals from misfolded proteins in the endoplasmic reticulum (ER), UPR signaling initially protects the cell from proteotoxic damage and then induces cellular changes that allow for increased protein load. Adaptations include pausing of further protein translation in the short-term through the inhibition of cap-dependent translation, followed by genetic changes that impact protein folding, degradation, glycosylation, ER-Golgi transport, and the induction of protein chaperone and heat shock protein (HSP) transcription [25]. On the other hand, UPR also signals apoptotic cell death with extensive and sustained ER stress. Studies have shown that UPR signaling is also an important component of the differentiation program of B and T cells. For example, the differentiation and maturation of B lymphocytes into antibody-secreting plasma cells is dependent upon the X-box Binding Protein-1 (XBP-1) arm of the UPR [26,27]. Likewise, evidence suggests that XBP-1 and its upstream activator, inositol-requiring enzyme 1α (IRE1α), are important for helper T-cell function and the differentiation of Th2 CD4+ helper T cells specifically [28,29]. Altogether, these reports support a role for UPR effector signaling in normal T-cell development, although, exactly how UPR signaling impacts the phenotypic diversity is not well understood. Furthermore, whether signaling is maintain or divergent in malignant CTCL T cells remains unknown.

In this study, we discovered that normal CD4+ T cells and CTCL cells exhibit distinct UPR signaling in response to pharmacological ER stressors. Using the proteasome inhibitor, bortezomib (Btz), as a model compound, we show that CTCL cells rely predominantly on the PERK-dependent induction of specific Hsp70 family members, HSPA5 (BiP; GRP78) and HSPA6, for protection against toxic levels of ER stress. We further demonstrate, using a selective inhibitor of HSPA5/6 (JG-023), that the production of Th2 cytokines, but not Th1 cytokines, is HSPA5/6-dependent in both normal and malignant T cells. Conditioned media from CTCL cells directly and indirectly suppress normal T-cell activation, and this immunosuppressive activity can be reversed using novel HSPA5/6-selective inhibitors. Our work offers new insight into the control of the Th2-dominant immunosuppressive effects of CTCL and suggests that selective HSPA5/6 inhibitors could be effective targeted therapies for the treatment of CTCL.

2. Materials and Methods

2.1. Cell Lines and Reagents

HH and Hut78 CTCL cell lines were purchased from the American Tissue Culture Collection (Manassas, VA, USA). Primary human peripheral blood mononuclear cells (PBMCs) were purchased from STEMCELL Technologies (Cambridge, MA, USA). Bortezomib (Catalog No. S1013), VER155008 (Catalog No. 57751), CB5083 (Catalog No. S8101), pevonedistat/MLN4924 (Catalog No. S7109), ricolinostat (Catalog No. S8001), alvespimycin/17-DMAG (Catalog No. S1142), GSK2606414 (Catalog No. S7307), and phorbol 12-myristate 13-acetate (Catalog No. S7791) were purchased from Selleck Chemicals (Houston, TX, USA). Ionomycin (Catalog No. 73722) was purchased from STEMCELL technologies (Cambridge, MA, USA). JG-023 was provided by Dr. Eli Chapman (U. of Arizona) and has been previously described [30].

2.2. Cell Culture

HH and Hut78 were serially passaged in RPMI 1640 with 2.05 mM L-Glutamine (Cytiva; Marlborough, MA, USA Catalog No. SH30027.01) supplemented with 10% heat-inactivated fetal bovine serum (FBS) (GeminiBio; Liverpool, UK Catalog No. 900-108) and antibiotics [1% penicillin (10,000 units/mL), 1% streptomycin (10,000 μg/mL), and 1% amphotericin B (25 μg/mL)] (Cytiva; Marlborough, MA, USA Catalog No. SV30079.01). PBMCs were stimulated with Immunocult Human CD3/CD28/CD2 T Cell Activator (STEMCELL technologies, Catalog No. 10970), and CD4+ T cells were isolated using the EasySep Human CD4+ T cell Enrichment Kit (STEMCELL technologies, Catalog No. 19052) according to the "The Big Easy" EasySep Magnet protocol (STEMCELL technologies, Catalog No. 18001). PBMCs and isolated CD4+ T cells were cultured in Immunocult- XF T cell Expansion Media (STEMCELL technologies, Catalog No. 10981) and serially passaged with the addition of CD3/CD28/CD2 T cell activator, 10 μL/1 × 10^6 cells and human IL2,

0.6 μL/1 × 10^6 cells (TECIN Teceleukin, National Cancer Institute; Frederick, MD, USA Catalog No. 23-6019).

2.3. Flow Cytometry

Stimulated PBMCs and CD4+ T-cell isolates were collected, washed with 1 mL of sterile flow buffer (1X PBS pH 7.4 with 150 μM of $CaCl_2$ and 1% FBS), and centrifuged at 2500 rpm 4 °C for 5 min. Cells were resuspended in 100 μL of flow antibodies prepared at a 1:50 dilution with FACs buffer and incubated on ice for 30 min protected from light. FITC mouse IgG1 kappa Isotype Control (Catalog No. 555748), FITC mouse anti-human CD4 (Catalog No. 555346), and FITC mouse anti-CD8 (Catalog No. 555634) were purchased from BD Pharmigen (San Jose, CA, USA). Live/dead staining was run using propidium iodide (Invitrogen, ThermoFisher Scientific; Waltham, MA, USA Catalog No. P3566). Following incubation, cells were washed with 1 mL of FACs buffer and resuspended in 300 μL of FACs buffer for analysis. For cleaved (active) caspase-3 staining, cells were washed with ice-cold PBS and fixed with the Fixation/Permeabilization Solution Kit from BD Biosciences (San Jose, CA, USA) according to the manufacturer's instructions. Cells were then treated with a 0.125 mg/mL final concentration of rabbit anti-active caspase-3 (BD Pharmigen; San Jose, CA, USA) and incubated at room temperature for 20 min. Cells were then washed with permeabilization/wash solution and resuspended in 50 μL of diluted Alexa Fluor 488 goat anti-rabbit IgG (Invitrogen; Carlsbad, CA, USA) according to the manufacturer's instructions. The cells were incubated for 20 min at room temperature while protected from light, washed with permeabilization/wash solution, and resuspended in 300 μL of permeabilization/wash solution and analyzed. All samples were analyzed with the NovoCyte flow cytometer (ACEA Biosciences; San Diego, CA, USA) and data were analyzed using FlowJo version 10.10.0 software.

2.4. Western Blot

HH, Hut78, and CD4+ T cells treated as indicated were collected on ice, rinsed with cold 1X PBS, centrifuged at 2500 rpm at 4 °C for 5 min, and lysed in 1X cell lysis buffer (Cell Signaling Technologies, Danvers, MA, USA; Catalog No. 9803) with the addition of protease (Catalog No. A32955) and phosphatase (Catalog No. A32957) inhibitors purchased from ThermoFisher Scientific (Waltham, MA, USA). Cell lysates were clarified and their relative protein concentrations were determined via a Bradford assay using Bio-Rad Protein Assay Dye Reagent Concentrate diluted 1:5 with MiliQ water (Bio-Rad Laboratories; Hercules, CA, USA Catalog No. 5000006). Values were normalized to the lowest average absorbance at 595 nm. Gel samples were prepared by mixing lysates with SDS sample buffer containing β-mercaptoethanol (final concentration 1X) and 1X cell lysis buffer. Samples were boiled for 10 min, loaded on NuPAGE Bis-Tris Gel 4–12% (Invitrogen Catalog No. NP0336BOX), and electrophoresed at 55 mA for 1 h and 45 min in 1X NuPAGE MOPS SDS Running Buffer (Invitrogen, ThermoFisher Scientific; Waltham, MA, USA Catalog No. NP0001). Gels were transferred to polyvinylidene difluoride (PDVF) membranes at 300 mA for 2 h in 1X Transfer Buffer containing 25 mM trizma base, 192 mM glycine, and 20% methanol. PDVF membranes were blocked for 1 h at room temperature with 5% MILK in TBS Tween prior to incubation with primary antibodies in 5% MILK TBS Tween overnight at 4 °C. Primary and secondary antibodies are listed below in Table 1. Detection was finalized using ECL (Catalog No. 32209) or Super Signal (Catalog No. 34094) detection reagents from ThermoFisher Scientific (Waltham, MA, USA).

Table 1. Primary and secondary antibodies.

Primary Antibodies		
Protein Target	**Vendor**	**Catalog #**
β-actin	Sigma	A5441
ATF3	Cell Signaling	33593S
ATF4	Cell Signaling	11815S
BiP	Cell Signaling	3183S
CHOP	Cell Signaling	2895S
eIF2α	Cell Signaling	5324T
p-eIF2α	Cell Signaling	3398S
HERPUD1	Cell Signaling	26730
HSC70/HSPA8	Santa Cruz	71270
HSPA6	Santa Cruz	374589
HSP70	Cell Signaling	4872S
Tubulin	Cell Signaling	3873S
Ubiquitin	Cell Signaling	3936S
XBP1S	Cell Signaling	12782S
Secondary Antibodies	**Vendor**	**Catalog #**
Goat anti-Mouse IgG-H + L	Invitrogen	31430
Goat anti-Rabbit IgG-H + L	Invitrogen	31480
Goat anti-Rat IgG-H + L	Invitrogen	9520

2.5. Rt-qPCR

Total RNA was extracted from treated cells for quantitative reverse transcriptase polymerase chain reaction (RT-qPCR) using the Qiagen RNeasy Plus Mini Kit (Catalog No. 74134) per the manufacturer's instructions. Isolated RNA concentrations were determined via NanoDrop, and RNA was reverse transcribed using the Luna Universal One-Step RT-qPCR Kit (New England BioLabs; Ipswich, MA, USA Catalog No. E3005L). Samples were run using the QuantStudio 3 Real-Time PCR System (AppliedBiosystems, ThermoFisher) and data were analyzed with QuantStudio3 qPCR Data Analysis. Triplicate raw C_t values for the housekeeping gene (GAPDH) and genes of interest were averaged and used to calculate the $\Delta\Delta C_t$ values. Changes in gene expression were evaluated as a measure of fold change ($2^{-\Delta\Delta Ct}$). The primer sequences used for each of the gene targets are shown below in Table 2.

Table 2. RT-qPCR primer sequences.

Target	**Fwd (5′-3′)**	**Rvs (5′-3′)**
ATF3	GGAGTGCCTGCAGAAAGAGT	CCATTCTGAGCCCGG ACAAT
ATF4	GACGGAGCGCTTTCCTCTT	TCCACAAAATGGACGCTCAC
DDIT3	GGAAACAGAGTGGTCATTCCC	CTGCTTGAGCCGTTCATTCTC
DNAJB1	CCAGTCACCCACGACCTTC	CCCTTCTTCACTTCGATGGTCA
HERPUD1	CCGGTTACACACCCTATGGG	TGAGGAGCAGCATTCTGATTG

Table 2. *Cont.*

Target	Fwd (5′-3′)	Rvs (5′-3′)
HSPA1A	GGCCTTGAGGACTTTGGGTTA	TGGGAATGCAAAGCACACG
HSPA1B	GGGAGGACTTCGACAACAGG	GACAAGGTTCTCTTGGCCCG
HSPA1L	AAAGCAGGTCAGGGAGAGCGA	GGAGGGATTCCAGTCAGGTCA
HSPA5	GGGAGGTGTCATGACCAAAC	GCAGGAGGAATTCCAGTCAG
HSPA6	GATGTGTCGGTTCTCTCCATTG	CTTCCATGAAGTGGTTCACGA
IL2	AGAATCCCAAACTCACCAGGATGC	AGATGTTTCAGTTCTGTGGCCTTC
IL4	ACAGCCTCACAGAGCAGAAGAC	TCTCATGGTGGCTGTAGAACTGC
IL5	GGCACTGCTTTCTACTCATCGA	AGTTGGTGATTTTTATGTACAGGAACA
IL6	TCTCCACAAGCGCCTTCG	CTCAGGGCTGAGATGCCG
IL10	GCTGGAGGACTTTAAGGGTTACCT	CTTGATGTCTGGGTCTTGGTTCT
IL12b	TCATCAAACCTGACCCACCCAAGA	TTTCTCTCTTGCTCTTGCCCTGGA
IL13	GAAGGCTCCGCTCTGCAAT	TCTGGGTCTTCTCGATGGCA
IL22	GCAGGCTTGACAAGTCCAACT	GCCTCCTTAGCCAGCATGAA
IL23a	ACTCAGCAGATTCCAAGCCTCAGT	TGGAGATCTGAGTGCCATCCTTGA
IFNg	TCCAAGTGATGGCTGAACTGTCG	ACCTCGAAACAGCATCTGACTCC
TGFb1	CAAGCAGAGTACACACAGCAT	TGCTCCACTTTTAACTTGAGCC
TNFa	CCAGGCAGTCAGATCATCTTCTCG	ATCTCTCAGCTCCACGCCATTG

2.6. Cell Viability Assays

CellTiter-Glo® (CTG) (Promega; Madison, WI, USA Catalog No. G9683) and a Spectramax L microplate luminometer (Molecular Devices; San Jose, CA, USA) were used to measure the cell viability of HH, Hut78, and CD4s treated as indicated. Following the 48 h incubation at 37 °C and 5% CO_2, CTG was added to each well, plates were incubated at 37 °C and 5% CO_2 for 5 min, and then, plates were read at 470 nm. Data were analyzed in Excel and the cytotoxicity of the compounds was expressed as the percentage cell viability compared to the control. The percent cell viability data were analyzed in GraphPad PRISM version 10.4.1. For experiments evaluating drug synergy, viability data were normalized to values in the absence of the drug 2 (Btz, carfilzomib) to account for cell death induced by single-agent drug 1 (JG-023, VER155008, AP-4-139B), as described previously [31,32]. By using this normalization method, any separation of the drug 2 dose response curves indicates a true synergistic effect of drug 1.

2.7. Production of HSPA5, HSPA6, and the Substrate-Binding Domain of HSPA5 and HSPA6

Full-length HSPA5 (26–636), HSPA6 (1–643), HSPA5 SBD (400–636), and HSPA6 SBD (396–643) were cloned into pSpeedET using ligation-independent cloning [33]. HSP70 constructs were expressed from pSpeedET vectors with a tobacco etch virus (TEV) protease cleavable N-terminal his tag. Codon plus *Escherichia coli* (Invitrogen) were transformed with the respective plasmid and grown on agar plates containing 35 μg/mL chloramphenicol and 50 μg/mL kanamycin. Colonies were washed into 2xYT media in baffled 2 L flasks and grown at 37 °C to an OD of 0.6. At this point, the bacteria were transferred to a room-temperature shaking incubator for 1 h before isopropyl β-d-1-thiogalactopyranoside (IPTG) was added to a final concentration of 500 μM. The bacteria were grown for 16 h and harvested by centrifugation. The pellets were resuspended in HKM buffer (50 mM HEPES pH 7.4, 150 mM KCl, 10 mM $MgCl_2$, 2 mM β-mercaptoethanol (BME)) and lysed by repeated passage through a microfluidizer (microfluidics corporation) at 12,000 psi. The lysate was clarified by centrifugation and incubated with cobalt talon agarose from Gold Biotech (Olivette, MO, USA) for 1 h. The slurry was then applied to a gravity column and washed with 20 CV HKM buffer. The protein of interest was eluted with HKM buffer containing 200 mM imidazole. TEV protease was added to cleave the Hisx6 tag for 16 h at 4 °C. During this time, the eluted protein was extensively dialyzed against HKM buffer. Once cleavage was complete, the TEV protease was recaptured using cobalt resin and the protein was aliquoted and flash-frozen in liquid nitrogen.

2.8. Fluorescence Polarization Assay

FP assays were conducted in black 384-well low-volume plates (corning) with 10 nM of fluorescent peptide (FAM-ALLLSAPRR from ABclonal, Woburn, MA, USA) or JG-023 in assay buffer (50 mM HEPES pH 7.4, 100 mM KCl, 10 mM $MgCl_2$, and 0.1% Triton X-100). Polarization was measured using the ID5 microplate reader (Molecular Devices, San Jose, CA, USA) with an excitation of 485 nm and emission of 535 nm after 4 h. For K_D determination, respective Hsp70s were serial diluted and the data were fit to a one-site binding model in PRISM 6.07 (GraphPad). For compound measurements, Hsp70 was added at a final concentration of 1 μM and % inhibition was calculated using unlabeled peptide as a positive control and DMSO as a negative control.

2.9. Determining Effect of CTCL Secreted Factors on Naïve T-Cell Activation via IFN-γ ELISA

IFN-γ levels in conditioned media were assessed via a human IFN-γ ELISA kit (Invitrogen Catalog No. 88-7316) Cells were treated as indicated, incubated at 37 °C 5% CO_2 for 6 h, collected and spun down at 1000 rpm for 5 min, washed with RPMI 1640 with 2.05 mM L-Glutamine (Cytiva Catalog No. SH30027.01) not supplemented, and then resuspended in Immunocult-XF T-cell expansion media and incubated at 37 °C 5% CO_2 for an additional 24 h. After 24 h, CTCL conditioned media were collected and spun down twice at 1500 rpm for 5 min to remove any remaining cells. PBMCs were plated with conditioned media and treated with human IL2 alone or in combination with Immunocult Human CD3/CD28/CD2 T-cell activator. Conditioned media alone were plated as a control. Plates were incubated at 37 °C 5% CO_2 for 72 h. After 72 h, conditioned media were collected from PBMCs and IFN-γ levels were assessed via ELISA according to the manufacturer's protocol. Results were analyzed in and compiled in GraphPad PRISM.

2.10. Cytokine Bead Array

Cells were stimulated with or without PMA and ionomycin for 6 h in the presence of drugs: JG-023 (10μM), Btz (5 nM), VER155008 (5μM), or combinations, followed by a wash-out. The cells were then incubated in fresh media for 24 h at 37 °C, 5% CO_2. The conditioned media from stimulated and unstimulated cells for each of the treatment conditions were collected for human Th1/Th2/Th17 cytokine bead array (CBA) (BD Biosciences; Franklin Lakes, NJ, USA; Cat #560484). CBA samples and standards were prepared according to the manufacturer's protocol and run on the LSR Fortessa flow cytometer (BD Biosciences; Franklin Lakes, NJ, USA). Samples were run undiluted or at 1:2 dilution in assay diluent. The data were analyzed using Flowjo software with the CBA plugin. Cytokine concentrations were calculated using the standard curve and values were normalized to cell counts.

3. Results

3.1. Divergent ER Stress Response Signaling in CTCL and Normal CD4+ T Cells

The role of UPR and ER stress signaling in the pathogenesis of T-cell malignancies, such as CTCL, is not known. To investigate this, we evaluated the effects of various pharmacological inducers of ER stress in a panel of CTCL cell lines and normal activated T cells. Cell models included HH cells, representative of the MF subtype, and Hut78, representative of the more aggressive SS form of CTCL. For normal T cells, we used activated PBMCs from healthy human donors and isolated CD4+ cells by immunomagnetic separation, as CD4+ is the dominant immunophenotype of CTCL [34,35] (Figure 1A). Specific ER stressors included drugs that inhibit vasolin-containing protein (VCP/p97) AAA ATPase (CB5083), Hsp90 (17-DMAAG), pan Hsp70 (VER155008), NEDD8-activating enzyme (NAE, MLN4924), HDAC6 (Ricolinostat), protein disulfide isomerase (PDI, LTI6426), and the 26S proteasome (bortezomib, Btz). These experiments showed that ER stress agents were widely cytotoxic to both CTCL and normal T cells (Figure 1B,C). They also revealed a statistically significant difference in sensitivity to Btz, with CTCL cells being >5-fold more sensitive than normal CD4+ in viability and apoptosis assays (Figure 1D,E). Upon investigation of signaling events in response to Btz, the difference between normal and malignant CD4+ T cells became even more evident. Btz induced a strong upregulation of classical UPR effectors such as ATF3, ATF4, CHOP, and HERPUD1 (Figure 1F). Normal CD4+ T cells, on the other hand, did not induce these pathways to any significant degree despite presenting with comparable levels of protein stress as indicated by the accumulation of high-molecular-weight poly-ubiquitinated protein smears (Figure 1F). Consistent with what was observed in viability and apoptosis assays, Btz induced apoptotic biomarkers such as the cleaved products of caspase substrates, PARP and Lamin A, in CTCL cells but not normal CD4+ T cells [36]. We confirmed that the observed UPR signaling was contributing to the apoptotic phenotype in experiments using the specific PERK inhibitor, GSK260614, which rescued HH and Hut78 cells from death induced by Btz (Figure S1). PERK is an apical stress sensor that triggers UPR signaling in response to ER stress. GSK260614 protected CTCL cells from Btz-induced death in a dose-dependent manner up to concentrations between 1 and 5 μM, concentrations above which it became cytotoxic as a single agent. Interestingly, PERK inhibition was unable to rescue cell death in normal primary T cells, suggesting different mechanisms of Btz-induced death between normal and malignant T cells. Together, these results reveal a divergence in the adaptive cellular response to ER stress induced by Btz between normal and malignant T cells that may indicate unique vulnerabilities of CTCL.

3.2. *HSPA5/6 Play a Dominant Role in the ER Stress Response in CTCL Cells*

To further characterize the UPR signaling profiles of malignant CTCL and normal CD4+ T cells, we used Btz as an ER stressor and evaluated the expression of a panel of UPR and oxidative stress genes that were selected previously [32]. Normal CD4+ cells showed a muted response to Btz with minimal activation of most gene targets in the set. CTCL, by comparison, showed a prominent activation of *ATF3*, the Hsp40 family member *DNAJB1*, and most notably, the inducible Hsp70 family member *HSPA6*, which was induced by 191- and 92-fold in CTCL cells versus only 5-fold in normal activated T cells ($p < 0.001$; Figure 2A). These hyper-responsive genes were further induced when Btz was combined with the PDI inhibitor LTI6426 (Figure S2A), which is highly synergistic with Btz in CTCL (Figure S2B) and a wide variety of other tumor types [31,37,38]. This suggests an important role of these genes and their products in CTCL's response to ER in a variety of stress-inducing drugs. Given the pronounced induction of *HSPA6*, we next evaluated the other stress-inducible members of the Hsp70 family (*HSPA1A*, *HSPA1B*, *HSPA1L*, *HSPA5*, *HSPA6*) [39] in response to a panel of ER stressors. *HSPA6* showed the most prominent induction in response to Btz and several other drug classes compared to the other Hsp70 isoforms (Figure 2B). UPR-responsive genes were continuously induced over a 16 h time period, and again, HSPA6 showed the most significant induction (Figure 2C). Additionally, HSPA6 induction was consistently higher in CTCL cells than normal CD4+ T cells (Figure 2D). Results were confirmed at the protein level, as Western blotting analysis showed high HSPA6 induction in Btz-treated CTCL but not normal CD4+ cells (Figure 2E). A similar trend was observed for HSPA5/BiP/Grp78, although the difference between normal and malignant cells was less striking. Total Hsp70 levels were induced to similar degrees in both CTCL and normal CD4+ T cells (Figure 2E). This suggests that while cellular fitness and the signaling response to Btz-induced stress was very different between cell types, the magnitude of stress incurred by the cells was similar. This rules out potential variables that affect drug target engagement (e.g., drug efflux pumps) between cell types that could be responsible for these observations. Importantly, these findings reveal a dominant role for HSPA6 and HSPA5 in the CTCL adaptive response to ER stress-inducing drugs, leading us to hypothesize that these Hsp70 family members are attractive drug targets in CTCL.

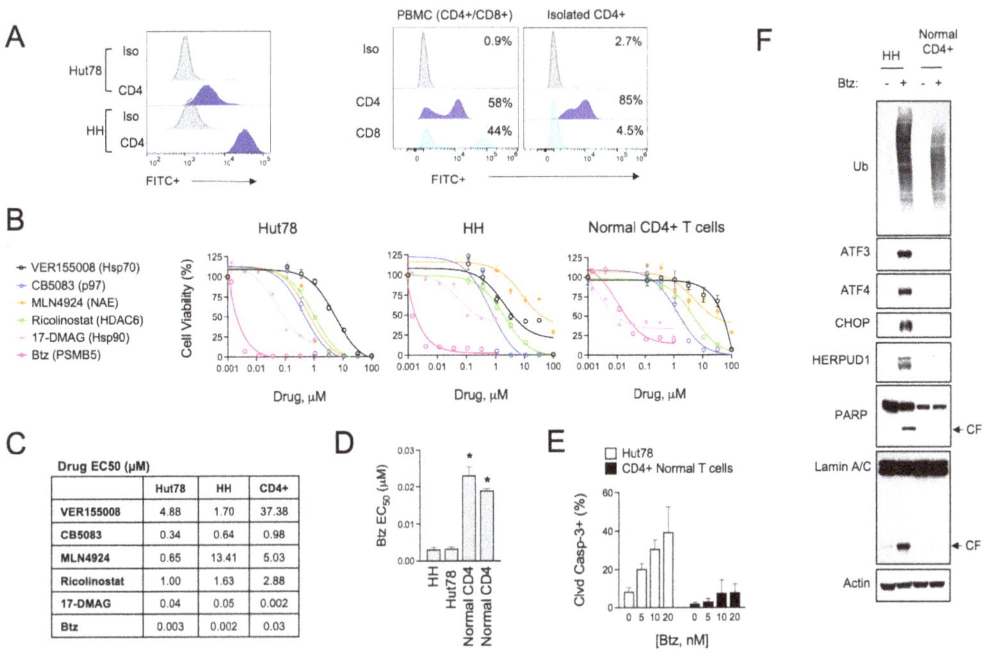

Figure 1. Btz selectively targets CTCL cells over normal CD4+ T cells. (**A**) (**Left**) Cell surface CD4 expression was measured by flow cytometry in Hut78 and HH CTCL cells. (**Right**) Non-malignant CD4+ T cells were isolated from PBMCs from normal human donors, which are a mixture of CD4+ and CD8+ T-cell populations. Flow cytometry data are shown for CD8- and CD4-stained cells. (**B**) Hut78, HH, and normal CD4+ T cells were treated with dose ranges of the indicated ER stress-inducing drugs for 48 h. Cell viability data are shown. (**C**) Effective concentration 50 (EC_{50}) values were extrapolated from dose curves shown in (**B**). EC50 values are shown in µM. (**D**) Effective concentration 50 (EC_{50}) values were extrapolated from Btz dose curves representing 2–3 experiments for each of the indicated cell lines. CD4+ T cells from 2 different human donors are shown. Statistical significance was determined using Student's *t*-test (* $p < 0.05$, N = 3). (**E**) Hut78 CTCL cells and normal CD4+ T cells were treated with the indicated concentrations of Btz for 24 h. Flow cytometry data for cleaved caspase-3 positive cells are shown (mean ± SEM, N = 3). (**F**) HH and normal CD4+ T cells were treated with Btz (20 nM) for 16 h. Western blot analysis of the indicated UPR and apoptotic markers is shown.

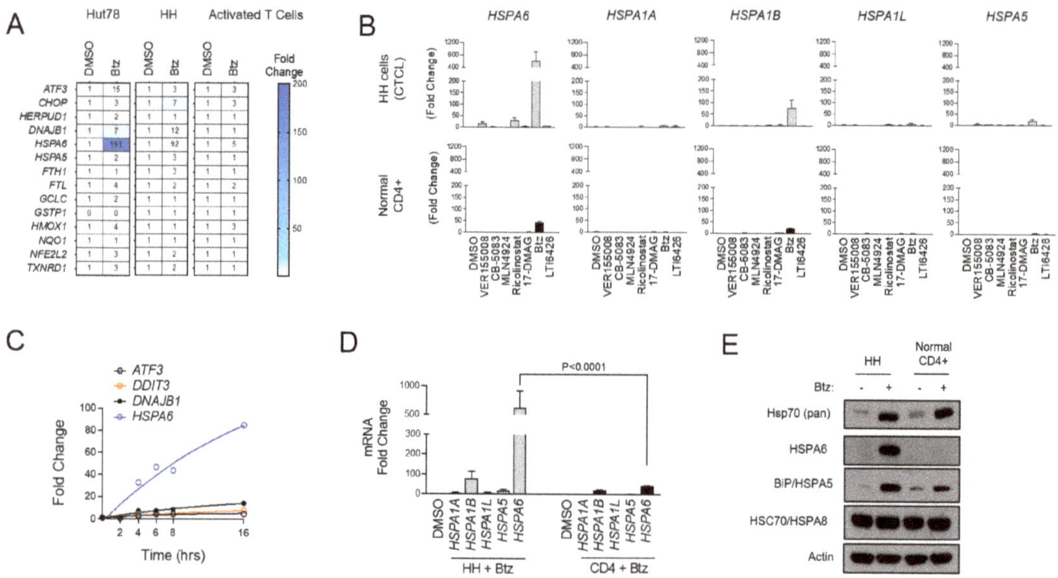

Figure 2. HSPA6 is a biomarker of response to Btz in CTCL cells. (**A**) Hut78, HH, and normal CD4+ T cells were treated with Btz for 16 h. RT-qPCR using primers for the indicated transcripts was conducted. Data were normalized to *GAPDH* (internal control) and DMSO-treated (treatment control) groups. (**B**) HH and normal CD4+ T cells were treated with an EC_{50} dose of the indicated ER stress-inducing drugs. RT-qPCR data for the indicated HSP70 isoform gene transcripts are shown. Data are expressed as average fold change ± SEM after normalization to *GAPDH* (internal control) and DMSO (treatment control). (**C**) HH cells were treated with Btz (10 nM) and expression levels of the indicated UPR genes were measured over time via RT-qPCR. Data are presented as fold change. (**D**) HH and normal CD4+ T cells were treated with Btz (20 nM) for 16 h. RT-qPCR data using primers for the indicated HSP70 isoform transcripts are shown. Statistical significance was determined using Student's *t*-test (N = 3). (**E**) HH and normal CD4+ T cells were treated with Btz (20 nM) for 24 h. Western blots for the indicated HSP70 proteins are shown.

3.3. HSPA5/6-Specific Inhibition Enhances ER Stress-Induced Death of CTCL Cells

Previous studies have identified Hsp70 isoform-selective inhibitors that target the substrate-binding pocket of the canonical human HSP70s with >100-fold selectivity for HSPA5 and HSPA6 compared to other isoforms [30]. Compound 35 from this series of analogs (herein referred to as JG-023; Figure 3A) showed selectivity for HSPA5 (also known as BiP or GRP78) and HSPA6. To further characterize the binding of JG-023 to HSPA5 and HSPA6, we generated a fluorescently labeled derivative, FAM-JG-023, using a copper-free click chemistry strategy (Schema in Figure 3B). We then titrated HSPA5 and HSPA6 against a fixed concentration (20 nM) of FAM-JG-023 in a fluorescence polarization (FP) assay to determine binding specificity and potency (Figure 3C). Competition binding assays were conducted using parent (unlabeled) JG-023 to displace FAM-JG-023 from full-length HSPA6 and HSPA5 and a truncated version of HSPA5 that encompasses the substrate-binding domain (SBD) mapping to amino acids 420–500 (Figure 3D). Unlabeled JG-023 successfully competed away FAM-JG-023 binding to HSPA6, HSPA5 FL, and HSPA5 SBD in a dose-dependent manner with IC50s of 0.5 ± 0.3, 1.3 ± 0.2, and 1.4 ± 0.06, respectively, confirming that JG-023 and FAM-JG-023 bound to the same binding pocket with specificity and good potency for an early-stage lead compound.

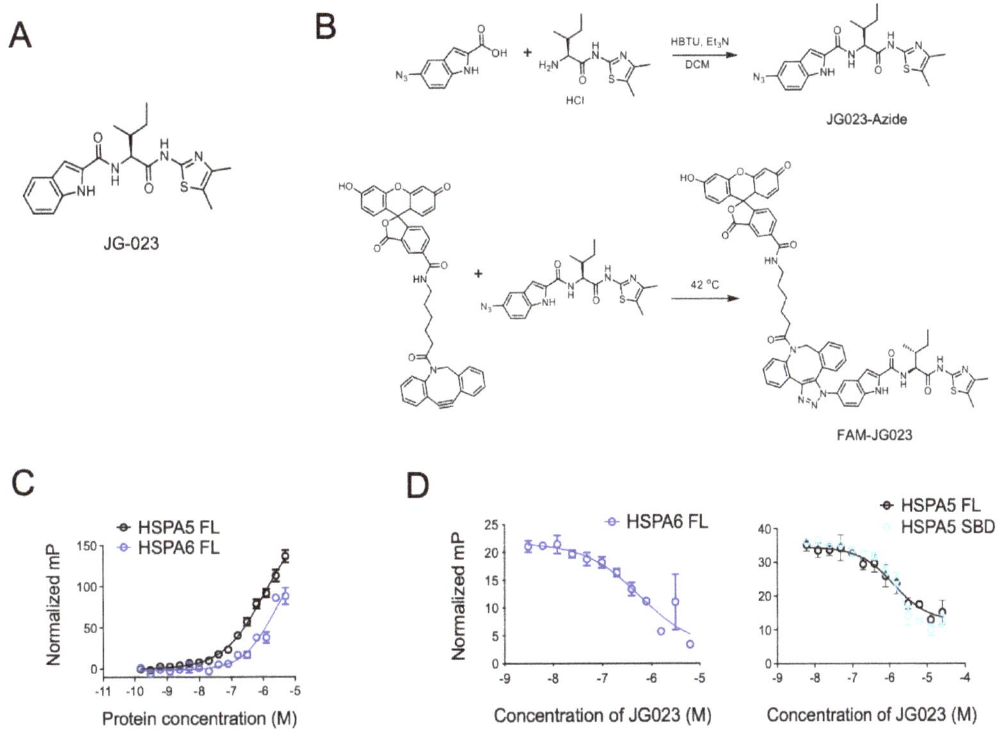

Figure 3. Selective HSPA5/6 inhibition using compound JG-023. (**A**) The chemical structure of JG-023 is shown. (**B**) A fluorescently labeled JG-023 (FAM-JG023) analog to be used in fluorescence polarization studies was synthesized using a copper-free click chemistry strategy. Chemical synthesis schema is shown. (**C**) Full-length (FL) HSPA5 and HSPA6 were titrated against 20 nM FAM-JG023. Fluorescence polarization binding data are shown. (**D**) Competition experiments were conducted using 1 mM HSPA6 FL, HSPA5 FL, and the HSPA5 substrate-binding domain (SBD) and 20 nM FAM-JG023. A dose range of unlabeled JG-023 was used to compete off FAM-JG023. Binding data from a fluorescence polarization assay are shown.

Given the hyper-responsiveness of the HSPA5 and A6 genes in the stress response of CTCL cells, we next evaluated the activity of JG-023 in our cell models. HSPA5/Bip/Grp78 is a chaperone that senses and regulates ER stress by engaging and inhibiting apical effectors of the UPR, including PERK, IRE1α, and ATF6 in non-stressed cells [40,41]. Misfolded proteins in the ER attract HSPA5 and sequester it from these effectors, thereby triggering their activation and UPR signaling. Because JG-023 inhibits HSPA5/6 function by binding to the SBD, we hypothesized that it would induce a UPR-like pseudo-response that mimics HSPA5/Bip/Grp78 displacement by misfolded proteins in the ER. Consistent with this, we observed the induction of key downstream UPR effectors, including ATF3, ATF4, and CHOP in a dose-dependent manner at the protein and mRNA levels upon treatment with JG-023 in CTCL cells (Figure 4A–C). HSPA6-specific inhibitors (Zxy-0028 and Zxy-0029) showed only modest activity with a slight induction of ATF4. In isobologram analyses, JG-023 showed synergistic killing when combined with Btz in CTCL cells, decreasing the EC50 by approximately 5-fold from 9.5 $\pm$ 2.3 to 1.9 $\pm$ 0.6 microM (Figure 4D). Similar synergy was observed with combinations of JG-023 and the NAE inhibitor MLN4924, and this was specific to CTCL cells and not normal CD4+ T cells (Figure S3). Similar synergies were detected with other commercially available non-selective HSP70 inhibitors, VER-155008 [42] and AP-4-139B [43], when combined with the second-generation proteasome inhibitor, carfilzomib (Figure 4E). These data suggest that HSP70 family members, including HSPA5 and A6, function in a protective role against ER stressors in CTCL cells. Furthermore, the combination of HSP70 inhibitors, such as JG-023, with pharmacological inducers of ER stress (e.g., proteasome and NAE inhibitors) is highly effective in killing CTCL cells.

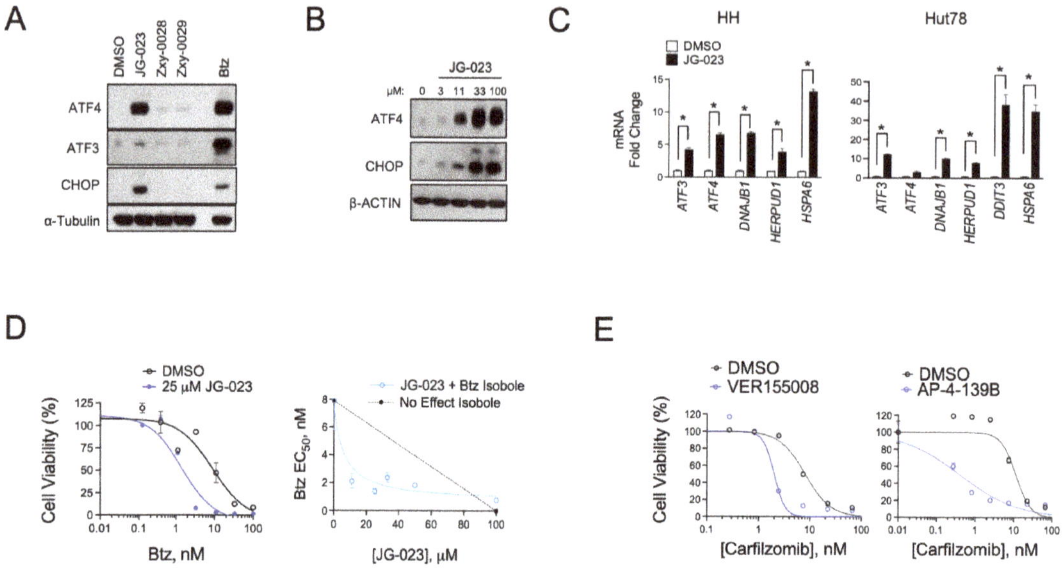

Figure 4. HSPA5/6-selective inhibitors induce the UPR and enhance proteasome inhibitor sensitivity in CTCL models. (**A**) Hut78 cells were treated with the indicated HSPA5/6- (JG-023) or HSPA6- (Zxy-0028 and Zxy-0029) selective inhibitors at 20 μM for 16 h. Western blots are shown. (**B**) HH cells were treated with a dose range of JG-023 for 16 h. Western blots are shown. (**C**) HH and Hut78 cells were treated with DMSO (control) or JG-023 (10 μM) for 16 h. RT-qPCR data are shown normalized to GAPDH (internal control) and DMSO (treatment control). Statistical significance was determined using Student's *t*-test (* $p < 0.05$, N = 3). (**D**) (**Left**) HH cells were treated with a dose range of Btz in the presence or absence (DMSO) of JG-023 (25 μM). Cell viability data are shown (DMSO: EC50 = 9.5 ± 2.3 μM; JG-023: EC50 = 1.9 ± 0.6 μM). (**Right**) Isobologram analysis is shown and confirms synergy between Btz and JG-023. The dotted line represents the no-effect isobole. (**E**) Hut78 cells were treated with a dose range of the second-generation proteasome inhibitor, carfilzomib, in the absence (DMSO) and presence of the pan HSP70 inhibitors, VER155008 (**Left**, 5 μM) and AP-4-139B (**Right**, 5 μM). Cell viability data are shown (carfilzomib EC50: DMSO = 8.1 ± 2.5 μM; VER155008 = 2.0 ± 0.6 μM; DMSO = 11.4 ± 3.3 μM; AP-4-139B = 0.42 ± 0.2 μM).

3.4. HSPA5/6 Inhibition Suppresses Th2 Cytokine Production by CTCL Cells

Malignant T cells from advanced CTCL show a Th2-dominant phenotype, characterized by the secretion of IL-4, IL-5, IL-10, and IL-13 with low IFNγ production [14,44]. We evaluated the expression of a panel of Th1 and Th2 cytokine genes in our cell models and confirmed that HH and Hut78 CTCL cells showed a Th2 bias (*IL4*, *IL6*, *IL10*, and *IL13*) with relatively lower induction of Th1 cytokine genes (*IFNG*, *TNFA*, and *IL2)* compared to primary CD4+ T cells from normal healthy donors (Figure 5A). Similar results were observed following T-cell stimulation using phorbol 12-myristate 13-acetate (PMA) and ionomycin (Figure 5B). These transcriptional profiles for Th1 and Th2 cytokines were also confirmed at the protein level (Figure S4). CTCL also induced high levels of the *IL22* gene, an observation that has been reported in patients and shown to influence the makeup and architecture of the CTCL TME [45,46]. As previous studies have shown that Hsp70 regulates the expression of Th2 cytokines such as IL-10 in CD4+ T cells [47,48], we next investigated the impact of HSPA5/6 inhibition on the production of cytokines in CTCL cells. In normal CD4+ T cells, JG-023 significantly reduced Th2-type cytokines, including *IL4, IL5,* and *IL6,* along with *IL22* (Figure 5C). However, JG-023 had little effect on the expression of Th1-type cytokines (*IFNG, TNFA,* and *IL2).* By comparison, in CTCL cells, JG-023 broadly suppressed the induction of cytokine genes with a prominent effect on Th2 cytokines including *IL4, IL10, IL13,* and *IL22* (Figure 5C). These results indicate that the inhibition of HSPA5 and HSPA6 suppresses the Th2 cytokine signature. Furthermore, in normal CD4+ T cells, JG-023 had no effect on Th1 cytokines, suggesting that HSPA5/6 may specifically regulate the Th2 differentiation program in normal T cells.

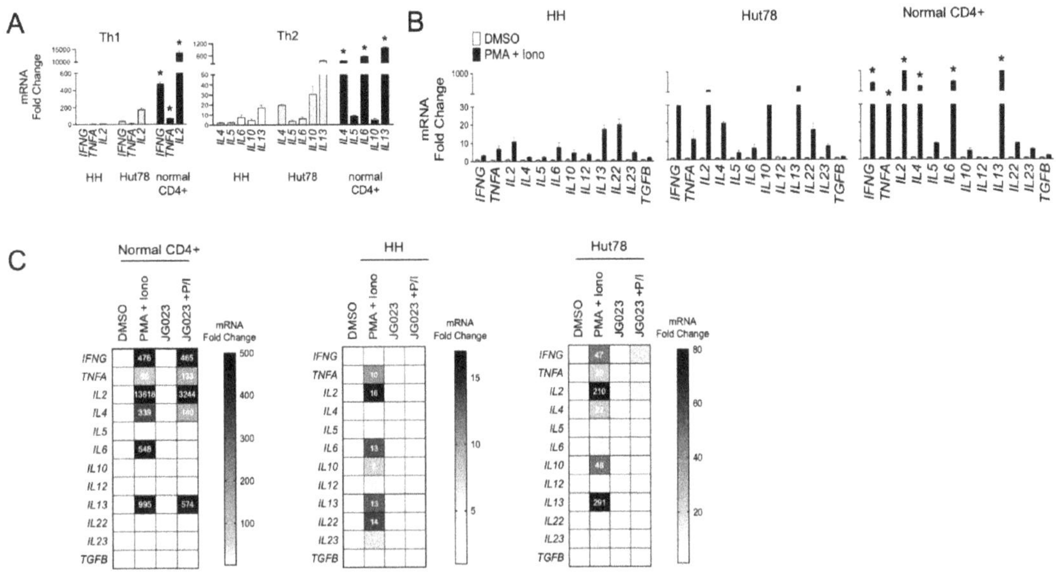

Figure 5. HSPA5/6 inhibition represses the Th2 phenotype in CTCL cells. (**A**) RT-qPCR was conducted on untreated HH, Hut78, and normal CD4+ T cells using primers targeting the indicated gene transcripts. Data were normalized to GAPDH to generate a relative fold change (mean ± SEM, N = 3). * $p < 0.01$ comparing normal CD4+ to both HH and Hut78 using a one-way ANOVA. (**B**) HH, Hut78, and normal CD4+ T cells were activated using phorbol ester (PMA, 1 μM) and ionomycin (Iono, 200 ng/mL) for 16 h and mRNA levels for the indicated cytokine gene panels were analyzed. RT-qPCR data are shown normalized to GAPDH (internal control) and DMSO (treatment control) and expressed as fold change (mean ± SEM, N = 3). * $p < 0.01$ comparing normal CD4+ to both HH and Hut78 using a one-way ANOVA. (**C**) The indicated T cells were activated with PMA + Iono in the presence or absence of JG-023 (20 μM) for 6 h. Cells were washed to remove treatments and incubated in fresh media for an additional 18 h. mRNA transcript levels for the indicated gene targets were then quantified. RT-qPCR data are shown in heat map format. Data were normalized to GAPDH (internal control) and DMSO (treatment control).

3.5. CTCL-Secreted Factors Suppress Normal T-Cell Function

Th2 cytokines play a key role in allergic responses and in defense and wound repair following helminth parasite infection [49,50]. They accomplish this in part by suppressing Th1 cytokines such as IFNγ and tipping the Th1/Th2 balance toward a Th2 bias [51]. Under normal conditions, Th2 immunity counterbalances Th1 immunity [52]. Under pathological conditions like CTCL, Th2 immunity has been proposed to induce global and local T-cell suppression in patients with CTCL [53]. Given the paucity of Th1 cells in CTCL lesions, we hypothesized that CTCL-derived secreted factors suppress normal Th1 cell activity. To test this, we activated naïve T cells from normal human donor PBMCs using a trimeric anti-CD3/CD28/CD2 antibody in the presence of conditioned media (CM) from HH and Hut78 CTCL cells. The activation of Th1-cell populations in PBMCs was measured using IFNγ production as a proxy. We found that CM from stimulated HH and Hut78 cells significantly repressed IFNγ production following the activation of normal T cells from three separate normal human donors in a dose-dependent manner (Figure 6A,B). Suppressive factors from CTCL cells exhibited high apparent potency with effects evident even when CM was diluted 50-fold (Figure 6B).

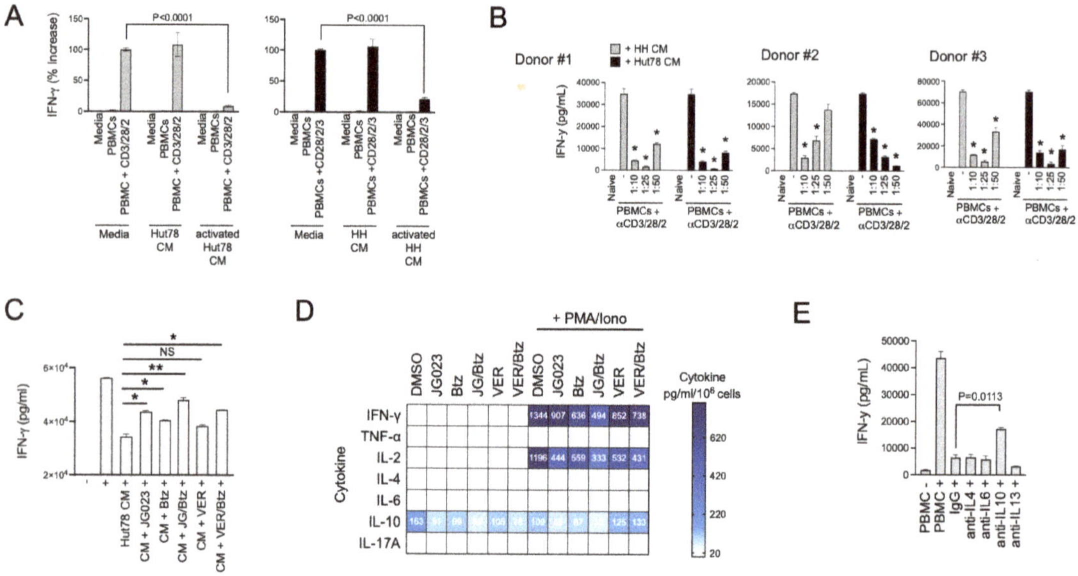

Figure 6. CTCL cells suppress normal T-cell activation. (**A**) Normal T cells from healthy human PBMCs were activated in the presence of conditioned media (CM) from resting and activated Hut78 and HH cells. IFNγ levels were determined by ELISA (mean ± SEM, N = 3). Normal T cells were activated using a trimeric anti-CD3/28/2 antibody. CM from activated CTCL cells were collected by treating Hut78 and HH cells with a combination of PMA + ionomycin for 6 h, after which, cells were washed to remove drug, and cells were finally incubated in fresh media for an additional 18 h. Statistical significance was determined using Student's t-test (N = 3). (**B**) Experiments described in (**A**) were conducted using normal T cells from 3 healthy human donors and varying dilutions of CM from Hut78 and HH cells. * $p < 0.0001$ compared to T-cell stimulation alone ("-") by Student's t-test (N = 3). (**C**) Normal T cells from healthy human donors were activated (+) in the absence or presence of Hut78 CM that were collected under the indicated conditions of treatment with JG-023 (10 μM), Btz (5 nM), VER155008 (5 mM), or combinations thereof. Hut78 cells were activated with PMA and ionomycin for 6 h in the presence of drugs followed by a wash-out. IFNg levels in the media from normal T-cell activation cultures were measured by ELISA after 48 h (mean ± SEM, N = 3). Statistical significance was determined by Student's t-test (* $p < 0.05$, ** $p < 0.01$, NS = no statistical significance). (**D**) Hut78 CM from the treatment groups described in (**A**) were analyzed by cytokine bead array for the indicated cytokines. To account for any effects of the drugs on cell viability and proliferation, cytokine levels were normalized to the number of viable cells counted at the time of CM collection. Data are expressed as pg/mL per 10^6 viable Hut78 cells. (**E**) Normal T cells from healthy human donors were activated in the presence of Hut78 CM and neutralizing antibodies to the indicated cytokines. IFNg levels in the media from activated normal T-cell cultures were analyzed by ELISA (mean ± SEM, N = 3). Statistical significance was determined using Student's t-test.

ER stress and the UPR play diverse roles in T-cell function, differentiation, and activation and TCR complex formation, as well as in T-cell pathologies [54–56]. Therefore, we next asked if the suppressive factors secreted by CTCL cells were dependent on HSP70 family members including HSPA5/6. To test this, we collected CM from activated CTCL cells that had been treated with JG-023 or the pan HSP70 inhibitor VER155008. We also tested the effects of Btz alone and in combination with HSP70 inhibitors given the synergy we observed between these two classes of drugs in prior experiments (Figure 4D,E). CTCL cells were activated with PMA + ionomycin for six hours in the presence and absence of drug. They were then washed to remove PMA/ionomycin and drug and incubated in basic T-cell media for an additional 24 h, after which, CM were collected. CM volumes were adjusted to viable cell count to normalize for any effects of the drugs on cell viability. Finally, normal T cells from human PBMCs were activated in the presence of CM from the different treatment groups to determine whether HSP70 inhibitors and ER stressors like Btz could block the immunosuppressive effects of CTCL. Indeed, we found that JG-023 monotherapy and the combination of JG-023 + Btz were able to rescue IFN-γ levels in normal T cells, indicating a restoration of normal Th1 cell activity (Figure 6C). VER155008 showed a modest but statistically insignificant rescue of IFN-γ as a single agent and did not add significantly to the activity of Btz alone.

To evaluate the contribution of select cytokines to this suppressive effect, we next quantified the concentrations of Th2 and Th1 cytokines in the CM from treated and untreated CTCL cells. Cytokine bead array analysis revealed that JG-023 and the combination of JG-023 and Btz reduced IL-10 levels in both resting and activated CTCL cells (Figure 6D). Similar results were observed for IFN-γ, IL-2, and to a lesser extent, TNFα, whereas minimal effects were observed for IL-4, IL-6, and IL-17a. The pan HSP70 inhibitor, VER155008, alone and in combination with Btz, had similar effects on Th1 cytokines, IFN-γ, and IL-2, but had no effect on IL-10 levels in either resting or activated CTCL cells. As JG-023 but not VER155008 blocked IL-10 production by CTCL cells and restored normal Th1 cell activation, we hypothesized that IL-10 was a key factor in the immunosuppressive phenotype of CTCL cells. To test this, we activated normal T cells in the presence of suppressive CM from CTCL cells and co-treated with neutralizing antibodies to IL-4, IL-6, IL-10, and IL-13. We found that only the IL-10-neutralizing antibody rescued normal T-cell activation (Figure 6E), indicating that IL-10 was indeed critical, at least in part, for the T-cell suppressing activity of CTCL cells. In summary, we show that CTCL-derived cytokines suppress normal Th1 cell activation. This repressive activity can be overcome by drugs that modulate the ER stress response, such as Btz-, pan HSP70-, and HSPA5/6-specific inhibitors. Lastly, IL-10 is a major inhibitory factor released by CTCL that contributes to the suppression of Th1 cells. Altogether, these data suggest that pharmacological interventions targeting the ER stress pathway, especially those that target HSPA5/6, not only induce anti-CTCL effects that include a strong stress and apoptotic signaling response but also inhibit their immunosuppressive Th2 phenotype.

4. Discussion

The ER stress response has been linked to various aspects of cancer initiation and progression, including cell growth and protection from apoptosis [57–59]. In the current study, we demonstrate a divergence in the ER stress response between CTCL and normal CD4+ T cells. Specifically, we demonstrate that CTCL cells are hyper-responsive to ER stress-inducing drugs (e.g., proteasome, NAE, and PDI inhibitors), exhibit the heightened activation of UPR effectors like HSPA6 in response to ER stressors, and present a Th2-suppressive phenotype bias that can be altered using inhibitors of specific UPR pathway components. This may suggest an important role of ER stress in CTCL disease initiation, progression, and T-cell biology and a new path for the treatment of CTCL in the clinic.

Our work highlights the significance of specific HSP70 isoforms, HSPA5 (BiP/Grp78) and HSPA6, in the CTCL immune phenotype and response to ER stress-inducing drugs. HSPA5, a glucose-regulated protein first identified in 1974 [60], is widely recognized for its overexpression in a variety of cancers and has emerged as a prognostic biomarker of disease and predictor of infiltrating immune cells in the tumor microenvironment [61–63]. By comparison, much less is known about HSPA6, which is at least partially due to it not being conserved in rodents [64,65]. HSPA6 (also known as HSP70B′) was initially sequenced and characterized as a heat- and stress-inducible HSP70 family member [65] and later described for its proinflammatory role in atherosclerotic plaque formation [66,67]. HSPA6 expression has been reported in certain tumor types and associated with outcomes, although its precise role in cancer is not well understood. Moreover, little is known about the expression and role of both HSPA5 and HSPA6 in CTCL. Our study highlights the unique role of HSPA5 and HSPA6 proteins in the ER stress response of CTCL, but not of normal CD4+ T cells. We observed the significant induction of HSPA5 and HSPA6 genes as well as increased protein expression in the ER stress response following treatment with Btz. We also observed enhanced ER stress-induced apoptosis in CTCL cells with the HSPA5/6-selective, inhibitor JG-023. Our findings are consistent with a recent study which showed Hsp70 to be upregulated in response to Btz as well as a synergistic relationship between Hsp70 inhibition and Btz [68]. The differential expression of HSPA5 and HSPA6 between CTCL and normal CD4+ T cells highlights an important role of ER stress signaling and the UPR in malignant T-cell biology and identifies these proteins as potential therapeutic targets for CTCL. Additionally, while Btz treatment has been utilized and proven effective as a single-agent therapeutic in CTCL, acquired resistance is a known barrier to its efficacy, and recent studies have linked the higher expression of heat shock proteins with acquired resistance to Btz [69,70]. Therefore, the induction of HSPA5 and HSPA6 in response to Btz in CTCL may play a cytoprotective role, and prolonged exposure and targeting HSPs such as HSPA5 and HSPA6 could enhance the efficacy of ER stress-inducing compounds like Btz. It is also important to note that pro-apoptotic ER stress signaling is a major component of the anti-cancer mechanism of Btz, but it is not the only one, as proteasome inhibition is known to have pleiotropic effects in cancer cells [71]. Our results show the clear convergence of HSPA5/6 inhibition and Btz on ER stress signaling, but we cannot rule out the possibility that HSPA5/6 inhibitors and Btz also synergize on non-ER stress pathways as well.

The ER stress response and UPR signaling pathways play a role in CD4+ T helper cell development, differentiation, and effector function [54,55,72]. In normal T-cell development, evidence supports a critical role for UPR effectors such as the chaperone gp96 (HSP90b1, grp94) in the thymus [73], while other HSP families, such as members of the heat shock protein 70 (HSP70) family, play a controversial and potentially stage-dependent role in T-cell differentiation and activation. For example, inducible HSP70 isoforms promote T-cell activity in the context of autoimmunity, whereas other reports have shown that HSP70 inhibits T-cell development early in thymic development [74–76]. In the current study, we demonstrate the Th2 skewed phenotype of HH and Hut78 CTCL could be normalized with selective HSPA5/6 inhibition. Interestingly, HSPA5/6 inhibition in normal CD4+ T cells suppressed Th2 cytokine production but not Th1 cytokines, suggesting a dominant role for HSPA5/6 in signaling in Th2 cell function and differentiation in both malignant and normal CD4+ T cells. While several studies have demonstrated the role of UPR signaling members such as IRE1α/XBP1 as important for promoting Th2 differentiation in normal T cells [48,49,77], less is known about the role of HSPA5 and HSPA6 in the function of CD4+ Th2 cells, specifically, or broadly in the overall differentiation and function of

T-cell subsets. One of the novel findings of this study is the spotlighting of HSPA5 and HSPA6 in the stress response of CTCL and the Th2 phenotype. Further studies are required to better understand this role, and the potential therapeutic implications that our work suggests warrant additional mechanistic studies and the optimization of HSPA5/6 isoform-selective inhibitors.

The Th2 immune phenotype of CTCL manifests as predominantly immunosuppressive [53,78,79]. By comparison, under normal conditions, Th2 immunity mediates a proinflammatory response against environmental allergens and helminth parasites and accelerates wound healing and damage repair at sites of infection [80–82]. The expansion of Th2-dominant malignant T cells in CTCL is thought to disrupt the balance between Th1 and Th2 immunity and favor a Th2-dominant TME which is immunosuppressive and resistant to immune surveillance and tumor control. Our findings support this as we found that CTCL-derived secreted factors suppressed the activation of normal naïve T cells. We were able to partially overcome the suppression using HSAP5/6 selective inhibitors and Btz, which we also showed could block Th2 cytokine production, demonstrating causality between ER stress and effectors of the UPR and the immunosuppressive phenotype of CTCL. Of the primary Th2 cytokines (i.e., IL-4, IL-5, IL-10, IL-13), we identified IL-10 as the dominant driver of CTCL-mediated immunosuppression, as we were able to rescue IFN-γ production in normal T cells with the addition of an IL-10-neutralizing antibody. As systemic immunosuppression and susceptibility is the leading cause of death amongst CTCL patients, identifying and targeting factors influencing immunosuppression is important for improving current standard of care and patient outcomes. While therapies exist that have been reported to decrease Th2 cytokine production in CTCL patients, such as bexarotene and IFN-α, resistance to these therapies has been reported, and no increase in overall survival has been observed [78]. Our work suggests that other targeted strategies, such as HSPA5/6-selective inhibitors alone or in combination with Btz, should be explored further for their potential to both overcome the immunosuppressive nature of CTCL and improve overall patient survival which the current standard of care is unable to do.

Future studies should further explore the potential of small-molecule inhibitors of HSPA5/6 for the treatment of CTCL in vivo and primary CTCL patient cells. Other methods of inhibition should also be explored, including the use of HSPA5/6-specific proteolysis targeting chimeras (PROTACs), which target proteins for proteasomal degradation and do not rely on the inhibition of target enzymatic function. Additional studies are also required to better understand the contribution of CTCL-derived secreted factors to the immune microenvironment and effects on global immune function. CTCL may affect extensive portions of the total body surface area (BSA), ranging anywhere from <10% of BSA in stage IA, to 10–25% of BSA in stage IB/IIA, and >25% of BSA in stage IB/IIA and more advanced disease forms [83]. The extent of disease burden coupled with the secretion of Th2 cytokines is likely to polarize the immune system away from Th1 immunity and have a significant impact on the ability of the host to fight infections. In fact, reports confirm that CTCL indeed suppresses Th1 responses, thereby reducing anti-tumor immunity in CTCL skin lesions and decreasing systemic immunity against microbial pathogens [20,84]. In addition, further evidence suggests that rebalancing toward the Th1 response correlates with a positive response to therapy in CTCL patients [53]. Our results suggest that therapeutic strategies to restore the Th1/Th2 immune balance should be explored in CTCL, and existing UPR-inducing drugs, like Btz, or novel HSPA5/6-selective inhibitors are potential candidates for pursuing that objective.

5. Conclusions

We identified divergent UPR signaling between malignant CTCL and normal CD4+ T cells that involves the hyper-responsiveness of Hsp70 family members, HSPA5 and HSPA6. These findings implicate HSA5/6 in CTCL biology, suggest a unique sensitivity of CTCL to proteasome inhibitors and isoform-selective HSPA5/6 inhibitors (JG-023), and connect Hsp70 and UPR signaling to the Th2-dominant immunosuppressive phenotype of CTCL.

Supplementary Materials: The following supporting information can be downloaded at: https://www.mdpi.com/article/10.3390/biom15010076/s1, Figure S1: Btz-induced killing of CTCL cells is PERK dependent; Figure S2: The combination of Btz and the PDI inhibitor, LTI6426, show synergistic killing of CTCL cells and induction of the *HSPA6* gene. Figure S3: HSPA5/6 selective inhibitor, JG-023, synergistically kills CTCL cells but not normal CD4+ T cells in combination with the NAE inhibitor, MLN4924. Figure S4: CM from resting and activated CTCL show a prominent Th2 cytokine profile. Figure S5: Original images of Western Blotting figures: Figure 1E; Figure 2D; Figure 4A,B.

Author Contributions: Conceptualization, N.S.T., E.C. and N.G.D.; methodology, N.S.T., B.N.C., L.R., R.M.R., L.G. and X.Z.; investigation, B.N.C., L.R., R.M.R., L.G. and X.Z.; writing—original draft, N.G.D.; writing—review and editing, N.S.T., E.C. and N.G.D.; supervision, N.G.D.; funding acquisition, N.G.D. All authors have read and agreed to the published version of the manuscript.

Funding: This research was funded by National Institutes of Health (R01CA245081).

Institutional Review Board Statement: Not applicable.

Informed Consent Statement: Not applicable.

Data Availability Statement: The original contributions presented in this study are included in the article. Further inquiries can be directed to the corresponding author.

Conflicts of Interest: The authors declare no conflict of interest.

References

1. Olsen, E.; Vonderheid, E.; Pimpinelli, N.; Willemze, R.; Kim, Y.; Knobler, R.; Zackheim, H.; Duvic, M.; Estrach, T.; Lamberg, S.; et al. Revisions to the staging and classification of mycosis fungoides and Sezary syndrome: A proposal of the International Society for Cutaneous Lymphomas (ISCL) and the cutaneous lymphoma task force of the European Organization of Research and Treatment of Cancer (EORTC). *Blood* **2007**, *110*, 1713–1722. [CrossRef] [PubMed]
2. Alpdogan, O.; Kartan, S.; Johnson, W.; Sokol, K.; Porcu, P. Systemic therapy of cutaneous T-cell lymphoma (CTCL). *Chin. Clin. Oncol.* **2019**, *8*, 10. [CrossRef] [PubMed]
3. Kim, Y.H.; Liu, H.L.; Mraz-Gernhard, S.; Varghese, A.; Hoppe, R.T. Long-term outcome of 525 patients with mycosis fungoides and Sezary syndrome: Clinical prognostic factors and risk for disease progression. *Arch. Dermatol.* **2003**, *139*, 857–866. [CrossRef] [PubMed]
4. Scarisbrick, J.J.; Bagot, M.; Ortiz-Romero, P.L. The changing therapeutic landscape, burden of disease, and unmet needs in patients with cutaneous T-cell lymphoma. *Br. J. Haematol.* **2021**, *192*, 683–696. [CrossRef]
5. Scarisbrick, J.J.; Prince, H.M.; Vermeer, M.H.; Quaglino, P.; Horwitz, S.; Porcu, P.; Stadler, R.; Wood, G.S.; Beylot-Barry, M.; Pham-Ledard, A.; et al. Cutaneous Lymphoma International Consortium Study of Outcome in Advanced Stages of Mycosis Fungoides and Sezary Syndrome: Effect of Specific Prognostic Markers on Survival and Development of a Prognostic Model. *J. Clin. Oncol.* **2015**, *33*, 3766–3773. [CrossRef]
6. Choi, J.; Goh, G.; Walradt, T.; Hong, B.S.; Bunick, C.G.; Chen, K.; Bjornson, R.D.; Maman, Y.; Wang, T.; Tordoff, J.; et al. Genomic landscape of cutaneous T cell lymphoma. *Nat. Genet.* **2015**, *47*, 1011–1019. [CrossRef]
7. da Silva Almeida, A.C.; Abate, F.; Khiabanian, H.; Martinez-Escala, E.; Guitart, J.; Tensen, C.P.; Vermeer, M.H.; Rabadan, R.; Ferrando, A.; Palomero, T. The mutational landscape of cutaneous T cell lymphoma and Sezary syndrome. *Nat. Genet.* **2015**, *47*, 1465–1470. [CrossRef]
8. McGirt, L.Y.; Jia, P.; Baerenwald, D.A.; Duszynski, R.J.; Dahlman, K.B.; Zic, J.A.; Zwerner, J.P.; Hucks, D.; Dave, U.; Zhao, Z.; et al. Whole-genome sequencing reveals oncogenic mutations in mycosis fungoides. *Blood* **2015**, *126*, 508–519. [CrossRef]
9. Liu, Z.; Wu, X.; Hwang, S.T.; Liu, J. The Role of Tumor Microenvironment in Mycosis Fungoides and Sezary Syndrome. *Ann. Dermatol.* **2021**, *33*, 487–496. [CrossRef]

10. Patil, K.; Kuttikrishnan, S.; Khan, A.Q.; Ahmad, F.; Alam, M.; Buddenkotte, J.; Ahmad, A.; Steinhoff, M.; Uddin, S. Molecular pathogenesis of Cutaneous T cell Lymphoma: Role of chemokines, cytokines, and dysregulated signaling pathways. *Semin. Cancer Biol.* **2022**, *86*, 382–399. [CrossRef]
11. Miyashiro, D.; Souza, B.d.C.e.; Torrealba, M.P.; Manfrere, K.C.G.; Sato, M.N.; Sanches, J.A. The Role of Tumor Microenvironment in the Pathogenesis of Sezary Syndrome. *Int. J. Mol. Sci.* **2022**, *23*, 936. [CrossRef] [PubMed]
12. Liu, X.; Jin, S.; Hu, S.; Li, R.; Pan, H.; Liu, Y.; Lai, P.; Xu, D.; Sun, J.; Liu, Z.; et al. Single-cell transcriptomics links malignant T cells to the tumor immune landscape in cutaneous T cell lymphoma. *Nat. Commun.* **2022**, *13*, 1158. [CrossRef] [PubMed]
13. Campbell, J.J.; Clark, R.A.; Watanabe, R.; Kupper, T.S. Sezary syndrome and mycosis fungoides arise from distinct T-cell subsets: A biologic rationale for their distinct clinical behaviors. *Blood* **2010**, *116*, 767–771. [CrossRef] [PubMed]
14. Vowels, B.R.; Lessin, S.R.; Cassin, M.; Jaworsky, C.; Benoit, B.; Wolfe, J.T.; Rook, A.H. Th2 cytokine mRNA expression in skin in cutaneous T-cell lymphoma. *J. Investig. Dermatol.* **1994**, *103*, 669–673. [CrossRef] [PubMed]
15. Rook, A.H.; Vowels, B.R.; Jaworsky, C.; Singh, A.; Lessin, S.R. The immunopathogenesis of cutaneous T-cell lymphoma. Abnormal cytokine production by Sezary T cells. *Arch. Dermatol.* **1993**, *129*, 486–489. [CrossRef]
16. Kural, Y.B.; Su, O.; Onsun, N.; Uras, A.R. Atopy, IgE and eosinophilic cationic protein concentration, specific IgE positivity, eosinophil count in cutaneous T Cell lymphoma. *Int. J. Dermatol.* **2010**, *49*, 390–395. [CrossRef]
17. Spellberg, B.; Edwards, J.E., Jr. Type 1/Type 2 immunity in infectious diseases. *Clin. Infect. Dis.* **2001**, *32*, 76–102. [CrossRef]
18. Murray, P.J. Macrophage Polarization. *Annu. Rev. Physiol.* **2017**, *79*, 541–566. [CrossRef]
19. Walker, J.A.; McKenzie, A.N.J. T(H)2 cell development and function. *Nat. Rev. Immunol.* **2018**, *18*, 121–133. [CrossRef]
20. Axelrod, P.I.; Lorber, B.; Vonderheid, E.C. Infections complicating mycosis fungoides and Sezary syndrome. *JAMA* **1992**, *267*, 1354–1358. [CrossRef]
21. Nguyen, V.; Huggins, R.H.; Lertsburapa, T.; Bauer, K.; Rademaker, A.; Gerami, P.; Guitart, J. Cutaneous T-cell lymphoma and *Staphylococcus aureus* colonization. *J. Am. Acad. Dermatol.* **2008**, *59*, 949–952. [CrossRef] [PubMed]
22. Talpur, R.; Bassett, R.; Duvic, M. Prevalence and treatment of *Staphylococcus aureus* colonization in patients with mycosis fungoides and Sezary syndrome. *Br. J. Dermatol.* **2008**, *159*, 105–112. [CrossRef] [PubMed]
23. Willerslev-Olsen, A.; Krejsgaard, T.; Lindahl, L.M.; Bonefeld, C.M.; Wasik, M.A.; Koralov, S.B.; Geisler, C.; Kilian, M.; Iversen, L.; Woetmann, A.; et al. Bacterial toxins fuel disease progression in cutaneous T-cell lymphoma. *Toxins* **2013**, *5*, 1402–1421. [CrossRef]
24. Walter, P.; Ron, D. The unfolded protein response: From stress pathway to homeostatic regulation. *Science* **2011**, *334*, 1081–1086. [CrossRef]
25. Travers, K.J.; Patil, C.K.; Wodicka, L.; Lockhart, D.J.; Weissman, J.S.; Walter, P. Functional and genomic analyses reveal an essential coordination between the unfolded protein response and ER-associated degradation. *Cell* **2000**, *101*, 249–258. [CrossRef]
26. Reimold, A.M.; Iwakoshi, N.N.; Manis, J.; Vallabhajosyula, P.; Szomolanyi-Tsuda, E.; Gravallese, E.M.; Friend, D.; Grusby, M.J.; Alt, F.; Glimcher, L.H. Plasma cell differentiation requires the transcription factor XBP-1. *Nature* **2001**, *412*, 300–307. [CrossRef]
27. Todd, D.J.; McHeyzer-Williams, L.J.; Kowal, C.; Lee, A.-H.; Volpe, B.T.; Diamond, B.; McHeyzer-Williams, M.G.; Glimcher, L.H. XBP1 governs late events in plasma cell differentiation and is not required for antigen-specific memory B cell development. *J. Exp. Med.* **2009**, *206*, 2151–2159. [CrossRef]
28. Henriksson, J.; Chen, X.; Gomes, T.; Ullah, U.; Meyer, K.B.; Miragaia, R.; Duddy, G.; Pramanik, J.; Yusa, K.; Lahesmaa, R.; et al. Genome-Wide CRISPR Screens in T Helper Cells Reveal Pervasive Crosstalk Between Activation and Differentiation. *Cell* **2019**, *176*, 882–896.e18. [CrossRef]
29. Kemp, K.L.; Lin, Z.; Zhao, F.; Gao, B.; Song, J.; Zhang, K.; Fang, D. The serine-threonine kinase inositol-requiring enzyme 1alpha (IRE1alpha) promotes IL-4 production in T helper cells. *J. Biol. Chem.* **2013**, *288*, 33272–33282. [CrossRef]
30. Ambrose, A.J.; Sivinski, J.; Zerio, C.J.; Zhu, X.; Godek, J.; Kumirov, V.K.; Brujas, T.C.; Garcia, J.T.; Annadurai, A.; Schmidlin, C.J.; et al. Discovery and Development of a Selective Inhibitor of the ER Resident Chaperone Grp78. *J. Med. Chem.* **2023**, *66*, 677–694. [CrossRef]
31. Duncan, R.M.; Reyes, L.; Moats, K.; Robinson, R.M.; Murphy, S.A.; Kaur, B.; Stessman, H.A.; Dolloff, N.G. ATF3 Coordinates Antitumor Synergy Between Epigenetic Drugs and Protein Disulfide Isomerase Inhibitors. *Cancer Res.* **2020**, *80*, 3279–3291. [CrossRef] [PubMed]
32. Robinson, R.M.; Basar, A.P.; Reyes, L.; Duncan, R.M.; Li, H.; Dolloff, N.G. PDI inhibitor LTI6426 enhances panobinostat efficacy in preclinical models of multiple myeloma. *Cancer Chemother. Pharmacol.* **2022**, *89*, 643–653. [CrossRef] [PubMed]
33. Klock, H.E.; Lesley, S.A. The Polymerase Incomplete Primer Extension (PIPE) method applied to high-throughput cloning and site-directed mutagenesis. *Methods Mol. Biol.* **2009**, *498*, 91–103. [CrossRef] [PubMed]
34. Boumsell, L.; Bernard, A.; Reinherz, E.L.; Nadler, L.M.; Ritz, J.; Coppin, H.; Richard, Y.; Dubertret, L.; Valensi, F.; Degos, L.; et al. Surface antigens on malignant Sezary and T-CLL cells correspond to those of mature T cells. *Blood* **1981**, *57*, 526–530. [CrossRef]
35. Rook, A.H.; Gottlieb, S.L.; Wolfe, J.T.; Vowels, B.R.; Sood, S.S.; Niu, Z.; Lessin, S.R.; Fox, F.E. Pathogenesis of cutaneous T-cell lymphoma: Implications for the use of recombinant cytokines and photopheresis. *Clin. Exp. Immunol.* **1997**, *107* (Suppl. S1), 16–20.

36. Slee, E.A.; Adrain, C.; Martin, S.J. Executioner caspase-3, -6, and -7 perform distinct, non-redundant roles during the demolition phase of apoptosis. *J. Biol. Chem.* **2001**, *276*, 7320–7326. [CrossRef]
37. Robinson, R.M.; Reyes, L.; Duncan, R.M.; Bian, H.; Reitz, A.B.; Manevich, Y.; McClure, J.J.; Champion, M.M.; Chou, C.J.; Sharik, M.E.; et al. Inhibitors of the protein disulfide isomerase family for the treatment of multiple myeloma. *Leukemia* **2019**, *33*, 1011–1022. [CrossRef]
38. Robinson, R.M.; Reyes, L.; Duncan, R.M.; Bian, H.; Strobel, E.D.; Hyman, S.L.; Reitz, A.B.; Dolloff, N.G. Tuning isoform selectivity and bortezomib sensitivity with a new class of alkenyl indene PDI inhibitor. *Eur. J. Med. Chem.* **2020**, *186*, 111906. [CrossRef]
39. Brocchieri, L.; de Macario, E.C.; Macario, A.J. hsp70 genes in the human genome: Conservation and differentiation patterns predict a wide array of overlapping and specialized functions. *BMC Evol. Biol.* **2008**, *8*, 19. [CrossRef]
40. Wang, M.; Wey, S.; Zhang, Y.; Ye, R.; Lee, A.S. Role of the unfolded protein response regulator GRP78/BiP in development, cancer, and neurological disorders. *Antioxid. Redox Signal.* **2009**, *11*, 2307–2316. [CrossRef]
41. Li, J.; Ni, M.; Lee, B.; Barron, E.; Hinton, D.R.; Lee, A.S. The unfolded protein response regulator GRP78/BiP is required for endoplasmic reticulum integrity and stress-induced autophagy in mammalian cells. *Cell Death Differ.* **2008**, *15*, 1460–1471. [CrossRef] [PubMed]
42. Massey, A.J.; Williamson, D.S.; Browne, H.; Murray, J.B.; Dokurno, P.; Shaw, T.; Macias, A.T.; Daniels, Z.; Geoffroy, S.; Dopson, M.; et al. A novel, small molecule inhibitor of Hsc70/Hsp70 potentiates Hsp90 inhibitor induced apoptosis in HCT116 colon carcinoma cells. *Cancer Chemother. Pharmacol.* **2010**, *66*, 535–545. [CrossRef] [PubMed]
43. Barnoud, T.; Leung, J.C.; Leu, J.I.-J.; Basu, S.; Poli, A.N.R.; Parris, J.L.; Indeglia, A.; Martynyuk, T.; Good, M.; Gnanapradeepan, K.; et al. A Novel Inhibitor of HSP70 Induces Mitochondrial Toxicity and Immune Cell Recruitment in Tumors. *Cancer Res.* **2020**, *80*, 5270–5281. [CrossRef] [PubMed]
44. Asadullah, K.; Docke, W.D.; Haeussler, A.; Sterry, W.; Volk, H.D. Progression of mycosis fungoides is associated with increasing cutaneous expression of interleukin-10 mRNA. *J. Investig. Dermatol.* **1996**, *107*, 833–837. [CrossRef] [PubMed]
45. Miyagaki, T.; Sugaya, M.; Suga, H.; Kamata, M.; Ohmatsu, H.; Fujita, H.; Asano, Y.; Tada, Y.; Kadono, T.; Sato, S. IL-22, but not IL-17, dominant environment in cutaneous T-cell lymphoma. *Clin. Cancer Res.* **2011**, *17*, 7529–7538. [CrossRef]
46. Gluud, M.; Pallesen, E.M.H.; Buus, T.B.; Gjerdrum, L.M.R.; Lindahl, L.M.; Kamstrup, M.R.; Bzorek, M.; Danielsen, M.; Bech, R.; Monteiro, M.N.; et al. Malignant T cells induce skin barrier defects through cytokine-mediated JAK/STAT signaling in cutaneous T-cell lymphoma. *Blood* **2023**, *141*, 180–193. [CrossRef]
47. Wang, Y.; Lin, F.; Zhu, X.; Leone, V.A.; Dalal, S.; Tao, Y.; Messer, J.S.; Chang, E.B. Distinct roles of intracellular heat shock protein 70 in maintaining gastrointestinal homeostasis. *Am. J. Physiol. Gastrointest. Liver Physiol.* **2018**, *314*, G164–G178. [CrossRef]
48. Wachstein, J.; Tischer, S.; Figueiredo, C.; Limbourg, A.; Falk, C.; Immenschuh, S.; Blasczyk, R.; Eiz-Vesper, B. HSP70 enhances immunosuppressive function of $CD4^+CD25^+FoxP3^+$ T regulatory cells and cytotoxicity in $CD4^+CD25^-$ T cells. *PLoS ONE* **2012**, *7*, e51747. [CrossRef]
49. Steinke, J.W.; Borish, L. Th2 cytokines and asthma. Interleukin-4: Its role in the pathogenesis of asthma, and targeting it for asthma treatment with interleukin-4 receptor antagonists. *Respir. Res.* **2001**, *2*, 66–70. [CrossRef]
50. Patel, N.; Kreider, T.; Urban, J.F., Jr.; Gause, W.C. Characterisation of effector mechanisms at the host: Parasite interface during the immune response to tissue-dwelling intestinal nematode parasites. *Int. J. Parasitol.* **2009**, *39*, 13–21. [CrossRef]
51. Webster, H.C.; Gamino, V.; Andrusaite, A.T.; Ridgewell, O.J.; McCowan, J.; Shergold, A.L.; Heieis, G.A.; Milling, S.W.; Maizels, R.M.; Perona-Wright, G. Tissue-based IL-10 signalling in helminth infection limits IFNgamma expression and promotes the intestinal Th2 response. *Mucosal Immunol.* **2022**, *15*, 1257–1269. [CrossRef] [PubMed]
52. Abbas, A.K.; Murphy, K.M.; Sher, A. Functional diversity of helper T lymphocytes. *Nature* **1996**, *383*, 787–793. [CrossRef] [PubMed]
53. Guenova, E.; Watanabe, R.; Teague, J.E.; Desimone, J.A.; Jiang, Y.; Dowlatshahi, M.; Schlapbach, C.; Schaekel, K.; Rook, A.H.; Tawa, M.; et al. TH2 cytokines from malignant cells suppress TH1 responses and enforce a global TH2 bias in leukemic cutaneous T-cell lymphoma. *Clin. Cancer Res.* **2013**, *19*, 3755–3763. [CrossRef] [PubMed]
54. Kemp, K.; Poe, C. Stressed: The Unfolded Protein Response in T Cell Development, Activation, and Function. *Int. J. Mol. Sci.* **2019**, *20*, 1792. [CrossRef] [PubMed]
55. Chen, S.; Wang, Q.; Wang, H.; Xia, S. Endoplasmic reticulum stress in T cell-mediated diseases. *Scand. J. Immunol.* **2023**, *98*, e13307. [CrossRef]
56. Thaxton, J.E.; Wallace, C.; Riesenberg, B.; Zhang, Y.; Paulos, C.M.; Beeson, C.C.; Liu, B.; Li, Z. Modulation of Endoplasmic Reticulum Stress Controls $CD4^+$ T-Cell Activation and Antitumor Function. *Cancer Immunol. Res.* **2017**, *5*, 666–675. [CrossRef]
57. Koong, A.C.; Chauhan, V.; Romero-Ramirez, L. Targeting XBP-1 as a novel anti-cancer strategy. *Cancer Biol. Ther.* **2006**, *5*, 756–759. [CrossRef]
58. Koumenis, C.; Naczki, C.; Koritzinsky, M.; Rastani, S.; Diehl, A.; Sonenberg, N.; Koromilas, A.; Wouters, B.G. Regulation of protein synthesis by hypoxia via activation of the endoplasmic reticulum kinase PERK and phosphorylation of the translation initiation factor eIF2alpha. *Mol. Cell. Biol.* **2002**, *22*, 7405–7416. [CrossRef]

59. Schleicher, S.M.; Moretti, L.; Varki, V.; Lu, B. Progress in the unraveling of the endoplasmic reticulum stress/autophagy pathway and cancer: Implications for future therapeutic approaches. *Drug Resist. Updates* **2010**, *13*, 79–86. [CrossRef]
60. Stone, K.R.; Smith, R.E.; Joklik, W.K. Changes in membrane polypeptides that occur when chick embryo fibroblasts and NRK cells are transformed with avian sarcoma viruses. *Virology* **1974**, *58*, 86–100. [CrossRef]
61. Dong, D.; Stapleton, C.; Luo, B.; Xiong, S.; Ye, W.; Zhang, Y.; Jhaveri, N.; Zhu, G.; Ye, R.; Liu, Z.; et al. A critical role for GRP78/BiP in the tumor microenvironment for neovascularization during tumor growth and metastasis. *Cancer Res.* **2011**, *71*, 2848–2857. [CrossRef] [PubMed]
62. Wang, Y.; Wang, J.; Liu, Y.; Wang, X.; Ren, M. Multidimensional pan-cancer analysis of HSPA5 and its validation in the prognostic value of bladder cancer. *Heliyon* **2024**, *10*, E27184. [CrossRef] [PubMed]
63. Luo, B.; Lee, A.S. The critical roles of endoplasmic reticulum chaperones and unfolded protein response in tumorigenesis and anticancer therapies. *Oncogene* **2013**, *32*, 805–818. [CrossRef] [PubMed]
64. Parsian, A.J.; Sheren, J.E.; Tao, T.Y.; Goswami, P.C.; Malyapa, R.; Van Rheeden, R.; Watson, M.S.; Hunt, C.R. The human Hsp70B gene at the HSPA7 locus of chromosome 1 is transcribed but non-functional. *Biochim. Biophys. Acta* **2000**, *1494*, 201–205. [CrossRef]
65. Leung, T.K.; Rajendran, M.Y.; Monfries, C.; Hall, C.; Lim, L. The human heat-shock protein family. Expression of a novel heat-inducible HSP70 (HSP70B′) and isolation of its cDNA and genomic DNA. *Biochem. J.* **1990**, *267*, 125–132. [CrossRef]
66. Hammad, S.M.; Twal, W.O.; Barth, J.L.; Smith, K.J.; Saad, A.F.; Virella, G.; Argraves, W.S.; Lopes-Virella, M.F. Oxidized LDL immune complexes and oxidized LDL differentially affect the expression of genes involved with inflammation and survival in human U937 monocytic cells. *Atherosclerosis* **2009**, *202*, 394–404. [CrossRef]
67. Smith, K.J.; Twal, W.O.; Soodavar, F.; Virella, G.; Lopes-Virella, M.F.; Hammad, S.M. Heat shock protein 70B′ (HSP70B′) expression and release in response to human oxidized low density lipoprotein immune complexes in macrophages. *J. Biol. Chem.* **2010**, *285*, 15985–15993. [CrossRef]
68. Huang, L.; Wang, Y.; Bai, J.; Yang, Y.; Wang, F.; Feng, Y.; Zhang, R.; Li, F.; Zhang, P.; Lv, N.; et al. Blockade of HSP70 by VER-155008 synergistically enhances bortezomib-induced cytotoxicity in multiple myeloma. *Cell Stress Chaperones* **2020**, *25*, 357–367. [CrossRef]
69. McConkey, D.J.; Zhu, K. Mechanisms of proteasome inhibitor action and resistance in cancer. *Drug Resist. Updates* **2008**, *11*, 164–179. [CrossRef]
70. Lu, S.; Wang, J. The resistance mechanisms of proteasome inhibitor bortezomib. *Biomark. Res.* **2013**, *1*, 13. [CrossRef]
71. Chauhan, D.; Catley, L.; Li, G.; Podar, K.; Hideshima, T.; Velankar, M.; Mitsiades, C.; Mitsiades, N.; Yasui, H.; Letai, A.; et al. A novel orally active proteasome inhibitor induces apoptosis in multiple myeloma cells with mechanisms distinct from Bortezomib. *Cancer Cell* **2005**, *8*, 407–419. [CrossRef] [PubMed]
72. Li, A.; Song, N.J.; Riesenberg, B.P.; Li, Z. The Emerging Roles of Endoplasmic Reticulum Stress in Balancing Immunity and Tolerance in Health and Diseases: Mechanisms and Opportunities. *Front. Immunol.* **2019**, *10*, 3154. [CrossRef] [PubMed]
73. Staron, M.; Yang, Y.; Liu, B.; Li, J.; Shen, Y.; Zúñiga-Pflücker, J.C.; Aguila, H.L.; Goldschneider, I.; Li, Z. gp96, an endoplasmic reticulum master chaperone for integrins and Toll-like receptors, selectively regulates early T and B lymphopoiesis. *Blood* **2010**, *115*, 2380–2390. [CrossRef]
74. Lee, W.H.; Park, Y.M.; Kim, J.; Jang, J.J.; Seo, J.S. Expression of heat shock protein 70 blocks thymic differentiation of T cells in transgenic mice. *Immunology* **1998**, *95*, 559–565. [CrossRef]
75. Mycko, M.P.; Cwiklinska, H.; Walczak, A.; Libert, C.; Raine, C.S.; Selmaj, K.W. A heat shock protein gene (Hsp70.1) is critically involved in the generation of the immune response to myelin antigen. *Eur. J. Immunol.* **2008**, *38*, 1999–2013. [CrossRef]
76. Mansilla, M.J.; Costa, C.; Eixarch, H.; Tepavcevic, V.; Castillo, M.; Martin, R.; Lubetzki, C.; Aigrot, M.-S.; Montalban, X.; Espejo, C. Hsp70 regulates immune response in experimental autoimmune encephalomyelitis. *PLoS ONE* **2014**, *9*, e105737. [CrossRef]
77. Pramanik, J.; Chen, X.; Kar, G.; Henriksson, J.; Gomes, T.; Park, J.-E.; Natarajan, K.; Meyer, K.B.; Miao, Z.; McKenzie, A.N.J.; et al. Genome-wide analyses reveal the IRE1a-XBP1 pathway promotes T helper cell differentiation by resolving secretory stress and accelerating proliferation. *Genome Med.* **2018**, *10*, 76. [CrossRef]
78. Weiner, D.M.; Durgin, J.S.; Wysocka, M.; Rook, A.H. The immunopathogenesis and immunotherapy of cutaneous T cell lymphoma: Current and future approaches. *J. Am. Acad. Dermatol.* **2021**, *84*, 597–604. [CrossRef]
79. Tendler, C.L.; Burton, J.D.; Jaffe, J.; Danielpour, D.; Charley, M.; McCoy, J.P.; Pittelkow, M.R.; Waldmann, T.A. Abnormal cytokine expression in Sezary and adult T-cell leukemia cells correlates with the functional diversity between these T-cell malignancies. *Cancer Res.* **1994**, *54*, 4430–4435.
80. Anthony, R.M.; Rutitzky, L.I.; Urban, J.F.; Jr Stadecker, M.J.; Gause, W.C. Protective immune mechanisms in helminth infection. *Nat. Rev. Immunol.* **2007**, *7*, 975–987. [CrossRef]
81. Chen, F.; Liu, Z.; Wu, W.; Rozo, C.; Bowdridge, S.; Millman, A.; Van Rooijen, N.; Urban, J.F., Jr.; Wynn, T.A.; Gause, W.C. An essential role for TH2-type responses in limiting acute tissue damage during experimental helminth infection. *Nat. Med.* **2012**, *18*, 260–266. [CrossRef] [PubMed]
82. Palm, N.W.; Rosenstein, R.K.; Medzhitov, R. Allergic host defences. *Nature* **2012**, *484*, 465–472. [CrossRef] [PubMed]

83. D'Agostino, P.; Kent, A.; Sharp, E.; Schmidt, F.; Turini, M. Mycosis fungoides-type cutaneous T-cell lymphoma (MF-CTCL) epidemiology and treatment pathway in Spain: New insights for an accurate description. *Drugs Context* **2020**, *9*, 1–10. [CrossRef] [PubMed]
84. Goldgeier, M.H.; Cohen, S.R.; Braverman, I.M.; Stenn, K.S. An unusual and fatal case of disseminated cutaneous herpes simplex. Infection in a patient with cutaneous T cell lymphoma (mycosis fungoides). *J. Am. Acad. Dermatol.* **1981**, *4*, 176–180. [CrossRef]

biomolecules

MDPI

Article

Modulating the CXCR2 Signaling Axis Using Engineered Chemokine Fusion Proteins to Disrupt Myeloid Cell Infiltration in Pancreatic Cancer

Benjamin N. Christopher [1], Lena Golick [1], Ashton Basar [1], Leticia Reyes [1], Reeder M. Robinson [1], Aaron O. Angerstein [1], Carsten Krieg [2], G. Aaron Hobbs [3,4], Denis C. Guttridge [4,5,6], John P. O'Bryan [3,4] and Nathan G. Dolloff [1,4,7,*]

1 Department of Pharmacology and Immunology, Medical University of South Carolina, Charleston, SC 29425, USA; christob@musc.edu (B.N.C.); golick@musc.edu (L.G.); basar@musc.edu (A.B.); reyesl@musc.edu (L.R.); robinree@musc.edu (R.M.R.); angerste@musc.edu (A.O.A.)
2 Department of Pathology and Laboratory Medicine, Medical University of South Carolina, Charleston, SC 29425, USA; kriegc@musc.edu
3 Department of Biochemistry, Medical University of South Carolina, Charleston, SC 29425, USA; hobbsg@musc.edu (G.A.H.); obryanjo@musc.edu (J.P.O.)
4 MUSC Hollings Cancer Center, Charleston, SC 29425, USA; guttridg@musc.edu
5 MUSC Darby Children's Research Institute, Charleston, SC 29425, USA
6 Department of Pediatrics, Medical University of South Carolina, Charleston, SC 29425, USA
7 Zucker Institute for Innovation Commercialization, Charleston, SC 29425, USA
* Correspondence: dolloffn@musc.edu; Tel.: +1-843-876-2204

Abstract: Pancreatic ductal adenocarcinoma (PDAC) has one of the lowest 5-year survival rates of all cancers, and limited treatment options exist. Immunotherapy is effective in some cancer types, but the immunosuppressive tumor microenvironment (TME) of PDAC is a barrier to effective immunotherapy. CXCR2+ myeloid-derived suppressor cells (MDSCs) are abundant in PDAC tumors in humans and in mouse models. MDSCs suppress effector cell function, making them attractive targets for restoring anti-tumor immunity. In this study, we show that the most abundant soluble factors released from a genetically diverse set of human and mouse PDAC cells are CXCR2 ligands, including CXCL8, CXCL5, and CXCL1. Expression of CXCR2 ligands is at least partially dependent on mutant KRAS and NFκB signaling, which are two of the most commonly dysregulated pathways in PDAC. We show that MDSCs are the most prevalent immune cells in PDAC tumors. MDSCs expressed high levels of CXCR2, and we found that myeloid cells readily migrate toward conditioned media (CM) prepared from PDAC cultures. We designed CXCR2 ligand-Fc fusion proteins to modulate the CXCR2 chemotactic signaling axis. Unexpectedly, these fusion proteins were superior to native chemokines in binding and activation of CXCR2 on myeloid cells. These "superkines" were potent inhibitors of PDAC CM-induced myeloid cell migration and were superior to CXCR2 small-molecule inhibitors and neutralizing antibodies. Our findings suggest that CXCR2 superkines may disrupt myeloid cell recruitment to PDAC tumors, ultimately improving immunotherapy outcomes in patients with PDAC.

Keywords: pancreatic ductal adenocarcinoma; tumor microenvironment; immunosuppression; myeloid cell migration; MDSCs; CXCR2; CXCL1

1. Introduction

Pancreatic ductal adenocarcinoma (PDAC) is a highly lethal disease with a 5-year survival rate of less than 10% [1]. PDAC exhibits significant resistance to therapy, mediated largely by a dense desmoplastic stroma and an immunosuppressive tumor microenvironment (TME) [2]. Immune checkpoint inhibitors (ICIs) have had success in a variety of cancer types, but their efficacy in PDAC has been disappointing. Moreover, ICIs are only approved for the 1–2% of patients that have high microsatellite instability (MSI-H) or DNA mismatch repair deficiency (dMMR) [3]. The PDAC TME supports an abundance of immunosuppressive immune and stromal cells such as myeloid-derived suppressor cells (MDSCs) [4], tumor-associated macrophages (TAMs) [5], regulatory T cells (Tregs) [6], and cancer-associated fibroblasts (CAFs) [7]. Consequently, infiltration and activation of anti-tumor immune cells such as CD8+ T cells and natural killer (NK) cells is limited [8–10]. The formation and maintenance of the PDAC TME are primarily driven by KRAS mutation, which is present in approximately 90% of PDAC cases. KRAS signaling in PDAC enhances the secretion of cytokines and chemokines, which recruit immunosuppressive cells and foster a supportive environment for immune evasion and tumor progression [11].

Myeloid cells play a distinctive role in shaping the PDAC TME due to their inflammatory properties, plasticity, and role in immunosuppression. In healthy individuals, the myeloid compartment plays a vital role in the inflammatory response, making it essential for proper immune system function and wound healing. However, the cytokines produced by PDAC have a profound effect on the myeloid compartment, which promotes a chronic inflammatory state where the inflammatory response never reaches a resolution phase [12,13]. This dysregulation underlies the characterization of tumors as "wounds that do not heal" [14].

One of the most significant consequences of this dysregulation is the expansion of MDSCs, which are potent suppressors of the anti-tumor immune response [15,16]. MDSCs employ multiple mechanisms to inhibit CD8+ T cells, such as depleting arginine, an essential amino acid for proper T cell proliferation and activation [17,18]; secreting reactive oxygen species, inducing stress in T cells and inactivating the T cell receptor [19,20]; and downregulating L-selectin on T cells, hindering T cell trafficking to lymphoid organs, which is essential for the T cell response [21]. Additionally, MDSCs suppress the activity of NK cells [22], promote the expansion of immunosuppressive regulatory T cells [23,24], and promote the immunosuppressive polarization of macrophages [25]. Beyond promoting the expansion of MDSCs, PDAC attracts MDSCs to the TME through chemokine secretion. Chemokine receptors play a central role in immune cell migration by directing cells to tissues that secrete their corresponding ligands. This migration relies on the ability of cells to sense a chemokine gradient and coordinate movement towards areas of higher chemokine concentration [26,27].

CXC motif chemokine receptor 2 (CXCR2) is one of many chemokine receptors implicated in establishing the immunosuppressive PDAC TME. In healthy individuals, the CXCR2 axis is critical for the homing of neutrophils and other cell types to sites of injury. This process is coordinated by the secretion of CXCR2 ligands from macrophages and mast cells in proximity to the site of injury [28]. However, PDAC and many other cancers hijack the normal physiological roles of CXCR2 ligands to recruit MDSCs and protect themselves from the anti-tumor immune response. The expression of CXCR2 on MDSCs is critical for their recruitment to the PDAC TME [29]. Since MDSCs are significant contributors to the immunosuppressive PDAC TME and CXCR2 is important for their recruitment to the PDAC TME, CXCR2 is an attractive target for pancreatic cancer. In our study we developed an IgG1 Fc-based fusion protein to inhibit myeloid cell migration towards PDAC-derived factors by hyperactivating the CXCR2 axis.

IgG Fc-based fusion proteins are a widely utilized class of biologics designed to enhance the stability, half-life, and efficacy of protein-based therapeutics. These molecules consist of a protein, usually a receptor or ligand, fused to the Fc domain of immunoglobulin G (IgG), which extends serum half-life by recycling through the neonatal Fc receptor (FcRn) and improves structural stability [30]. The Fc domain may retain its native function, engaging Fc receptors on immune cells to modulate immune responses, or it may be engineered to eliminate Fc receptor interactions [31]. Several clinically relevant IgG Fc fusion proteins, including etanercept (TNFR-Fc) and aflibercept (VEGFR-Fc), have been developed for inflammatory diseases and oncology, where they modulate immune activation, block immunosuppressive pathways, or reshape the tumor microenvironment [32,33]. These therapeutics work by acting as cytokine traps to sequester pro-inflammatory or pro-angiogenic ligands, thereby preventing their interaction with cell surface receptors and dampening pathological signaling within the inflammatory or tumor microenvironment [34]. Other Fc-based fusion proteins incorporate ligands as their fusion partners and act as activators or suppressors of immune responses [31]. In this study, we identified an unanticipated consequence of Fc fusion, where the addition of the Fc domain enhanced the binding and signaling activity of CXCL1, indicating that Fc fusion can actively influence cytokine function rather than serving solely as a pharmacokinetic enhancer, thereby revealing new opportunities for therapeutic development.

2. Materials and Methods

2.1. Cell Lines and Reagents

Cell lines were purchased from the American Tissue Culture Collection (Manassas, VA, USA) and cultured according to the manufacturer's specifications. The cell lines were PANC-1, CFPAC-1, BxPC-3, Capan-1, Capan-2, HEK293, THP-1, HL60, and K562. HPNE cells with doxycycline-inducible KRAS mutations were a gift from Dr. Aaron Hobbs (Medical University of South Carolina). Cells were grown at 37 °C and in 5% CO_2 in the recommended base media supplemented with heat-inactivated fetal bovine serum (FBS) (900–108, GeminiBio, Sacramento, CA, USA), and the antibiotics 1% penicillin (10,000 units/mL) and 1% streptomycin (10,000 μg/mL) were purchased from Selleck Chemicals (Houston, TX, USA). Other materials were as follows: SX-682 (S8947), Navarixin (S8506), and AZD5069 (S6645) were purchased from Selleck Chemicals (Houston, TX, USA), αCXCR2 (MA1-24669, Thermo Fisher Scientific, Waltham, MA, USA), doxycycline (D9891, Sigma-Aldrich, St. Louis, MO, USA), IκB-SR retrovirus was a gift from Dr. Dennis Guttridge (Medical University of South Carolina), and TNF-α (1006-050, CellGenix, Freiburg, Germany). A Proteome Profiler Human XL Cytokine Array Kit (ARY022B, R & D Systems, Minneapolis, MN, USA) was used according to the manufacturer's specifications. All ELISAs were performed according to the manufacturer's specifications. Kits were as follows: Human CXCL1 (88-52122-22, Invitrogen, Waltham, MA, USA), Human CXCL5 (dy254-05, R & D Systems, Minneapolis, MN, USA), Human CXCL8 (88-8086-22, Invitrogen, Waltham, MA, USA), Mouse CXCL1 (P352782, R & D Systems, Minneapolis, MN, USA), and Mouse CXCL5 (MX000, R & D Systems, Minneapolis, MN, USA).

2.2. Subcloning

Genes for CXCL1 (35–107), CXCL5 (37–114), IgG1 Fc (216–447 Eu numbering), and CXCR2 were codon-optimized for human cells by GenScript Biotech (Piscataway, NJ, USA). CXCL1 or CXCL5 were fused to Fc with DNA encoding GGGGS. CXCR2 was cloned into pLJM1 using Afe1 and EcoR1. All other genes were subcloned into a custom pcDNA3.4 using NheI and XhoI with an N-terminal secretion signal and a C-terminal 6x-His tag. Proper ligation and fusion of each gene was confirmed by DNA sequencing at Eurofins Genomics (Louisville, KY, USA).

2.3. Protein Expression and Purification

CXCL1, CXCL5, Fc, and all fused genes were expressed in the expi293F expression system (Thermo Fisher Scientific, Waltham, MA, USA) following the manufacturer's protocol. Briefly, 1 µg/mL pcDNA3.4 containing each gene with a secretion signal and C-terminal 6x-His tag was transfected into expi293F cells using expifectamine 293. Medium was collected by centrifugation of cells 5–7 days later when cell viability was <60%. Medium was then loaded at 5 mL/min onto a 5 mL HisTrap Excel column (Cytiva, Marlborough, MA, USA) equilibrated with buffer A (20 mM sodium phosphate, 500 mM NaCl, pH 7.4). After loading, the column was washed 3 subsequent times with buffer A containing 10 mM, 25 mM, and then 50 mM imidazole (A10221, Thermo Fisher Scientific, Waltham, MA, USA). The bound protein was subsequently eluted with buffer A containing 500 mM imidazole. Eluted proteins were then dialyzed overnight in PBS, at pH 7.4 at 4 °C, sterile-filtered, flash-frozen in liquid nitrogen, and stored at −80 °C. Protein purity was determined using SDS-PAGE. Concentrations were determined using a Bradford assay using Bio-Rad Protein Assay Dye Reagent Concentrate (Bio-Rad Laboratories, Hercules, CA, USA) diluted 1:5 in MilliQ water.

2.4. Western Blotting

Cells were collected on ice, rinsed with cold 1× PBS, centrifuged at 2500 rpm at 4 °C for 5 min, and lysed in 1x cell lysis buffer (Cell Signaling Technology, Danvers, MA, USA) with the addition of protease (Thermo Fisher Scientific, Waltham, MA, USA) and phosphatase (Thermo Fisher Scientific, Waltham, MA, USA) inhibitors. For Western blots analyzing the nuclear fraction, cytoplasmic and nuclear fractions were isolated with an NE-PER Nuclear and Cytoplasmic Extraction Kit (78833, Thermo Fisher Scientific, Waltham, MA, USA). Cell lysates were clarified, and their relative protein concentrations were determined via a Bradford assay. Values were normalized to the lowest average absorbance at 595 nm. Gel samples were prepared by mixing lysates with SDS sample buffer containing β-mercaptoethanol (bp176-100, Thermo Fisher Scientific, Waltham, MA, USA) (final concentration 1×) or without β-mercaptoethanol for non-reducing conditions and 1x cell lysis buffer. Samples were boiled for 10 min, loaded on NuPAGE Bis-Tris Gel 4–12% (Invitrogen, Waltham, MA, USA), and subjected to gel electrophoresis at 55 mA for 1 h and 30 min in 1× NuPAGE MOPS SDS Running Buffer (Invitrogen, Waltham, MA, USA). Gels were transferred to polyvinylidene difluoride (PDVF) membranes at 300 mA for 2 h in 1× transfer buffer containing 25 mM trizma base, 192 mM glycine, and 20% methanol. PVDF membranes were blocked for 1 h at room temperature with 5% milk in TBS-Tween prior to incubation with primary antibodies in 5% milk in TBS-Tween overnight at 4 °C. Primary antibodies were as follows: KRAS was a gift from Dr. Aaron Hobbs (Medical University of South Carolina) (WH0003845MI, Sigma-Aldrich, St. Louis, MO, USA), β-Actin (A5441, Sigma-Aldrich, St. Louis, MO, USA), IκBα (9242, Cell Signaling Technology, Danvers, MA, USA), α-Tubulin (3873, Cell Signaling Technology, Danvers, MA, USA), NF-κB p65 (sc-8008, Santa Cruz Biotechnology, Dallas, TX, USA) Phospho-NF-κB p65 (3033, Cell Signaling Technology, Danvers, MA, USA), Lamin A/C (2032 Cell Signaling Technology, Danvers, MA, USA), ERK1/2 (4695 Cell Signaling Technology, Danvers, MA, USA), Phospho-ERK1/2 (4377 Cell Signaling Technology, Danvers, MA, USA), AKT (pan) (4685 Cell Signaling Technology, Danvers, MA, USA), and Phospho-AKT (4060 Cell Signaling Technology, Danvers, MA, USA). Membranes were washed with 1× TBS-Tween and subsequently incubated with secondary antibodies in 5% milk in TBS-Tween for 2 h at room temperature. Secondary antibodies conjugated to horseradish peroxidase were goat anti-mouse IgG-H+L (31430, Invitrogen, Waltham, MA, USA) and goat anti-rabbit IgG-H+L. Detection was finalized using ECL (T32209, Thermo Fisher Scientific, Waltham,

MA, USA) or Super Signal (34094, Thermo Fisher Scientific, Waltham, MA, USA) detection reagents.

2.5. Migration Assays

CM for migration assays was obtained by plating 10×10^6 CFPAC-1 or PANC-1 cells in 10 mL of fully supplemented medium in 10 cm^2 petri dishes, incubating for 48 h, and collecting the medium. THP-1 cells were collected and pelleted at $300\times g$ for 5 min. Cells were resuspended in fully supplemented RPMI-1640, and 5×10^5 cells were added in 0.5 mL with or without treatment to ThinCert Cell Culture Inserts (24 well, 8 μm pore size) (662638, Greiner Bio-One, Kremsmünster, Austria). Next, 0.75 mL of PDAC CM or fully supplemented DMEM was added the lower chamber in each well. Plates were allowed to incubate for 2 h at 37 °C with 5% CO_2. After incubation, the inserts were removed, and 0.25 mL of CellTiter-Glo (Promega, Madison, WI, USA) was added to each well. Plates were shaken and incubated at 37 °C for 10 min. After incubation, 0.2 mL of the mixture in each well was transferred to black 96-well plates in technical duplicate. Luminescence was recorded on a SpectraMax L Microplate Reader (Molecular Devices, San Jose, CA, USA) at 470 nm with a 1 s integration time.

2.6. Thermodynamic Stability Assay

The thermodynamic stability of Fc, CXCL1-Fc, and each CXCL1-Fc variant was determined using a Tycho NanoTemper (Munich, Germany). Proteins were diluted to 100 μg/mL in PBS and drawn into glass capillary tubes. The protein in the glass capillary tubes was then subjected to a thermal ramp, and the absorbance at 350 nm and 330 nm was monitored. The first derivative of the ratio of 350 nm/330 nm was then plotted as a function of temperature by the onboard software, where the peak was defined as the melting temperature (T_m) of the sample.

2.7. Aggregation Assay

Proteins were diluted to 86 μg/mL and analyzed using a PROTEOSTAT Protein aggregation assay following the manufacturer's instructions (Enzo, Cat #ENZ-51023-KP050). The plate was read using an excitation setting of 550 nm and an emission filter of 600 nm.

2.8. Production of Lentiviral Particles and Infection of K562 Cells

A total of 2.5×10^6 HEK-293T cells were plated in 10 mL of DMEM with 10% FBS. Cells were infected with 15 μg pLJM1 CXCR2 DNA, 7.5 μg pCMV-dR8.91, and 1.5 μg pCMV-VSV-G diluted in Opti-MEM and Lipofectamine 2000. Seventy-two hours post-transfection, the supernatant containing viral particles was collected and stored at −80 °C. A total of 5×10^5 K562 cells were infected with 2 mL of supernatant and 1 mL of RPMI with 10% FBS with 8 μg/mL polybrene. Cells were treated with 2 μg/mL puromycin (A1113803, Thermo Fisher Scientific, Waltham, MA, USA) for selection of infected cells. Experiments were performed once stable expression was confirmed by flow cytometry.

2.9. Flow Cytometry

Cells were collected and washed with ice-cold FACS buffer (PBS with 1% FBS and 150 μM $CaCl_2$). Fc receptors were blocked using Human SeroBlock (BUF070, Bio-Rad Laboratories, Hercules, CA, USA) or Mouse SeroBlock (BUF041 Bio-Rad Laboratories, Hercules, CA, USA) according to the manufacturer's instructions. Cells were then treated with fluorophore-conjugated antibodies at dilutions indicated by the manufacturers and incubated at 4 °C for 30 min. Cells were then washed with 1 mL of FACS buffer and resuspended in 300 μL of FACS buffer and analyzed with either the FITC or PE channel or both of a NovoCyte flow cytometer (Acea Biosciences; San Diego, CA, USA). Antibodies used for flow cytometry were CD11b-FITC (24442, Cell Signaling Technology, Danvers, MA, USA), Rat IgG2b, κ-FITC (400634, BioLegend, San Diego, CA, USA), GR-1-PE (78355S, Cell Signaling Technology, Danvers, MA, USA), Rat IgG2b-PE (27426, Cell Signaling Technology, Danvers, MA, USA) mouse CXCR2-FITC (149310, BioLegend, San Diego, CA, USA), Rat IgG2a, κ-FITC (553929, BD Pharmingen, Franklin Lakes, NJ, USA), His-Tag-Alexa Fluor 488 (14930 Cell Signaling Technology, Danvers, MA, USA), human CXCR2-PE (12-1892-42, Thermo Fisher Scientific, Waltham, MA, USA), and Mouse IgG1, κ-PE (400114, BioLegend, San Diego, CA, USA).

2.10. MDSC Isolation from Mouse Spleens

C57BL/6 mice were injected subcutaneously with 100,000 KPC908 cells in 0.1 mL of a 1:1 mixture of DMEM and Matrigel (354234, Corning, Corning, NY, USA) or a mixture with no cells as a control. Tumors were implanted in the left and right flanks of each mouse for a total of two tumors per mouse. Spleens were harvested from the C57BL/6 mice once tumors were near 2 cm^3. Spleens were disrupted in PBS containing 2% FBS by the rubber head of a syringe plunger and passed through a 70 μm mesh nylon strainer (542070 Greiner Bio-One, Kremsmünster, Austria). Cells were then centrifuged at 300× g for 10 min and resuspended in PBS containing 2% FBS and 1 mM EDTA (J15694.AE Thermo Fisher Scientific, Waltham, MA, USA). Mouse MDSCs were then isolated by negative selection from the single-cell suspension using an EasySep™ Mouse MDSC (CD11b+Gr1+) Isolation Kit (STEMCELL Technologies; Vancouver, BC, Canada). Proper isolation was verified by flow cytometry. Mice were handled according to the Medical University of South Carolina (MUSC) institutional animal care and use committee (IACUC).

2.11. CyTOF

To comprehensively characterize immune cell populations within the KPC tumor microenvironment, mass cytometry by time-of-flight (CyTOF) was performed on barcoded CD45+ immune cells isolated from subcutaneous tumors. Single-cell suspensions were prepared and stained with a 20+ metal-tagged antibody panel designed to distinguish key immune subsets, including myeloid-derived suppressor cells (MDSCs), tumor-associated macrophages (TAMs), neutrophils, and T cells. Cells were then fixed, permeabilized for nuclear staining, and acquired on a Helios™ mass cytometer (Standard Biotools, South San Francisco, CA, USA). Data were normalized using bead-based calibration and analyzed using the CATALYST workflow [35,36]. Dimensionality reduction techniques, including UMAP, were applied to visualize immune cell clustering and heterogeneity.

3. Results

3.1. CXCR2 Ligands Are the Most Abundant Factors Secreted by PDAC Cells

The PDAC tumor microenvironment is composed of multiple immune cell types and a dense network of stromal fibroblasts and proteins [9,37,38]. The formation of this desmoplastic stroma is coordinated by a host of signaling molecules including cytokines, growth factors, and chemokines that are released from tumor cells, stromal fibroblasts, and infiltrating immune cells [13,39]. To identify the secreted factors produced solely by PDAC tumor cells, we screened the secretome of a genetically diverse panel of human PDAC cells. We measured the relative concentration of 105 soluble factors in the conditioned medium of four cell lines (PANC-1, BxPC3, Capan1, and CFPAC) using an antibody-based multiplex screening strategy. We found the highest levels of CXCL8/IL-8, CXCL5/ENA-78, GDF-15, VEGF, and CXCL1/GROα to be present in the conditioned media (CM) of all four cell lines examined (Figure 1A,B and Figure S1). Interestingly, three of the five secreted factors that were detected in all four cell lines were CXCR2 ligands (CXCL1, CXCL5, and CXCL8). In addition to these five, we observed a high expression of several other potentially relevant molecules in CM from two out of four or three out of four but not all of the cell lines tested. These included angiogenin, lipocalin-2, serpinE1, Dkk1, uPAR, and IL-17a (Figure 1B). Given the high levels and consistent expression of CXCR2 ligands in PDAC cells, we focused on these chemokines and confirmed the initial screening findings by quantitative ELISA (Figure 1C). Lower levels of all three were detected in immortalized normal pancreatic epithelial cells (HPNE) and HEK cells, demonstrating that the release of CXCR2 ligands is related to the specific genotype and phenotype of PDAC cells rather than a consequence of tissue culture conditions or methods (e.g., serum, media supplements, etc.). Elevated levels of CXCL8 were detected in Capan-1, CFPAC-1, and PANC-1; CXCL5 in Capan-1 and CFPAC-1; and CXCL1 in Capan-1, CFPAC-1, and PANC-1. Mouse-specific CXCR2 ligands, CXCL5 and CXCL1 (mice lack the CXCL8 gene or a human homolog), were also abundant in the CM from murine PDAC cell lines including Panc02 cells and two cell lines derived from PDAC tumors in Kras+/LSL-G12D;Trp53+/LSL-R172H;Pdx-1-Cre (KPC) mice (Figure 1D), further supporting the observation that CXCR2 ligand secretion is a characteristic of the PDAC phenotype.

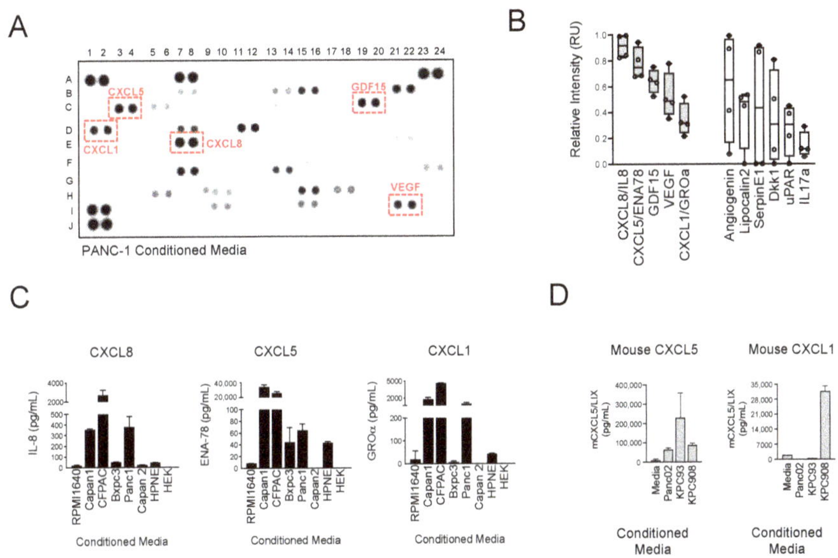

Figure 1. CXCR2 ligands are the most abundant soluble factors secreted by PDAC cells. (**A**) Cytokine arrays were used to identify relative concentrations of 105 secreted factors from four PDAC cell lines (BxPC-3, Capan-1, CFPAC-1, and PANC-1). Concentrations were determined relative to reference spots (A1,2,23,24 and J1,2). A representative cytokine array image using PANC-1 CM is shown. (**B**) Quantification of secreted factors that were upregulated in four out of four (gray) or three out of four (white) cell lines are shown. (**C**) CM from PDAC cell lines, HEK cells, and HPNE cells was analyzed by ELISA for CXCL8, CXCL5, and CXCL1 (n = 2–10). (**D**) CM from cell lines established from PDAC mouse models Panc02 and KPC was analyzed by ELISA for mouse CXCL5 and mouse CXCL1 (n = 2).

3.2. CXCR2 Ligand Production in PDAC Cells Is Induced by Mutant KRAS and NFκB Signaling

KRAS mutations are one of the most common genomic abnormalities in PDAC and are known to be driver events [40]. The cell autonomous effects of KRAS mutations have been extensively studied, but their impact effects on the secretome are less understood. Therefore, we next conducted experiments to determine if KRAS mutations also contribute to heightened CXCR2 ligand secretion by PDAC. We first used a model of immortalized normal human pancreatic epithelial (HPNE) cells, which were engineered with a doxycycline (DOX)-inducible system to express KRAS G12D and G12R mutants (Figure 2A). CM from HPNE cultures revealed that the KRAS mutations led to significant increases in CXCL8 and CXCL5 secretion but not to changes in CXCL1 production (Figure 2B). We note that the absolute magnitude of CXCL5 expression induced by KRAS mutations was 10-fold lower than the levels detected in human PDAC cell lines like Capan-1 (75–100 pg/mL versus ~1000 pg/mL). In human PDAC cell lines, which harbor KRAS mutations as well as a host of other genomic alterations, we found that treatment with the KRAS G12D mutant specific inhibitor, MRTX1133, significantly suppressed CXCR2 ligand secretion in G12D mutant cells (PANC-1 [41]; Figure 2C). In support of the KRAS G12D specificity of MRTX1133, the drug had no effect on KRAS G12V mutant cells (CFPAC-1 [42]; Figure 2C). These findings partially implicate mutant KRAS signaling in the induction and secretion of CXCR2 ligands, with the caveat that KRAS mutations contribute to, but are not sufficient to, induce all CXCR2 ligands.

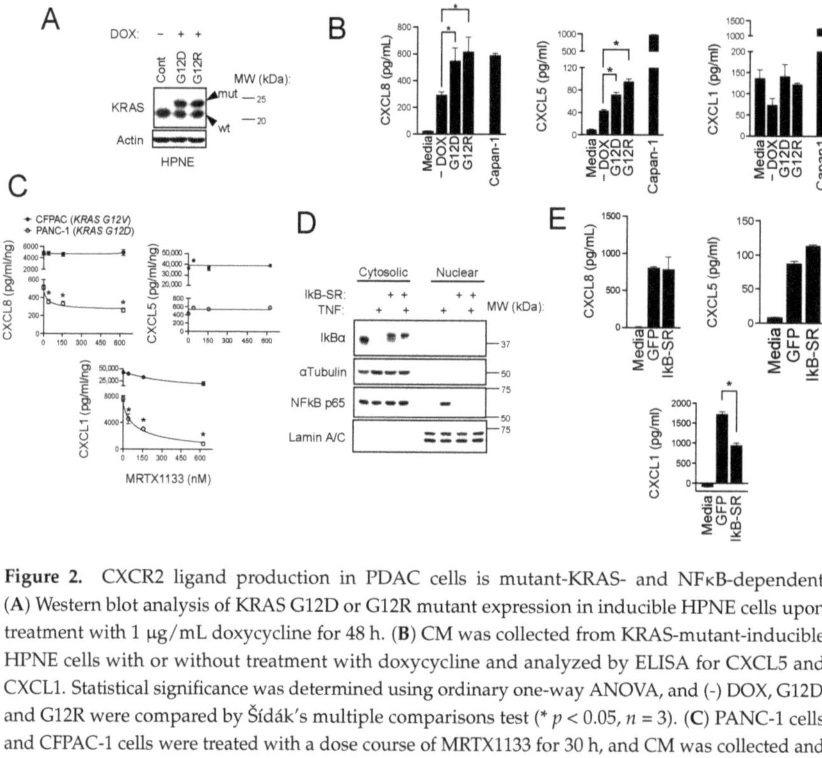

Figure 2. CXCR2 ligand production in PDAC cells is mutant-KRAS- and NFκB-dependent. (**A**) Western blot analysis of KRAS G12D or G12R mutant expression in inducible HPNE cells upon treatment with 1 μg/mL doxycycline for 48 h. (**B**) CM was collected from KRAS-mutant-inducible HPNE cells with or without treatment with doxycycline and analyzed by ELISA for CXCL5 and CXCL1. Statistical significance was determined using ordinary one-way ANOVA, and (-) DOX, G12D, and G12R were compared by Šídák's multiple comparisons test (* $p < 0.05$, $n = 3$). (**C**) PANC-1 cells and CFPAC-1 cells were treated with a dose course of MRTX1133 for 30 h, and CM was collected and analyzed by ELISA for CXCL8 and CXCL1. Statistical significance was determined using Student's *t*-test (* $p < 0.05$, $n = 2$). (**D**) PANC-1 cells expressing IκB-SR were treated with 10 ng/mL TNFα for 30 min. Cells were lysed and separated into the cytosolic and nuclear fractions. Western blotting analysis is shown. (**E**) CM was collected from PANC-1 cells expressing GFP or IκB-SR and analyzed by ELISA for CXCL1 and CXCL5. Statistical significance was determined using ordinary one-way ANOVA, and GFP and IkB-SR were compared by Šídák's multiple comparisons test (* $p < 0.05$, $n = 3$). Original images of (**A**,**D**) can be found in Supplementary Materials.

We also investigated the role of nuclear factor kappa light chain enhancer of activated B cells (NF-κB) in CXCR2 ligand production. NF-κB is a linchpin regulator of pro-inflammatory cytokine and chemokine production [43,44], and dysregulation of NF-κB signaling has been reported in approximately 70% of PDAC cases [45,46]. We explored the role of NF-κB by overexpressing an IκB super repressor (IκB-SR), which is a degradation-resistant form of IκBα. Activation of NFκB is set in motion by cytokines such as TNFα that induce IκBα degradation and the release and nuclear translocation of NFκB transcription factor subunits. IκB-SR is stable following TNFα treatment and prevents NFκB nuclear translocation by binding and sequestering it in the cytoplasm (Figure 2D). In PANC-1 cells, we found that IκB-SR significantly reduced CXCL1 levels in CM (Figure 2E) but had no effect on CXCL8 or CXCL5 production and secretion. Similar to what we observed with KRAS mutations, these findings suggest a multifactorial regulation of CXCR2 chemokines by a variety of factors and parallel signaling pathways. Our results show that these include mutant KRAS and NFκB signaling, which are some of the most commonly deregulated pathways in PDAC.

3.3. CXCR2 Is Part of the Immunophenotype of Myeloid-Derived Suppressor Cells (MDSCs) in Mice with PDAC

We next analyzed the cellular components of the PDAC tumor immune microenvironment in KPC mice. To achieve this, we conducted CyTOF on CD45-positive immune cells isolated from subcutaneous tumors established from a KPC clone. The vast majority (>80%, N = 4) of infiltrating immune cells were myeloid in origin (Figure 3A,B), as indicated by T cell-negative (CD4, CD8), CD45+, Ly6C+, and CD11b+ cells with varying levels of F4/80, Ly6G, CD11c, and MHC-II (Figure S2). Three distinct myeloid subsets were defined based on differential marker expression: Myeloid 1 cells exhibited low CD11b, Ly6G, CD11c, and f4/80 with intermediate Ly6C and high MHC-II, consistent with inflammatory monocytes or immature monocytic cells; Myeloid 2 cells expressed low levels of Ly6C and MHC-II with intermediate Ly6G, CD11c, CD11b, and F4/80, indicative of tumor-associated macrophages (TAMs) or differentiated macrophages; and Myeloid 3 cells demonstrated high CD11c, MHC-II, CD11b, Ly6C, Ly6G, and F4/80 consistent with antigen-presenting dendritic-like cells. The most abundant of those myeloid cell clusters (Myeloid 3) corresponded to an immunophenotype consistent with MDSCs [CD11b- and GR1 (Ly6C/G)-positive] [47].

We confirmed the presence of tumor-infiltrating MDSCs in PDAC tumors growing in mice by analyzing CD11b+GR1+ cells by flow cytometry (Figure 3C). We also analyzed the spleens from tumor-naïve and tumor-bearing mice, and in line with previous reports by others [47], we found that spleens from PDAC-tumor-bearing mice contained a significantly higher portion of MDSCs (Figure 3D). Importantly, splenic MDSCs also expressed CXCR2 (Figure 3E). Altogether, these findings implicate CXCR2 and its ligands in the high density of MDSCs found in the PDAC TME.

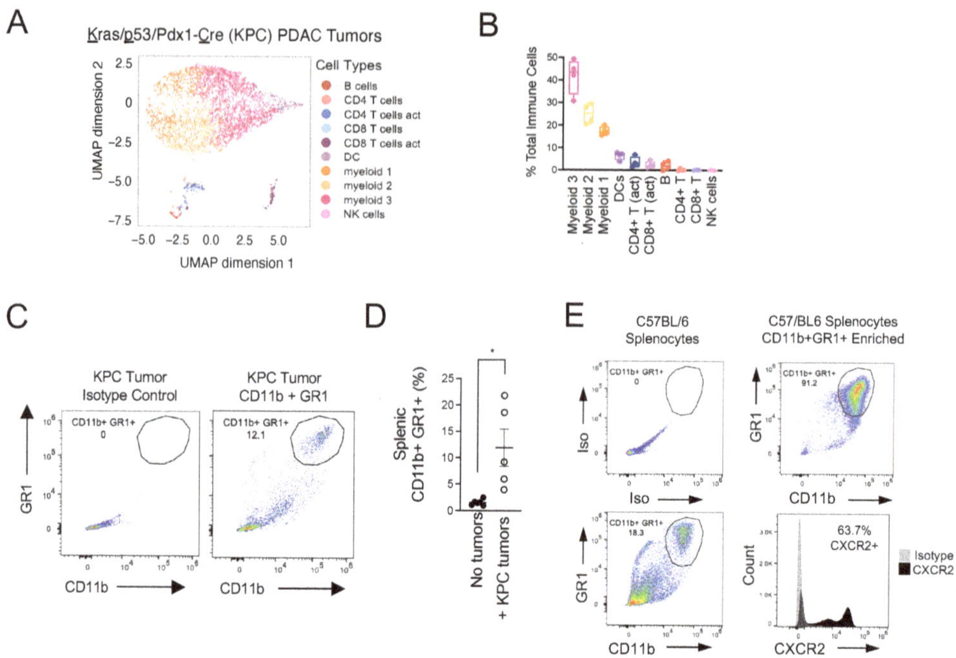

Figure 3. MDSCs in mice with PDAC are CXCR2-positive. (**A**) C57BL/6 mice were injected subcutaneously with KPC cells, and tumors and spleens were harvested. The composition of tumor-infiltrating immune cells was analyzed by CyTOF. (**B**) Quantification of each cell type in (**A**) is shown. (**C**) Single-cell suspensions were prepared from harvested KPC tumors and were analyzed by flow cytometry for tumor infiltrating MDSCs (CD11b+/GR1+). (**D**) Spleens from healthy mice (No tumors) and mice bearing subcutaneous KPC tumors (+KPC tumors) were analyzed by flow cytometry for splenic MDSCs. Statistical significance was determined using Student's *t*-test (* $p < 0.05$, $n = 6$ [No tumors], $n = 5$ [+KPC tumors]). (**E**) MDSCs were isolated from spleens by negative selection and analyzed by flow cytometry for CD11b, GR1, and CXCR2.

3.4. Design and Characterization of CXCR2 Ligand Fusion Proteins

We next set out to engineer CXCR2-targeted biotherapeutic molecules that could modulate this pathway and disrupt myeloid cell trafficking to PDAC tumor sites. We designed, expressed, and purified CXCL5 and CXCL1 fusion proteins that consisted of the mature chemokines conjugated to the N-terminal hinge domain of an IgG1 fragment crystallizable (Fc) region. A flexible GGGGS linker was used, and the Fc domain was engineered with multiple site-specific mutations. These included (1) C220S, which eliminates crosslinking of Fc domains by removing the cysteine that is normally involved in an interchain disulfide bond with the light chain of IgG1, (2) N297G, which eliminates antibody-dependent cellular cytotoxicity (ADCC) by removing a glycosylation site that is required for FcγR binding [48], and (3) the combination of M428L and N434S, which increases serum half-life in vivo through enhanced affinity for neonatal Fc receptors (FcRn) [49]. We gave priority to CXCL5 and CXCL1 over CXCL8, since the former are selective CXCR2 ligands, whereas the latter binds with similar affinity for both CXCR1 and CXCR2 and therefore offered less specificity. The layouts of the proteins produced for this study are shown in the insert of Figure 4A. Analysis of the protein products by SDS-PAGE revealed that the Fc fusion proteins formed dimers, presumably due to Fc domain coupling (Figure 4A). The CXCL5-Fc and CXCL1-Fc fusion proteins showed greater binding than the native chemokines to THP-1 cells (Figure 4B), which are acute myeloid leukemia (AML) cells of myeloid origin and CXCR2-positive (Figure S3A). Similar results were observed with HL60 AML cells and primary human neutrophils (Figure 4C), which were CXCR2-positive (Figure S3B). Binding was CXCR2-dependent, as CXCL5-Fc bound to CXCR2-positive K562 human chronic myelogenous leukemia (CML) cells but not to isogenic CXCR2-negative K562 cells (Figure S4). Ultimately, CXCL1-Fc was chosen for further analysis due to its greater binding to THP-1 cells over a range of concentrations (Figure 4D). In line with superior cell surface binding of the chemokine-fusion proteins compared to the monomeric non-fusions, the Fc fusions also induced the most intense cell signaling in THP-1 cells. For example, CXCL1 induced weak and transient induction of the p42/44 MAPK (ERK) and Akt pathways, whereas the CXCL1-Fc super chemokine (dubbed "superkines" by others [50]) stimulated these pathways with greater intensity and for sustained time periods in both THP-1 and HL60 cells (Figure 4E). Similar results were observed in dose range experiments, where CXCL1-Fc generated stronger ERK signaling than the native CXCL1 across a range of concentrations (Figure 4F).

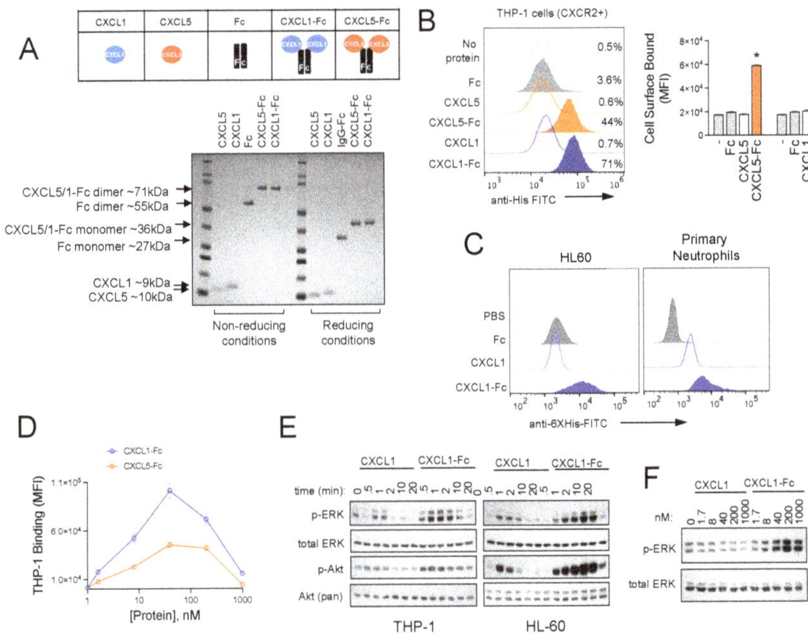

Figure 4. Design and characterization of CXCR2 ligand fusion proteins. (**A**) Purified CXCL1, CXCL5, Fc, and fusion proteins CXCL1-Fc and CXCL5-Fc were analyzed by SDS-PAGE. Proteins were separated under reducing and non-reducing conditions and visualized using Coomassie Blue staining. (**B**) THP-1 cells were incubated on ice with proteins for 30 min, and binding was analyzed by flow cytometry using a FITC-conjugated anti-His tag antibody. Statistical significance was determined using Student's *t*-test (* $p < 0.05$, $n = 3$). (**C**) HL60 and primary human neutrophils were incubated with proteins as in (**B**) and analyzed by flow cytometry, as in (**B**). (**D**) Binding of CXCL1-Fc and CXCL5-Fc to THP-1 cells was analyzed as in (**B**) across a range of concentrations. (**E**) THP-1 cells and HL60 cells were incubated with CXCL1 or CXCL1-Fc for the indicated times. Cell lysates were analyzed by Western blotting for p-ERK, total ERK, p-AKT, and total AKT (pan). (**F**) THP-1 cells were incubated with the indicated concentrations of CXCL1 or CXCL1-Fc for 5 min. Cell lysates were analyzed by Western blotting for p-ERK and total ERK. Original images of (**E**,**F**) can be found in Supplementary Materials.

3.5. CXCR2 Ligand Fusion Protein Disrupts Chemotaxis of Myeloid Cells

Previous studies have shown that knockout of the mouse homolog of CXCR2 (Cxcr2) reduces infiltration of MDSCs, neutrophils, and tumor-associated macrophages into mouse PDAC tumors [51,52]. CXCR1/2 small-molecule inhibitors were reported to reduce myeloid cell infiltration into tumors in mice [53,54]. However, others observed compensatory effects that counteract this response [55], which is perhaps not unexpected given the multifactorial composition of PDAC-secreted factors and the receptor/ligand promiscuity common among chemokines [56]. Given these challenges, we took an alternative approach to disrupting CXCR2 signaling. Rather than inhibiting CXCR2, we hypothesized that we could disrupt chemotactic gradients by hyperactivating the CXCR2 signaling axis with our ligand fusion superkine. Because the binding and signaling through these molecules was superior to the native chemokines, we reasoned that these signals would cancel out directional chemotactic signals. We first demonstrated that CM from multiple PDAC cell lines indeed induced migration of myeloid cells (THP-1) in a Boyden chamber transwell assay (Figure 5A). When CXCL1-F_c was added to the top well of the chamber with cells, it significantly reduced their ability to migrate towards the PDAC CM chemotactic gradient (Figure 5B). Interestingly, native CXCL1 had no effect on migration, further demonstrating the inferiority of the natural chemokine compared to the engineered fusion protein. Dose–response analyses were also conducted to quantify these differences. The migration Inhibitory Concentration 50 (IC50) for CXCL1-Fc was calculated at 8.5 nM compared to >1000 nM for CXCL1 and the Fc control protein (Figure 5C). In addition, we found that an anti-CXCR2 neutralizing antibody, like the native chemokine, failed to inhibit THP-1 myeloid cell migration toward PDAC CM. Further to this point, we found that multiple small-molecule CXCR2 inhibitors (SX-682, AZD5069, and Navarixin) also failed to inhibit THP-1 cell migration at concentrations that inhibited CXCL8- and CXCL5-induced signaling (Figure 5D and Figure S5). Together, these findings support our hypothesis that hyperactivation of the CXCR2 signaling axis with a CXCL1 fusion superkine is able to overcome chemotactic forces in PDAC CM, whereas CXCR2 inhibitors are not. We also found that inhibition by CXCL1-Fc could be neutralized by a CXCR2 antibody (Figure 5E). This demonstrates that there are CXCR2-independent chemotactic forces in PDAC CM, and while CXCR2 inhibition is not sufficient to inhibit migration, the effects of CXCL1-Fc remain dependent on binding and signaling through CXCR2.

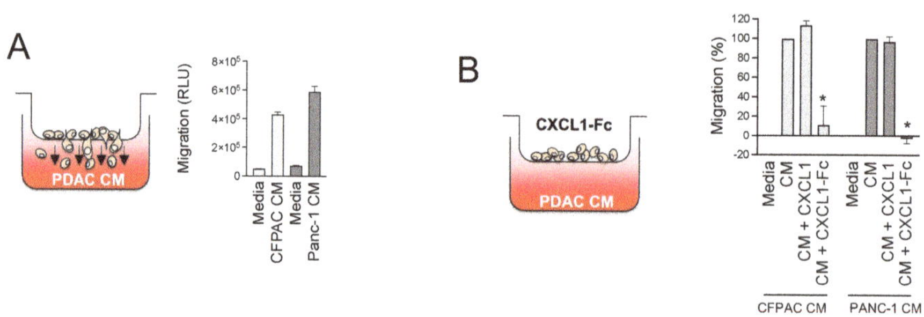

Figure 5. *Cont.*

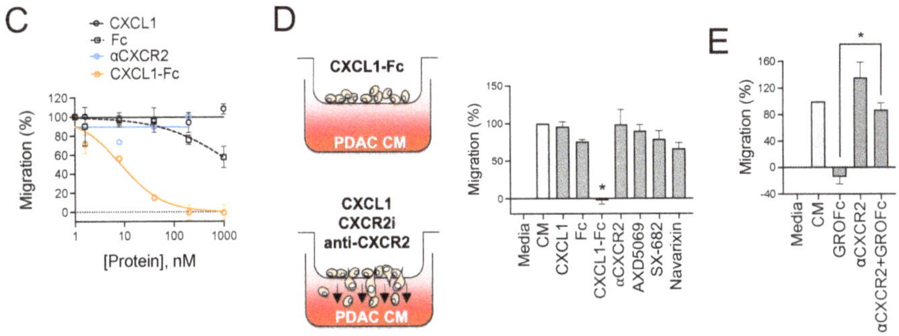

Figure 5. CXCL1-Fc superkine disrupts chemotaxis of myeloid cells. (**A**) Migration assays using a Boyden chamber design were performed to analyze the migration of myeloid cells (THP-1) towards PDAC CM. THP-1 myeloid cells were seeded in the top well of an 8 μm porous membrane insert. The bottom well contained PANC-1 CM, CFPAC-1 CM, or DMEM + 10% FBS as a control. Cells were incubated for 2 h, and migrated cells were quantified using a CellTiter Glo assay. Data are displayed as relative luminescence units (RLUs). (**B**) Either 100 nM CXCL1 or CXCL1-Fc was added to the top well with cells, and migration was analyzed as in (**A**). Data are shown as % migration normalized to untreated controls. Statistical significance was determined using a one-sample *t*-test against a hypothetical mean of 100% (* $p < 0.05$, $n = 3$). (**C**) Cells in the top well were treated with a dose range of CXCL1, Fc, anti-CXCR2 neutralizing antibody, or CXCL1-Fc. Migration towards PANC-1 CM was analyzed as in (**A**). (**D**) Cells in the top well were treated with 5 μM CXCR1/2 inhibitors AZD5069, SX-682, and Navarixin, and migration towards PANC-1 CM was analyzed as in (**C**). Statistical significance was determined using a one-sample *t*-test against a hypothetical mean of 100% (* $p < 0.05$, $n = 3$). (**E**) Cells in the top well were treated with CXCL1-Fc, anti-CXCR2 neutralizing antibody, or both, and migration was analyzed as in (**C**). Statistical significance was determined using ordinary one-way ANOVA, and CXCL1-Fc and CXCL1-Fc + αCXCR2 were compared by Šídák's multiple comparisons test (* $p < 0.05$, $n = 4$).

3.6. Structural Investigation of CXCL1-Fc

We initially constructed CXCR2 ligand Fc fusions to improve the stability of the chemokines and ultimately improve the pharmacokinetic properties, as reported by others [57]. However, having observed stark differences in binding, signaling, and cellular effects between the native and fusion proteins, we next asked why these differences existed. We considered two hypotheses. One was that the forced dimerization of the chemokine by virtue of Fc conjugation could enhance signaling in an unexpected non-linear fashion, or, secondly, the Fc domain of the fusion could impart enhanced signaling independent of the chemokine dimer. To test this, we made two structural variants of CXCL1-Fc by introducing knobs-into-holes technology, where complementary mutations (T366Y and Y407T) were made in the CH3 domain of the Fc region to prevent Fc homodimerization, as is typically carried out for bispecific antibody construction [58,59]. The first ($CXCL1^1$-Fc^2) consisted of a CXCL1-Fc monomer with a knob mutation matched to an Fc monomer with a hole mutation, forming an intact Fc dimer with only one CXCL1 protein per molecule (Figure 6A). The second ($CXCL1^1$-Fc^1) consisted of a CXCL1-Fc monomer with a knob mutation but no complementary hole mutation counterpart, resulting in a dimerization-deficient CXCL1-Fc with monomeric chemokine and Fc region (Figure 6A). With these constructs in hand, we were able to test the importance of CXCL1 bivalency ($CXCL1^1$-Fc^2) and Fc bivalency ($CXCL1^1$-Fc^1) in the superior activity of the CXCL1-Fc superkine. In the binding assays, CXCL1-Fc bound significantly more effectively than either of the variants with an approximately three- to four-fold higher mean fluorescence intensity per cell, with no significant difference in binding between the two monomeric CXCL1 variants. However, both variants bound better than CXCL1 alone (Figure 6B,C). Signaling experiments revealed that CXCL1-Fc was the most potent molecule of the three, as it induced sustained ERK signaling compared to the two variants, which were transient, peaking at 1–2 min and rapidly returning to baseline. Again, both variants were superior to CXCL1 alone (Figure 6D). However, we noted that the CXCL1-Fc knob/hole variant achieved a significantly higher signal amplitude than the CXCL1-Fc knob that was comparable to the full CXCL1-Fc protein. This suggested that the intact Fc dimer had an impact on the downstream signaling activity of the molecule even though it did not necessarily increase its binding properties. Finally, the migration experiments revealed that CXCL1-Fc was a superior inhibitor of myeloid cell migration toward a PDAC CM gradient (Figure 6E), as it had a significantly lower IC50 compared to CXCL1^1-Fc^2 (8.5 nM vs. 28 nM). CXCL1^1-Fc^1 was the least effective inhibitor of myeloid cell migration, with an IC50 25-fold and 7.5-fold higher than CXCL1-Fc (8.5 nM vs. 211 nM) and CXCL1^1-Fc^2 (28 nM vs. 211 nM, Figure 6E). This demonstrates that fusion of the CXCL1 chemokine to an Fc domain, and to an Fc dimer in particular, is important for CXCL1-Fc-mediated inhibition of myeloid cell migration. To better understand the contribution of the Fc domain to the potency of the CXCL1-Fc superkine, we investigated the biochemical properties of the different protein variants. In the thermal stability studies, we found that the melting temperature (T_m) of CXCL1^1-Fc^1 was ~2 °C lower than CXCL1-Fc and CXCL1^1-Fc^2 (Figure S6A). Additionally, assessments of protein aggregation revealed that the CXCL1^1-Fc^1 variant was more prone to aggregation than the other proteins (Figure S6B). These results suggest that the inferiority of CXCL1^1-Fc^1 may be due to decreased stability from the loss of Fc dimerization. Taken together, these results demonstrate that fusion of CXCL1 to an Fc domain is sufficient to enhance CXCL1 binding and signaling, independent of Fc or CXCL1 dimerization, thereby enabling CXCL1-Fc to function as an inhibitor of myeloid cell migration.

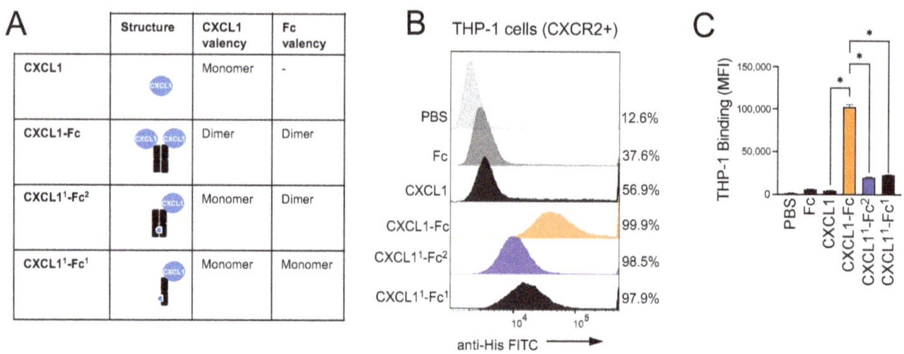

Figure 6. *Cont.*

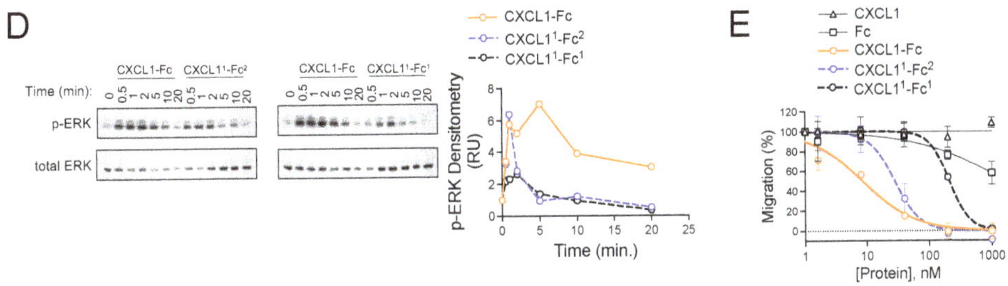

Figure 6. Fc dimerization is critical for the superkine activity of CXCL1-Fc. (**A**) Schematic and characteristics of the CXCL1-Fc variants are shown. Compared to the CXCL1-Fc superkine, which consists of two CXCL1 ligands and an Fc dimer, CXCL1^1-Fc^2 consisted of a CXCL1 monomer with an Fc dimer, and CXCL1^1-Fc^1 was composed of a CXCL1 monomer and an Fc monomer. (**B**) Binding of Fc, CXCL1, CXCL1-Fc, CXCL1^1-Fc^2, and CXCL1^1-Fc^1 to THP-1 cells was analyzed by flow cytometry using an FITC-conjugated anti-6X His antibody. (**C**) Mean fluorescence intensity of flow cytometry in (**B**) was quantified. Statistical significance was determined using ordinary one-way ANOVA, and CXCL1, CXCL1-Fc, CXCL1^1-Fc^2, and CXCL1^1-Fc^1 were compared by Šídák's multiple comparisons test (* $p < 0.05$, $n = 3$). Statistical tests not shown in (**C**) are CXCL1-Fc vs. CXCL1^1-Fc^2 ($p < 0.05$), CXCL1-Fc vs. CXCL1^1-Fc^1 ($p < 0.05$), and CXCL1^1-Fc^2 vs. CXCL1^1-Fc^1 ($p = 0.1163$). (**D**) THP-1 cells were incubated with CXCL1, CXCL1-Fc, CXCL1^1-Fc^2, or CXCL1^1-Fc^1 for the indicated times, ranging from 0.5–20 min. Downstream cell signaling events were analyzed by Western blotting using p-ERK as the read-out. Total ERK is included as a loading control. p-ERK signal was normalized to total ERK and plotted over time in an x-y plot. (**E**) Migration assays were performed to analyze the inhibition of THP-1 cell migration toward PDAC CM by CXCL1, Fc, CXCL1-Fc, CXCL1^1-Fc^2, and CXCL1^1-Fc^1 ($n = 2$). Original images of (**D**) can be found in Supplementary Materials.

4. Discussion

MDSCs are a major constituent of the PDAC TME and represent a significant obstacle to anti-tumor immunity and immunotherapies due to their suppressive effects on immune effector cell types. In our study, we identified CXCR2 ligands as the most abundant soluble factors secreted from PDAC cells and that CD11b+/GR1+ MDSCs express the receptor for these ligands, CXCR2. Additionally, we showed that mutant KRAS and NFκB signaling, which are two of the major aberrant signaling pathways in PDAC, contributed to the production and secretion of CXCR2 ligands from human and mouse PDAC cells. In addition to the production of these ligands from PDAC cells, previous studies have reported the secretion of CXCR2 ligands from cancer-associated fibroblasts (CAFs) [39,60] and tumor-associated macrophages (TAMs) [61] in the PDAC TME. Our data indicate that myeloid cells, and MDSCs specifically, were the most prominent immune cell type in KPC tumors. These MDSCs expressed CXCR2 and were present in the tumors and enriched in the spleens of these mice. Taken together, our data and previously published data strongly demonstrate that CXCR2 ligands are produced in high quantity from PDAC tumors and promote the recruitment and/or retention and differentiation of CXCR2+ MDSCs, which counteract anti-tumor immunity.

Several CXCR1/2 small-molecule inhibitors have shown efficacy in preclinical models of inflammatory diseases and in combination treatments for cancer [62,63]. However, their results in clinical trials have been mixed with few positive signals of efficacy [64,65]. The development of therapeutics that inhibit MDSCs or myeloid cell recruitment to PDAC tumors presents several challenges. Myeloid cell migration is driven by multiple chemokines and chemokine receptors. While CXCR2 ligands are the most prevalent soluble factors secreted by a genetically diverse set of PDAC cells, CXCR2 is not the only chemokine receptor that regulates myeloid cell migration. For example, work by others has shown that CC Chemokine Receptor 2 (CCR2) also mediates MDSC recruitment in PDAC [66]. Chemokine networks also show a high degree of binding promiscuity between ligands and receptors [56]. CXCL8, for example, which is one of the most prevalent PDAC ligands, binds to both CXCR2 and CXCR1, and CXCR2 is also a receptor for CXCL1, -2, -3, -5, and -7 [67]. Therefore, pharmacological inhibition of a single chemokine receptor like CXCR2 using a small molecule or neutralizing antibody is likely to be offset by compensatory recruitment of other chemokine receptor signals.

To address this challenge, we hypothesized that hyperactivating CXCR2, rather than inhibiting it, could circumvent the complexity and compensatory nature of chemokine networks by disrupting directional signaling created by chemokine gradients. This was made possible by the serendipitous discovery of the CXCL1-Fc superkine that demonstrated enhanced CXCR2 signaling properties. Chemokines induce cell migration by concentrating chemokine receptor signaling at the cellular pole facing a chemokine gradient. This enriched signaling biases cytoskeletal dynamics at the leading edge of the cell and propels the cell in the direction of the chemokine source [68]. We propose that the hyperactivation of CXCR2 by the CXCL1-Fc superkine neutralizes chemotactic signals originating from PDAC factors and inhibits migration by preventing cytoskeletal polarization and directional movement. Because of this CXCR2 hyperactivation effect, CXCL1-Fc could impair not just movement toward CXCR2 ligand gradients but also broadly inhibit chemotaxis toward chemokine gradients in general. In support of this hypothesis, we showed that CXCL1-Fc is the only molecule that could inhibit myeloid cell migration towards PDAC CM. CXCR1/2 small-molecule inhibitors were unable to block migration, as was a CXCR2-neutralizing antibody. The native CXCL1 chemokine was also unable to inhibit migration, demonstrating the importance of the Fc domain in the design of the fusion superkine.

Beyond PDAC, CXCL1-Fc may have broader therapeutic applications. Given its superior ability to inhibit myeloid cell migration compared to small-molecule inhibitors, it may be beneficial in treating inflammatory diseases. In addition, CXCL1-Fc may have applications in wound healing. The CXCR2 axis is critical for neutrophil recruitment and epithelial resurfacing mediated by keratinocytes in the wound healing process, and loss of CXCR2 expression delays wound healing [69]. Various strategies have been explored to enhance wound healing by modulating chemokine signaling. Notably, a topical therapy utilizing lactic acid bacteria engineered to express CXCL12 was recently found to accelerate wound healing in a phase 1 clinical trial [70]. Given that CXCL1-Fc exhibits superior CXCR2 activation compared to CXCL1 alone, it may represent an alternative or complementary strategy for enhancing neutrophil recruitment and keratinocyte migration during wound healing. Furthermore, other chemokines may benefit from Fc fusion conjugation, broadening the potential for therapeutic intervention across various conditions.

IgG Fc fusion proteins are widely used in biomedical applications to enhance stability, solubility, and functional activity in diverse biomedical applications [71]. In our study, we leveraged the IgG Fc region with the intent to improve pharmacokinetic properties, and in the process, we unexpectedly created a CXCL1 superkine with enhanced binding to CXCR2 and enhanced activation of CXCR2 downstream signaling effectors like ERK (p42/44 MAPK) and Akt. Our data show that this is not due simply to the forced dimerization of CXCL1, as a CXCL1-dimerization-deficient variant still demonstrated superior function. This finding suggests that the Fc region itself contributes to the enhanced activity of the fusion. Potential explanations for this enhanced function include Fc-mediated alteration of CXCL1 steric positioning, stabilization of a CXCL1 conformation that promotes CXCR2 activation, or enabling CXCL1 to behave as a partial agonist, selectively activating specific components of the CXCR2 signaling pathway while sparing others. Further studies are needed to elucidate the precise mechanism and to explore the translational potential of CXCL1-Fc in preclinical studies of PDAC and other disease models.

5. Conclusions

Collectively, our data implicate the CXCR2 signaling axis in the migration of myeloid cells toward soluble factors released by PDAC tumors (Figure 7A). CXCR2+ myeloid cells infiltrate PDAC tumors along a chemotactic gradient that is composed predominantly of CXCR2 ligands, including CXCL1, CXCL5, and CXCL8. These myeloid cells appear to have suppressor function (i.e., MDSCs) prior to entering the PDAC environment, as our data show that PDAC tumors increase MDSC numbers in secondary lymphoid organs such as the spleen. Once in the PDAC tumor, MDSCs suppress immune effector cell function and shield PDAC cells from host immune surveillance. Through design and evaluation of a CXCL1-Fc fusion "superkine" with unexpectedly potent signaling activity, we found that hyperstimulation of CXCR2 of MDSCs could completely block migration toward PDAC-secreted factors (Figure 7B). This approach was, somewhat paradoxically, more effective than inhibiting CXCR2 with small-molecule inhibitors or anti-CXCR2 antibodies. This suggests that there are additional chemotactic factors in PDAC CM that can compensate for CXCR2 inhibition and attract myeloid cells in the presence of the inhibitors. The CXCL1-Fc superkine, on the other hand, had the ability to drown out all other chemotactic gradients due to its superior signaling ability, and it completely inhibited myeloid cell migration.

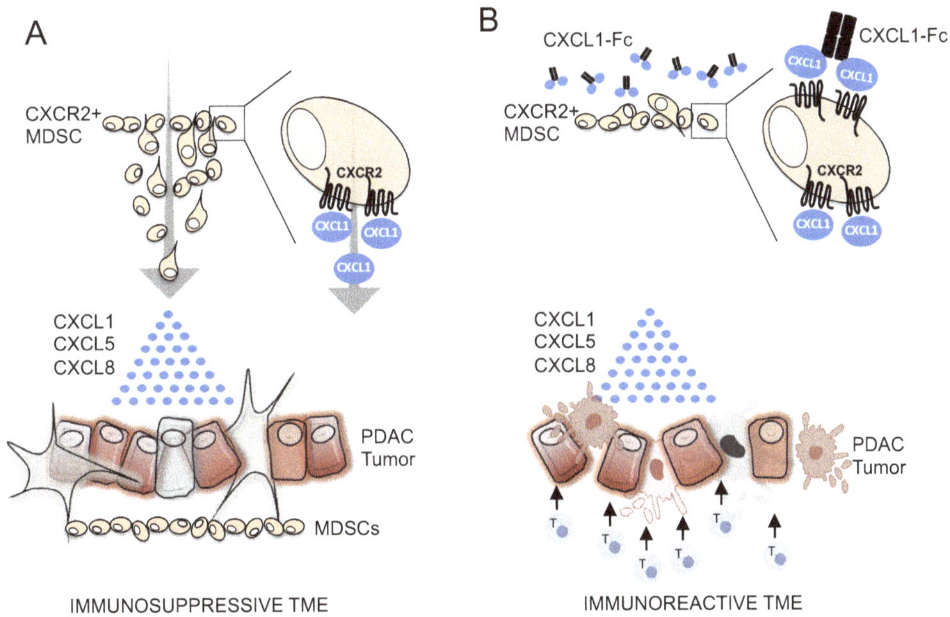

Figure 7. Model of CXCL1-Fc superkine inhibition of myeloid cell infiltration of PDAC tumors. (**A**) MDSCs migrate into PDAC tumors along chemotactic gradients composed primarily of CXCR2 ligands. In the PDAC TME, MDSCs suppress immune effector cells and protect tumors from host immune surveillance. (**B**) CXCL1-Fc superkine induces robust CXCR2 signaling and drowns out chemotactic signals from PDAC tumors and prevents MDSC infiltration.

Supplementary Materials: The following supporting information can be downloaded at: https://www.mdpi.com/article/10.3390/biom15050645/s1, Figure S1: soluble factors secreted by PDAC cell lines detected by cytokine array; Figure S2: Characterization of immune cell subsets based on immunophenotype in CyTOF analysis; Figure S3: CXCR2 is expressed on the surface of THP-1, HL60, and primary human neutrophils; Figure S4: CXCL5-Fc binds to CXCR2-positive K562 cells but not isogenic CXCR2-negative K562 cells; Figure S5: Inhibition of IL-8- and CXCL5-mediated activation of CXCR2 downstream effector ERK by CXCR2 small-molecule inhibitors SX682, Navarixin, and AZD5069; Figure S6: $CXCL1^1$-Fc^1 has a lower melting temperature and a higher susceptibility to aggregation than CXCL1-Fc and $CXCL1^1$-Fc^2. Figures S7–S12: Original Western blot images.

Author Contributions: Conceptualization, B.N.C. and N.G.D.; investigation, B.N.C., A.B. and N.G.D.; methodology, B.N.C., L.G., A.B., L.R., R.M.R., A.O.A., C.K., G.A.H., D.C.G., J.P.O. and N.G.D.; writing—original draft, B.N.C. and N.G.D.; writing—review and editing, B.N.C. and N.G.D. All authors have read and agreed to the published version of the manuscript.

Funding: This research was funded by the National Cancer Institute (NCI) of the National Institutes of Health (NIH), grant number R01CA245081 to N.G.D. and Hollings Cancer Center Team Science Award funding to N.G.D. and J.O.O.

Institutional Review Board Statement: Not applicable.

Informed Consent Statement: Not applicable.

Data Availability Statement: The original contributions presented in this study are included in the article. Further inquiries can be directed to the corresponding author.

Conflicts of Interest: The authors declare no conflicts of interest.

References

1. Society, A.C. *Cancer Facts & Figures 2024*; American Cancer Society: Atlanta, GA, USA, 2024.
2. Murphy, K.J.; Chambers, C.R.; Herrmann, D.; Timpson, P.; Pereira, B.A. Dynamic Stromal Alterations Influence Tumor-Stroma Crosstalk to Promote Pancreatic Cancer and Treatment Resistance. *Cancers* **2021**, *13*, 3481. [CrossRef] [PubMed]
3. Luchini, C.; Brosens, L.A.A.; Wood, L.D.; Chatterjee, D.; Shin, J.I.; Sciammarella, C.; Fiadone, G.; Malleo, G.; Salvia, R.; Kryklyva, V.; et al. Comprehensive characterisation of pancreatic ductal adenocarcinoma with microsatellite instability: Histology, molecular pathology and clinical implications. *Gut* **2021**, *70*, 148–156. [CrossRef] [PubMed]
4. Khaled, Y.S.; Ammori, B.J.; Elkord, E. Increased Levels of Granulocytic Myeloid-Derived Suppressor Cells in Peripheral Blood and Tumour Tissue of Pancreatic Cancer Patients. *J. Immunol. Res.* **2014**, *2014*, 879897. [CrossRef] [PubMed]
5. Liou, G.-Y.; Döppler, H.; Necela, B.; Edenfield, B.; Zhang, L.; Dawson, D.W.; Storz, P. Mutant KRAS–Induced Expression of ICAM-1 in Pancreatic Acinar Cells Causes Attraction of Macrophages to Expedite the Formation of Precancerous Lesions. *Cancer Discov.* **2015**, *5*, 52–63. [CrossRef]
6. Hiraoka, N.; Onozato, K.; Kosuge, T.; Hirohashi, S. Prevalence of FOXP3+ Regulatory T Cells Increases During the Progression of Pancreatic Ductal Adenocarcinoma and Its Premalignant Lesions. *Clin. Cancer Res.* **2006**, *12*, 5423–5434. [CrossRef]
7. Öhlund, D.; Handly-Santana, A.; Biffi, G.; Elyada, E.; Almeida, A.S.; Ponz-Sarvise, M.; Corbo, V.; Oni, T.E.; Hearn, S.A.; Lee, E.J.; et al. Distinct populations of inflammatory fibroblasts and myofibroblasts in pancreatic cancer. *J. Exp. Med.* **2017**, *214*, 579–596. [CrossRef]
8. Bailey, P.; Chang, D.K.; Forget, M.-A.; Lucas, F.A.S.; Alvarez, H.A.; Haymaker, C.; Chattopadhyay, C.; Kim, S.-H.; Ekmekcioglu, S.; Grimm, E.A.; et al. Exploiting the neoantigen landscape for immunotherapy of pancreatic ductal adenocarcinoma. *Sci. Rep.* **2016**, *6*, 35848. [CrossRef]
9. Ino, Y.; Yamazaki-Itoh, R.; Shimada, K.; Iwasaki, M.; Kosuge, T.; Kanai, Y.; Hiraoka, N. Immune cell infiltration as an indicator of the immune microenvironment of pancreatic cancer. *Br. J. Cancer* **2013**, *108*, 914–923. [CrossRef]
10. Lim, S.A.; Kim, J.; Jeon, S.; Shin, M.H.; Kwon, J.; Kim, T.-J.; Im, K.; Han, Y.; Kwon, W.; Kim, S.-W.; et al. Defective Localization With Impaired Tumor Cytotoxicity Contributes to the Immune Escape of NK Cells in Pancreatic Cancer Patients. *Front. Immunol.* **2019**, *10*, 496. [CrossRef]
11. Huang, L.; Guo, Z.; Wang, F.; Fu, L. KRAS mutation: From undruggable to druggable in cancer. *Signal Transduct. Target. Ther.* **2021**, *6*, 386. [CrossRef]
12. Väyrynen, S.A.; Zhang, J.; Yuan, C.; Väyrynen, J.P.; Dias Costa, A.; Williams, H.; Morales-Oyarvide, V.; Lau, M.C.; Rubinson, D.A.; Dunne, R.F.; et al. Composition, Spatial Characteristics, and Prognostic Significance of Myeloid Cell Infiltration in Pancreatic Cancer. *Clin. Cancer Res.* **2021**, *27*, 1069–1081. [CrossRef] [PubMed]

13. Bayne, L.J.; Beatty, G.L.; Jhala, N.; Clark, C.E.; Rhim, A.D.; Stanger, B.Z.; Vonderheide, R.H. Tumor-Derived Granulocyte-Macrophage Colony-Stimulating Factor Regulates Myeloid Inflammation and T Cell Immunity in Pancreatic Cancer. *Cancer Cell* **2012**, *21*, 822–835. [CrossRef] [PubMed]
14. Dvorak, H.F. Tumors: Wounds That Do Not Heal—Redux. *Cancer Immunol. Res.* **2015**, *3*, 1–11. [CrossRef] [PubMed]
15. Markowitz, J.; Brooks, T.R.; Duggan, M.C.; Paul, B.K.; Pan, X.; Wei, L.; Abrams, Z.; Luedke, E.; Lesinski, G.B.; Mundy-Bosse, B.; et al. Patients with pancreatic adenocarcinoma exhibit elevated levels of myeloid-derived suppressor cells upon progression of disease. *Cancer Immunol. Immunother.* **2015**, *64*, 149–159. [CrossRef]
16. Sharma, V.; Sachdeva, N.; Gupta, V.; Nada, R.; Jacob, J.; Sahni, D.; Aggarwal, A. IL-6 is associated with expansion of myeloid-derived suppressor cells and enhanced immunosuppression in pancreatic adenocarcinoma patients. *Scand. J. Immunol.* **2021**, *94*, e13107. [CrossRef]
17. Zhang, J.; Xu, X.; Shi, M.; Chen, Y.; Yu, D.; Zhao, C.; Gu, Y.; Yang, B.; Guo, S.; Ding, G.; et al. CD13^{hi} Neutrophil-like myeloid-derived suppressor cells exert immune suppression through Arginase 1 expression in pancreatic ductal adenocarcinoma. *OncoImmunology* **2017**, *6*, e1258504. [CrossRef]
18. Zea, A.H.; Rodriguez, P.C.; Culotta, K.S.; Hernandez, C.P.; DeSalvo, J.; Ochoa, J.B.; Park, H.-J.; Zabaleta, J.; Ochoa, A.C. l-Arginine modulates CD3ζ expression and T cell function in activated human T lymphocytes. *Cell. Immunol.* **2004**, *232*, 21–31. [CrossRef]
19. Corzo, C.A.; Cotter, M.J.; Cheng, P.; Cheng, F.; Kusmartsev, S.; Sotomayor, E.; Padhya, T.; McCaffrey, T.V.; McCaffrey, J.C.; Gabrilovich, D.I. Mechanism Regulating Reactive Oxygen Species in Tumor-Induced Myeloid-Derived Suppressor Cells. *J. Immunol.* **2009**, *182*, 5693–5701. [CrossRef]
20. Huang, J.; Zhao, Y.; Zhao, K.; Yin, K.; Wang, S. Function of reactive oxygen species in myeloid-derived suppressor cells. *Front. Immunol.* **2023**, *14*, 1226443. [CrossRef]
21. Hanson, E.M.; Clements, V.K.; Sinha, P.; Ilkovitch, D.; Ostrand-Rosenberg, S. Myeloid-Derived Suppressor Cells Down-Regulate L-Selectin Expression on CD4+ and CD8+ T Cells. *J. Immunol.* **2009**, *183*, 937–944. [CrossRef]
22. Li, H.; Han, Y.; Guo, Q.; Zhang, M.; Cao, X. Cancer-Expanded Myeloid-Derived Suppressor Cells Induce Anergy of NK Cells through Membrane-Bound TGF-β1. *J. Immunol.* **2009**, *182*, 240–249. [CrossRef] [PubMed]
23. Huang, B.; Pan, P.-Y.; Li, Q.; Sato, A.I.; Levy, D.E.; Bromberg, J.; Divino, C.M.; Chen, S.-H. Gr-1+CD115+ Immature Myeloid Suppressor Cells Mediate the Development of Tumor-Induced T Regulatory Cells and T-Cell Anergy in Tumor-Bearing Host. *Cancer Res.* **2006**, *66*, 1123–1131. [CrossRef] [PubMed]
24. Serafini, P.; Mgebroff, S.; Noonan, K.; Borrello, I. Myeloid-Derived Suppressor Cells Promote Cross-Tolerance in B-Cell Lymphoma by Expanding Regulatory T Cells. *Cancer Res.* **2008**, *68*, 5439–5449. [CrossRef]
25. Beury, D.W.; Parker, K.H.; Nyandjo, M.; Sinha, P.; Carter, K.A.; Ostrand-Rosenberg, S. Cross-talk among myeloid-derived suppressor cells, macrophages, and tumor cells impacts the inflammatory milieu of solid tumors. *J. Leukoc. Biol.* **2014**, *96*, 1109–1118. [CrossRef]
26. Olson, T.S.; Ley, K. Chemokines and chemokine receptors in leukocyte trafficking. *Am. J. Physiol.-Regul. Integr. Comp. Physiol.* **2002**, *283*, R7–R28. [CrossRef]
27. Wang, F. The Signaling Mechanisms Underlying Cell Polarity and Chemotaxis. *Cold Spring Harb. Perspect. Biol.* **2009**, *1*, a002980. [CrossRef]
28. De Filippo, K.; Dudeck, A.; Hasenberg, M.; Nye, E.; Van Rooijen, N.; Hartmann, K.; Gunzer, M.; Roers, A.; Hogg, N. Mast cell and macrophage chemokines CXCL1/CXCL2 control the early stage of neutrophil recruitment during tissue inflammation. *Blood* **2013**, *121*, 4930–4937. [CrossRef]
29. Highfill, S.L.; Cui, Y.; Giles, A.J.; Smith, J.P.; Zhang, H.; Morse, E.; Kaplan, R.N.; Mackall, C.L. Disruption of CXCR2-mediated MDSC tumor trafficking enhances anti-PD1 efficacy. *Sci. Transl. Med.* **2014**, *6*, 237ra67. [CrossRef]
30. Unverdorben, F.; Richter, F.; Hutt, M.; Seifert, O.; Malinge, P.; Fischer, N.; Kontermann, R.E. Pharmacokinetic properties of IgG and various Fc fusion proteins in mice. *mAbs* **2016**, *8*, 120–128. [CrossRef]
31. Czajkowsky, D.M.; Hu, J.; Shao, Z.; Pleass, R.J. Fc-fusion proteins: New developments and future perspectives. *EMBO Mol. Med.* **2012**, *4*, 1015–1028. [CrossRef]
32. Giordano, G.; Febbraro, A.; Venditti, M.; Campidoglio, S.; Olivieri, N.; Raieta, K.; Parcesepe, P.; Imbriani, G.C.; Remo, A.; Pancione, M. Targeting Angiogenesis and Tumor Microenvironment in Metastatic Colorectal Cancer: Role of Aflibercept. *Gastroenterol. Res. Pract.* **2014**, *2014*, 526178. [CrossRef] [PubMed]
33. Shirmohammadi, E.; Ebrahimi, S.-E.S.; Farshchi, A.; Salimi, M. The efficacy of etanercept as anti-breast cancer treatment is attenuated by residing macrophages. *BMC Cancer* **2020**, *20*, 836. [CrossRef]
34. Tan, J.K.; Aphale, A.; Malaviya, R.; Sun, Y.; Gottlieb, A.B. Mechanisms of Action of Etanercept in Psoriasis. *J. Investig. Dermatol. Symp. Proc.* **2007**, *12*, 38–45. [CrossRef]
35. Nowicka, M.; Krieg, C.; Crowell, H.L.; Weber, L.M.; Hartmann, F.J.; Guglietta, S.; Becher, B.; Levesque, M.P.; Robinson, M.D. CyTOF workflow: Differential discovery in high-throughput high-dimensional cytometry datasets. *F1000Research* **2017**, *6*, 748. [CrossRef]

36. Krieg, C.; Nowicka, M.; Guglietta, S.; Schindler, S.; Hartmann, F.J.; Weber, L.M.; Dummer, R.; Robinson, M.D.; Levesque, M.P.; Becher, B. Author Correction: High-dimensional single-cell analysis predicts response to anti-PD-1 immunotherapy. *Nat. Med.* **2018**, *24*, 1773–1775. [CrossRef]
37. Feig, C.; Gopinathan, A.; Neesse, A.; Chan, D.S.; Cook, N.; Tuveson, D.A. The Pancreas Cancer Microenvironment. *Clin. Cancer Res.* **2012**, *18*, 4266–4276. [CrossRef]
38. Clark, C.E.; Hingorani, S.R.; Mick, R.; Combs, C.; Tuveson, D.A.; Vonderheide, R.H. Dynamics of the Immune Reaction to Pancreatic Cancer from Inception to Invasion. *Cancer Res.* **2007**, *67*, 9518–9527. [CrossRef]
39. Velez-Delgado, A.; Donahue, K.L.; Brown, K.L.; Du, W.; Irizarry-Negron, V.; Menjivar, R.E.; Lasse Opsahl, E.L.; Steele, N.G.; The, S.; Lazarus, J.; et al. Extrinsic KRAS Signaling Shapes the Pancreatic Microenvironment Through Fibroblast Reprogramming. *Cell. Mol. Gastroenterol. Hepatol.* **2022**, *13*, 1673–1699. [CrossRef]
40. Ying, H.; Kimmelman, A.C.; Lyssiotis, C.A.; Hua, S.; Chu, G.C.; Fletcher-Sananikone, E.; Locasale, J.W.; Son, J.; Zhang, H.; Coloff, J.L.; et al. Oncogenic Kras maintains pancreatic tumors through regulation of anabolic glucose metabolism. *Cell* **2012**, *149*, 656–670. [CrossRef]
41. Loukopoulos, P.; Kanetaka, K.; Takamura, M.; Shibata, T.; Sakamoto, M.; Hirohashi, S. Orthotopic transplantation models of pancreatic adenocarcinoma derived from cell lines and primary tumors and displaying varying metastatic activity. *Pancreas* **2004**, *29*, 193–203. [CrossRef]
42. Moore, P.S.; Sipos, B.; Orlandini, S.; Sorio, C.; Real, F.X.; Lemoine, N.R.; Gress, T.; Bassi, C.; Kloppel, G.; Kalthoff, H.; et al. Genetic profile of 22 pancreatic carcinoma cell lines. Analysis of K-ras, p53, p16 and DPC4/Smad4. *Virchows Arch.* **2001**, *439*, 798–802. [CrossRef] [PubMed]
43. Baldwin, A.S., Jr. The NF-kappa B and I kappa B proteins: New discoveries and insights. *Annu. Rev. Immunol.* **1996**, *14*, 649–683. [CrossRef] [PubMed]
44. Richmond, A. Nf-kappa B, chemokine gene transcription and tumour growth. *Nat. Rev. Immunol.* **2002**, *2*, 664–674. [CrossRef] [PubMed]
45. Wang, W.; Abbruzzese, J.L.; Evans, D.B.; Larry, L.; Cleary, K.R.; Chiao, P.J. The nuclear factor-kappa B RelA transcription factor is constitutively activated in human pancreatic adenocarcinoma cells. *Clin. Cancer Res.* **1999**, *5*, 119–127.
46. Fujioka, S.; Sclabas, G.M.; Schmidt, C.; Niu, J.; Frederick, W.A.; Dong, Q.G.; Abbruzzese, J.L.; Evans, D.B.; Baker, C.; Chiao, P.J. Inhibition of constitutive NF-kappa B activity by I kappa B alpha M suppresses tumorigenesis. *Oncogene* **2003**, *22*, 1365–1370. [CrossRef]
47. Youn, J.-I.; Nagaraj, S.; Collazo, M.; Gabrilovich, D.I. Subsets of Myeloid-Derived Suppressor Cells in Tumor-Bearing Mice. *J. Immunol.* **2008**, *181*, 5791–5802. [CrossRef]
48. Leabman, M.K.; Meng, Y.G.; Kelley, R.F.; Deforge, L.E.; Cowan, K.J.; Iyer, S. Effects of altered FcγR binding on antibody pharmacokinetics in cynomolgus monkeys. *mAbs* **2013**, *5*, 896–903. [CrossRef]
49. Zalevsky, J.; Chamberlain, A.K.; Horton, H.M.; Karki, S.; Leung, I.W.L.; Sproule, T.J.; Lazar, G.A.; Roopenian, D.C.; Desjarlais, J.R. Enhanced antibody half-life improves in vivo activity. *Nat. Biotechnol.* **2010**, *28*, 157–159. [CrossRef]
50. Levin, A.M.; Bates, D.L.; Ring, A.M.; Krieg, C.; Lin, J.T.; Su, L.; Moraga, I.; Raeber, M.E.; Bowman, G.R.; Novick, P.; et al. Exploiting a natural conformational switch to engineer an interleukin-2 'superkine'. *Nature* **2012**, *484*, 529–533. [CrossRef]
51. Sano, M.; Ijichi, H.; Takahashi, R.; Miyabayashi, K.; Fujiwara, H.; Yamada, T.; Kato, H.; Nakatsuka, T.; Tanaka, Y.; Tateishi, K.; et al. Blocking CXCLs-CXCR2 axis in tumor-stromal interactions contributes to survival in a mouse model of pancreatic ductal adenocarcinoma through reduced cell invasion/migration and a shift of immune-inflammatory microenvironment. *Oncogenesis* **2019**, *8*, 8. [CrossRef]
52. Chao, T.; Furth, E.E.; Vonderheide, R.H. CXCR2-Dependent Accumulation of Tumor-Associated Neutrophils Regulates T-cell Immunity in Pancreatic Ductal Adenocarcinoma. *Cancer Immunol. Res.* **2016**, *4*, 968–982. [CrossRef] [PubMed]
53. Prajapati, D.R.; Molczyk, C.; Purohit, A.; Saxena, S.; Sturgeon, R.; Dave, B.J.; Kumar, S.; Batra, S.K.; Singh, R.K. Small molecule antagonist of CXCR2 and CXCR1 inhibits tumor growth, angiogenesis, and metastasis in pancreatic cancer. *Cancer Lett.* **2023**, *563*, 216185. [CrossRef] [PubMed]
54. Yang, J.; Yan, C.; Vilgelm, A.E.; Chen, S.C.; Ayers, G.D.; Johnson, C.A.; Richmond, A. Targeted Deletion of CXCR2 in Myeloid Cells Alters the Tumor Immune Environment to Improve Antitumor Immunity. *Cancer Immunol. Res.* **2021**, *9*, 200–213. [CrossRef] [PubMed]
55. Nywening, T.M.; Belt, B.A.; Cullinan, D.R.; Panni, R.Z.; Han, B.J.; Sanford, D.E.; Jacobs, R.C.; Ye, J.; Patel, A.A.; Gillanders, W.E.; et al. Targeting both tumour-associated CXCR2(+) neutrophils and CCR2(+) macrophages disrupts myeloid recruitment and improves chemotherapeutic responses in pancreatic ductal adenocarcinoma. *Gut* **2018**, *67*, 1112–1123. [CrossRef]
56. Hughes, C.E.; Nibbs, R.J.B. A guide to chemokines and their receptors. *FEBS J.* **2018**, *285*, 2944–2971. [CrossRef]
57. Jazayeri, J.A.; Carroll, G.J. Fc-based cytokines: Prospects for engineering superior therapeutics. *BioDrugs* **2008**, *22*, 11–26. [CrossRef]

58. Merchant, A.M.; Zhu, Z.; Yuan, J.Q.; Goddard, A.; Adams, C.W.; Presta, L.G.; Carter, P. An efficient route to human bispecific IgG. *Nat. Biotechnol.* **1998**, *16*, 677–681. [CrossRef]
59. Ridgway, J.B.; Presta, L.G.; Carter, P. 'Knobs-into-holes' engineering of antibody CH3 domains for heavy chain heterodimerization. *Protein Eng.* **1996**, *9*, 617–621. [CrossRef]
60. Awaji, M.; Saxena, S.; Wu, L.; Prajapati, D.R.; Purohit, A.; Varney, M.L.; Kumar, S.; Rachagani, S.; Ly, Q.P.; Jain, M.; et al. CXCR2 signaling promotes secretory cancer-associated fibroblasts in pancreatic ductal adenocarcinoma. *FASEB J.* **2020**, *34*, 9405–9418. [CrossRef]
61. Caronni, N.; La Terza, F.; Vittoria, F.M.; Barbiera, G.; Mezzanzanica, L.; Cuzzola, V.; Barresi, S.; Pellegatta, M.; Canevazzi, P.; Dunsmore, G.; et al. IL-1β+ macrophages fuel pathogenic inflammation in pancreatic cancer. *Nature* **2023**, *623*, 415–422. [CrossRef]
62. Sitaru, S.; Budke, A.; Bertini, R.; Sperandio, M. Therapeutic inhibition of CXCR1/2: Where do we stand? *Intern. Emerg. Med.* **2023**, *18*, 1647–1664. [CrossRef] [PubMed]
63. Xie, Y.; Kuang, W.; Wang, D.; Yuan, K.; Yang, P. Expanding role of CXCR2 and therapeutic potential of CXCR2 antagonists in inflammatory diseases and cancers. *Eur. J. Med. Chem.* **2023**, *250*, 115175. [CrossRef] [PubMed]
64. Lazaar, A.L.; Miller, B.E.; Donald, A.C.; Keeley, T.; Ambery, C.; Russell, J.; Watz, H.; Tal-Singer, R.; Bardin, P.; Bremner, P.; et al. CXCR2 antagonist for patients with chronic obstructive pulmonary disease with chronic mucus hypersecretion: A phase 2b trial. *Respir. Res.* **2020**, *21*, 149. [CrossRef] [PubMed]
65. Armstrong, A.J.; Geva, R.; Chung, H.C.; Lemech, C.; Miller, W.H.; Hansen, A.R.; Lee, J.-S.; Tsai, F.; Solomon, B.J.; Kim, T.M.; et al. CXCR2 antagonist navarixin in combination with pembrolizumab in select advanced solid tumors: A phase 2 randomized trial. *Investig. New Drugs* **2024**, *42*, 145–159. [CrossRef]
66. Gu, H.; Deng, W.; Zheng, Z.; Wu, K.; Sun, F. CCL2 produced by pancreatic ductal adenocarcinoma is essential for the accumulation and activation of monocytic myeloid-derived suppressor cells. *Immun. Inflamm. Dis.* **2021**, *9*, 1686–1695. [CrossRef]
67. Korbecki, J.; Kupnicka, P.; Chlubek, M.; Goracy, J.; Gutowska, I.; Baranowska-Bosiacka, I. CXCR2 Receptor: Regulation of Expression, Signal Transduction, and Involvement in Cancer. *Int. J. Mol. Sci.* **2022**, *23*, 2168. [CrossRef]
68. Jin, T.; Xu, X.; Hereld, D. Chemotaxis, chemokine receptors and human disease. *Cytokine* **2008**, *44*, 1–8. [CrossRef]
69. Devalaraja, R.M.; Nanney, L.B.; Qian, Q.; Du, J.; Yu, Y.; Devalaraja, M.N.; Richmond, A. Delayed Wound Healing in CXCR2 Knockout Mice. *J. Investig. Dermatol.* **2000**, *115*, 234–244. [CrossRef]
70. Öhnstedt, E.; Vågesjö, E.; Fasth, A.; Lofton Tomenius, H.; Dahg, P.; Jönsson, S.; Tyagi, N.; Åström, M.; Myktybekova, Z.; Ringstad, L.; et al. Engineered bacteria to accelerate wound healing: An adaptive, randomised, double-blind, placebo-controlled, first-in-human phase 1 trial. *eClinicalMedicine* **2023**, *60*, 102014. [CrossRef]
71. Yang, H.; Liu, L.; Xu, F. The promises and challenges of fusion constructs in protein biochemistry and enzymology. *Appl. Microbiol. Biotechnol.* **2016**, *100*, 8273–8281. [CrossRef]

biomolecules

MDPI

Article

A Novel Ashwagandha (*Withania somnifera*) Formulation Mitigates Sleep Deprivation-Induced Cognitive Impairment and Oxidative Stress in a Rat Model

Besir Er [1], Busra Ozmen [2], Emre Sahin [3], Cemal Orhan [2], Nurhan Sahin [2], Abhijeet A. Morde [4], Muralidhara Padigaru [4] and Kazim Sahin [2,*]

1 Department of Biology, Faculty of Science, Firat University, Elazig 23119, Turkey; ber@firat.edu.tr
2 Department of Animal Nutrition, Faculty of Veterinary Medicine, Firat University, Elazig 23119, Turkey; busraagzikucuk01@gmail.com (B.O.); corhan@firat.edu.tr (C.O.); nsahin@firat.edu.tr (N.S.)
3 Department of Animal Nutrition, Faculty of Veterinary Medicine, Bingol University, Bingol 12100, Turkey; esahin@bingol.edu.tr
4 Research and Development, OmniActive Health Technologies Co., Ltd., Mumbai 400013, India; a.morde@omniactives.com (A.A.M.); m.padigaru@omniactives.com (M.P.)
* Correspondence: ksahin@firat.edu.tr; Tel.: +90-532-747-3506 or +90-424-237-0000 (ext. 3938)

Abstract: Ashwagandha (*Withania somnifera*) is a well-known adaptogenic herb traditionally used to enhance sleep quality and mitigate stress-induced cognitive decline. This study investigated the effects of different doses of ashwagandha root extract (AE) formulations on cognitive function, oxidative stress, and neuronal plasticity in a rat model of sleep deprivation (SD). Forty-nine rats were randomly assigned to seven groups: control, wide platform (WP), SD, SD + A1 (15 mg/kg AE 1.5%), SD + A2 (30 mg/kg AE 1.5%), SD + A3 (5.5 mg/kg AE 8.0%), and SD + A4 (11 mg/kg AE 8.0%). The extract was administered orally for four weeks. SD induced via a modified wide platform model significantly impaired spatial memory, increased oxidative stress, and suppressed GABA receptor activity. Treatment with all AE doses, except 15 mg/kg AE 1.5%, considerably reduced serum corticosterone (12% for SD + A2, 15% for SD + A3, and 32% for SD + A4), CRH (11% for SD + A2, 14% for SD + A3, and 17% for SD + A4), ACTH (22% for SD + A2, 26% for SD + A3, and 38% for SD + A4), and MDA levels (31% for SD + A2, 34% for SD + A3, and 46% for SD + A4) ($p < 0.05$). All doses improved antioxidant enzyme activity and memory performance, while AE 8.0% doses notably increased serotonin (19% for SD + A3 and 33% for SD + A4) and dopamine levels (40% for SD + A3 and 50% for SD + A4). Moreover, AE treatment enhanced markers of neuronal plasticity and partially improved GABAergic function. These findings suggest that AE formulations, particularly at higher concentrations, exert neuroprotective effects against SD-induced cognitive impairment by modulating oxidative stress, neurotransmitter balance, and neuroplasticity, indicating their potential application in managing stress-related neurological disorders.

Keywords: ashwagandha; withanolides; memory; sleep deprivation; stress; GABAergic pathway

1. Introduction

Cognitive health depends on a complex interplay of physiological and neurological processes, with sleep playing a pivotal role in maintaining memory and brain function [1]. Sleep deprivation (SD)—whether due to insufficient duration or poor quality—is increasingly prevalent in modern society and is known to impair various aspects of cognition [2].

Evidence from both human and animal studies indicates that SD detrimentally affects motor performance, cognitive processing, and emotional regulation, often exacerbating anxiety and depressive behaviors [3]. Moreover, sleep disturbances such as insomnia and fragmented sleep are frequently observed in patients with neurodegenerative diseases like Alzheimer's, as well as in aging animals with cognitive dysfunction [4]. Therefore, elucidating the intricate molecular and hormonal relationship between cognitive function, sleep disturbance, and stress is vital for developing effective interventions to mitigate the adverse effects of SD-related stress for humans and animals [3,5].

SD is closely linked to the activation of the hypothalamic–pituitary–adrenal (HPA) axis, leading to elevated levels of corticotropin-releasing hormone (CRH) and adrenocorticotropic hormone (ACTH), which can interfere with memory consolidation [6,7]. Additionally, SD induces oxidative stress by reducing total antioxidant capacity and the activity of critical enzymes such as superoxide dismutase (SOD), catalase (CAT), and glutathione peroxidase (GSH-Px), thereby increasing neuronal vulnerability [8].

Alterations in neurotransmitter systems—particularly serotonin, dopamine, and gamma-aminobutyric acid (GABA)—also disrupt synaptic function and neuroplasticity, essential components of memory encoding and retrieval [9]. Neuroplasticity and cellular communication markers, such as the neural cell adhesion molecule (NCAM) [10], intercellular adhesion molecule-1 (ICAM-1) [11], brain-derived neurotrophic factor (BDNF), and nerve growth factor (NGF), are also adversely affected, impairing synaptic connectivity and plasticity in sleep-deprived states [12]. It possibly mitigates stress and detrimental effects of SD by regulating cortisol release [13] and protein expression of $GABA_A$, $GABA_B1$, and serotonin receptors in the brain [14].

Ashwagandha (*Withania somnifera*) is a medicinal plant widely recognized for its therapeutic properties, primarily attributed to a group of naturally occurring compounds known as withanolides. These steroidal lactones, predominantly found in the roots and, to a lesser extent, in the leaves and stems, possess a C28 ergostane backbone with a lactone ring structure. In addition to withanolides, the plant contains other bioactive components such as withaferin A, withanone, alkaloids, and steroidal lactones, which contribute to its broad pharmacological potential. Major identified withanolides include withanolide A–D, withanoside IV and V, withanone, withaferin A, and sitoindosides VII–X [15]. The ashwagandha extract (AE) used in this study is a standardized water–alcoholic preparation containing 1.5% total withanolides, quantified by high-performance liquid chromatography (HPLC) in accordance with the United States Pharmacopeia (USP) guidelines, which define seven specific peaks for withanolide analysis [16]. This standardized method ensures consistency and reliability, addressing limitations found in earlier studies that often relied on non-standardized commercial extracts with variable and unverified withanolide content.

Withanolides have been extensively studied for their diverse biological effects. These include antioxidant activity through the reduction in oxidative stress, anti-inflammatory effects via modulation of cytokine pathways, immunomodulatory actions that enhance or balance immune responses, and neuroprotective effects that support neuronal health and function. Moreover, several withanolides have shown promising antitumor activity and have been linked to cognitive enhancement and stress resilience [17,18]. Ashwagandha extract (AE) is a natural sleep-promoting agent known for its favorable safety profile and efficacy in improving sleep onset and quality in individuals with insomnia. Its beneficial effects are thought to involve modulation of GABAergic pathways and brain-derived neurotrophic factor (BDNF), contributing to reduced stress and preserved neuroplasticity [17,18]. However, the poor solubility of ashwagandha's key bioactive components—particularly polar withanolide glycosides—may limit its bioavailability and clinical effectiveness [19]. Furthermore, the dose-dependent effects of AE on SD-related cognitive and molecular changes remain insufficiently understood.

This study aimed to investigate the potential neuroprotective effects of different doses of a water-soluble, highly bioavailable AE formulation on memory impairment induced by SD in a rat model. We focused on evaluating biochemical and molecular markers of oxidative stress [e.g., malondialdehyde (MDA), antioxidant enzymes], hormonal responses (e.g., corticosterone, CRH, ACTH), and neural plasticity including BDNF, NGF, NCAM, and growth-associated protein 43 (GAP-43), as well as the expression of GABAergic receptor subtypes ($GABA_AR2$, $GABA_BR1$, $GABA_BR2$) in the brain. This comprehensive approach aimed to clarify the dose-dependent impact of AE on SD-related stress and cognitive dysfunction.

2. Materials and Methods

2.1. Animals and Experimental Design

A total of 49 male Sprague–Dawley rats (n = 7), 8 weeks old, were used in this study. The animals were housed under standard laboratory conditions, including a controlled temperature of 22 ± 2 °C, relative humidity of 55 ± 5%, and a 12 h light/dark cycle. All rats had ad libitum access to food and water. The experimental protocol was approved by the Animal Ethics Committee of Fırat University (Approval No.: 06.05.2024-23993) and conducted in accordance with the National Institutes of Health Guide for the Care and Use of Laboratory Animals and relevant EU directives (approval date 6 May 2024).

The rats were randomly assigned into seven groups (n = 7 per group), each designed to assess the effects of SD and AE at different concentrations and dosages: (1) Control (C): Rats were maintained under standard conditions without exposure to SD. They received daily oral gavage of normal saline throughout the experimental period. (2) Wide platform (WP): This group served as a procedural control for the SD setup. Rats were placed on large platforms that allowed for normal sleep while controlling for the environmental stress of platform housing. They also received daily oral saline. (3) Sleep deprivation (SD): Rats were subjected to SD for 8 h per day (08:30–16:30) for four weeks using a modified multiple-platform method to induce stress. These rats received oral saline daily and served as the negative control group. (4) SD + A1: Rats were sleep-deprived as in the SD group and treated with 15 mg/kg of ashwagandha 1.5% extract (low dose of standard concentration) via oral gavage once daily for four weeks. (5) SD + A2: Rats were sleep-deprived and received a higher dose of ashwagandha 1.5% extract (30 mg/kg) daily to assess potential dose-dependent effects of the standard concentration formulation. (6) SD + A3: This group was sleep-deprived and treated with a low dose (5.5 mg/kg) of a concentrated ashwagandha 8.0% extract formulation to examine the efficacy of a more potent preparation. (7) SD + A4: Sleep-deprived rats in this group received 11 mg/kg of the ashwagandha 8.0% extract (high dose of the concentrated formulation). All treatments were administered via oral gavage once daily for four weeks, starting simultaneously across all groups (Figure 1).

Samples of AE containing 1.5% (item code: 301273) and 8% (item code: 301274) withanolides, the active phytochemical constituents, were obtained from OmniActive Health Technologies Ltd. (Thane, India). These formulations are designed for improved solubility and bioavailability, utilizing cellulose polymers as hydrophilic carriers. The varying concentrations of *Withania somnifera* extract were incorporated into the formulations to create water-dispersible products suitable for oral administration in experimental settings.

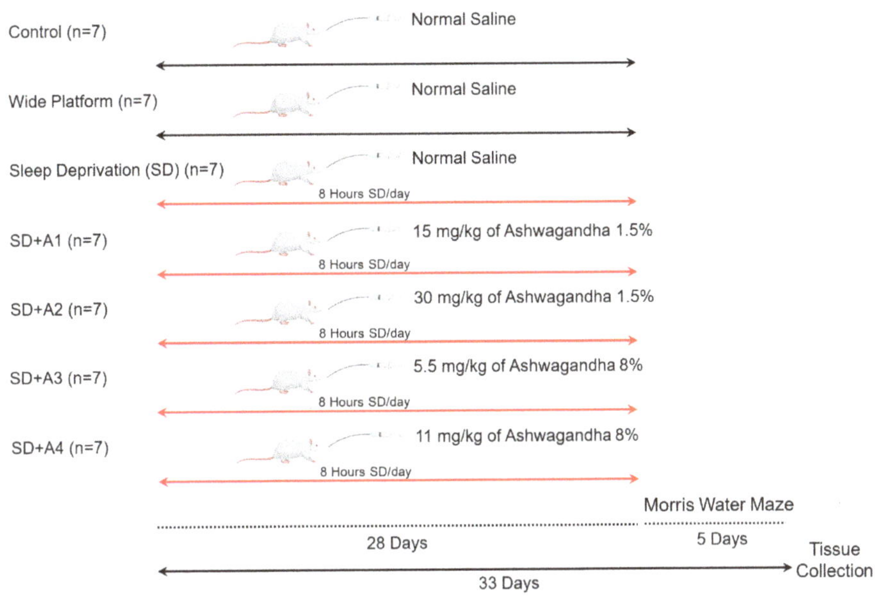

Figure 1. Experimental procedures.

2.2. Induction of Sleep Deprivation

SD was induced using multiple columns in the modified platform (in water) model [20]. Rats were positioned in a large tank measuring 170 cm in length, 40 cm in width, and 55 cm in depth, which contained tap water at a temperature of 24 °C. The tank included 20 small platforms, each with a diameter of 5 cm, positioned 10 cm apart from edge to edge and organized into two rows (Figure 2). The tank contained water up to 2 cm under the platform surface. In this tank, animals were able to traverse freely between platforms. Muscle atonia occurred during the paradoxical sleep period [21], causing the animals to fall into the water and wake up. Following that, they ascended to the platform and seated themselves upon it again. To determine the effects of possible stresses that may occur in the tank environment, WP with a diameter of 12 cm was used to ensure that the rats slept uninterruptedly and to prevent them from falling into the water.

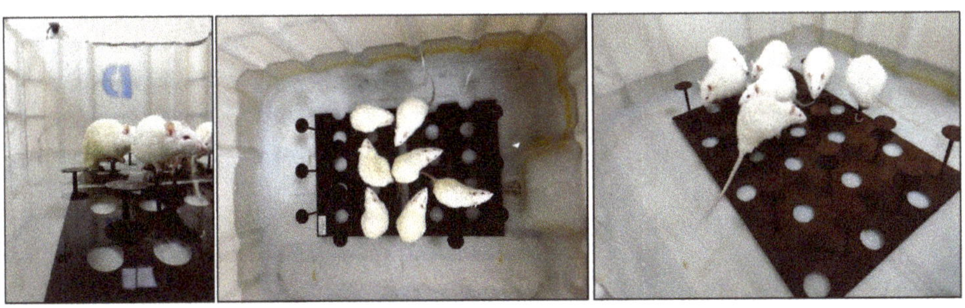

Figure 2. Sleep deprivation (SD) platform images for rats.

2.3. Morris Water Maze

The Morris water maze (MWM) test was used to test spatial learning and memory among all groups of animals at the end of the study for 5 days. This model and the detailed procedure were previously described [22]. A circular platform with a diameter of 10 cm was positioned at the center of one quadrant of the tank, situated 1.5 cm below the water's surface. Initially, the platform was discreetly placed in a randomly selected area of the reservoir to ensure its concealment throughout the testing period. The pool was divided into four equal quadrants by the designated starting positions (north, south, east, and west) arranged along its perimeter. Each rat was allowed a maximum of 60 s to locate the submerged platform and was required to remain on it for 30 s. In instances where a rat was unable to find the hidden platform within this timeframe, it was gently placed on the platform for a duration of 10 s. Following this, the rat was returned to a heated cage for a five-minute recuperation period between trials and subsequently released into the water from a different starting point to repeat the procedure. Upon completion of the fifth day of trials, a probe test was conducted. During this assessment, the platform was removed from the water, and the rats were positioned at the center of the tank to monitor their movements for 60 s. The duration each rat spent in proximity to the area where the platform was previously located was recorded and analyzed.

2.4. Sample Collection

All rats were overnight-fasted after the MWM test (33rd day), and blood was taken from decapitated animals through cervical dislocation. Blood samples were collected in biochemical tubes, and the serum samples were subsequently subjected to centrifugation at 4 °C with a force of 2300× g for 10 min in a refrigerated centrifuge. Brain tissues were promptly extracted and stored in a deep freezer at −80 °C until their analysis was conducted.

2.5. Biochemical Analysis

The levels of serum biochemical markers such as serum glucose, cholesterol, triglycerides, blood urea nitrogen (BUN), and creatinine and the activities of aspartate aminotransferase (AST) and alanine aminotransferase (ALT) were assessed using a portable automated chemistry analyzer (Samsung LABGEO PT10, Samsung Electronics Co., Suwon, Republic of Korea) with rat-specific kits (IVR-PT06, Samsung LABGEO PT Biochemistry Test 15, Samsung Electronics Co., Suwon, Republic of Korea).

Corticosterone, CRH, ACTH (Cayman Chemical Co., Ann Arbor, MI, USA), serotonin, dopamine (Elabscience Biotechnology, Wuhan, China), total antioxidant activity (TAC), and antioxidant enzymes (SOD, CAT, GSHPx) were quantified using appropriate commercial kits (BT-LABS, Shanghai, China) following the manufacturer's instructions in a microplate reader (Elx-800, Bio-Tek Instruments Inc., Winooski, VT, USA).

The levels of serum, liver, and brain MDA were assessed using a high-performance liquid chromatography system (HPLC, Shimadzu, Kyoto, Japan). A system incorporating a UV–vis SPD-10 AVP detector and a CTO-10 AS VP column was chosen for the analysis. The mobile phase consisted of a mixture of 30 mM KH_2PO_4 and methanol (82.5:17.5, v/v, pH 3.6), with a flow rate established at 1.2 mL/min.

2.6. Western Blot Analysis

Brain samples were homogenized in a cold Tris-HCl buffer (10 mM, pH 7.4) with the addition of protease inhibitors (Sigma, St. Louis, MO, USA). Following homogenization, the samples were centrifuged at 15,000× g for 30 min at 4 °C. The supernatants obtained were then heated and mixed with 2x Laemmli sample buffer to prepare for analysis. To facilitate the separation of proteins, the Mini-PROTEAN Tetra Cell system (Bio-Rad, Hercules, CA, USA) was utilized for the 12% SDS-PAGE stage. Subsequently, proteins were transferred to nitrocellulose membranes via a semi-dry transfer method using a Power Blotter (Thermo Fisher, Waltham, MA, USA). The membranes underwent an incubation with a 5% bovine serum albumin solution at room temperature for two hours to block non-specific binding to proteins. They were then incubated overnight at 4 °C with antibodies specific to the brain, including NCAM, BDNF, NGF, GAP-43, $GABA_AR2$, $GABA_BR1$, $GABA_BR2$, and 5-HT1A (Santa Cruz Biotechnology, Dallas, TX, USA). Following the blocking and incubation phase, membranes were treated with an HRP-linked secondary antibody at room temperature for two hours. To assess the presence of proteins, an anti-β-actin antibody (Sigma, St. Louis, MO, USA) was employed, followed by the application of the diaminobenzidine substrate method to observe antibody interactions. Densitometric analyses of the protein bands were conducted using ImageJ software (version 1.54g, National Institutes of Health, Bethesda, MD, USA).

2.7. Statistical Analysis

The SPSS statistical package program (IBM SPSS Version 22.0) was used for data analysis. The sample size (N = 49) was determined using the G*Power program (Version 3.1.9.3) with a power of 85%, an effect size of 0.65, and a significance level of 0.05. All data are presented as the mean ± standard error of the mean (SEM). A one-way analysis of variance (ANOVA), followed by the Tukey post hoc test, was employed for multiple comparisons of parametric data. Conversely, non-parametric data (probe trial) were analyzed using the Kruskal–Wallis and Mann–Whitney U tests. The threshold for statistical significance was established at $p < 0.05$.

3. Results

3.1. Body Weight and Serum Biochemical Parameters

Rats exposed to chronic stress induced by SD showed a significant reduction in final body weight compared to both the control and WP groups ($p < 0.001$). Among the ashwagandha-treated groups, only the highest dose of the 8% extract (11 mg/kg; SD + A4 group) significantly prevented this weight loss, resulting in a final body weight (BW) that was higher than that in the SD group ($p < 0.05$), though still lower than the control group.

Chronic stress led to increased serum glucose and AST levels in the SD group compared to healthy control rats ($p < 0.001$). Notably, the higher dose of AE at 11 mg/kg (8% ashwagandha) decreased serum glucose and AST levels in comparison to those in the SD group ($p < 0.01$). We observed that serum BUN levels decreased due to stress induced by SD ($p < 0.001$), and only the higher dose of AE at 11 mg/kg of 8% ashwagandha was effective in partially reversing this decrease ($p < 0.01$) (Table 1).

Table 1. Effects of different doses of ashwagandha on body weight (BW, g) and serum biochemical parameters (mg/dL) in sleep-deprived (SD) rats (n = 7).

Items	C	WP	SD	SD + A1	SD + A2	SD + A3	SD + A4
Initial BW, g	235.30 ± 5.51	232.90 ± 4.16	233.90 ± 6.116	231.90 ± 5.587	230.70 ± 5.768	233.00 ± 3.823	234.70 ± 5.172
Final BW, g	297.00 ± 4.99 ^{a}	293.90 ± 3.06 ^{a}	234.6 ± 3.54 ^{c}	244.90 ± 6.32 ^{bc}	250.1 ± 5.36 ^{bc}	253.3 ± 7.22 ^{bc}	258.6 ± 5.08 ^{b}
Glucose	97.23 ± 2.57 ^{d}	101.1 ± 3.66 ^{d}	146.40 ± 2.78 ^{a}	136.1 ± 3.04 ^{ab}	128.2 ± 1.83 ^{bc}	130.7 ± 1.56 ^{bc}	121.9 ± 2.42 ^{c}
Cholesterol	67.22 ± 1.97	66.96 ± 2.29	65.13 ± 1.21	64.59 ± 1.52	66.74 ± 1.36	67.35 ± 2.07	65.26 ± 2.72
Triglyceride	73.93 ± 2.65	74.39 ± 2.40	75.98 ± 1.97	73.96 ± 1.80	73.76 ± 2.06	74.45 ± 2.86	70.01 ± 1.85
AST, U/L	115.10 ± 1.97 ^{ab}	116.00 ± 1.90 ^{ab}	122.90 ± 2.25 ^{a}	115.40 ± 2.64 ^{ab}	112.70 ± 1.57 ^{b}	114.00 ± 1.78 ^{ab}	110.40 ± 2.15 ^{b}
ALT, U/L	83.71 ± 2.18	85 ± 2.16	89.71 ± 1.95	86.14 ± 2.40	87 ± 2.29	84.14 ± 1.68	81.86 ± 2.20
BUN	31.07 ± 0.92 ^{a}	30.72 ± 0.92 ^{a}	23.1 ± 0.79 ^{c}	24.29 ± 0.41 ^{bc}	24.55 ± 0.53 ^{bc}	25.31 ± 0.41 ^{bc}	26.98 ± 0.27 ^{b}
Creatinine	0.61 ± 0.016	0.60 ± 0.015	0.62 ± 0.021	0.58 ± 0.019	0.58 ± 0.017	0.56 ± 0.022	0.56 ± 0.018

Data are expressed as means ± SEM. Data were analyzed using one-way ANOVA followed by Tukey's post hoc test for multiple comparisons. Different superscript letters (a–d) in the same row indicate statistically significant differences between groups ($p < 0.05$). Groups not sharing the same letter are significantly different. C: control, WP: wide platform, SD: sleep deprivation, SD + A1: SD + ashwagandha 1.5% (15 mg/kg), SD + A2: SD + ashwagandha 1.5% (30 mg/kg), SD + A3: SD + ashwagandha 8% (5.5 mg/kg), SD + A4: SD + ashwagandha 8% (11 mg/kg). ALT: alanine transaminase, AST: aspartate transaminase, BUN: blood urea nitrogen.

3.2. Serum Hormone Levels

SD significantly elevated serum levels of stress-related hormones, including corticosterone (panel A), CRH (panel B), and ACTH (panel C), compared to both the control and WP groups ($p < 0.001$, Figure 3). These increases confirm the successful induction of chronic stress through SD. Treatment with AE effectively modulated these stress markers. Specifically, 30 mg/kg of ashwagandha 1.5% (SD + A2, $p < 0.01$) and both doses of ashwagandha 8% (SD + A3 and SD + A4, $p < 0.001$ for all) significantly reduced corticosterone, CRH, and ACTH levels compared to those in the untreated SD group. Notably, the 11 mg/kg dose of ashwagandha 8% (SD + A4) demonstrated the most pronounced effect, bringing hormone levels closer to those of the control group. Furthermore, even the low dose of ashwagandha 1.5% (SD + A1) significantly reduced ACTH levels, although its impact on corticosterone and CRH was limited.

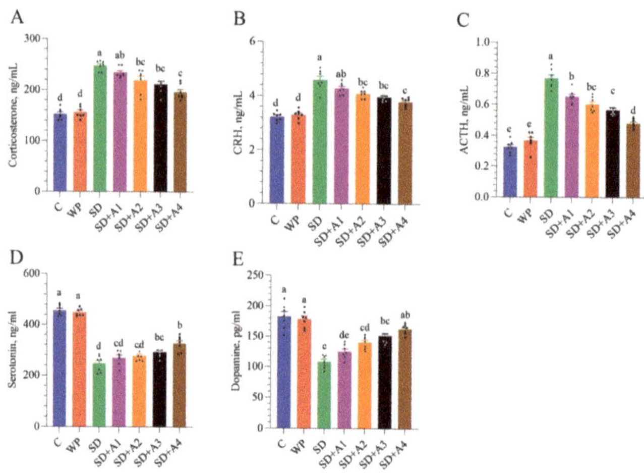

Figure 3. Effects of different ashwagandha doses on corticosterone (**A**), CRH (**B**), ACTH (**C**), serotonin (**D**), and dopamine (**E**) in sleep-deprived rats. Data were analyzed using one-way ANOVA followed by Tukey's post hoc test for multiple comparisons. Different superscript letters (a–e) indicate statistically significant differences between groups ($p < 0.05$). The groups not sharing the same letter are significantly different. The error lines indicate SEM, and symbols show individual values. CRH: corticotropin-releasing hormone, ACTH: adrenocorticotropic hormone, C: control, WP: wide platform, SD: sleep deprivation, SD + A1: SD + ashwagandha 1.5% (15 mg/kg), SD + A2: SD + ashwagandha 1.5% (30 mg/kg), SD + A3: SD + ashwagandha 8% (5.5 mg/kg), SD + A4: SD + ashwagandha 8% (11 mg/kg).

In terms of neurotransmitters, SD led to a significant decline in serum serotonin (panel D) and dopamine (panel E) levels, indicating disrupted mood and cognitive signaling. AE treatment, especially at higher doses, ameliorated these effects (Figure 3). Serotonin levels were partially improved by both doses of ashwagandha 8% ($p < 0.05$ for SD + A3 and $p < 0.001$ for SD + A4), while dopamine levels were significantly enhanced in the SD + A2, SD + A3, and SD + A4 groups ($p < 0.001$). Notably, the rats receiving 11 mg/kg ashwagandha 8% had serum dopamine levels similar to those of the control group rats ($p > 0.05$).

3.3. Malondialdehyde and Antioxidant Enzymes

Chronic stress induced by SD led to a marked increase in MDA levels, a key marker of lipid peroxidation and oxidative stress, in the serum (panel A), liver (panel B), and brain (panel C) ($p < 0.001$, Figure 4). In contrast, the control and WP groups maintained significantly lower MDA levels, confirming the impact of SD on systemic and central oxidative damage. All doses of AE (SD + A1 to SD + A4) reduced serum, liver, and brain MDA levels compared to those in non-supplemented rats ($p < 0.01$, Figure 4). The SD + A4 group showed the most significant decrease in serum and brain MDA levels compared to the other treatments ($p < 0.001$), suggesting dose-dependent protection. Simultaneously, brain antioxidant enzymes (SOD, CAT, GSH-Px) and TAC were suppressed under these stress conditions in rats ($p < 0.001$). All doses of AE (SD + A1 to SD + A4) improved the levels of SOD, CAT, GSH-Px, and TAC in the brain compared to SD group ($p < 0.05$). Notably, the highest dose of AE (11 mg/kg ashwagandha 8%) substantially improved brain SOD and CAT levels compared to other AE formulations ($p < 0.001$). Although the SD + A4 and SD + A3 groups had similar brain GSH-Px and TAC levels ($p > 0.05$), the SD + A4 group exhibited slightly higher values, indicating a more pronounced antioxidant effect at the highest AE dose.

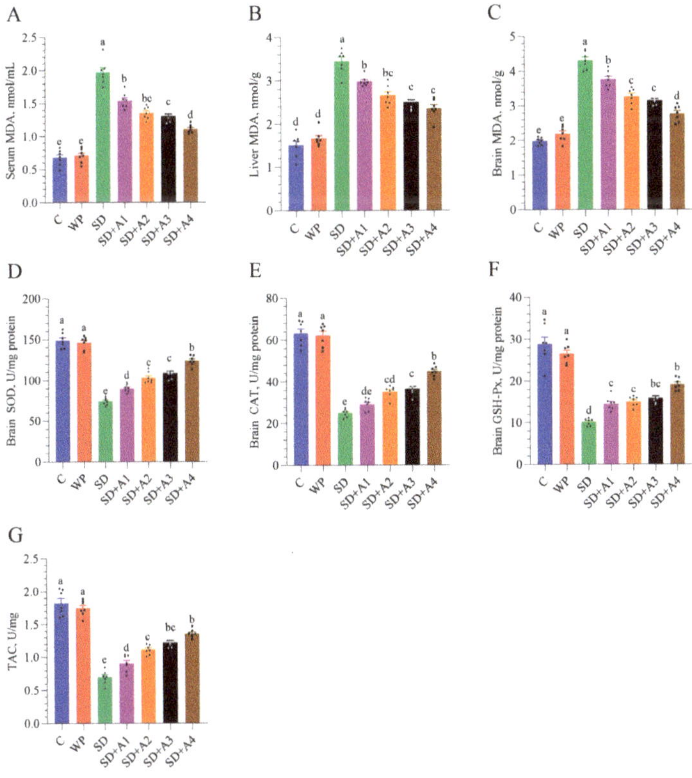

Figure 4. Effects of different ashwagandha doses on oxidative stress markers in sleep-deprived rats. (**A**) serum MDA, (**B**) liver MDA, (**C**) brain MDA, (**D**) brain superoxide dismutase (SOD), (**E**) brain catalase (CAT), (**F**) brain glutathione peroxidase (GSH-Px), and (**G**) total antioxidant capacity (TAC). Data were analyzed using one-way ANOVA followed by Tukey's post hoc test for multiple comparisons. Different superscript letters (a–e) indicate statistically significant differences between groups ($p < 0.05$). The groups not sharing the same letter are significantly different. The error lines indicate SEM, and symbols show individual values. MDA: malondialdehyde, SOD: superoxide dismutase, CAT: catalase, GSH-Px: glutathione peroxidase, TAC: total antioxidant capacity. C: control, WP: wide platform, SD: sleep deprivation, SD + A1: SD + ashwagandha 1.5% (15 mg/kg), SD + A2: SD + ashwagandha 1.5% (30 mg/kg), SD + A3: SD + ashwagandha 8% (5.5 mg/kg), SD + A4: SD + ashwagandha 8% (11 mg/kg).

3.4. Morris Water Maze

The Morris water maze test confirmed that SD-induced stress may impair spatial memory performance. Across all groups, the latency time to reach the target gradually decreased from day 1 to day 5 (Figure 5). On day 5, the latency time was highest in the SD group compared to all groups ($p < 0.01$, except SD + A1), indicating impaired learning ability. All doses of AE showed varying levels of improvement in latency time ($p < 0.01$, except 15 mg/kg ashwagandha 1.5%), and a dose of 11 mg/kg ashwagandha 8% demonstrated the most significant effect, approaching the control and WP groups' performance ($p > 0.05$). The time of entries to the target quadrant was reduced in the SD group ($p < 0.001$) compared to that in the control and WP groups. Treatment groups SD + A2 ($p < 0.05$), SD + A3 ($p < 0.01$), and SD + A4 ($p < 0.001$) showed significant improvement in the time of entries to the target quadrant as compared to the SD group. Moreover, the time of entries to the target quadrant of the SD + A4 group was similar to the control and WP groups' levels ($p > 0.05$). The probe trial was not changed between the SD and SD + A1–A3 groups ($p > 0.05$); however, the SD + A4 group showed more probe trials than the untreated SD group ($p < 0.05$).

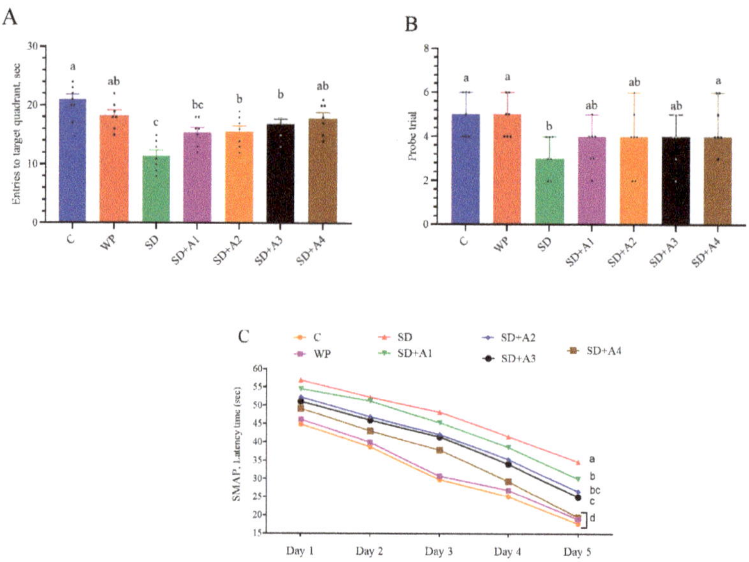

Figure 5. Effects of different ashwagandha doses on entries to target quadrant (**A**), probe trial (**B**), and spatial learning memory acquisition phase (SMAP) (**C**) in sleep-deprived rats. One-way ANOVA was used for group differences, and Tukey post hoc analysis for multiple comparisons. Kruskal–Wallis and Mann–Whitney U tests were used to compare the probe trials. Different superscripts (a–d) indicate the mean differences between the groups ($p < 0.05$). The groups not sharing the same letter are significantly different. The error lines indicate SEM for entries to the target quadrant and 95% confidence interval for the probe trial. Symbols show individual values. C: control, WP: wide platform, SD: sleep deprivation, SD + A1: SD + ashwagandha 1.5% (15 mg/kg), SD + A2: SD + ashwagandha 1.5% (30 mg/kg), SD + A3: SD + ashwagandha 8% (5.5 mg/kg), SD + A4: SD + ashwagandha 8% (11 mg/kg).

3.5. Brain Neurotrophic Factors

Stress induced by SD significantly decreased brain levels of NCAM, BDNF, NGF, and GAP-43 while simultaneously increasing brain ICAM-1 levels compared to those of healthy control rats ($p < 0.001$). All doses of AE (except for 15 mg/kg ashwagandha 1.5% for BDNF) partially restored the brain activity of NCAM, BDNF, NGF, GAP-43, and ICAM-1 compared to that in the SD group ($p < 0.05$), but 11 mg/kg of ashwagandha 8% formulation demonstrated the most significant partial improvement across all doses ($p < 0.001$, Figure 6). However, the SD + A2–A4 group had similar brain GAP-43 levels ($p > 0.05$).

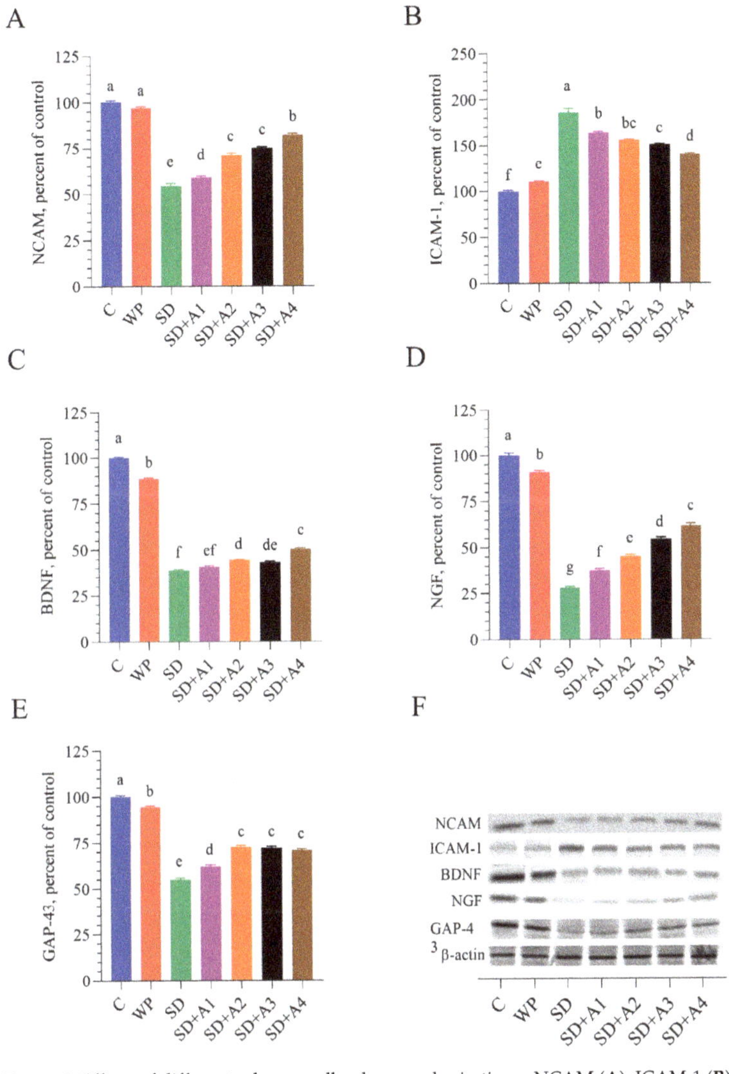

Figure 6. Effects of different ashwagandha doses on brain tissue NCAM (**A**), ICAM-1 (**B**), BDNF (**C**), NGF (**D**), and GAP-43 (**E**) protein levels and representative Western blot bands ((**F**) and Figure S1) in sleep-deprived rats. Western blot analysis was performed with incorporated β-actin to ensure equal protein loading. Representative bands are shown in panel (**F**). Data were analyzed using one-way ANOVA followed by Tukey's post hoc test for multiple comparisons. Different superscript letters (a–g) indicate statistically significant differences between groups ($p < 0.05$). The groups not sharing the same letter are significantly different. The error lines indicate SEM. NCAM: neural cell adhesion molecule, ICAM: intercellular adhesion molecule-1, BDNF: brain-derived neurotrophic factor, NGF: nerve growth factor. C: control, WP: wide platform, SD: sleep deprivation, SD + A1: SD + ashwagandha 1.5% (15 mg/kg), SD + A2: SD + ashwagandha 1.5% (30 mg/kg), SD + A3: SD + ashwagandha 8% (5.5 mg/kg), SD + A4: SD + ashwagandha 8% (11 mg/kg). The original Western blot image can be found in the Supplementary Materials.

3.6. Brain GABAergic and Serotonergic Receptors

Chronic SD caused a significant reduction in the expression of key neurotransmitter-related receptors in the brain, including $GABA_AR2$, $GABA_BR1$, $GABA_BR2$, and 5-HT1A compared to the control and WP groups ($p < 0.001$, Figure 7). AE treatments demonstrated varying levels of protection against SD-induced suppression, with SD + A2, SD + A3, and SD + A4 showing significant effects compared to the SD group ($p < 0.01$). While the lowest dose (15 mg/kg of ashwagandha 1.5%, SD + A1) failed to impact $GABA_AR2$ and 5-HT1A levels ($p > 0.05$ vs. SD), higher doses, namely, 30 mg/kg of ashwagandha 1.5% (SD + A2) and both 5.5 mg/kg (SD + A3) and 11 mg/kg (SD + A4) of ashwagandha 8%, resulted in notable improvement of all four receptors compared to the untreated SD group ($p < 0.001$). Among all treatment groups, SD + A4 (11 mg/kg of ashwagandha 8%) showed the most substantial improvement in the expression of all measured receptors ($p < 0.05$ for the SD + A3, $p < 0.001$ for the SD + A1 and SD + A2 groups for all receptors), indicating a stronger regulatory effect on neurotransmission under chronic stress conditions.

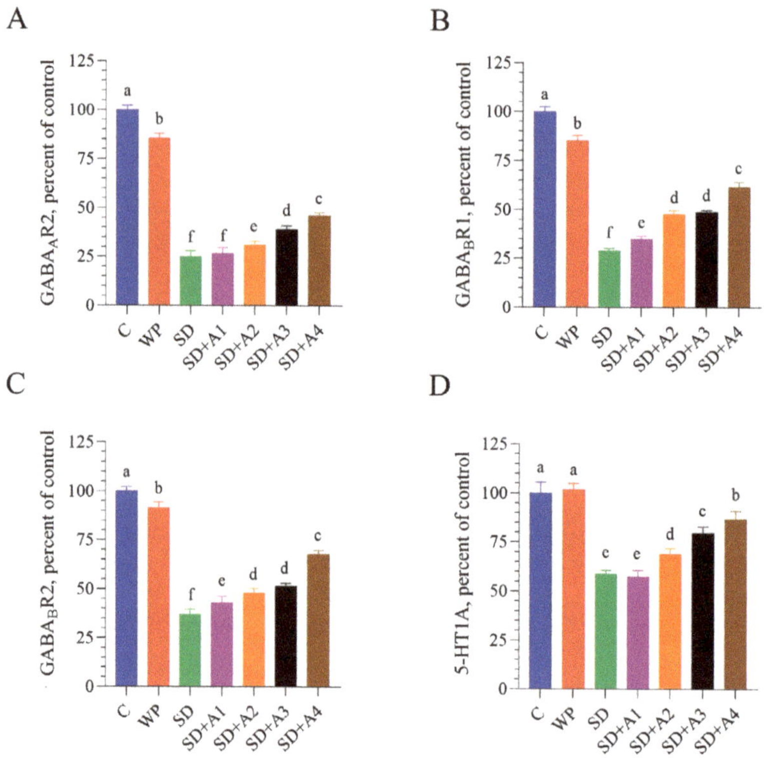

Figure 7. *Cont.*

E

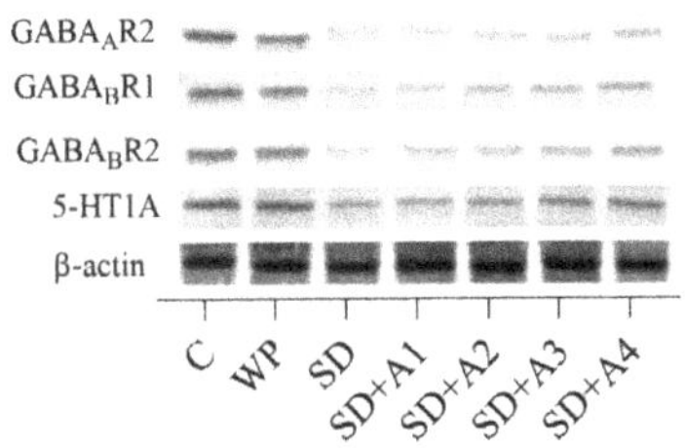

Figure 7. Effects of different ashwagandha doses on brain tissue $GABA_AR2$ (**A**), $GABA_BR1$ (**B**), $GABA_BR2$ (**C**), and 5-HT1A (**D**) protein levels and representative Western blot bands ((**E**) and Figure S2) in sleep-deprived rats. Western blot analysis was performed with incorporated β-actin to ensure equal protein loading. Representative bands are shown in panel E. Data were analyzed using one-way ANOVA followed by Tukey's post hoc test for multiple comparisons. Different superscript letters (a–f) indicate statistically significant differences between groups ($p < 0.05$). The groups not sharing the same letter are significantly different. The error lines indicate SEM. $GABA_A$: gamma-aminobutyric acid type A receptor subunit alpha, GAP-43: growth-associated protein-43. C: control, WP: wide platform, SD: sleep deprivation, SD + A1: SD + ashwagandha 1.5% (15 mg/kg), SD + A2: SD + ashwagandha 1.5% (30 mg/kg), SD + A3: SD + ashwagandha 8% (5.5 mg/kg), SD + A4: SD + ashwagandha 8% (11 mg/kg). The original Western blot image can be found in the Supplementary Materials.

4. Discussion

This study comprehensively evaluated the effects of different doses and formulations of AE on physiological, biochemical, and neurobiological parameters in rats exposed to chronic stress induced by SD. The findings demonstrated that AE exerted protective effects across a range of stress-induced impairments, including disruptions in body weight, metabolic biomarkers, oxidative stress, cognitive performance, and neuroplasticity. The most notable improvements were observed with the higher-dose AE formulation (11 mg/kg of 8% extract), suggesting a dose-dependent efficacy consistent with the plant's adaptogenic and neuroprotective profile. No dose-dependent adverse effects were observed in animals at higher concentrations of AE during the study period. All administered doses were well tolerated, and no signs of toxicity, behavioral abnormalities, or weight loss were noticed. We emphasize that the AE used in this study is a chemically defined, water–alcoholic extract of *Withania somnifera*, standardized to either 1.5% or 8.0% total withanolides. These formulations were analyzed and quantified using high-performance liquid chromatography (HPLC) following the United States Pharmacopeia (USP) guidelines, which specify seven characteristic chromatographic peaks for withanolide quantification. This standardization ensures consistency and enhances the interpretability and reproducibility of the results, addressing limitations seen in previous studies using non-standardized commercial extracts.

SD-induced chronic stress significantly reduced body weight and altered serum glucose, AST, and BUN levels, indicating metabolic disturbances in both liver and kidney function. These findings align with the understanding that chronic stress enhances catabolic activity [23], hepatic gluconeogenesis [24], and renal dysfunction [25]. It may explain increased serum glucose, AST levels, and lowered BUN levels due to SD. While all AE-treated groups showed improvement, the high-dose 8% AE formulation was particularly effective in normalizing these biochemical parameters, consistent with previous reports on its hepatoprotective and renoprotective actions [26]. This effect was likely related to AE's withanolide content, which previously led to a significant decrease in serum bilirubin, ALP, AST, and ALT levels [27].

Stress led to a marked increase in serum corticosterone, CRH, and ACTH levels, reflecting activation of the HPA axis. These results are consistent with prior studies demonstrating HPA axis hyperactivity under SD-related stress conditions [28]. The ability of AE to lower cortisol levels has been previously reported and is thought to be linked to its adaptogenic properties, primarily attributed to bioactive compounds such as withaferin A, as well as its calming and sleep-promoting effects [29]. Moreover, withaferin A can potentiate the anti-stress effect of acylsterylglucosides, namely, sitoindoside VII and sitoindoside VIII. In addition, their combination exhibited significant anti-stress activity [30]. Additionally, glycowithanolides, sitoindoside IX, and sitoindoside X produced significant anti-stress activity in albino mice and rats [31]. In this study, AE—particularly at higher doses—significantly reduced the elevated stress hormone levels, suggesting its effectiveness in modulating excessive HPA axis activity [32].

While SD may increase serotonin release, this response is often independent of stress [33]. In contrast, certain stressors—such as immobility—have been shown to suppress serum serotonin levels in rats [34–36]. Sahin et al. [37] demonstrated that caffeine-induced SD impairs both serotonergic and dopaminergic signaling, as evidenced by reductions in serum serotonin and dopamine. Similarly, in the present study, SD-induced stress significantly lowered serotonin and dopamine levels, suggesting disruption of neurotransmitter homeostasis. AE may modulate neurotransmitter activity partly due to its antioxidant and anti-inflammatory effects; however, certain bioactive compounds like withanolides could also directly affect neurotransmitter pathways. Withanolide A has been demonstrated to bind to serotonin receptors and serotonin transporters in both humans and Caenorhabditis elegans through molecular investigations [38]. A methanolic extract of ashwagandha roots (rich in Withaferin and sitoindosides VII-X) was found to inhibit the effects of morphine and ethanol on the dopamine-producing neurons in the ventral tegmental area of the rodent brain [39]. Treatment with AE effectively reversed these declines, underscoring its role in stabilizing neurotransmitter balance. These findings are in agreement with Dawane et al. [34], who reported that AE dose-dependently improved serotonin levels in rats subjected to immobility stress, supporting the current evidence of AE's neuromodulatory and adaptogenic effects.

The administration of AE exhibited a clear dose-dependent effect in reducing MDA levels, a marker of lipid peroxidation and oxidative stress, in the serum, liver, and brain. The most pronounced reduction was observed at the highest tested dose (11 mg/kg of 8% ashwagandha), suggesting strong antioxidant efficacy. Alongside this, AE, particularly in its high-dose formulation, significantly enhanced the activity of key antioxidant enzymes (SOD, CAT, GSH-Px) and increased TAC in the brain tissue of sleep-deprived rats. The potent antioxidant effects of AE are likely attributed to its rich content of flavonoids, alkaloids, steroidal lactones, saponins, phenolic compounds, withanolides, and other phytochemicals that help mitigate reactive oxygen species (ROS), repair oxidative cellular damage, and reduce lipid peroxidation [40]. These findings are consistent with those of Suganya et al. [41], who reported that AE significantly boosts antioxidant enzyme levels while decreasing oxidative damage and lipid peroxidation in sleep-deprived rats. The bioactive components, sitoindosides VII–X and withaferin A, are considered key contributors to these effects.

Oxidative stress has been closely linked to cognitive deficits, particularly in memory-related tasks [22]. In the current study, rats exposed to SD-induced stress showed impaired performance in the MWM test, as evidenced by fewer entries into the target quadrant and delayed platform acquisition. These impairments are consistent with prior research suggesting that increased oxidative stress in the brain, particularly in the hippocampus, is associated with memory dysfunction and anxiety-like behaviors [42].

Treatment with AE, especially at higher doses, significantly improved spatial memory performance in stressed rats. This cognitive enhancement is likely due to the combined neuroprotective, antioxidant, and anxiolytic properties of AE. The hydrophobic core of β-amyloid 1–42 was shown to interact in the form of oligomers with withanolide A, withanolide B, withanoside IV, withanoside V, and sominone. As a result, further interaction with monomers was prevented, leading to reduced aggregation [43,44]. Specifically, compounds such as withanolide A [45] and withanoside IV may support synaptic function and promote neurogenesis, contributing to improved learning and memory [44]. These results are in agreement with Gladen-Kolarsky et al. [46], who demonstrated that AE improved spatial memory in a dose-dependent manner and correlated these improvements with enhanced brain antioxidant activity in an Alzheimer's disease mouse model.

The present study highlights the profound impact of chronic stress induced by SD on brain function, mainly through its detrimental effects on neurotrophic factors and inflammatory markers. Key neurotrophic factors essential for neuroplasticity, synaptic connectivity, and neuronal survival, including NCAM, BDNF, NGF, and GAP-43, were significantly downregulated in the brains of SD-exposed rats. Concurrently, levels of ICAM-1, a pro-inflammatory marker associated with neuroinflammation and potential neurodegeneration, were markedly elevated. AE treatment effectively counteracted these alterations, particularly at higher doses. Previous in vitro studies have shown AE's potential to promote neurodifferentiation in glioma cells [47] and suppress cell migration in neuroblastoma models [48], supporting its neurodegenerative properties. Withanoside IV can prevent Aβ(25–35)-induced axonal, dendritic, and synaptic losses and memory deficits in mice [44]. In line with these findings, the current study observed an upregulation of NCAM in AE-treated animals, suggesting a reversal of SD-induced suppression.

Mechanistic insights from earlier studies further support AE's neurotrophic potential. Konar et al. [17] demonstrated that AE enhances BDNF expression by modulating cyclic AMP response element-binding protein (CREB) activity and increasing intracellular calcium levels. Similarly, Kim et al. [35] reported that a high dose of AE containing withanolide A elevated BDNF levels in a chronic stress-induced depression model, consistent with our observations. In mice that were treated for spinal cord injury, the expression of BDNF was significantly increased by withaferin A [49]. The neuroprotective effect of BDNF is dependent on GAP-43 activity, which is essential for neuroprotection and neuronal plasticity [50]. Withaferin A- and withanone-treated neuroblastoma cells may exhibit GAP-43 upregulation [17]. Our study, aligned with the findings of [17], demonstrated that AE could enhance the production of brain GAP-43 in memory-impaired mice.

A molecular docking study by Mitra et al. [51] explained the NGF-promoting effect of AE. The study revealed that AE might be acting NGF-mimetic due to interacting with TrkA [51]. Withaferin A-related anti-inflammatory properties of AE [35] can suppress tumor necrosis factor-α-induced expression of nuclear factor-κB and their lower cascade, ICAM-1 [52]. This mechanism may explain the suppressive effect of AE on ICAM-1 in the brain in SD rats.

According to Park et al. [14], AE may improve sleep by increasing the levels of GABA in the brain and boosting the protein levels of $GABA_A$, $GABA_{B1}$, and serotonin receptors through the GABAergic system. Biochemical studies suggested that AE's GABA-mimetic activity originated from some withanolide derivatives (not from withaferin A and withanolide A), which had a GABA-mimetic activity [53]. Sitoindosides VII and VIII, in particular, have been implicated in producing anxiolytic and anti-stress effects, likely by potentiating GABA receptor activity, which leads to reduced neuronal excitability and anxiolysis [54]. This GABA-mimetic action is crucial in mediating ashwagandha's calming and neuroprotective properties. In a mouse restraint stress model, AE prevented a stress-induced decrease in hippocampal serotonin levels, similar to the present findings [55]. The serotonergic activity of AE was likely associated with the regulation of mRNA expression of serotonin receptors and transporters by withanolide A [38]. Additionally, this GABA-mimetic and serotonergic activity can be improved by administering an elevated dose of AE [14]. In this study, the observed suppression of GABA and serotonin receptors in SD rats may have been mitigated by AE's dose-dependent GABA-mimetic and serotonergic effects. The combined antioxidant properties of AE and its GABAergic modulation may have synergistically enhanced its neuroprotective and cognitive effects by reducing oxidative stress and improving GABA receptor function. This dual action likely protected neuronal integrity and contributed to improved cognitive performance.

While the present study provides compelling evidence for the neuroprotective, antioxidant, and cognitive-enhancing effects of ashwagandha extract (AE) under chronic stress conditions, several limitations should be acknowledged. First, the absence of a positive control group, such as a standard therapeutic agent (e.g., benzodiazepines, antidepressants, or known adaptogens), represents a limitation. The primary aim of this study was to evaluate the dose-dependent efficacy of AE formulations standardized to specific withanolide concentrations (1.5% and 8.0%), rather than to directly compare AE's effects with those of established pharmacological treatments. Nonetheless, including a positive control group in future studies would strengthen the comparative value of the findings and help position AE more precisely within the therapeutic landscape of stress-related disorders. Second, the study was conducted solely on male rats, which limits the generalizability of the results to both sexes. Given the known hormonal and neurochemical differences between males and females, it is possible that females may exhibit distinct responses to AE treatment. Additionally, biochemical analyses were conducted on whole-brain homogenates, which precludes the identification of region-specific effects. This is particularly important considering that neurotransmitter receptors and neurotrophic factors exhibit region-dependent expression and function; for example, presynaptic receptors are more prominent in the midbrain, while postsynaptic receptors are abundant in the cortex and hippocampus. Future studies should include region-specific assessments to elucidate the precise neural substrates and molecular pathways modulated by AE. Furthermore, cognitive performance in this study was evaluated using a single behavioral test, which may not fully capture the multifaceted nature of cognition. Employing a comprehensive battery of behavioral assays in future research would enhance the robustness and interpretability of the cognitive outcomes.

5. Conclusions

The findings of this study indicate that water-soluble AE, particularly at higher doses, may enhance spatial memory and cognitive performance—likely due to improved bioavailability of its active phytochemicals, especially withanolides. In rats exposed to SD-induced stress, these cognitive benefits were associated with increased brain antioxidant capacity and the GABA-mimetic effects of AE. These results highlight the potential of high-dose AE as a therapeutic option for managing cognitive impairments in clinical contexts. AE may also be beneficial in other models of cognitive impairment, such as aging-related decline and neurodegenerative disorders (e.g., Alzheimer's or Parkinson's disease). However, further preclinical investigations and well-designed clinical trials are essential to confirm and extend these observations.

Supplementary Materials: The following supporting information can be downloaded at https://www.mdpi.com/article/10.3390/biom15050710/s1, Figure S1: Full immunoblots related to Figure 5 in the main text. Figure S2: Full immunoblots related to Figure 6 in the main text.

Author Contributions: Conceptualization, K.S. and B.E.; methodology, K.S. and B.E.; formal analysis, investigation, data curation, B.E., B.O., C.O. and N.S.; writing—original draft preparation, E.S., A.A.M. and M.P.; writing—review and editing, A.A.M., M.P. and K.S. All authors have read and agreed to the published version of the manuscript.

Funding: This research was funded by OmniActive Health Technologies (No.: 2024-1; Mumbai, India) and partially by the Turkish Academy of Sciences (2024-2; K.S., Ankara, Turkey). The funders were not involved in the project design, collection, analysis, and interpretation of data; the writing of this article; or the decision to submit it for publication.

Institutional Review Board Statement: The experimental procedures were approved by the Animal Experiments Local Ethics Committee of Firat University (protocol number: 06.05.2024-23993) following EU directives (approval date 6 May 2024).

Informed Consent Statement: Not applicable.

Data Availability Statement: The original contributions presented in this study are included in the article/Supplementary Material. Further inquiries can be directed to the corresponding author(s).

Acknowledgments: The authors thank OmniActive Health Technologies (Mumbai, India) and the Turkish Academy of Science (Ankara, Turkey, KS).

Conflicts of Interest: A.M.M. and M.P. are employees of OmniActive Health Technologies, which funded this research. The funders had no role in the design, execution, or analysis of this study. Independent validation of data integrity was conducted to ensure unbiased results. The other authors declare no conflicts of interest.

References

1. Gohari, A.; Baumann, B.; Jen, R.; Ayas, N. Sleep Deficiency: Epidemiology and Effects. *Clin. Chest Med.* **2022**, *43*, 189–198. [CrossRef]
2. Krause, A.J.; Simon, E.B.; Mander, B.A.; Greer, S.M.; Saletin, J.M.; Goldstein-Piekarski, A.N.; Walker, M.P. The Sleep-Deprived Human Brain. *Nat. Rev. Neurosci.* **2017**, *18*, 404–418. [CrossRef] [PubMed]
3. Alrousan, G.; Hassan, A.; Pillai, A.A.; Atrooz, F.; Salim, S. Early Life Sleep Deprivation and Brain Development: Insights From Human and Animal Studies. *Front. Neurosci.* **2022**, *16*, 833786. [CrossRef]
4. Prpar Mihevc, S.; Majdič, G. Canine Cognitive Dysfunction and Alzheimer's Disease—Two Facets of the Same Disease? *Front. Neurosci.* **2019**, *13*, 604. [CrossRef] [PubMed]
5. Taylor, T.L.; Fernandez, E.J.; Handley, K.N.; Hazel, S.J. Non-Pharmacological Interventions for the Treatment of Canine Cognitive Dysfunction: A Scoping Review. *Appl. Anim. Behav. Sci.* **2023**, *269*, 106097. [CrossRef]
6. Sapolsky, R.M. Why Stress Is Bad for Your Brain. *Science* **1996**, *273*, 749–750. [CrossRef]
7. Grillon, C.; Duncko, R.; Covington, M.F.; Kopperman, L.; Kling, M.A. Acute Stress Potentiates Anxiety in Humans. *Biol. Psychiatry* **2007**, *62*, 1183–1186. [CrossRef]
8. Neculicioiu, V.S.; Colosi, I.A.; Costache, C.; Toc, D.A.; Sevastre-Berghian, A.; Colosi, H.A.; Clichici, S. Sleep Deprivation-Induced Oxidative Stress in Rat Models: A Scoping Systematic Review. *Antioxidants* **2023**, *12*, 1600. [CrossRef]
9. Vaseghi, S.; Arjmandi-Rad, S.; Eskandari, M.; Ebrahimnejad, M.; Kholghi, G.; Zarrindast, M.-R. Modulating Role of Serotonergic Signaling in Sleep and Memory. *Pharmacol. Rep.* **2022**, *74*, 1–26. [CrossRef]
10. Mishra, R.; Manchanda, S.; Gupta, M.; Kaur, T.; Saini, V.; Sharma, A.; Kaur, G. Tinospora Cordifolia Ameliorates Anxiety-like Behavior and Improves Cognitive Functions in Acute Sleep Deprived Rats. *Sci. Rep.* **2016**, *6*, 25564. [CrossRef]
11. Frey, D.J.; Fleshner, M.; Wright, K.P. The Effects of 40 Hours of Total Sleep Deprivation on Inflammatory Markers in Healthy Young Adults. *Brain. Behav. Immun.* **2007**, *21*, 1050–1057. [CrossRef]
12. Sahin, K.; Korkusuz, A.K.; Sahin, E.; Orhan, C.; Er, B.; Morde, A.; Padigaru, M.; Kilic, E. The Effect of Water-Soluble Alpinia Galanga Extract on Sleep and the Activation of the GABAAergic/Serotonergic Pathway in Mice. *Pharmaceuticals* **2024**, *17*, 1649. [CrossRef]

13. Chandrasekhar, K.; Kapoor, J.; Anishetty, S. A Prospective, Randomized Double-Blind, Placebo-Controlled Study of Safety and Efficacy of a High-Concentration Full-Spectrum Extract of Ashwagandha Root in Reducing Stress and Anxiety in Adults. *Indian J. Psychol. Med.* **2012**, *34*, 255–262. [CrossRef] [PubMed]
14. Park, C.W.; Hong, K.-B.; Suh, H.J.; Ahn, Y. Sleep-Promoting Activity of Amylase-Treated Ashwagandha (*Withania somnifera* L. Dunal) Root Extract via GABA Receptors. *J. Food Drug Anal.* **2023**, *31*, 278–288. [CrossRef] [PubMed]
15. Abdelwahed, M.T.; Hegazy, M.A.; Mohamed, E.H. Major Biochemical Constituents of Withania Somnifera (*Ashwagandha*) Extract: A Review of Chemical Analysis. *Rev. Anal. Chem.* **2023**, *42*, 20220055. [CrossRef]
16. United States Pharmacopeia Ashwagandha Root Dry Extract. 2020. Available online: https://doi.usp.org/USPNF/USPNF_M2789_08_01.html (accessed on 2 May 2025).
17. Konar, A.; Shah, N.; Singh, R.; Saxena, N.; Kaul, S.C.; Wadhwa, R.; Thakur, M.K. Protective Role of Ashwagandha Leaf Extract and Its Component Withanone on Scopolamine-Induced Changes in the Brain and Brain-Derived Cells. *PLoS ONE* **2011**, *6*, e27265. [CrossRef]
18. Langade, D.; Kanchi, S.; Salve, J.; Debnath, K.; Ambegaokar, D. Efficacy and Safety of Ashwagandha (*Withania somnifera*) Root Extract in Insomnia and Anxiety: A Double-Blind, Randomized, Placebo-Controlled Study. *Cureus* **2019**, *11*, e5797. [CrossRef] [PubMed]
19. Dudhat, K.; Bhalodiya, M.; Dudhrejiya, A.; Shah, S.; Parmar, R.; Baldaniya, L.; Dhaval, M. Application of Amorphous Solid Dispersion Technology for Improving the Physicochemical Properties, Saturation Solubility, and In Vitro Dissolution of *Withania somnifera* Methanolic Root Powder Extract. *J. Pharm. Innov.* **2023**, *18*, 1338–1349. [CrossRef]
20. Alzoubi, K.H.; Khabour, O.F.; Salah, H.A.; Abu Rashid, B.E. The Combined Effect of Sleep Deprivation and Western Diet on Spatial Learning and Memory: Role of BDNF and Oxidative Stress. *J. Mol. Neurosci.* **2013**, *50*, 124–133. [CrossRef]
21. Grahnstedt, S.; Ursin, R. Platform Sleep Deprivation Affects Deep Slow Wave Sleep in Addition to REM Sleep. *Behav. Brain Res.* **1985**, *18*, 233–239. [CrossRef]
22. Sahin, K.; Orhan, C.; Karatoprak, S.; Tuzcu, M.; Deeh, P.B.D.; Ozercan, I.H.; Sahin, N.; Bozoglan, M.Y.; Sylla, S.; Ojalvo, S.P.; et al. Therapeutic Effects of a Novel Form of Biotin on Propionic Acid-Induced Autistic Features in Rats. *Nutrients* **2022**, *14*, 1280. [CrossRef] [PubMed]
23. Chrousos, G.P. Stress and Disorders of the Stress System. *Nat. Rev. Endocrinol.* **2009**, *5*, 374–381. [CrossRef]
24. Nikolic, A.; Fahlbusch, P.; Riffelmann, N.-K.; Wahlers, N.; Jacob, S.; Hartwig, S.; Kettel, U.; Schiller, M.; Dille, M.; Al-Hasani, H.; et al. Chronic Stress Alters Hepatic Metabolism and Thermodynamic Respiratory Efficiency Affecting Epigenetics in C57BL/6 Mice. *iScience* **2024**, *27*, 109276. [CrossRef] [PubMed]
25. Bruce, M.A.; Griffith, D.M.; Thorpe, R.J. Stress and the Kidney. *Adv. Chronic Kidney Dis.* **2015**, *22*, 46–53. [CrossRef]
26. Jain, V.; Chaturvedi, S.; Jamil, S.; Tyagi, R.; Arya, S.; Madan, S. Ashwagandha: Botanic Occurrence, Conventional Uses, and Significance in Heart, Metabolic, Renal and Hepatic Disorder. *Nutr. Food Sci.* **2024**, *54*, 1337–1355. [CrossRef]
27. Devkar, S.T.; Kandhare, A.D.; Zanwar, A.A.; Jagtap, S.D.; Katyare, S.S.; Bodhankar, S.L.; Hegde, M.V. Hepatoprotective Effect of Withanolide-Rich Fraction in Acetaminophen-Intoxicated Rat: Decisive Role of TNF-α, IL-1β, COX-II and iNOS. *Pharm. Biol.* **2016**, *54*, 2394–2403. [CrossRef]
28. Vgontzas, A.N.; Mastorakos, G.; Bixler, E.O.; Kales, A.; Gold, P.W.; Chrousos, G.P. Sleep Deprivation Effects on the Activity of the Hypothalamic-Pituitary-Adrenal and Growth Axes: Potential Clinical Implications. *Clin. Endocrinol.* **1999**, *51*, 205–215. [CrossRef]
29. Della Porta, M.; Maier, J.A.; Cazzola, R. Effects of *Withania somnifera* on Cortisol Levels in Stressed Human Subjects: A Systematic Review. *Nutrients* **2023**, *15*, 5015. [CrossRef]
30. Bhattacharya, S.K.; Goel, R.K.; Kaur, R.; Ghosal, S. Anti-Stress Activity of Sitoindosides VII and VIII, New Acylsterylglucosides from *Withania somnifera*. *Phytother. Res.* **1987**, *1*, 32–37. [CrossRef]
31. Ghosal, S.; Lal, J.; Srivastava, R.; Bhattacharya, S.K.; Upadhyay, S.N.; Jaiswal, A.K.; Chattopadhyay, U. Immunomodulatory and CNS Effects of Sitoindosides IX and X, Two New Glycowithanolides from *Withania somnifera*. *Phytother. Res.* **1989**, *3*, 201–206. [CrossRef]
32. Wiciński, M.; Fajkiel-Madajczyk, A.; Kurant, Z.; Kurant, D.; Gryczka, K.; Falkowski, M.; Wiśniewska, M.; Słupski, M.; Ohla, J.; Zabrzyński, J. Can Ashwagandha Benefit the Endocrine System?-A Review. *Int. J. Mol. Sci.* **2023**, *24*, 16513. [CrossRef] [PubMed]
33. Longordo, F.; Kopp, C.; Lüthi, A. Consequences of Sleep Deprivation on Neurotransmitter Receptor Expression and Function. *Eur. J. Neurosci.* **2009**, *29*, 1810–1819. [CrossRef]
34. Dawane, J.; Seok, S.; Dhande, P.; Langade, D.; Han, H.; Kim, S.-B.; Ju, J.-Y. Evaluation of the Anxiolytic and Antidepressant Effects of Standardized Ashwagandha (*Withania somnifera*) Root Extract in Wistar Rats. *Prev. Nutr. Food Sci.* **2024**, *29*, 414–421. [CrossRef] [PubMed]
35. Kim, H.; Choi, H.-S.; Han, K.; Sim, W.; Suh, H.J.; Ahn, Y. Ashwagandha (*Withania somnifera* (L.) Dunal) Root Extract Containing Withanolide a Alleviates Depression-like Behavior in Mice by Enhancing the Brain-Derived Neurotrophic Factor Pathway under Unexpected Chronic Mild Stress. *J. Ethnopharmacol.* **2025**, *340*, 119224. [CrossRef]

36. Tyagi, A.; Choi, Y.-Y.; Shan, L.; Vinothkanna, A.; Lee, E.-S.; Chelliah, R.; Barathikannan, K.; Raman, S.T.; Park, S.J.; Jia, A.-Q.; et al. Limosilactobacillus Reuteri Fermented Brown Rice Alleviates Anxiety Improves Cognition and Modulates Gut Microbiota in Stressed Mice. *NPJ Sci. Food* **2025**, *9*, 5. [CrossRef]
37. Sahin, K.; Gencoglu, H.; Korkusuz, A.K.; Orhan, C.; Aldatmaz, İ.E.; Erten, F.; Er, B.; Morde, A.; Padigaru, M.; Kilic, E. Impact of a Novel Valerian Extract on Sleep Quality, Relaxation, and GABA/Serotonin Receptor Activity in a Murine Model. *Antioxidants* **2024**, *13*, 657. [CrossRef] [PubMed]
38. Naß, J.; Abdelfatah, S.; Efferth, T. Induction of Stress Resistance and Extension of Lifespan in Chaenorhabditis Elegans Serotonin-Receptor Knockout Strains by Withanolide A. *Phytomedicine* **2021**, *84*, 153482. [CrossRef]
39. Bassareo, V.; Talani, G.; Frau, R.; Porru, S.; Rosas, M.; Kasture, S.B.; Peana, A.T.; Loi, E.; Sanna, E.; Acquas, E. Inhibition of Morphine- and Ethanol-Mediated Stimulation of Mesolimbic Dopamine Neurons by *Withania somnifera*. *Front. Neurosci.* **2019**, *13*, 545. [CrossRef]
40. Guo, S.; Rezaei, M.J. The Benefits of Ashwagandha (*Withania somnifera*) Supplements on Brain Function and Sports Performance. *Front. Nutr.* **2024**, *11*, 1439294. [CrossRef]
41. Suganya, K.; Kayalvizhi, E.; Yuvaraj, R.; Chandrasekar, M.; Kavitha, U.; Konakanchi Suresh, K. Effect of *Withania somnifera* on the Antioxidant and Neurotransmitter Status in Sleep Deprivation Induced Wistar Rats. *Bioinformation* **2020**, *16*, 631–637. [CrossRef]
42. Konakanchi, S.; Raavi, V.; Ml, H.K.; Shankar MS, V. Impact of Chronic Sleep Deprivation and Sleep Recovery on Hippocampal Oligodendrocytes, Anxiety-like Behavior, Spatial Learning and Memory of Rats. *Brain Res. Bull.* **2023**, *193*, 59–71. [CrossRef] [PubMed]
43. Dubey, S.; Kallubai, M.; Subramanyam, R. Improving the Inhibition of β-Amyloid Aggregation by Withanolide and Withanoside Derivatives. *Int. J. Biol. Macromol.* **2021**, *173*, 56–65. [CrossRef]
44. Kuboyama, T.; Tohda, C.; Komatsu, K. Withanoside IV and Its Active Metabolite, Sominone, Attenuate Abeta(25-35)-Induced Neurodegeneration. *Eur. J. Neurosci.* **2006**, *23*, 1417–1426. [CrossRef] [PubMed]
45. Mishra, L.C.; Singh, B.B.; Dagenais, S. Scientific Basis for the Therapeutic Use of *Withania somnifera* (Ashwagandha): A Review. *Altern. Med. Rev. J. Clin. Ther.* **2000**, *5*, 334–346.
46. Gladen-Kolarsky, N.; Monestime, O.; Bollen, M.; Choi, J.; Yang, L.; Magaña, A.A.; Maier, C.S.; Soumyanath, A.; Gray, N.E. *Withania somnifera* (Ashwagandha) Improves Spatial Memory, Anxiety and Depressive-like Behavior in the 5xFAD Mouse Model of Alzheimer's Disease. *Antioxidants* **2024**, *13*, 1164. [CrossRef]
47. Kumar, P.; Singh, R.; Nazmi, A.; Lakhanpal, D.; Kataria, H.; Kaur, G. Glioprotective Effects of Ashwagandha Leaf Extract against Lead Induced Toxicity. *BioMed Res. Int.* **2014**, *2014*, 182029. [CrossRef]
48. Kataria, H.; Wadhwa, R.; Kaul, S.C.; Kaur, G. *Withania somnifera* Water Extract as a Potential Candidate for Differentiation Based Therapy of Human Neuroblastomas. *PLoS ONE* **2013**, *8*, e55316. [CrossRef]
49. Yan, X.; Huang, G.; Liu, Q.; Zheng, J.; Chen, H.; Huang, Q.; Chen, J.; Huang, H. Withaferin A Protects against Spinal Cord Injury by Inhibiting Apoptosis and Inflammation in Mice. *Pharm. Biol.* **2017**, *55*, 1171–1176. [CrossRef]
50. Gupta, S.K.; Dongare, S.; Mathur, R.; Mohanty, I.R.; Srivastava, S.; Mathur, S.; Nag, T.C. Genistein Ameliorates Cardiac Inflammation and Oxidative Stress in Streptozotocin-Induced Diabetic Cardiomyopathy in Rats. *Mol. Cell. Biochem.* **2015**, *408*, 63–72. [CrossRef] [PubMed]
51. Mitra, S.; Munni, Y.A.; Dash, R.; Sultana, A.; Moon, I.S. Unveiling the Effect of *Withania somnifera* on Neuronal Cytoarchitecture and Synaptogenesis: A Combined in Vitro and Network Pharmacology Approach. *Phytother. Res.* **2022**, *36*, 2524–2541. [CrossRef] [PubMed]
52. Oh, J.H.; Kwon, T.K. Withaferin A Inhibits Tumor Necrosis Factor-Alpha-Induced Expression of Cell Adhesion Molecules by Inactivation of Akt and NF-kappaB in Human Pulmonary Epithelial Cells. *Int. Immunopharmacol.* **2009**, *9*, 614–619. [CrossRef]
53. Candelario, M.; Cuellar, E.; Reyes-Ruiz, J.M.; Darabedian, N.; Feimeng, Z.; Miledi, R.; Russo-Neustadt, A.; Limon, A. Direct Evidence for GABAergic Activity of *Withania somnifera* on Mammalian Ionotropic GABAA and GABAϱ Receptors. *J. Ethnopharmacol.* **2015**, *171*, 264–272. [CrossRef] [PubMed]
54. Bhattacharya, S.K.; Bhattacharya, A.; Sairam, K.; Ghosal, S. Anxiolytic-Antidepressant Activity of *Withania somnifera* Glycowithanolides: An Experimental Study. *Phytomedicine* **2000**, *7*, 463–469. [CrossRef]
55. Bhatnagar, M.; Sharma, D.; Salvi, M. Neuroprotective Effects of *Withania somnifera* Dunal.: A Possible Mechanism. *Neurochem. Res.* **2009**, *34*, 1975–1983. [CrossRef]

biomolecules

MDPI

Article

Antimicrobial and Antiradical Activity of Extracts from Leaves of Various Cultivars of *Pyrus communis* and *Pyrus pyrifolia*

Beata Żbikowska [1], Magdalena Kotowska [2], Andrzej Gamian [3], Katarzyna Patek [4], Katarzyna Matuła [4], Daria Augustyniak [5], Kamila Korzekwa [6] and Zbigniew Sroka [1,*]

1 Department of Pharmacognosy and Herbal Medicine, Wroclaw Medical University, Borowska 211a, 50-556 Wrocław, Poland; beata.zbikowska@umw.edu.pl
2 Laboratory of Molecular Biology of Microorganisms, Hirszfeld Institute of Immunology and Experimental Therapy, Polish Academy of Sciences, Rudolfa Weigla 12, 53-114 Wrocław, Poland
3 Department of Immunology of Infectious Diseases, Hirszfeld Institute of Immunology and Experimental Therapy, Polish Academy of Sciences, Rudolfa Weigla 12, 53-114 Wrocław, Poland; andrzej.gamian@hirszfeld.pl
4 Faculty of Pharmacy, Wroclaw Medical University, Borowska 211a, 50-556 Wrocław, Poland
5 Department of Pathogen Biology and Immunology, University of Wroclaw, S. Przybyszewskiego 63, 51-148 Wrocław, Poland; daria.augustyniak@uwr.edu.pl
6 Department of Microbiology, University of Wroclaw, S. Przybyszewskiego 63, 51-148 Wrocław, Poland; kamila.korzekwa@uwr.edu.pl
* Correspondence: zbigniew.sroka@umw.edu.pl

Abstract: Certain plant raw materials are rich in antioxidant and antimicrobial compounds, which are highly valued in modern medicine. These include the leaves of various species and cultivars of pears. For our research, we chose the leaves of the common pear (*Pyrus communis*) and Asian pear (*Pyrus pyrifolia*). Four different extracts were obtained from all raw materials and were investigated for their antimicrobial and antioxidant activity. The content of total phenolics and flavonoids was measured using colorimetric methods, and antiradical activity was measured using DPPH and ABTS radical probes. The antimicrobial activity of extracts was measured using the disc diffusion method, and the amount of major antimicrobial components (hydroquinone and arbutin) was measured using the HPLC method. The highest amount of general phenols and flavonoids was found in ethyl acetate extracts in all cultivars, and the lowest amount of phenols was found in the remaining aqueous solution. The amount of general phenols positively correlated with the antiradical activity of extracts. The strongest antimicrobial activity against Gram-positive and Gram-negative pathogens corresponded to the highest content of hydroquinone and arbutin in ethyl acetate extracts. Extracts obtained from pear leaves showed an average content of phenolic compounds and average antiradical activity compared to extracts from other raw materials, especially green tea or bergenia leaves. The amount of hydroquinone was moderate, lower than that of arbutin. The antimicrobial activity of the extracts was moderate due to the average amount of hydroquinone, which is the main antimicrobial compound.

Keywords: pear leaf extracts; arbutin; hydroquinone; phenolic compounds; antibacterial activity; antioxidant activity

1. Introduction

The golden era of antibiotics was between 1940 and 1965, when they were discovered and introduced into modern medicine [1]. The improper usage of antibiotics has led to growing bacterial resistance, which presents an increasing threat to contemporary medicine. This increases the risk of life-threatening infections in hospital wards and facilitates, the spread of resistant bacteria from hospital units to the wider society, leading to an escalating prevalence of antibiotic-resistant pathogens [2–12]. The other reasons for bacterial resistance are the inappropriate use of antibiotics in the population in general, as well as the use of these drugs as feed additives in the nutrition of farm animals.

Particularly dangerous are infections with highly and multidrug-resistant Gram-negative bacteria such as *Pseudomonas aeruginosa, Klebsiella pneumoniae,* and *Acinetobacter baumannii,* which belong to the so-called ESCAPE group of pathogens (*Enterococcus faecium, Staphylococcus aureus, Klebsiella pneumoniae, Acinetobacter baumannii, Pseudomonas aeruginosa,* and *Enterobacter* spp.), causing most nosocomial infections [1].

The problem of antibiotic resistance is complicated by too few new effective antimicrobial drugs being developed, while antibiotics' effectiveness is falling sharply [13]. In addition, economic problems will increase due to the rising costs of treatment, health care, and mortality. This significantly undermines the effectiveness of current antimicrobial therapies, raising serious concerns for the future fight against bacterial and fungal diseases in both developed and developing countries.

To alleviate these urgent problems, funding should be given to research laboratories engaged in the search for new antimicrobial drugs, either by chemical synthesis or by the search for naturally occurring antimicrobial compounds [14,15] or by looking for inhibitors of enzymes that break down antibiotics, e.g., carbapenemases [16]. A potential alternative or complement to antibiotics may be the use of bacteriophages, i.e., viruses that can kill bacteria [17–19], or phage-derived enzymes, including endolysins and depolymerases [20]. Phytotherapy is also of great interest based on the search for naturally occurring antimicrobial compounds in medicinal and other plants [14,15]. There are many such plants whose extracts are effective against clinically relevant pathogens. For example, the antimicrobial activities of cranberry extract inhibiting the growth and limiting the biofilm formation of *Enterococcus faecalis* isolated from urinary tract infections have been confirmed [21]. Likewise, extracts from *Syzygium aromaticum, Punica granatum,* and *Thymus vulgaris* were effective against bacteria such as *Bacillus cereus, Staphylococcus aureus, Escherichia coli, Pseudomonas aeruginosa,* and *Salmonella typhi,* which cause food poisoning diseases [22]. Finally, extracts from *Pistacia lentiscus, Brassica oleracea, Glycyrrhiza glabra, Camellia sinensis, Cinnamomum cassia, Allium sativum, Nigella sativa,* and the leaves of various species of the genus *Bergenia* and natural compounds such as phenols—resveratrol, curcumin, quercetin, arbutin, and hydroquinone—and the organosulfur compound allicin demonstrated high antimicrobial activity against *H. pylori* [23,24]. Many similar examples could be provided, but we will stop here, as this is not a review work.

In addition to antibiotic resistance, another very important task for modern medicine is combating free radicals and oxidative stress resulting from an imbalance between the generation and neutralization of reactive oxygen and nitrogen species (ROS, RNS), as their excess occurs during various metabolic disorders and chronic diseases [25–29]. Excess of free radicals, ROS, and RNS leads to damage of important cellular structures such as lipids (the major components of membranes), causing the peroxidation of polyunsaturated fatty acids and oxidation of active and structural proteins, as well as DNA or RNA [30–32]. This, in turn, may lead to tissue damage, initiating many pathological processes, such as metabolic disorders (type II diabetes mellitus, obesity) [33], cancers [34], and chronic inflammatory disorders such as autoimmune diseases [35], which can also additionally generate an increase in ROS levels [36–38]. ROS affect wound healing, cell proliferation, granulation formation, and extracellular matrix formation and also harm wound healing [39]. Phytochemicals are increasingly used to treat inflammation-associated diseases by reducing the production of pro-inflammatory cytokines or regulating molecular mechanisms that synergize with the increased production of anti-inflammatory cytokines [40]. It has been documented that plant extracts from flax can inhibit human skin cell inflammation and cause remodeling of the extracellular matrix and wound closure activation [41]. The intensification of processes that generate free radicals destroys the body and accelerates the aging of organisms [42]. Plants and plant-derived compounds are therefore cited as a promising alternative or supportive components in strategies used to reduce microbial load or excessive oxidative stress [43]. The major groups of antimicrobial and antioxidant compounds made by plants comprise phenolics and polyphenols, terpenoids, alkaloids, and flavonoids. Among these compounds, phenols and polyphenols have the greatest power to reduce oxidative processes, and their activity strongly depends on their chemical structure [44]. Furthermore, mixtures of plant bioactive compounds are more effective than purified ones due to combination interactions [45].

The *Pyrus* species has been used in traditional medicine for over 2000 years. The genus *Pyrus* includes many species, such as Chinese pear (*Pyrus ussuriensis*; *Pyrus bretschneideri*), Asian pear (*Pyrus pyrifolia*), and the common pear (*Pyrus communis*) [46]. Different parts of pears, including leaves, contain both antimicrobial (arbutin and hydroquinone) [47,48] and antioxidant (polyphenols) compounds [24].

In this work, extracts prepared from leaves of different cultivars of *Pyrus pyrifolia* and *Pyrus communis* were tested for antimicrobial and antioxidant activity. The general amount of phenols was determined with colorimetric methods, and the amounts of hydroquinone and arbutin were measured with high-pressure liquid chromatography (HPLC).

2. Materials and Methods

2.1. Raw Material

The leaves used as research raw material were obtained from an experimental orchard of Wrocław University of Environmental and Life Sciences. There were three cultivars of Asian pear (*Pyrus pyrifolia* (Burm.f.) Nak.—cultivars *Chojuro, Kosui, and Shu-Li*) and three cultivars of common pear (*Pyrus communis* L.—cultivars *Dicolor, Clapp's favourite*, and *Radana*). Voucher specimens were deposited in the Department of Horticulture (Wrocław University of Environmental and Life Sciences, Poland).

2.2. Reagents

The following reagents were used: analysis-grade methanol, Chempur Poland; gradient-grade methanol, Merck, Darmstadt, Germany; analytical-grade ethyl acetate, Chempur, Poland; Folin–Ciocalteu's reagent, Fluka, Switzerland, analytical-grade sodium carbonate, Chempur, Poland; DPPH (2,2-diphenyl-1-picrylhydrazyl) free radical, Sigma-Aldrich, Steinheim, Germany; ABTS (2,2′-azino-bis(3-ethylbenzothiazoline-6-sulfonic acid) diammonium salt) and potassium persulfate Sigma-Aldrich Steinheim, Germany; analytical-grade anhydrous aluminum chloride, Chempur, Poland; arbutin and hydroquinone, Sigma-Aldrich, Steinheim, Germany.

2.3. Apparatus

Spectrophotometer Cecil 3021, Cambridge, UK;

Spectrophotometer Hitachi U5100, Tokyo, Japan;

The DIONEX UHPLC Ultimate 3000 apparatus (Thermo Fisher Scientific, Waltham, MA, USA).

2.4. Preparation of Extracts

Extracts were prepared according to the method described by Sroka et al. [24], with modifications. A sample of 50 g of dried and crushed leaves from various cultivars of *Pyrus pyrifolia* and *Pyrus communis* was extracted with 900 mL of 60% solution of methanol in water at 45 °C for 48 h. Then, the extract was filtered with filter paper (Filtrak 388). A total of 180 mL of the extract was condensed under reduced pressure and lyophilized to obtain **EA** extract. The remaining part of the solution, with a volume of 720 mL, was condensed under reduced pressure to dryness and dissolved in 600 mL of water at 45 °C. The obtained solution was stored in a refrigerator at 4 °C for two days. The precipitate formed was separated using a filter (Filtrak 388) and freeze-dried to obtain residue **EB**. After **EB** separation, the aqueous solution was exhaustively extracted with ethyl acetate using a separatory funnel. The ethyl acetate solution was condensed under reduced pressure to obtain residue **EC**. After ethyl acetate extraction, the remaining aqueous solution was condensed under reduced pressure and then freeze-dried to obtain residue **ED**. Extracts obtained from different cultivars of *Pyrus communis* were labeled by adding the last letter corresponding to the cultivar. Extracts obtained from the *Radana* cultivar were identified as **EAR**, **EBR**, **ECR**, and **EDR** (weights of extracts [g] 2.943, 0.831, 1.894, and 8.851 g); extracts from *Clapp's favourite* as **EAF**, **EBF**, **ECF**, and **EDF** (3.573, 0.729, 1.53, and 8.088 g); and extracts from *Dicolor* as **EAD**, **EBD**, **ECD**, and **EDD** (2.858, 0.517, 2.286, and 8.803 g). Extracts from the *Pyrus pyrifolia Chojuro* cultivar were identified as **EAC**, **EBC**, **ECC**, and **EDC** (4.067, 0.126, 0.966, and 7.007 g); extracts from *Kosui* as **EAK**, **EBK**, **ECK**, and **EDK** (5.812, 0.369, 1.736, and 9.518 g); and extracts from *Shu li* as **EAS**, **EBS**, **ECS**, and **EDS** (5.278, 0.271, 2.072, and 6.166 g).

2.5. Colorimetric Measurement of Total Phenol Content

The total amount of phenolic compounds in extracts was measured using the colorimetric method outlined by Singleton and Rossi [49] with Folin–Ciocalteu phenol reagent [50,51], which is a mixture of phosphotungstate and phosphomolybdate.

To the test tubes, 7 mL of water and 0.5 mL of methanol solution of extract at 1 mg/mL for **EA**, **EB**, and **EC** and 2 mg/mL for extract **ED** were added, followed by 0.5 mL of Folin–Ciocalteu phenol reagent. After three minutes, 2 mL of a 20% solution of Na_2CO_3 in water was added to the samples. The samples were heated for 1 min in a boiling water bath. After cooling, the samples were centrifuged at 5500 RCF. The absorbance of supernatants was measured at a wavelength of 685 nm. Measurements were repeated five times. The amount of phenolic compounds was expressed in mg of phenols per mg of extract (gallic acid equivalents per mg of extract, GAE/mg) and per g of raw material (GAE/g).

2.6. Measurement of Flavonoids in Extracts

Flavonoid content in extracts was measured according to the method first proposed by Christ and Muller [52] using the aluminum chloride ($AlCl_3$) method. Flavonoids form with aluminum chloride (III) complexes with a yellow color, which absorb at a wavelength λ = 430 nm.

A total of 1 mL of methanolic solution of extracts at 1 mg/mL for extracts **EA**, **EB**, and **EC** and 2 mg/mL for extract **ED** was added to the test tube, and then 1 mL of 2% aluminum chloride solution in methanol was added. After 30 min of storage of samples in a dark place, absorbance was measured at a wavelength λ = 430 nm in glass cuvettes with an optical path length of 1 cm. Measurements for each sample were repeated five times.

The amount of flavonoids was expressed in mg of flavonoids per mg of extracts (as quercetin equivalents per mg of extracts, QE/mg) and per g of raw material (QE/g).

2.7. Investigation of Antioxidant Activity of Extracts

2.7.1. Investigation of Antioxidant Activity of Extracts Using $ABTS^{\bullet+\cdot}$ Radical

The reduction in the cationic radical $ABTS^{\bullet+}$ determines the activity [53], which is obtained by the oxidation of ABTS (2,2′-azino-bis(3-ethylbenzothiazoline-6-sulfonic acid) diammonium salt) with potassium persulfate ($K_2S_2O_8$). The color of the cationic radical is green-blue, the intensity of which decreases in the presence of the reductants, such as phenolic compounds. This is partly achieved through HAT (hydrogen atom transfer) and ET (electron transfer) types of reaction. The ABTS radical solution was prepared as follows. The solution ABTS in water at 7 mM was mixed with 2.45 mM aqueous solution of $K_2S_2O_8$ in a volume ratio of 1:1, then left in a dark place for 16 h for the cation radical $ABTS^{\bullet+}$ to form. The final solution was diluted with 50% methanol solution so that the absorbance was equal to 1.

To measure antiradical activity, the extract was dissolved in a 1:1 methanol–water mixture (**EC** extract) or water **EA**, **EB**, and **ED** extracts. A reagent control sample was prepared by adding 20 µL of 50% methanol in water to 2 mL of $ABTS^{\bullet+}$ radical solution. The absorbance was measured at the beginning and after 1 min of the reaction at 25 °C using glass cuvettes with an optical path length equal to 1 cm. The maximal error was calculated using the total differential method.

The antiradical activity was calculated using the method described by Sroka et al. [54].

The antiradical activity was presented as the number of antiradical units per mg of extract $TAU_{ABTS/mg}$ and g of raw material $TAU_{ABTS/g}$. One unit of antiradical activity is the amount of extract that causes the elimination of 1 mmole of $ABTS^{\bullet+}$ radical within 1 min, at the temperature of 25 °C, measured at wavelength λ = 734 nm.

2.7.2. Calculation of the Number of $TAU_{ABTS/mg}$ and $TAU_{ABTS/g}$ Units

To calculate the number of antiradical units, the molar absorption coefficient ε for $ABTS^{\bullet+}$ equal to 1.5×10^4 dm^3 mol^{-1} cm^{-1} was used.

The number of antiradical units per mg of extracts was calculated according to the following equation:

$$TAU_{ABTS/mg} = 6.8 \times 10^{-2} \frac{(A_{s0} - A_{s1}) - (A_{c0} - A_{c1})}{m}$$

where $TAU_{ABTS/mg}$ is the number of antiradical units per mg of extract; A_{s0} is the absorbance of the sample at the beginning of the reaction; $A_{s1}$is the absorbance of the sample after 1 min of the reaction; A_{c0} is the absorbance of the control sample at the beginning of the reaction; A_{c1} is the absorbance of the control sample after 1 min of the reaction; m is the amount of extract [mg] in 1 mL of the sample.

Because $A_{c0} - A_{c1}$ is always equal to 0, the equation was simplified to

$$TAU_{ABTS/mg} = 6.8 \times 10^{-2} \frac{A_{s0} - A_{s1}}{m}$$

The amount of antiradical units per g of raw material was calculated according to the following equation:

$$TAU_{ABTS/g} = \frac{(TAU_{\frac{ABTS}{mgEA}} \cdot m_{EA}) + (TAU_{\frac{ABTS}{mgEB}} \cdot m_{EB}) + (TAU_{\frac{ABTS}{mgEC}} \cdot m_{EC}) + (TAU_{\frac{ABTS}{mgED}} \cdot m_{ED})}{W_R}$$

where $TAU_{ABTS/g}$ is number of antiradical units per g of raw material; $TAU_{ABTS/mgEA}$ is the number of antiradical units per mg of extract **EA**; m_{EA} is total mass of extract **EA** [mg]; $TAU_{ABTS/mgEB}$ is the number of antiradical units per mg of extract **EB**; m_{EB} is total mass of extract **EB** [mg]; $TAU_{ABTS/mgEC}$ is the number of antiradical units per mg of extract **EC**; m_{EC} is total mass of **EC** extract [mg]; $TAU_{ABTS/mgED}$ is the number of antiradical units per mg of extract **ED**; m_{ED} is total mass of extract **ED** [mg]; W_R is total weight of raw material used for extraction [g].

2.7.3. Measurement of Antiradical Activity of Extracts Using DPPH Radical

Extracts' antiradical features were measured using the DPPH (2,2-diphenyl-picrylhydrazyl) radical, whose blue color changes in the presence of compounds with reducing properties such as phenols [55]. This is a HAT type of reaction involving the transfer of hydrogen atoms. DPPH in the radical form absorbs at λ = 515 nm; in the presence of antioxidants, absorption decreases.

A total of 2 mg of DPPH was dissolved in 54 mL of methanol. The solution was left until the absorbance stabilized at 515 nm. The absorbency of the DPPH solution was adjusted with methanol to a value of 1. The test was performed by adding 50 µL of extract solution to 2 mL of DPPH solution in methanol. The control sample was prepared by adding 50 µL of methanol, instead of extract solution, to DPPH. The absorbency was measured at the beginning of the reaction and after 1 min. Measurements were repeated 5 times. The maximal error was calculated using the total differential method.

The antiradical activity was presented as the number of antiradical units per mg of extract $TAU_{DPPH/mg}$ and g of raw material $TAU_{DPPH/g}$. One unit of antiradical activity is the amount of extract that causes the elimination of 1 µmole of DPPH radical within 1 min, at the temperature of 25 °C, measured at wavelength λ = 515 nm.

2.7.4. Calculation of the Number of $TAU_{DPPH/mg}$ Units

To calculate the number of antiradical units per mg of extract, the molar absorption coefficient ε for DPPH equal to 1.25×10^4 dm^3 mol^{-1} cm^{-1} was used.

The amount of antiradical units per mg of extracts was calculated according to the following equation:

$$TAU_{DPPH/mg} = 8.0 \times 10^{-2} \frac{(A_{D0} - A_{D1}) - (A_{C0} - A_{C1})}{m}$$

where $TAU_{515/\mathrm{mg}}$ is the number of antioxidant units calculated per mg of extract; A_{D0} is the absorbance DPPH solution at the beginning of the reaction; A_{D1} is the absorbance of DPPH solution after 1 min of reaction; A_{C0} is the absorbance of the control sample at the beginning of the reaction; A_{C1} is the control sample after 1 min of reaction; m is the amount of extract in mg in 1 cm^3 of sample.

Because $A_{C0} - A_{C1}$ is always equal to 0, the equation was simplified to

$$TAU_{DPPH/mg} = 8.0 \times 10^{-2} \frac{A_{D0} - A_{D1}}{m}$$

The amount of antiradical units per g of raw material was calculated according to the following equation:

$$TAU_{DPPH/g} = \frac{(TAU_{\frac{DPPH}{mgEA}} \cdot m_{EA}) + (TAU_{\frac{DPPH}{mgEB}} \cdot m_{EB}) + (TAU_{\frac{DPPH}{mgEC}} \cdot m_{EC}) + (TAU_{\frac{DPPH}{mgED}} \cdot m_{ED})}{W_R}$$

where $TAU_{DPPH/g}$ is the number of antiradical units per g of raw material; $TAU_{DPPH/mgEA}$ is the number of antiradical units per mg of extract **EA**; m_{EA} is the total mass of extract **EA** [mg]; $TAU_{DPPH/mgEB}$ is the number of antiradical units for mg of extract **EB**; m_{EB} is the total mass of extract **EB** [mg]; $TAU_{DPPH/mgEC}$ is the number of antiradical units per mg of extract **EC**; m_{EC} is the total mass of **EC** extract [mg]; $TAU_{DPPH/mgED}$ is the number of antiradical units per mg of extract **ED**; m_{ED} is the total mass of extract **ED** [mg]; W_R is the total weight of raw material used for extraction [g].

2.8. Measurement of Arbutin and Hydroquinone in Extracts Obtained from Leaves of Pyrus communis and Pyrus pyrifolia

Analysis of the amounts of arbutin and hydroquinone in extracts from leaves of *Pyrus communis* and *Pyrus pyrifolia* was performed with the Dionex UHPLC Ultimate 3000 apparatus (Thermo Fisher Scientific, USA) equipped with degasser (UltiMate 3000), pump (LPG-3400 SD), autosampler (WPS 3000) with dynamic mixing chamber (Reodyne 7770-362), temperature-controlled (10 °C) diode array detector (DAD-3000) and column thermostat (TCC 3000), Hypersil GOLD C18 column (250 × 4.6 mm, 5 μm), and C18 precolumn (10 × 4.6 mm, 5 μm), both from Thermo Scientific, USA. The analysis was carried out using a gradient as described by Fecka et al. [56]: A (water: formic acid, 98.5: 1.5, *v/v*) and B) acetonitrile: formic acid, 98.5: 1.5, *v/v*). The following gradient was applied: 0–30 min (3–30%, B in A), 30–33 min (30–70%, B in A), 33–45 min. (70%, B in A isocratic), and 45–48 min (70–100%, B in A). Then, the system was equilibrated to the starting condition of 3% B in A. The flow rate was 1.2 mL/min, the injection volume was 5 μL, and the temperature of analysis was 22 ± 1 °C. Spectral measurements were made at the wavelength of 190–800 nm, in steps of 1 nm and bunch width of 1 nm. Chromatograms were recorded at wavelengths 254, 271, 280, 283, 289, and 350 nm.

2.9. Investigation of Antimicrobial Activity of Extracts

2.9.1. Bacterial Strains and Storage

The following Gram-positive and Gram-negative bacteria were used: (I) reference strains including *S. aureus* ATCC 25923, *E. faecalis* ATCC 25212, *E. coli* ATCC 25922, *P. aeruginosa* ATCC 27853, *H. pylori* ATCC 43504, and nontypeable *Haemophilus influenzae* (NTHi) ATCC 49247; (II) clinical strains including *S. aureus* MRSA 112, *E. faecalis* HLAR 78, *E. coli* ESBL 79, *P. aeruginosa* 256, nontypeable *Haemophilus influenzae* (NTHi 6), and *Bacillus subtilis* 21.

Most of the tested strains belong to the collection of the Department of Microbiology, University of Wroclaw. *H. influenzae* strains and *P. aeruginosa 256* belong to the Department of Pathogen Biology and Immunology, University of Wroclaw. *H. pylori* and *E. faecalis* were kindly lent from the Department of Microbiology, Wroclaw Medical University. Strains were stored frozen at −70 °C in relevant media, as described below, supplemented with 20% glycerol.

2.9.2. Cultivation of Microorganisms

For experiments, the aerobic/facultative anaerobic bacterial strains (*S. aureus*, *E. faecalis*, *E.coli*, *P. aeruginosa*, and *B. subtilis*) were refreshed on Trypticase Soy Agar (TSA) (Becton Dickinson and Company, Cockeysville, MD, USA) at 37 °C for 18 h, whereas the microaerophilic strains were cultivated on Mueller–Hinton with 5% horse blood and NAD (*H. influenzae*, 24 h) or Columbia Agar with 5% sheep blood plus (*H. pylori*, 48 h) (Thermo Scientific, Wesel, Germany). *H. influenzae* and *H. pylori* were incubated in a humidified atmosphere with 5% CO_2 and in a gas pack (GENbag microaer) with atmosphere of 80% N_2, 10% H_2, and 10% CO_2 (bioMerieux, Craponne, France), respectively.

2.9.3. Measurement of Antimicrobial Activity of Extracts

Plant extracts were diluted to a concentration of 100 mg/mL in DMSO, mixed vigorously on a shaker until completely dissolved, portioned, and stored in a refrigerator until use. For microbiological purposes, the disc diffusion method was used as recommended by EUCAST (Disc Diffusion Method for Antimicrobial Susceptibility Testing—Version 9.0, January 2021). Aerobic/facultative anaerobic bacteria and *H. influenzae* were inoculated at a concentration of ~1.5×10^8 cfu/mL, referring to McFarland, 0.5 in 0.085% NaCl on

Mueller–Hinton Agar 2. *H. pylori* was inoculated at a concentration of ~12 × 10^8 cfu/mL, corresponding to McFarland 4, in Schaedler broth + vit. K3 (bioMerieux, Marcy-l'Etoile, France). Ten microliters with 1 mg of each extract were applied to 6 mm diameter blotting paper discs placed on Mueller–Hinton Agar 2 for aerobic/facultative anaerobic strains and Mueller–Hinton agar supplemented with 5% horse blood and NAD for microaerophilic strains. The plates were incubated for 18–24 h at 37 °C, except for those with *H. pylori*, which were incubated for 48 h. After incubation, growth inhibition zones were measured (as diameter given in millimeters). Analyses were performed in duplicate in three independent experiments. The positive control was a relevant synthetic antibiotic. The negative control was DMSO alone.

2.9.4. Investigation of Inhibition of Growth of Extracts on *Campylobacter Jejuni* NCTC 11168

Due to increasingly frequent infections with Gram-negative *Campylobacter jejuni*, especially in the elderly and immunocompromised [57], the effect of this bacterium was additionally examined as described below.

The *Campylobacter jejuni* strain NCTC 11168 was obtained from the collection of the Laboratory of Molecular Biology of Microorganisms, Hirszfeld Institute of Immunology and Experimental Therapy, Polish Academy of Sciences.

C. jejuni was cultured in Brain Heart Infusion (Oxoid) (BHI) to OD_{600} = ~0.6 and then diluted to OD_{600} = 0.1 in BHI. Next, diluted bacteria culture was spread evenly with a cotton swab on the surface of square plates (12 cm × 12 cm) containing 40 mL of Columbia blood agar base medium supplemented with 10% fetal bovine serum and an antibiotic mix according to Contreras et al. [58].

Plant extracts were dissolved in DMSO to a concentration of 100 mg/mL and applied on Whatman GF/F sterile discs with a diameter of 6 mm (10 µL/disc). Discs were then placed on the surface of the inoculated plates. H_2O_2 (5 µL of 2% solution/disc) was used as a positive control.

C. jejuni was cultivated for 48 h at 42 °C under microaerobic conditions (3.5% H_2, 6% O_2, 7% CO_2, and 85% N_2) generated using the jar evacuation–replacement method (the Anaerobic Gas System PetriSphere).

Tests were perfromed in triplicate. Results are shown as mean ± standard deviation.

3. Results

3.1. Amounts of Phenolic Compounds and Flavonoids in Extracts and Raw Materials

The amounts of phenolic compounds in extracts and raw materials are presented in Tables 1 and 2 and Figures 1–3. The highest total phenol amount was measured in extract **ECK** (0.416 ± 0.011, GAE/mg) obtained from *P. pyrifolia* cultivar *Kosui* leaves. The highest amounts of phenolic compounds were generally detected in ethyl acetate extracts in all *P. pyrifolia* and *P. communis* cultivars with the following GAE/mg values: 0.401 ± 0.0015, 0.377 ± 0.0019, 0.347 ± 0.0008, 0.338 ± 0.009, and 0.334 ± 0.009 for extracts **ECF**, **ECD**, **ECR**, **ECC**, and **ECS**, respectively.

The lowest amount of total phenols was found for lyophilized aqueous residues: GAE/mg values 0.108 ± 0.003, 0.092 ± 0.002, 0.092 ± 0.0018, 0.083 ± 0.0004, 0.077 ± 0.0004, and 0.067 ± 0.002, for extracts **EDS**, **EDK**, **EDF**, **EDD**, **EDR**, and **EDC**, respectively.

Table 1. The amount of general phenols (GAE/mg, GAE/g) and flavonoids (QE/mg, QE/g) in mg of extracts and g of dry leaves of *Pyrus communis* L. Antiradical activity of extracts from leaves of *Pyrus communis* measured with the use of $ABTS^{\bullet+}$ and DPPH radicals and expressed as the number of antiradical units per mg of extract ($TAU_{ABTS/mg}$, $TAU_{DPPH/mg}$) and grams of raw material ($TAU_{ABTS/g}$, $TAU_{DPPH/g}$) (±maximal error − ME).

Raw Material	Extracts	GAE/mg	GAE/g	QE/mg	QE/g	$TAU_{ABTS/mg}$	$TAU_{ABTS/g}$	$TAU_{DPPH/mg}$	$TAU_{DPPH/g}$
Pyrus communis "Dicolor"	**EAD**	0.211 ± 0.0010	45.61 ± 1.16	0.019 ± 0.0001	5.61 ± 0.08	2.29 ± 0.17	450.3 ± 35.2	1.10 ± 0.06	169.4 ± 13.9
	EBD	0.171 ± 0.0018		0.028 ± 0.0001		1.38 ± 0.07		1.02 ± 0.05	
	ECD	0.377 ± 0.0019		0.068 ± 0.0002		2.91 ± 0.17		1.24 ± 0.04	
	EDD	0.083 ± 0.0004		0.006 ± 0.0001		0.98 ± 0.04		0.22 ± 0.03	
Pyrus communis "Clapp's favourite"	**EAF**	0.197 ± 0.0017	43.73 ± 2.07	0.022 ± 0.0001	6.42 ± 0.05	2.64 ± 0.14	501.6 ± 35.0	0.98 ± 0.05	169.9 ± 15.4
	EBF	0.170 ± 0.0025		0.039 ± 0.0001		2.73 ± 0.22		0.96 ± 0.05	
	ECF	0.401 ± 0.0015		0.100 ± 0.0002		2.81 ± 0.15		1.02 ± 0.04	
	EDF	0.092 ± 0.0018		0.007 ± 0.0001		1.16 ± 0.04		0.34 ± 0.04	
Pyrus communis "Radana"	**EAR**	0.152 ± 0.0010	38.58 ± 0.59	0.016 ± 0.0001	4.57 ± 0.10	1.31 ± 0.07	378.3 ± 27.9	0.80 ± 0.05	186.5 ± 11.0
	EBR	0.174 ± 0.0003		0.029 ± 0.0002		1.74 ± 0.09		0.95 ± 0.05	
	ECR	0.347 ± 0.0008		0.059 ± 0.0003		2.93 ± 0.19		1.12 ± 0.03	
	EDR	0.077 ± 0.0004		0.005 ± 0.0001		0.91 ± 0.04		0.46 ± 0.01	

Table 2. The amount of general phenols (GAE/mg, GAE/g) and flavonoids (QE/mg, QE/g) in mg of extracts and g of dry leaves of *Pyrus pyrifolia* L. Antiradical activity of extracts from leaves of *Pyrus pyrifolia* measured with the use of $ABTS^{\bullet+}$ and DPPH radicals and expressed as the number of antiradical units per mg of extract ($TAU_{ABTS/mg}$, $TAU_{DPPH/mg}$) and grams of raw material ($TAU_{ABTS/g}$, $TAU_{DPPH/g}$) (±ME).

Raw Material	Extracts	GAE/mg	GAE/g	QE/mg	QE/g	$TAU_{ABTS/mg}$	$TAU_{ABTS/g}$	$TAU_{DPPH/mg}$	$TAU_{DPPH/g}$
Pyrus pyrifolia "Chojuro"	EAC	0.108 ± 0.003	28.88 ± 0.77	0.017 ± 0.001	4.31 ± 0.12	1.17 ± 0.07	294.7 ± 29.6	0.595 ± 0.028	127.47 ± 8.95
	EBC	0.126 ± 0.003		0.035 ± 0.001		0.96 ± 0.070		0.821 ± 0.040	
	ECC	0.338 ± 0.009		0.075 ±0.002		2.277 ± 0.055		1.257 ± 0.052	
	EDC	0.067 ± 0.002		0.006 ± 0.001		0.807 ± 0.089		0.254 ± 0.013	
Pyrus pyrifolia "Kosui"	EAK	0.159 ± 0.004	52.37 ± 1.39	0.016 ± 0.001	6.81 ± 0.19	1.466 ± 0.054	447.59 ± 26.79	0.881 ± 0.024	244.86 ± 12.42
	EBK	0.168 ± 0.005		0.036 ± 0.001		1.332 ± 0.072		1.092 ± 0.034	
	ECK	0.416 ± 0.011		0.103 ± 0.003		2.466 ± 0.095		1.661 ± 0.047	
	EDK	0.092 ± 0.002		0.006 ± 0.001		0.929 ± 0.039		0.389 ± 0.010	
Pyrus pyrifolia "Shu Li"	EAS	0.154 ± 0.004	44.31 ± 1.17	0.020 ± 0.001	7.39 ± 0.21	1.411 ± 0.058	369.93 ± 24.94	1.006 ± 0.064	212.22 ± 14.03
	EBS	0.159 ± 0.004		0.037 ± 0.001		1.181 ± 0.069		1.114 ± 0.034	
	ECS	0.334 ± 0.009		0.099 ± 0.003		2.278 ± 0.115		1.334 ± 0.040	
	EDS	0.108 ± 0.003		0.008 ± 0.001		0.977 ± 0.051		0.364 ± 0.0010	

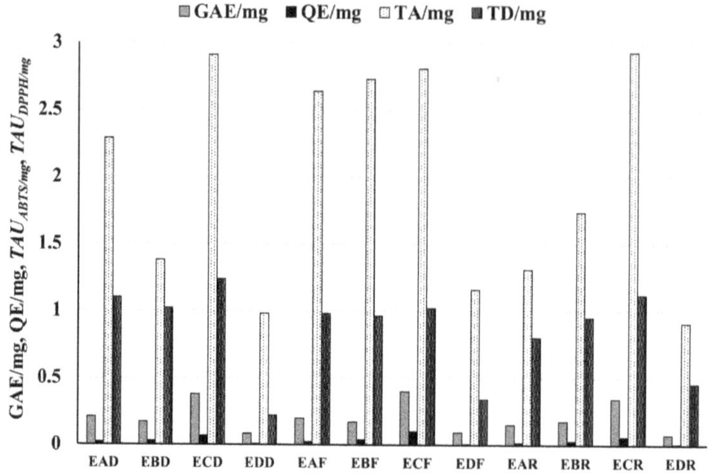

Figure 1. Total phenolic (GAE/mg) and flavonoid (QE/mg) content and antiradical activity ($TAU_{ABTS/mg}$ presented in the figure as TA/mg, $TAU_{DPPH/mg}$ as TD/mg) of extracts from different cultivars of *Pyrus communis*. Antiradical activity is expressed as the number of antiradical units per mg of extracts with the use of cation radical $ABTS^{\bullet+}$ and radical DPPH. An explanation of abbreviations of extracts is provided in the section on preparation of extracts.

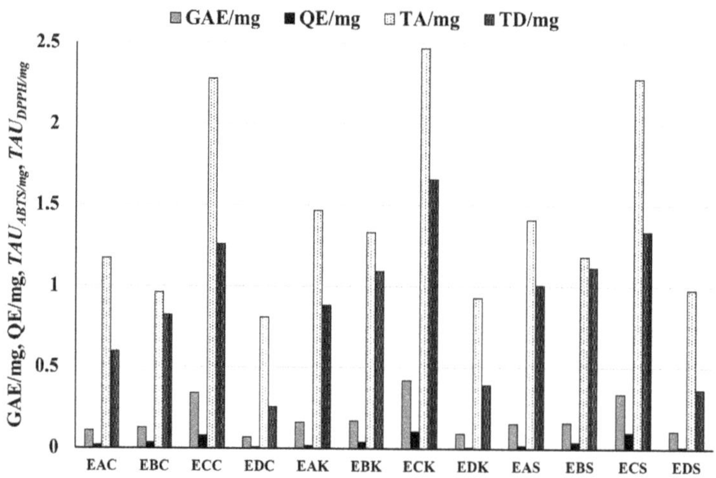

Figure 2. Total phenolic (GAE/mg) and flavonoid (QE/mg) content and antiradical activity ($TAU_{ABTS/mg}$, presented in the figure as TA/mg, $TAU_{DPPH/mg}$ as TD/mg) of extracts from different cultivars of *Pyrus pyrifolia*. Antiradical activity is expressed as the number of antiradical units per mg of extracts with the use of cation radical $ABTS^{\bullet+}$ and radical DPPH. An explanation of abbreviations of extracts is provided in the section on preparation of extracts.

The raw material with the highest amount of total phenols was the leaves of *P. pyrifolia Kosui* (52.37 ± 1.39, GAE/g). The smallest amount of phenolic compounds was found in the leaves of *P. pyrifolia Chojuro* (28.88 ± 0.77, GAE/g) (see Table 2, Figure 3).

The amount of flavonoids was much lower than the total phenols. As with the total phenol content, the highest amount of flavonoids was detected in the ethyl acetate extracts **ECD**, **ECF**, **ECR**, **ECC**, **ECK**, and **ECS**, and the lowest amount of flavonoids was measured for the lyophilized water residues **EDD**, **EDF**, **EDR**, **EDC**, **EDK**, and **EDS**. The highest content of flavonoids was found in the leaves of *P. pyrifolia Shu Li* (7.39 ± 0.21 QE/g), and the lowest in the leaves of *P. pyrifolia Chojuro* (4.31 ± 0.12 QE/g) (see Tables 1 and 2, Figures 1–3).

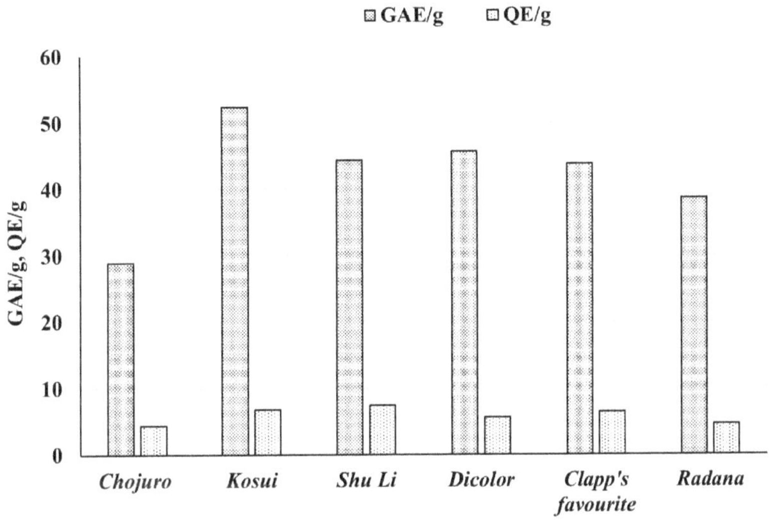

Figure 3. Total phenol (GAE/g) and flavonoid (QE/g) content in leaves of different cultivars of *Pyrus communis* and *Pyrus pyrifolia* expressed in mg per g of raw material. An explanation of abbreviations of extracts is provided in the section on preparation of extracts.

3.2. Antiradical Potential of Extracts and Leaves of *P. pyrifolia* and *P. communis*

The antiradical features of extracts and leaves are presented in Tables 1 and 2 and Figures 1, 2 and 4.

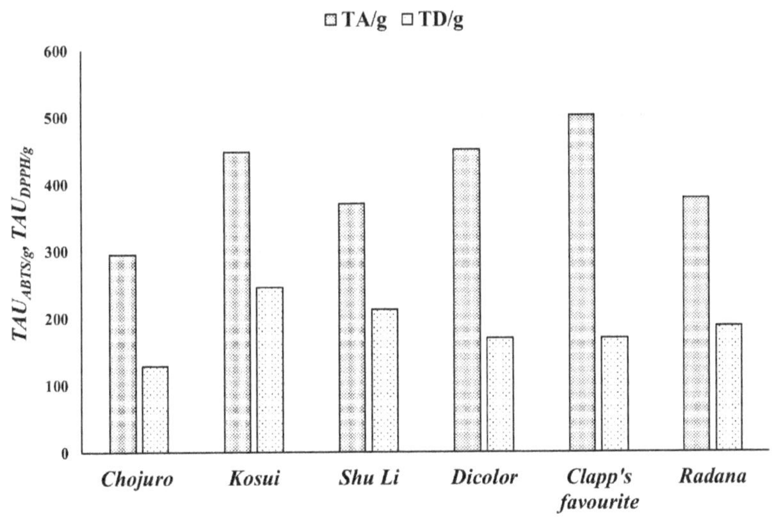

Figure 4. The number of antiradical units per g of raw material ($TAU_{ABTS/g}$ $TAU_{DPPH/g}$) measured using $ABTS^{\bullet+}$ cation radical and DPPH radical. An explanation of abbreviations of extracts is provided in the section on preparation of extracts.

Antiradical activity of extracts was measured using special probes, namely, cation radical ABTS$^{\bullet+}$ (2,2′- azino-bis(3-ethylbenzothiazoline-6-sulfonic acid) diammonium salt) and radical DPPH (2,2-diphenyl-1-picrylhydrazyl). When ABTS was used, the activity was presented as the number of antiradical units per mg of extract and grams of raw materials ($TAU_{ABTS/mg}$, $TAU_{ABTS/g}$). When the DPPH radical was used, the antiradical activity was presented as $TAU_{DPPH/mg}$ and $TAU_{DPPH/g}$.

As with the content of phenolic compounds, the greatest antiradical activity was detected for the ethyl acetate extracts in both the DPPH and ABTS tests.

Namely, for ABTS tests for extracts **ECR**, **ECD**, **ECF**, **ECK**, **ECS**, and **ECC**, the activity was, respectively, 2.93 ± 0.19, 2.91 ± 0.17, 2.81 ± 0.15, 2.47 ± 0.095, 2.278 ± 0.115, and 2.277 ± 0.055 $TAU_{ABTS/mg}$. For DPPH tests, the activity was, respectively, 1.12 ± 0.03, 1.24 ± 0.04, 1.02 ± 0.04, 1.661 ± 0.047, 1.334 ± 0.04, and 1.257 ± 0.052 $TAU_{DPPH/mg}$. The lowest antiradical activity levels were observed for the aqueous residues labeled as **EDD**, **EDF**, **EDR**, **EDC**, **EDK**, and **EDS**. Pearson's correlation coefficient (r) between the amount of total phenols and antiradical activity was 0.85 for the ABTS test and 0.82 for the DPPH test.

When we calculated the antiradical potential per g of raw material for the ABTS test, the highest number of antiradical units was observed for leaves from *P. communis Clapps favoritae* cultivar (501.6 ± 35.0 $TAU_{ABTS/g}$), and leaves from *P. communis Dicolor* (450.3 ± 35.2 $TAU_{ABTS/g}$). The lowest activity was noted for leaves from *P. pyrifolia Chojuro* (294.7 ± 29.6 $TAU_{ABTS/g}$). In the DPPH tests, the highest number of antiradical units was observed for *P. pyrifolia Kosui* (244.86 ± 12.42 $TAU_{DPPH/g}$), and the lowest was observed for *P. pyrifolia Chojuro* (127.47 ± 8.95 $TAU_{DPPH/g}$).

3.3. *Antibacterial Activity of Extracts Obtained from Leaves of Different P. pyrifolia and P. communis Cultivars*

During this study, eleven types of extracts were tested in the *Pyrus pyrifolia* group, and twelve extracts were tested in the *Pyrus communis* group. Analyzing the antibacterial activity of ethyl acetate extracts **ECD**, **ECF**, and **ECR** of *Pyrus communis* (Table 3) and **ECC**, **ECK**, **ECS**, and water residue **EDS** of *Pyrus pyrifolia* (Table 4), a wide spectrum of activity against both aerobic/facultative anaerobic and microaerophilic strains belonging to Gram-positive and Gram-negative pathogens was recorded. All bacterial reference strains from the American Type Culture Collection (ATCC), except *E. faecalis* ATCC 25212, were sensitive to most of these specified extracts, although the level of sensitivity was different. Most of these extracts possess activity against clinical multidrug-resistant *P. aeruginosa* 256 and nontypeable *H. influenzae* NTHi6.

The most susceptible to the plant extract tested, as documented by the largest inhibition zones, were *H. pylori* ATCC 43504 and, to a lesser degree, *B. subtilis* 21. For both strains, growth inhibition was observed for most of the extract formulations derived from *Pyrus pyrifolia* and for *H. pylori* ATCC 43504, and also in the case of the extract from *Pyrus communis*. The above-mentioned most potent extracts from both *Pyrus* species also act against *S. aureus* ATCC 25 923 and *Haemophilus influenzae* (reference and clinical strains).

As documented, the ethyl acetate extracts containing the most hydroquinone (Tables 5 and 6) showed the strongest antibacterial potency (Tables 3 and 4), suggesting that hydroquinone is responsible for the observed antimicrobial effect. This is consistent with our previous research documenting that extracts rich in hydroquinone usually exhibited high antimicrobial activity, where the dependence of activity on the concentration of hydroquinone is proportional [24,59,60].

Table 3. Inhibiting zones [mm] of bacterial growth for extracts from leaves of *Pyrus communis*.No antibacterial activity was found for extracts EDD, EDF, and EDR, and extract EBD was not investigated. S.a.—*Staphylococcus aureus*, *E.c.*—*Escherichia coli*, *P.a.*—*Pseudomonas aeruginosa*, *B.s.*—*Bacillus subtilis*, *NTHi*—nontypeable *Haemophilus influenza*, AMC—augmentin, C—chloramphenicol, CIP—ciprofloxacin, CPX—ciclopirox, DO—doxycillin, FOX—cefoxitin, IMP—imipenem, SXT—sulfamethoxazole, TZP—tazobactum, *SZD*—the sum of inhibition zone diameter for all bacterial strains for extract (±standard deviation—SD). 0 means no antibacterial activity.

Extract	*S.a.*ATCC 25923	*S.a.* MRSA	*E.c.*ATCC 25922	*E.c.*ESBL	*P.a.*ATCC 27853	*P.a.* 256	*B.s.*	*NTHi* ATCC 47249	*NTHi6*	*H.p.* ATCC 43504	*SZD*
EAD	0	0	0	0	0	0	0	0	0	9.2 ± 0.68	9.2 ± 0.68
ECD	11 ± 0.9	0	8.7 ± 0.5	0	9.3 ± 1	7.2 ± 0.6	8 ± 0.9	9.7 ± 0.6	8.5 ± 0.6	10 ± 0.52	72.4 ± 5.62
EAF	0	0	0	0	0	0	0	0	0	10.5 ± 1.2	10.5 ± 1.2
EBF	0	0	0	0	0	0	0	0	0	8.3 ± 0.52	8.3 ± 0.52
ECF	12.2 ± 0.4	0	7.1 ± 0.5	0	9 ± 0.7	6.9 ± 0.2	7.9 ± 1.4	9.2 ± 1	7.9 ± 0.9	9.3 ± 1.03	69.5 ± 6.13
EAR	0	0	0	0	0	0	0	0	0	8.7 ± 0.93	8.7 ± 0.93
EBR	10.3 ± 0.8	0	0	0	0	0	0	0	0	9.7 ± 1.03	20 ± 1.83
ECR	9.8 ± 1.6	0	7.6 ± 0.7	0	8.8 ± 0.8	8 ± 0.9	7.6 ± 0.5	8.7 ± 0.8	8.9 ± 0.9	10 ± 0.89	69.4 ± 7.09
Antibiotic	CPX (16) FOX(30)	CIP(25)	CIP(25) AMC(25)	AMC(16) SXT(30)		CIP(30) IMP(22)	AMC(23)		IMP(25) AMC(23)	DO(31) C(20)	
	CIP(27)		SXT(34)	TZP(21)		TZP(22)			SXT(28)		
DMSO	0	0	0	0	0	0	0	0	0	0	

Table 4. Inhibiting zones [mm] of bacterial growth for extracts from leaves of *Pyrus pyrifolia*. No antibacterial activity was found for extracts EBC, EDC, and EDK. S.a.—*Staphylococcus aureus*, *E.c.*—*Escherichia coli*, *P.a.*—*Pseudomonas aeruginosa*, *B.s.*—*Bacillus subtilis*, *N.T.H.i.*—nontypeable *Haemophilus influenza*, *H.p.*—*Helicobacter pylori*. *SZD*—the sum of inhibition zone diameter for all bacterial strains for extract [mm] (±SD). 0 means no antibacterial activity.

Extract	*S.a.*ATCC 25923	*S.a.* MRSA	*E.c.*ATCC 25922	*E.c.*ESBL	*P.a.*ATCC 27853	*P.a.* 256	*B.s.*	*N.T.H.i.* ATCC 47249	*N.T.H.i.*6	*H.p.* ATCC 43504	*SZD*
EAC	0	0	0	0	0	0	8.2 ± 0.6	0	0	8.4 ± 0.8	16.6 ± 1.4
ECC	13 ± 2	7.6 ± 0.5	9 ± 0.9	0	8.7 ± 1	8 ± 0.6	11.8 ± 1.5	8.9 ± 0.7	8.4 ± 0.9	8.8 ± 1.6	84.2 ± 9.7
EAK	0	0	0	0	0	0	9.8 ± 0.8	0	0	7.9 ± 0.8	17.7 ± 1.6
EBK	0	0	0	0	0	0	8.5 ± 1.2	0	0	11 ± 1.7	19.5 ± 2.9
ECK	14.4 ± 3	0	7.3 ± 0.9	0	7.4 ± 0.5	7.7 ± 0.8	12.1 ± 0.8	9.8 ± 0.8	9.4 ± 0.5	10.0 ± 1.6	78.1 ± 8.9
EAS	0	0	0	0	0	0	0	0	0	9.2 ± 0.29	9.2 ± 0.29
EBS	0	0	0	0	0	0	8.8 ± 1.2	0	0	10 ± 3.5	18.8 ± 4.7
ECS	11.2 ± 1	0	0	0	8.6 ± 1.1	7.8 ± 0.7	8.7 ± 0.7	9.1 ± 0.5	8.8 ± 0.8	11 ± 0.8	65.2 ± 5.6
EDS	11.4 ± 2	0	0	0	0	0	11.5 ± 1	9.3 ± 0.8	9.8 ± 0.3	7.6 ± 0.4	40.4 ± 4.5

Table 5. Amount of hydroquinone and arbutin in µg per mg of extract and in mg per g of raw material for *Pyrus communis* (± ME).

Cultivar	Extract	Hydrochinon [µg/mg Dry Weight]	Arbutin [µg/mg dr Weight]	Hydroquinone [mg/g Raw Material]	Arbutin [mg/g Raw Material]
Dicolor	EAD	2.54 ± 0.004	37.48 ± 0.141	0.565 ± 0.014	9.849 ± 0.213
	EBD	2.48 ± 0.007	24.56 ± 0.111		
	ECD	3.89 ± 0.033	62.34 ± 0.034		
	EDD	1.23 ± 0.005	26.14 ± 0.031		
Clapp's favourite	EAF	2.30 ± 0.013	54.26 ± 0.076	0.542 ± 0.013	14.361 ± 0.314
	EBF	2.64 ± 0.026	28.71 ± 0.108		
	ECF	4.95 ± 0.020	78.79 ± 0.163		
	EDF	1.16 ± 0.004	47.27 ± 0.093		
Radana	EAR	2.32 ± 0.001	63.73 ± 0.204	0.536 ± 0.014	16.717 ± 0.369
	EBR	2.31 ± 0.012	15.49 ± 0.126		
	ECR	4.03 ± 0.045	145.91 ± 0.112		
	EDR	1.18 ± 0.009	40.57 ± 0.091		

Table 6. Amount of hydroquinone and arbutin in µg per mg of extract and in mg per g of raw material for *Pyrus pyrifolia* (±ME).

Cultivar	Extract	Hydrochinon [µg/mg Extract]	Arbutin [µg/mg Extract]	Hydroquinone [mg/g Raw Material]	Arbutin [mg/g Raw Material]
Chojuro	EAC	2.37 ± 0.005	48.30 ± 0.509	0.461 ± 0.012	11.899 ± 0.260
	EBC	2.39 ± 0.025	26.59 ± 0.085		
	ECC	4.63 ± 0.014	92.68 ± 0.161		
	EDC	1.23 ± 0.017	43.62 ± 0.014		
Kosui	EAK	3.11 ± 0.016	70.69 ± 0.188	0.753 ± 0.021	24.754 ± 0.533
	EBK	2.37 ± 0.018	41.85 ± 0.132		
	ECK	4.37 ± 0.040	148.24 ± 0.144		
	EDK	1.17 ± 0.009	58.21 ± 0.051		
Shu Li	EAS	2.59 ± 0.030	47.17 ± 0.150	1.519 ± 0.061	9.502 ± 0.216
	EBS	2.32 ± 0.010	15.94 ± 0.081		
	ECS	5.71 ± 0.041	46.05 ± 0.192		
	EDS	8.08 ± 0.204	20.50 ± 0.016		

The research described in this study showed (Tables 3 and 4) that the strongest inhibition of bacterial growth was observed for *Helicobacter pylori*, both for extracts obtained from leaves of *Pyrus communis* and *Pyrus pyrifolia*, where the greatest activity was observed for ethyl acetate extracts (**ECD**, **ECF**, **ECR**, **ECC**, **ECK**, and **ECS**). In the case of *H. pylori*, the growth inhibition zone diameter for the **ECS** extract was 11 ± 0.8 mm. In addition to ethyl acetate extracts, *Helicobacter pylori* growth was strongly inhibited by **EBK** extract (11 ± 1.7 mm). The aqueous residues (**EDD**, **EDF**, **EDR**, **EDC**, and **EDK**) did not show any activity against any strain, except for the **EDS** extract (Table 6), which contained a significant amount of hydroquinone. Some antibacterial activity was observed for *Bacillus subtilis*, *Haemophilus influenzae*, and *Staphylococcus aureus* ATCC 25923, especially for extracts from *Pyrus pyrifolia* leaves.

Low antimicrobial activity was observed against *Escherichia coli* (reference and clinical strains) and *Pseudomonas aeruginosa* (reference and clinical strains). No antimicrobial activity was observed against the strains *Staphylococcus aureus* MRSA and *Enterococcus faecalis* (reference and clinical strains).

The sum of diameters of inhibition zones for all strains (*SZD*) was highest for ethyl acetate extracts in the following decreasing order: **ECC** (84.2) > **ECK** (78.1) > **ECD** 72.4 > **ECF** (69.5) > **ECR** (69.4) > **ECS** (65.2).

In general, extracts from leaves of *Pyrus pyrifolia* were more active than extracts from leaves of *Pyrus communis*. The correlation coefficient *r* between the sum of inhibition zones (*SZD*) for different bacterial species and the hydroquinone content for *Pyrus communis* extracts was 0.9, while for *P. pyrifolia* extracts, it was 0.72. For arbutin, the corresponding correlation coefficient values were 0.68 and 0.59, respectively.

Although the correlation coefficient between arbutin content and antimicrobial effect is relatively high, arbutin is not an active substance, as our previous control research has shown. Our research demonstrated that the most active substance is hydroquinone [60].

Due to the increasing clinical significance of infections caused by *Campylobacter jejuni*, we additionally performed tests on the ability of extracts to inhibit the growth of this bacterium. These tests were performed independently in the Hirszfeld Institute of Immunology and Experimental Therapy laboratory. The results are graphically displayed in Figure 5.

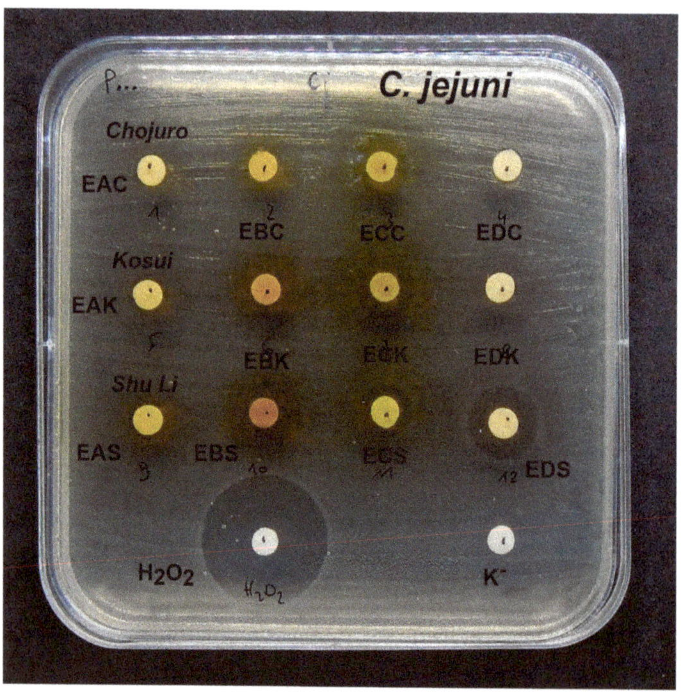

Figure 5. Figure showing the plate with the zones of inhibition of growth of *Campylobacter jejuni* in the presence of extracts from *Chojuro* cultivar, *Pyrus pyrifolia*, with the strongest antibacterial activity. An explanation of abbreviations of extracts is given in the section on preparation of extracts.

The diameters of inhibition zones (mm) for extracts from both *P. communis* and *P. pyrifolia* were **EAD** (7.5 ± 1.2), **EBD** (7.2 ± 0.3), **ECD** (7.5 ± 0.2), **EDD** (no activity), **EAF** (7.4 ± 0.9), **EBF** (8.1 ± 0.5), **ECF** (9.5 ± 0.4), **EDF** (6.6 ± 0.6), **EAR** (7.3 ± 0.4), **EBR** (7.1 ± 0.7), **ECR** (7.4 ± 0.7), **EDR** (6.5 ± 0.5), **EAC** (8.3 ± 1.1), **EBC** (6.5 ± 0.5), **ECC** (11.2 ± 0.8), **EDC** (7.6 ± 0.4), **EAK** (9.4 ± 1.4), **EBK** (9.0 ± 1.6), **ECK** (13.5 ± 2.3), **EDK** (7.8 ± 1.1), **EAS** (9.8 ± 1.7), **EBS** (8.7 ± 2.2), **ECS** (13.2 ± 3.5), and **EDS** (15.8 ± 2.3).

P. pyrifolia extracts showed significantly greater activity than extracts from *P. communis* against *Campylobacter jejuni*. The most potent inhibitors of *C. jejuni* growth were **ECC**, **ECK**, **ECS**, and **EDS** extracts. The correlation coefficient between the activity of extracts against *C. jejuni* and the content of hydroquinone and arbutin was, respectively, $r = 0.85$, $r = 0.24$ (Figure 6). The high correlation between the hydroquinone content in the extracts and the inhibition of the growth of *C. jejuni* indicates that the main antimicrobial component is hydroquinone. Arbutin seems to be less important. Our previous research led to a similar conclusion [60].

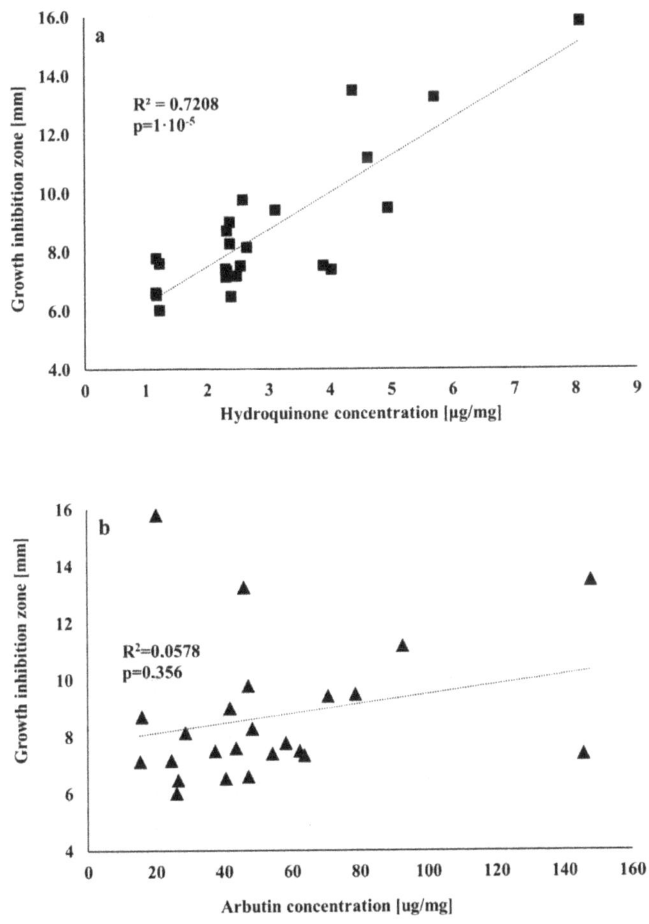

Figure 6. Graphical demonstration of the dependence of bacterial growth inhibition of *Campylobacter jejuni* on the concentration of hydroquinone (**a**) and arbutin (**b**) in extracts. The differences between samples are significant when $p < 0.05$. The dotted line indicates a trend line.

3.4. Amounts of Arbutin and Hydroquinone in Extracts and Raw Materials

As described above, hydroquinone significantly inhibits the growth of different species of bacteria.

The amounts of hydroquinone and arbutin in extracts investigated in this research are demonstrated in Table 5 for *P. communis* and Table 6 for *P. pyrifolia* cultivars. A graphical presentation of the results is shown in Figure 7 for hydroquinone and Figure 8 for arbutin. Selected chromatograms are presented in Figure 9. The highest amount of arbutin was measured in **ECK** and **ECR** extracts from *P. pyrifolia Kosui* (148.2 ± 0.14) and *P. communis Radana* (145.9 ± 0.11) leaves, respectively. A lower value was obtained for leaf extract **ECC** of *P. pyrifolia Chojuro* (92.7 ± 0.014) and then, in decreasing order, **ECF, EAK, EAR, ECD, EDK, EAF, EAC, EDF, EAS, ECS, EDC, EBK, EDR, EAD, EBF, EBC, EDD, EBD, EDS, EBS**, and **EBR**.

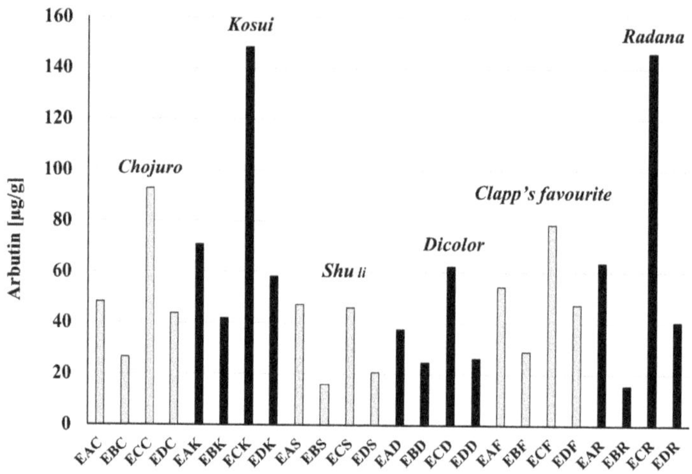

Figure 7. Comparison of arbutin content in extracts. An explanation of abbreviations of extracts is given in the section on preparation of extracts.

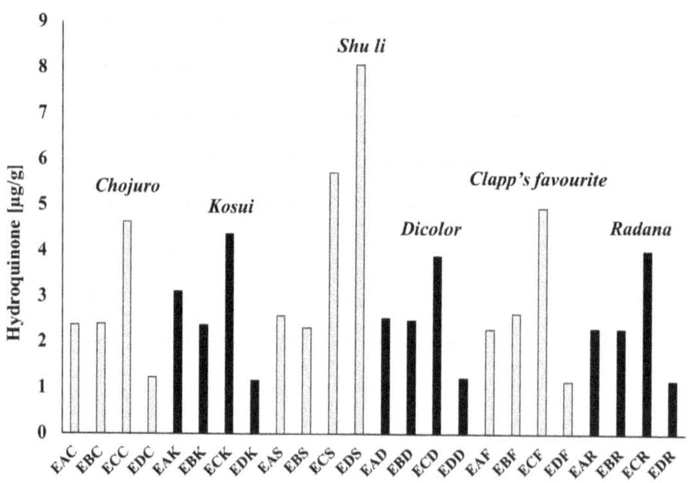

Figure 8. Comparison of hydroquinone content in extracts. An explanation of abbreviations of extracts is given in the section on preparation of extracts.

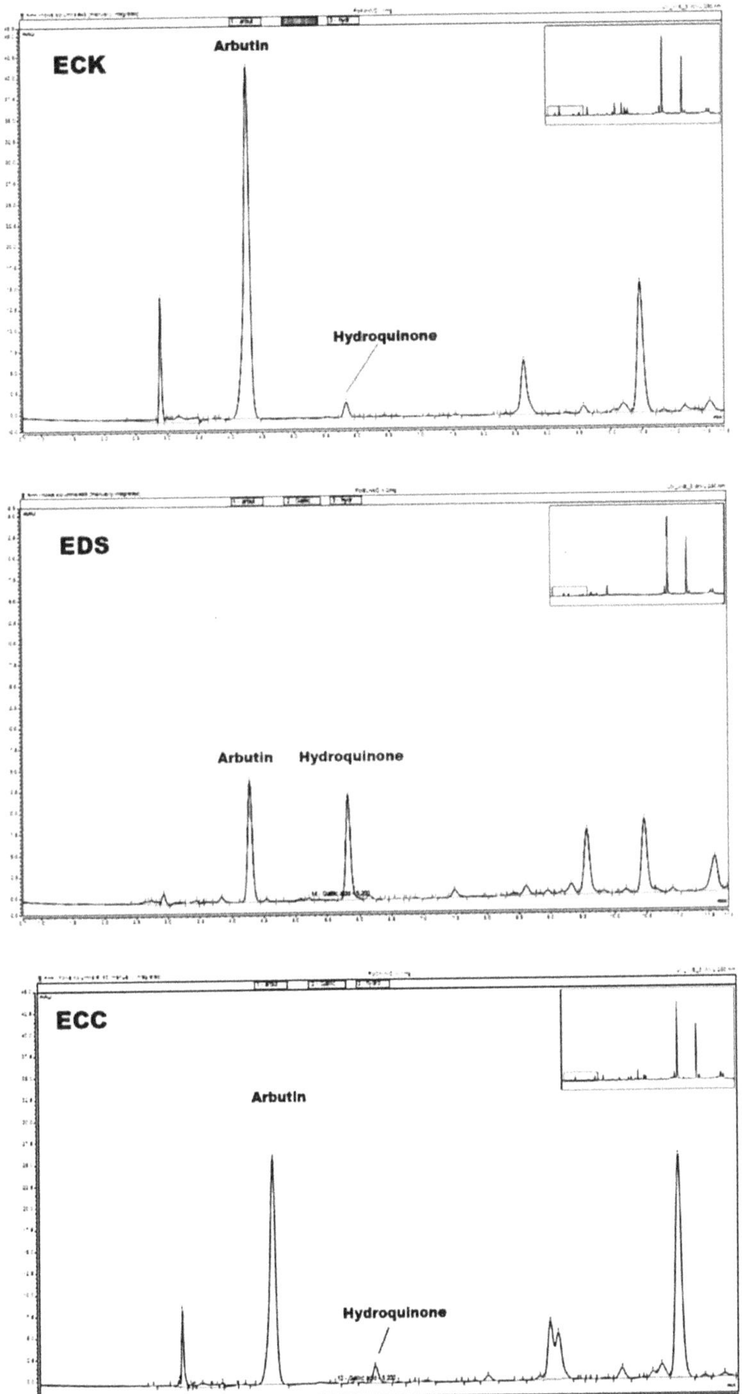

Figure 9. HPLC chromatograms of selected extracts with marked peaks of arbutin and hydroquinone. ECK—extract with the highest amount of arbutin, EDS—extract with the highest amount of hydroquinone, ECC—extract with the highest antimicrobial activity (*SZD* value).

The highest amount of hydroquinone was observed for extract **EDS** (8.08 ± 0.204), then **ECS** (5.71 ± 0.041), **ECF** (4.95 ± 0.020), and **ECC** (4.63 ± 0.014), and, in decreasing order, **ECK, ECR, ECD, EAK, EBF, EAS, EAD, EBD, EBC, EAC = EBK, EBS = EAR, EBR, EAF, EDC = EDD, EDR, EDK,** and **EDF**.

The contents of hydroquinone and arbutin were also calculated per g of raw material (Tables 5 and 6). The highest amount of hydroquinone was found in the leaves of the *Shu li* cultivar (1.517 mg/g raw material), which also contained the least arbutin (9.503 mg/g). The lowest amount of hydroquinone was found in the leaves of *Chojuro* (0.461 mg/g), and the most arbutin (24.751 mg/g) was present in *Kosui* leaves.

4. Discussion

Since increasing bacterial resistance to antibiotics is a global threat, combating microbial resistance requires coordinated efforts that include the development of new conventional antibiotics and alternative antimicrobials. One of these is plant-derived compounds that can serve as phytotherapeutic agents used in good treatment alternatives.

We investigated the antimicrobial activity of a variety of plant extracts (leaves of *Bergenia crassifolia, Bergenia cordifolia,* and some species of the *Pyrus* genus), and we found high inhibition of the growth of dangerous and/or clinically relevant pathogens, including *Pseudomonas aeruginosa*, MRSA, *Helicobacter pylori, Enterococcus faecalis,* and *Escherichia coli* ESBL [24,60].

In our previous studies, we obtained highly active methanolic and ethyl acetate extracts from leaves of different cultivars of the pear (Conference and Shinseiki) containing arbutin, hydroquinone, and a large amount of phenols, mainly against Gram-positive strains such as *Staphylococcus aureus*, including MRSA strains, and *Bacillus subtilis* [24,60].

Some ethyl acetate extracts rich in hydroquinone showed high antibacterial activity against Gram-negative *Helicobacter pylori* that was similar to antibiotics such as erythromycin, chloramphenicol, and tetracycline [24,60].

In some cases, we obtained good effects of extracts containing a high content of hydroquinone against clinically relevant pathogens [60] such as Gram-positive *Enterococcus faecalis*, which may cause endocarditis, urinary tract infections, prostatitis, and intra-abdominal infection [61,62], and Gram-negative *Pseudomonas aeruginosa*, which may infect patients with carcinomas, immunodeficiency, cystic fibrosis, and chronic obstructive pulmonary disorder (COPD and serious infection requiring ventilation [63–65]. Antibacterial activity against Gram-negative *E. coli* ATCC 25922 and *E. coli* ESBL has also been demonstrated. It is worth adding that numerous *E. coli* strains can cause gastrointestinal tract infections [66], urinary tract infections [67], and Lesniowski–Crohn's disease [68].

It can be found in the literature that hydroquinone and hydroquinone-rich extracts are inactive against Gram-negative bacteria [69]. Our study showed that extracts containing large amounts of hydroquinone are active against Gram-negative bacteria, but their activity is lower than that of Gram-positive bacteria [24].

In this study, the extracts showed moderate antimicrobial activity and an average content of hydroquinone, and these extracts contained a large amount of arbutin, which is microbiologically inactive. Thanks to the high content of arbutin (hydroquinone β D-glucopyranoside), the extracts have a high antimicrobial potential because arbutin relatively easily hydrolyzes (e.g., in alkaline conditions in urine), resulting in the release of active hydroquinone. Such hydrolysis usually leads to an increase in the antimicrobial activity of arbutin-rich extracts [70].

Further studies will be conducted to increase the antimicrobial potential of the extracts.

Another important problem for modern medicine is chronic and degenerative diseases [71], the severity of which may be related to oxidative and free radical processes that occur in the body, especially if the intensity of these processes is high [72].

Due to their chemical structure, phenolic compounds, abundant in plants, usually have strong antioxidant properties and inhibit free radical processes [73]. Their activity strongly depends on their chemical structure [54]. Flavonoids are an important group of plant phenols with significant antioxidant potential, which can weaken inflammatory processes by interfering with the arachidonic acid metabolism pathways [74].

In this study, we measured the total content of phenols and flavonoids in plant extracts and raw materials and evaluated their antiradical and antioxidant features.

Our research showed that the highest antiradical activity was found in ethyl acetate extracts with the highest content of phenolic compounds, while the weakest antioxidant activity and the lowest amount of phenols were found for water residues. The correlation coefficient (r) between the amount of total phenols and antiradical activity was 0.85 for ABTS tests and 0.82 for DPPH tests.

The leaves of the *Kosui* cultivar showed the strongest antiradical activity and contained the highest amount of phenolic compounds, while leaves from the *Chojuro* cultivar had the weakest antiradical properties and contained the least phenolic compounds.

In general, both phenolic content and antiradical properties of both extracts and raw materials are high, although weaker than in the case of extracts from bergenia leaves [60] and green and black tea leaves [75].

5. Conclusions

1. Extracts obtained from leaves of different *Pyrus communis* and *Pyrus pyrifolia* cultivars exhibited antioxidant and antiradical activity.
2. The antiradical activity of extracts was positively correlated with the amount of total phenolic compounds in them.
3. The highest antibacterial activity was observed for ethyl acetate extracts.
4. Antibacterial activity positively correlated with the hydroquinone content and, to a lesser extent, with the amount of arbutin.

Author Contributions: Z.S.: conceptualization, data curation, formal analysis, funding acquisition, supervision, validation, writing—original draft, writing—review and editing. B.Ż.: data curation, investigation, methodology, supervision, writing—review and editing. K.P. and K.M.: students, investigation. D.A., K.K., and M.K.: data curation, formal analysis, investigation, methodology, writing—review and editing. A.G.: resources, writing—review and editing. All authors have read and agreed to the published version of the manuscript.

Funding: This work was supported by a grant to Wrocław Medical University, number SUBZ. D 110.24.085, from the Hirszfeld Institute of Immunology and Experimental Therapy, PAS, number 501-15.

Institutional Review Board Statement: Not applicable.

Informed Consent Statement: Not applicable.

Data Availability Statement: The original contributions presented in this study are included in the article. Further inquiries can be directed to the corresponding author.

Conflicts of Interest: This manuscript has not been published or presented in part or entirety and is not under consideration by another journal. There are no conflicts of interest to declare.

References

1. Jadimurthy, R.; Mayegowda, S.B.; Nayak, S.C.; Mohan, C.D.; Rangappa, K.S. Escaping mechanisms of ESKAPE pathogens from antibiotics and their targeting by natural compounds. *Biotechnol. Rep.* **2022**, *34*, e00728. [CrossRef] [PubMed]
2. Nimer, N.A. Nosocomial infection and antibiotic-resistant threat in the Middle East. *Infect. Drug. Resist.* **2022**, *15*, 631–639. [CrossRef] [PubMed]
3. Sahoo, S.; Mohanty, J.N.; Routray, S.P.; Khandia, R.; Das, J.; Shah, S.; Swarnkar, T. Colistin the last resort drug in 21st century antibiotics to combat multidrug resistance superbugs. *J. Exp. Biol. Agric. Sci.* **2023**, *11*, 919–929. [CrossRef]
4. Huy, T.X.N. Overcoming *Klebsiella pneumonia* antibiotic resistance: New insights into mechanisms and drug discovery. *Beni Suef. Univ. J. Basic Appl. Sci.* **2024**, *13*, 13. [CrossRef]
5. Gauba, A.; Rahman, K.M. Evaluation of antibiotic resistance mechanisms in Gramm-negative bacteria. *Antibiotics* **2023**, *12*, 1590. [CrossRef] [PubMed]
6. Singha, B.; Murmu, S.; Nair, T.; Rawat, R.S.; Sharma, A.K.; Soni, V. Metabolic rewiring of *Mycobacterium tuberculosis* upon drug treatment and antibiotics resistance. *Metabolites* **2024**, *4*, 63. [CrossRef]
7. Martins-Santana, L.; Rezende, C.P.; Rossi, A.; Martinez-Rossi, N.M.; Almeida, F. AddressingMicrobial ResistanceWorldwide: Challenges over controlling life-threatening fungal infections. *Pathogens* **2023**, *12*, 293. [CrossRef]
8. Aiken, A.M.; Rehman, A.M.; Kraker, M.E.A.D.; Madrid, L.; Kebede, M.; Labi, A.K.; Obeng-Nkrumah, N.; Nyamwaya, B.; Kagucia, E.; Cocker, D.; et al. Mortality associated with third-generation cephalodsporin resistance in Enterobacterales bloodstream infections at eight sub-Saharan African hospitals (MBIRA): A prospective cohort study. *Lancet Infect. Dis.* **2023**, *23*, 1280–1290. [CrossRef]
9. Wheatley, R.M.; Botelho, J. Chasing resistance: Analysing the fight against hospital infections. *Lancet Reg. Health Eur.* **2023**, *34*, 100740. [CrossRef]
10. Nouwen, J.L. Controlling antibiotic use and resistance. *Clin. Infect. Dis.* **2006**, *42*, 776–777. [CrossRef]
11. Del Fiol, F.S. Antibiotic use and resistance. *Basic. Clin. Pharmacol. Toxicol.* **2014**, *114*, 439–440. [CrossRef] [PubMed]
12. Solano, T.R. Antimicrobial resistance and antibiotic use. *Anesth. Intensive Care* **2010**, *38*, 987–988. [CrossRef] [PubMed]
13. Widmer, A.F. Emerging antibiotic resistance: Why we need new antibiotics! *Swiss Med. Wkly.* **2022**, *152*, 40032. [CrossRef] [PubMed]
14. Zazharskyi, W.; Davydenko, P.O.; Kulishenko, M.; Borovik, I.V.; Brygadyrenko, V.V. Antimicrobial activity of 50 plant extracts. *Biosyst. Divers.* **2019**, *27*, 163–169. [CrossRef]
15. Oulahal, N.; Degraeve, P. Phenolic rich plant extracts with antimicrobial activity: An alternative to food preservatives and biocides. *Front. Microbiol.* **2022**, *12*, 753518. [CrossRef]
16. Lee, S.Y.; Gill, C.M.; Nicolau, D.P. Activity of novel β-lactam. β-lactamase inhibitor combinations against serine carbapenemase-producing carbapenem resistant *Pseudomonas Aeruginosa*. *J. Antimicrob. Chemoth.* **2023**, *78*, 2795–2800. [CrossRef]
17. Lin, D.M.; Koskella, B.; Lin, H.C. Phage therapy: An alternative to antibiotics in the age of multidrug resistance. *World J. Gastrointest Pharmacol. Ther.* **2017**, *8*, 162–173. [CrossRef]
18. Strathdee, S.A.; Hatfull, G.F.; Mutalik, V.K.; Schooley, R.T. Phage therapy: From biological mechanisms to future directions. *Cell* **2023**, *186*, 17–31. [CrossRef]
19. Dorotkiewicz-Jach, A.; Augustyniak, D.; Olszak, T.; Drulis-Kawa, Z. Modern therapeutic approaches against *Pseudomonas aeruginosa* infections. *Curr. Med. Chem.* **2015**, *22*, 1642–1664. [CrossRef]
20. Maciejewska, B.; Olszak, T.; Drulis-Kawa, Z. Applications of bacteriophages versus phage enzymes to combat and cure bacterial infections: An ambitious and also a realistic application? *Appl. Microbiol. Biotechnol.* **2018**, *102*, 2563–2581. [CrossRef]
21. Wojnicz, D.; Tichaczek-Goska, D.; Korzekwa, K.; Kicia, M.; Hendrich, A.B. Study of the impact of cranberry extract on the virulence factors and biofilm formation by *Enterococcus faecalis* strains isolated from urinary tract infection. *Int. J. Food Sci. Nutr.* **2016**, *67*, 1005–1016. [CrossRef] [PubMed]
22. Mostafa, A.A.; Al-Askar, A.A.; Almaary, K.S.; Dawoud, T.M.; Sholkamy, E.N.; Bakri, M.M. Antimicrobial activity of some plants extracts against bacterial strains causing food poisoning diseases. *Saudi J. Biol. Sci.* **2018**, *25*, 361–366. [CrossRef] [PubMed]
23. Sathianarayanan, S.; Ammanath, A.V.; Biswas, R.; Anita, B.; Anita Sukumaran, S.; Venkidasamy, B. A new approach against *Helicobacter pylori* using plants and its constituents: A review study. *Microb. Pathog.* **2022**, *168*, 105594. [CrossRef] [PubMed]
24. Sroka, Z.; Zgórka, G.; Żbikowska, B.; Sowa, A.; Franiczek, R.; Wychowaniec, K.; Krzyżanowska, B. High antimicrobial efficacy, antioxidant activity, and a novel approach to phytochemical analysis of bioactive polyphenols in extracts from leaves *Pyrus communis* and *Pyrus pyrifolia* collected during one vegetative season. *Microb. Drug Resist.* **2019**, *25*, 582–593. [CrossRef]
25. Caiati, C.; Stanca, A.; Lepera, M.E. Free radicals and obesity related chronic inflammation contrasted by antioxidants: A new perspective in coronary artery disease. *Metabolites* **2023**, *13*, 712. [CrossRef]
26. Sakaguchi, K.; Yishikawa, M.; Fujita, Y.; Yamamoto, Y.; Ibaraki, T.; Fujioka, N.; Osa, T.; Yamauchi, M.; Ichikawa, H.; Muro, S. Serum free radical scavenging capacity profiles of patients with chronic obstructive pulmonary disease. *Int. J. Chronic. Obstr.* **2023**, *18*, 613–624. [CrossRef]

27. Yang, H.M.; Leng, J.P.; Liu, N.; Huang, L. Editorial: Free radicals and antioxidants in diseases associated with immune dysfunction, inflammatory process, and aberrant metabolism. *Front. Endocrinol.* **2024**, *15*, 1363854. [CrossRef]
28. Chaudhary, P.; Janmeda, P.; Docea, A.O.; Yeskaliyeva, B.; Razis Afa Modu, B.; Calina, D.; Sharifi-Rad, J. Oxidative stress, free radicals and antioxidants: Potential crosstalk in pathophysiology of human diseases. *Front. Chem.* **2023**, *11*, 1158198. [CrossRef]
29. Hajam, Y.A.; Rani, R.; Ganie, S.Y.; Sheikh, T.A.; Javaid, D.; Qadri, S.S.; Pramodh, S.; Alsulimani, A.; Alkhanani, M.; Harakeh, S.; et al. Oxidative stress in human pathology and aging: Molecular mechanisms and perspectives. *Cells* **2022**, *11*, 552. [CrossRef]
30. Ramos-Tovar, E.; Muriel, P. Free radicals, antioxidants, nuclear factor-E2- related factor-2 and liver damage. *J. Appl. Toxicol.* **2020**, *40*, 151–168. [CrossRef]
31. Li, S.; Tan, H.Y.; Wang, N.; Zhang, Z.J.; Lao, L.X.; Wong, C.W.; Feng, Y.B. The role of oxidative stress and antioxidants in liver diseases. *Int. J. Mol. Sci.* **2015**, *16*, 26087–26124. [CrossRef]
32. Juan, C.A.; de la Lastra, J.M.P.; Plou, F.J.; Pérez-Lebeña, E. The chemistry of reactive oxygen species (ROS) revisited: Outlining their role in biological macromolecules (DNA, lipids, and proteins) and induced pathologies. *Int. J. Mol. Sci.* **2021**, *22*, 4642. [CrossRef] [PubMed]
33. Krawczyk, M.; Burzynska-Pedziwiatr, I.; Wozniak, L.A.; Bukowiecka-Matusiak, K. Impact of polyphenols on inflammatory and oxidative stress factors in diabetes mellitus: Nutritional antioxidants and their application in improving antidiabetic therapy. *Biomolecules* **2023**, *13*, 1402. [CrossRef] [PubMed]
34. Yalçin, A.S.; Attaallah, W.; Yilmaz, A.M.; Aktan, A.Ő. Free radicals, whey proteins and colorectal cancers. *Marmara Med. J.* **2014**, *27*, 1–6. [CrossRef]
35. Nielsen, O.H.; Ahnfelt-Ronne, I. Involvement of oxygen-derived free-radicals in the pathogenesis of chronic inflammatory bowel-disease. *Klin. Wochenschr.* **1991**, *69*, 995–1000. [CrossRef]
36. Simpson, D.S.A.; Oliver, P.l.L. ROS generation in microglia: Understanding oxidative stress and inflammation in neurodegenerative disease. *Antioxidants* **2020**, *9*, 743. [CrossRef]
37. El-Kenawi, A.; Ruffell, B. Inflammation, ROS, and mutagenesis. *Cancer Cell* **2017**, *32*, 727–729. [CrossRef]
38. Wang, Y.Q.; Han, D.; Huang, Y.J.; Dai, Y.L.; Wang, Y.; Liu, M.; Wang, N.; Yin, T.Y.; Du, W.Q.; He, K. Oral administration of punicalagin attenuates imiquimod-induced psoriasis by reducing ROS generation and inflammation via MAPK/ERK and NFκB signaling pathways. *Phytother. Res.* **2024**, *38*, 713–726. [CrossRef] [PubMed]
39. Wang, G.; Yang, F.; Zhou, W.; Xiao, N.; Luo, M.; Tang, Z. The initiation of oxidative stress and therapeutic strategies in wound healing. *Biomed. Pharmacother.* **2023**, *157*, 114004. [CrossRef]
40. Nisar, A.; Jagtap, S.; Vyavahare, S.; Deshpande, M.; Harsulkar, A.; Ranjekar, P.; Prakash, O. Phytochemicals in the treatment of inflammation-associated diseases: The journey from preclinical trials to clinical practice. *Front. Pharmacol.* **2023**, *14*, 1177050. [CrossRef]
41. Styrczewska, M.; Kostyn, A.; Kulma, A.; Majkowska-Skrobek, G.; Augustyniak, D.; Prescha, A.; Czuj, T.; Szopa, J. Flax fiber hydrophobic extract inhibits human skin cells inflammation and Causes remodeling of extracellular matrix and wound closure activation. *BioMed Res. Int.* **2015**, *2015*, 862391. [CrossRef] [PubMed]
42. Jyoti, A.; Mishra, N.; Dhas, Y. Ageing consequences of excessive free radicals and inflammation. *Curr. Sci. India* **2017**, *111*, 1787–1793. [CrossRef]
43. Awad, A.M.; Kumar, P.; Ismail-Fitry, M.R.; Jusoh, S.; Ab Aziz, M.F.; Sazili, A.Q. Green extraction of bioactive compounds from plant biomass and their application in meat as natural antioxidant. *Antioxidants* **2021**, *10*, 1465. [CrossRef] [PubMed]
44. Lang, Y.X.; Gao, N.X.; Zang, Z.H.; Meng, X.J.; Lin, Y.; Yang, S.F.; Yang, Y.Y.; Jn, Z.; Li, B. Classification and antioxidant assays of polyphenols: A review. *J. Future Foods* **2024**, *4*, 193–204. [CrossRef]
45. Vaou, N.; Stavropoulou, E.; Voidarou, C.; Tsakris, Z.; Rozos, G.; Tsigalou, C.; Bezirtzoglou, E. Interactions between plant-derived bioactive compounds: Focus on antimicrobial combination effect. *Antibiotics* **2022**, *11*, 1014. [CrossRef]
46. Singh, S.; Verma, R.; Sharma, H. Exploring the therapeutic potential and bioactive compounds in *Pyrus* species. *Pharmacol. Res. -Mod. Chin. Med.* **2024**, *10*, 100342. [CrossRef]
47. Cho, J.Y.; Park, K.Y.; Lee, K.H.; Lee, H.J.; Lee, S.H.; Cho, J.A.; Kim, W.S.; Shin, S.C.; Park, K.H.; Moon, J.H. Recovery of arbutin in high purity from fruit peals of pear (*Pyrus pyrifolia* Nakai). *Food Sci. Biotechnol.* **2011**, *20*, 801–807. [CrossRef]
48. Sasaki, C.; Ichitani, M.; Kunimoto, K.K.; Asada, C.; Nakamura, Y. Extraction of arbutin and its comparative content in branches, leaves, stems, and fruits of Japanese pear *Pyrus pyrifolia* cv. Kousui. *Biosci. Biotechnol. Biochem.* **2014**, *78*, 874–877. [CrossRef]
49. Singleton, V.L.; Rossi, J.A. Colorimetry of total phenolics with phosphomolybdic-phosphotungustic acid reagents. *Am. J. Enol. Vitic.* **1965**, *16*, 144–158. [CrossRef]
50. Folin, O.; Ciocalteu, V. On tyrosine and tryptophane determination in proteins. *J. Biol. Chem.* **1927**, *27*, 239–243. [CrossRef]
51. Folin, O.; Denis, W. Tyrosine in proteins as determined by a new colorimetric method. *J. Biol. Chem.* **1912**, *12*, 245–251. [CrossRef]
52. Christ, B.; Müller, K.H. Zur serienmaessigen Bestimmung des an Flavonol-Derivaten in Drogen. *Arch. Pharm.* **1960**, *293*, 1033–1042. [CrossRef] [PubMed]

53. Re, R.; Pellegrini, N.; Proteggente, A.; Pannala, A.; Yang, M.; Rice-Evans, C. Antioxidant activity applying an improved ABTS radical cation decolorization assay. *Free Radic. Bio Med.* **1999**, *26*, 1231–1237. [CrossRef] [PubMed]
54. Sroka, Z.; Żbikowska, B.; Hładyszowski, J. The antiradical activity of some selected flavones and flavonols. Experimental and quantum mechanical study. *J. Mol. Model.* **2015**, *21*, 307. [CrossRef]
55. Brand-Williams, W.; Cuvelier, M.E.; Berset, C. Use of the free radical method to evaluate antioxidant activity. *LWT–Food Sci. Technol.* **1995**, *28*, 25–30. [CrossRef]
56. Fecka, I.; Włodarczyk, M.; Starzec, A. Isolation and structure elucidation of cistusin: A new ellagitannin from *Cistus* x *incanus* L. leaves. *Ind. Crops Prod.* **2020**, *158*, 112971. [CrossRef]
57. Ternhag, A.; Törner, A.; Svensson, Å.; Giesecke, J.; Ekdahl, K. Mortality following *Campylobacter infection*: A registry-based linkage study. *BMC Infect. Dis.* **2005**, *5*, 1–5. [CrossRef]
58. Contreras, M.; Thiberge, J.M.; Mandrand-Berthelot, M.A.; Labigne, A. Characterization of the roles of NikR, a nickel-responsive pleiotropic autoregulator of *Helicobacter Pylori*. *Mol. Microbiol.* **2003**, *49*, 947–963. [CrossRef]
59. Sroka, Z.; Żbikowska, B.; Janicki, K.; Franiczek, R.; Krzyżanowska, B.; Dryś, A. Antimicrobial and antiradical activity of extracts obtained from leaves of tree species of the genus *Pyrus*. *Microb. Drug Resist.* **2014**, *20*, 337–343.
60. Żbikowska, B.; Franiczek, R.; Sowa, A.; Połukord, G.; Krzyżanowska, B.; Sroka, Z. Antimicrobial and antiradical activity of extracts obtained from leaves of five species of the genus *Bergenia*: Identification of antimicrobial compounds. *Microb. Drug Resist.* **2017**, *23*, 771–780. [CrossRef]
61. Saleh, R.O.; Dheyab, A.S.; Hadi, B.H.; Hasan, R.N.; Jasim, S.A. Effect of ethanolic extract of *Syzygium aromaticum* plant against *Enterococcus faecalis* isolated from woman with urinary tract infections. *Arch. Clin. Infect. Dis.* **2024**, *19*, 134924. [CrossRef]
62. Said, M.S.; Tirthani, E.; Lesho, E. Enterococcus infections. In *StatPearls [Internet]*; StatPearls Publishing: Treasure Island, FL, USA, 2024. Available online: https://www.ncbi.nlm.nih.gov/books/NBK567759/ (accessed on 12 February 2024).
63. Sanya, D.R.A.; Onésime, D.; Vizzarro, G.; Jacquier, N. Recent advances in therapeutic targets identification and development of treatment strategies towards *Pseudomonas aeruginosa* infections. *BMC Microbiol.* **2023**, *23*, 86. [CrossRef] [PubMed]
64. Sathe, N.; Beech, P.; Croft, L.; Suphioglu, C.; Kapat, A.; Athan, E. *Pseudomonas aeruginosa*: Infections and novel approaches to tratment "Knowing the enemy" the treat of *Pseudomonas aeruginosa* and exploring novel approaches to treatment. *Infect. Med.* **2023**, *2*, 178–194. [CrossRef] [PubMed]
65. Quin, S.; Xiao, W.; Zhou, C.; Pu, Q.; Deng, X.; Lan, L.; Liang, H.; Song, X.; Wu, M. *Pseudomonas aeruginosa*: *Pathogenesis*, virulence factors, antibiotic resistance, interaction with host, technology advances and emerging therapeutics. *Signal Transduct. Target. Ther.* **2022**, *7*, 199. [CrossRef]
66. Monte, D.F.M.; Sellera, F.P.; Lincopan, N.; Landgraf, M. Genome-based diagnostic of MDR *Escherichia coli* ONT: H19ST10955 causing human gastrointestinal infection. *Diagn. Microbiol. Infect. Dis.* **2024**, *110*, 116340. [CrossRef]
67. Zhou, Y.; Zhou, Z.; Zheng, L.; Gong, Z.; Li, Y.; Jin, Y.; Huang, Y.; Chi, M. Urinary tract infections caused by uropathogenic *Escherichia coli*: Mechanisms of infection and treatment options. *Int. J. Mol. Sci.* **2023**, *24*, 10537. [CrossRef]
68. Siniagina, M.N.; Markelova, M.I.; Boulygina, E.A.; Laikov, A.V.; Odintsova, A.H.; Abdulkhakov, R.A.; Grigoryeva, T.V. Diversity and adaptations of *Escherichia coli* strains: Exploring the intestinal community in Crohn's disease patients and healthy individuals. *Microorganisms* **2021**, *9*, 1299. [CrossRef]
69. Ma, C.; He, N.; Zhao, Y.; Xia, D.; Wei, J.; Kang, W. Antimicrobial mechanism of hydroquinone. *Appl. Biochem. Biotechnol.* **2019**, *189*, 1291–1303. [CrossRef]
70. Sroka, Z.; Kolasa, A. Polish Patent 2011. PL 208366, 29 April 2011.
71. Santana-Gálvez, J.; Cisneros-Zevallos, L.; Jacobo-Velázquez, D.A. A practical guide for designing effective nutraceutical combinations in the form of foods beverages, and dietary supplements against chronic degenerative diseases. *Trends Foods Sci. Technol.* **2019**, *88*, 179–193. [CrossRef]
72. Pham-Huy, L.A.; He, H.; Pham-Huy, C. Free Radicals, Antioxidants in disease and health. *Int. J. Biomed. Sci.* **2008**, *4*, 89–96. [CrossRef]
73. Nisa, R.U.; Nisa, A.U.; Tantray, A.Y.; Shah, A.H.; Jan, A.T.; Shah, A.A.; Wani, I.A. Plant phenolics with promising therapeutic applications against skin disorders: A miechanistic review. *J. Agric. Food Res.* **2024**, *16*, 101090. [CrossRef]
74. Chen, G.Y.; Tao, Q.W.; Liu, X.Y.; Yan, X.E.; Yu, X.; Liu, Y.; Luo, J. Total flavonoids of *rhizoma Drynariae* treat osteoarthritis by inhibiting arachidonic acid metabolites through AMPK/NFκB pathway. *J. Inflamm. Res.* **2023**, *16*, 4123–4140. [CrossRef] [PubMed]
75. Wojciechowski, D.; Sroka, Z.; Gamian, A. Investigation of antiradical potential of different kinds of teas and extracts from these teas using antiradical activity units (TAU). *Postep. Hig. I Med. Dosw.* **2011**, *65*, 796–803. [CrossRef] [PubMed]

biomolecules

MDPI

Article

Uric Acid, the End-Product of Purine Metabolism, Mitigates Tau-Related Abnormalities: Comparison with DOT, a Non-Antibiotic Oxytetracycline Derivative

Bianca Andretto de Mattos [1,2], Rodrigo Hernán Tomas-Grau [1,3], Thaís Antonia Alves Fernandes [1,2], Florencia González-Lizárraga [1,3], Aurore Tourville [1], Ismaila Ciss [1,4], Jean-Michel Brunel [5], Rosana Chehin [3], Annie Lannuzel [1,6], Laurent Ferrié [4], Rita Raisman-Vozari [1], Bruno Figadère [4], Elaine Del Bel [2,7] and Patrick Pierre Michel [1,*]

1 Paris Brain Institute-ICM, Inserm, CNRS, APHP, Hôpital de la Pitié Salpêtrière, Sorbonne Université, 75013 Paris, France; biancaandretto@usp.br (B.A.d.M.); rodrigo.tomasgrau@fbqf.unt.edu.ar (R.H.T.-G.); thais.alves.fernandes@usp.br (T.A.A.F.); mflorenciagl@hotmail.com.ar (F.G.-L.); cissismaila87@gmail.com (I.C.); annie.lannuzel@chu-guadeloupe.fr (A.L.); ritaraisman@gmail.com (R.R.-V.)

2 Medical School of Ribeirão Preto, Department of Physiology, University of Sao Paulo, Ribeirão Preto 14040-904, Brazil; eadelbel@usp.br

3 IMMCA, CONICET-UNT-SIPROSA, Tucumán 4000, Argentina; rosana.chehin@fbqf.unt.edu.ar

4 BioCIS, CNRS, Université Paris-Saclay, 91400 Orsay, France; laurent.ferrie@universite-paris-saclay.fr (L.F.); bruno.figadere@universite-paris-saclay.fr (B.F.)

5 Inserm, Membranes et Cibles Thérapeutiques, Service de Santé des Armées, Aix Marseille Université, 13385 Marseille, France; bruneljm@yahoo.fr

6 Department of Neurology, CIC1424, University Hospital of Guadeloupe, Antilles University, French West Indies, 97159 Pointe-à-Pitre, France

7 Dentistry School, Basic and Oral Biology, University of Sao Paulo, Ribeirão Preto 14040-904, Brazil

* Correspondence: patrick.michel@icm-institute.org

Abstract: We aimed to simulate tau abnormalities—specifically hyperphosphorylation and aggregation—that are hallmarks of tauopathies, including Alzheimer's disease, to evaluate tau-targeting therapies. To model pathological p-tau accumulation at early disease stages, we exposed mouse cortical cultures to redox-active iron from hemin (Hm), a breakdown product of hemoglobin, or challenged them with the excitatory neurotransmitter glutamate. Using the AT8 phospho-specific antibody, we demonstrate that a subtoxic concentration of Hm (3 μM) promotes pathological p-tau accumulation in a subpopulation of cultured cortical neurons and their proximal neurites. Uric acid (UA; 0.1–200 μM), the metabolic end-product of purines in humans, prevented p-tau build-up. Neither xanthine, the immediate precursor of UA, nor allantoin, its oxidized product, reproduced this effect. Live cell imaging studies revealed that UA operates by repressing iron-driven lipid peroxidation. DOT (3 μM), a brain-permeant tetracycline (TC) without antibiotic activity, mimicked UA's anti-tau and antioxidant effects. Interestingly, both UA and DOT remained effective in preventing p-tau accumulation induced by glutamate (10 μM). To simulate tau aggregation at more advanced disease stages, we conducted a Thioflavin-T aggregation assay. Our findings revealed that UA and DOT prevented tau aggregation seeded by heparin. However, only DOT remained effective when heparin-assembled tau fibrils were used as the seeding material. In summary, our results indicate that UA-elevating agents may hold therapeutic utility for tauopathies. The non-purine compound DOT could serve as an effective alternative to UA-related therapies.

Keywords: aggregation; Alzheimer disease; excitotoxicity; hemin; iron; non-antibiotic tetracyclines; oxidative stress; phospho-tau; tauopathies

https://doi.org/10.3390/biom15070941

1. Introduction

Tauopathies belong to a heterogeneous class of age-related neurodegenerative disorders primarily characterized by the accumulation of insoluble proteinaceous aggregates made of abnormally phosphorylated tau [1,2]. Aggregates of misfolded tau protein forming neurofibrillary tangles (NFTs) are believed to be implicated in neuronal dysfunction and death, leading to a range of symptoms including cognitive decline, behavioral changes, dementia, and motor deficits [3,4]. It has been hypothesized that neurodegeneration in tauopathies results from a loss of tau's ability to stabilize microtubule assemblies, but it is conceivable that other cellular functions are also perturbed during tau pathological processes [5].

Tauopathies can be classified as primary or secondary depending on whether the tau pathology appears as the main driver of neurodegeneration. Primary tauopathies include corticobasal degeneration, progressive supranuclear palsy, frontotemporal lobar degeneration with tau pathology, Pick's disease, and argyrophilic grain disease [3,6,7]. Alzheimer's disease (AD), the most well-studied tauopathy, is classified as a secondary tauopathy as tau NFT formation follows the appearance of amyloid-β protein deposits. Secondary tauopathies may also have an environmental cause, such as trauma in chronic traumatic encephalopathy [7,8] or toxins in atypical forms of degenerative Parkinsonism [9,10].

In AD, it is believed that tau aggregation and the formation of NFTs are preceded by a pre-tangle stage during which non-fibrillary hyperphosphorylated tau accumulates in the soma and dendrites of vulnerable neurons [11–13]. This suggests that tau hyperphosphorylation may be an initiating step for tau aggregation into NFTs [14–16].

In this study, we aimed to simulate tau abnormalities occurring in the early and advanced stages of AD and other tauopathies to identify new tau-directed therapies. To model the early stages of pathological tau build-up, we established mouse primary cortical cultures and exposed them to redox-active iron from hemin (Hm), a breakdown product of hemoglobin (Hb) [17,18], or to low excitotoxic levels of the neurotransmitter glutamate [19]. The selection of these two cellular stressors was motivated by the fact that iron dyshomeostasis [20,21] and neuronal hyperexcitability [22,23] are believed to actively contribute to the pathophysiology of tauopathies. For cellular immunodetection of pathological tau, we used the AT8 phospho-specific antibody, which enables tracking of the progression of a pathological tau burden in the human brain by specifically detecting phospho-tau (p-tau) Ser202/Thr205 residues [16,24,25].

Our findings revealed that p-tau accumulates in certain subsets of neuronal cell bodies and their proximal neurites, enabling us to test molecules that could potentially prevent these pathological events. Specifically, we evaluated the effects of uric acid (UA), which is the metabolic end-product of purines. Investigating UA in this context is of interest because lower circulating levels of this purine may represent a risk factor for AD dementia and other tauopathies [26,27]. We compared the effect of UA to that of a newly designed non-antibiotic TC having potent neuroprotective/antioxidant effects in experimental paradigms that model Parkinson's disease (PD) neurodegeneration [28].

To reproduce tau abnormalities at a more advanced disease stage, we monitored the aggregation of 2N4R tau monomers in a Thioflavin-T fluorescent assay. Specifically, test compounds were evaluated when tau aggregation was seeded by heparin or heparin-assembled tau fibrils [29,30].

2. Materials and Methods

2.1. Use of Animals

The mice used were housed, handled, and cared for in strict accordance with the European Union Council Directives (2010/63/EU). The Committee on the Ethics of Animal

Experiments Charles Darwin no. 5 approved experimental protocols under the authorization number Ce5/2024/001.

2.2. Cortical Neuronal Cultures

We established primary cultures of cortical neurons from Swiss mouse embryos at day 13.5 of gestation (Janvier LABS; Le Genest St Isle, France). Briefly, brain cortices were dissected out and then incubated for 20 min at 37 °C in an EDTA (2 mM)–trypsin (0.05%) solution for brain tissue digestion before trituration. Then, trypsin was neutralized with Dulbecco's Modified Eagle's Medium (DMEM; Thermo Fisher Scientific, Courtaboeuf, France) containing 10% fetal calf serum (FCS; Biowest LLC, Les Ulis, France), and mechanical trituration performed in Leibovitz L15 culture medium (Thermo Fisher Scientific; Courtaboeuf, France) following protocols previously described in detail for midbrain cultures [28,31]. Dissociated cells in suspension were seeded at a density of 40–60 $\times$ 10^3 cells/cm^2 onto Nunc 48-well multiwell plates (Roskilde, Denmark) or 8-well glass bottom µ-slides (#80807; Ibidi, Gräfelfing, Germany) pre-coated with 1 mg/mL polyethylenimine (PEI; P3143; Sigma Aldrich, L'Isle-d'Abeau Chesnes, France) dissolved in a pH = 8.3 borate buffer as described before [32,33].

Cell culture was initiated in Neurobasal-A medium (Nb; #10888022; Gibco, Saint Aubin, France) supplemented with a B27 cocktail without antioxidants (#10889038; Gibco), an N2 mix (#17502048; Gibco), 100 IU/mL of penicillin/streptomycin (#15140122; Gibco), and 1% FCS. This medium is referred to as modified Nb (mNb). After plating, the cultures were treated daily with 1.5 µM of the antimitotic agent cytosine arabinoside (Ara-C; Sigma Aldrich) to reach a cumulative concentration of 4.5 µM at day in vitro (div)3. This treatment regimen, which is not toxic for neuronal cells, allows us to eliminate >95% of glial cells from the cultures [34,35]. At the end of div3, the plating medium was completely substituted by mNb medium lacking FCS, which had been previously conditioned in pure astrocyte cultures [31,36]. Cortical cultures were then maintained in astrocyte-conditioned medium (ACM) until termination of the cultures.

To model tau-related neurodegenerative changes, we treated the cultures with hemin (Hm), a breakdown product of Hb [17,18], which contains redox-active iron, or with the excitatory neurotransmitter glutamate [19]. When exposing cortical cultures to Hm, treatments were carried out on div7, 8, and 9, and the consequences of such treatments were evaluated at div10. When tau neurodegenerative changes were provoked by a challenge with glutamate, we used div14 cortical cultures and a concentration of the excitatory neurotransmitter of 10 µM, causing the partial neurodegeneration of cortical neurons. In that case, cultures were taken for analysis at div15. The two protocols used to evaluate the anti-tau effects of test treatments are described in Figure 1.

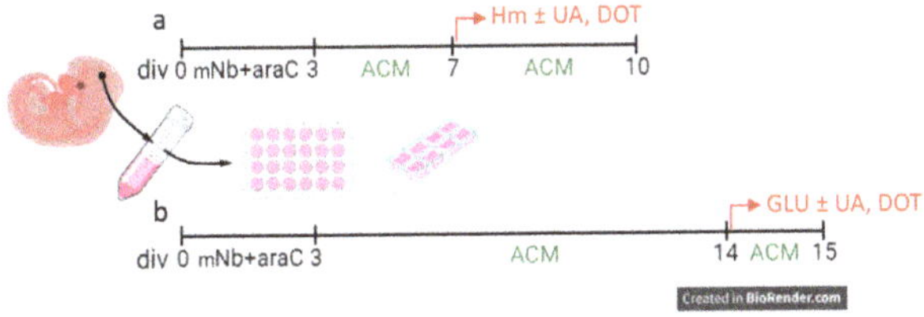

Figure 1. Experimental conditions implemented to promote tau abnormalities in mouse cortical neurons in culture. (**a**) Cortical cultures treated with hemin (Hm) on div7, 8, and 9, and then processed at div10 for analysis. (**b**) Div14 cortical cultures challenged with 10 µM of glutamate (GLU) for 24 h and then processed for analysis. In each experimental paradigm, we evaluated the capacity of UA, DOT, and the other test compounds to prevent p-tau accumulation. ACM: astrocyte-conditioned medium. mNb: modified neurobasal medium. Created with Biorender.com.

2.3. The Non-Antibiotic TC Compound, DOT

2.3.1. Synthesis of DOT

The non-antibiotic TC compound, 4-dedimethylamino-12a-deoxy oxytetracycline (DOT), was synthesized in-house using a protocol previously described [28]. The synthesis strategy is a modification of that initially developed by Golub and colleagues [37] to generate non-antibiotic TC derivatives. Specifically, we removed the dimethylamino substituent at position 4 to eliminate the antimicrobial activity, along with the hydroxy group at position 12a on ring A of oxytetracycline [28]. Stock solutions of DOT dissolved at 50 mM in DMSO were kept at −20 °C for less than 6 months. Intermediate dilutions used for cell culture treatments were made in distilled water and stored for 7 days at 4 °C, protected from light.

2.3.2. Capacity of DOT to Penetrate the Brain

The efficacy of DOT in penetrating the brain was assessed using adult Swiss mice receiving a single subcutaneous injection of 30 mg/kg of DOT diluted in saline with 5% DMSO and 5% Tween 80. This formulation allowed administration in a volume of 4 mL/kg. After being sacrificed 30 min, 1 h, 8 h, and 24 h after treatment, brain and serum samples (n = 3/time point) were collected and mixed with acetonitrile for compound extraction. After vortexing and sonication, proteins and solid residues were removed by centrifugation (15,000× *g*, 5 min), and supernatants were analyzed using a UHPLC system coupled with a triple quadrupole mass spectrometer LCMS-8030 (Shimadzu Corporation, Kyoto, Japan). The brain-to-plasma ratio calculated from the area under the concentration–time curves in the brain and plasma was 0.6 ± 0.2 (n = 3), which demonstrates the good brain penetration of this compound.

2.4. Uric Acid and Related Purine Derivatives

UA (#U2625), its immediate precursor xanthine (XANT; #X7375), its oxidation product allantoin (ALTN; #93791), and its synthetic analog 1, 7-dimethyluric acid (DMUA; #40407) were obtained from Sigma Aldrich (L'Isle-d'Abeau Chesnes, France). Stock solutions of UA, XANT, DMUA, and ALTN were all made at 10 mM. UA was first solubilized in 1N NaOH before bringing the pH of the solution to ~7 using 1N HCl. XANT and DMUA were diluted using the same procedure, whereas ALTN was diluted in sterile distilled water.

2.5. Other Pharmacological Reagents

Hemin (Hm; #51280) was diluted in DMSO to obtain a 20 mM stock solution and intermediate dilutions were made fresh in distilled water just before use. Stock solutions of inhibitors of lipid peroxidation/ferroptosis Trolox-C (TROL; #238813) and Liproxstatin-1 (LIP; #6113) were made at 50 mM in pure ethanol and DMSO, respectively. The iron chelator desferoxamine (DESF; #D9533) and vitamin C (VitC; #255564) were diluted at 10 mM in distilled water. The iron-chelating glycoprotein (APO, #T1428) was dissolved in distilled water at a concentration of 10 mg/mL.

2.6. Immunocytochemical Procedures

After the termination of test treatments, cortical cultures were fixed for 12 min in Dulbecco's phosphate-buffered saline (PBS) containing 4% formaldehyde (#252549; Sigma Aldrich), washed with PBS, and then incubated for 18 h with a mouse p-tau monoclonal antibody (AT8) (MN1020, Thermofisher Scientific, Courtaboeuf, France; 1:500 in 0.2% Triton X-100/PBS), which was revealed with an anti-mouse IgG (H + L) conjugated to Alexa Fluor 488 (#A11001 Thermofisher Scientific). Then, cortical cultures were incubated for 18 h with a chicken anti-microtubule-associated protein-2 (MAP-2) antibody (ab5392,

Abcam, Cambridge, UK; 1:1000 in PBS), which was revealed with an anti-chicken IgG (H + L) conjugated to Alexa Fluor 555 (#A32932, Thermofisher Scientific).

2.7. Cell Counting Procedures

For cell-counting operations, we used a Nikon Eclipse Ti-U fluorescence inverted microscope (Nikon France, Champigny sur Marne, France) equipped with an ORCA-Flash digital camera (Hamamatsu Photonics, Massy, France) and the NIS-Elements software Version 5.41 (Nikon). The number of p-tau$^+$ cell bodies and MAP-2$^+$ neurons was estimated by taking microphotographs with a 40× objective of 5–10 visual fields that were randomly selected for each treatment condition. Cell-counting operations were performed with the open-source FIJI software (version 2.1.0/1.54p) [38]. No blinding procedure was undertaken.

2.8. Confocal Imaging

Confocal imaging was performed using a Nikon A1R HD25 microscope (Nikon, Amstelveen, The Netherlands) equipped with the NIS-Elements software (Version 5.41; Nikon). Images from cultures grown on glass-bottom Ibidi μ-slides were acquired every 0.38 μm in the Z direction using a 40× water immersion objective (NA 1.3; WD 0.2 mm). All acquisitions were performed under the resonant scanner mode. Image reconstructions were performed with the FIJI software.

2.9. Assessment of ROS Emission and Changes in Mitochondrial Membrane Potential

To assess changes in reactive oxygen species (ROS) and mitochondrial membrane potential ($\Delta\Psi$m), the culture medium was removed and immediately replaced by warm PBS–glucose (5 mM). Then, the cultures were exposed to tetramethylrhodamine methyl ester, perchlorate (TMRM; 50 nM; ab228569; Abcam, Cambridge, UK), and dihydrorhodamine 123 (DHR-123; 25 μM; D23806; Thermo Fisher Scientific) 10 min later. After 35 min, the cultures were washed extensively (3×) with PBS–glucose to remove fluorescent probes in excess before carrying out live cell imaging [35,39,40]. In some experiments, ROS production was induced with the prooxidant compound H_2O_2 (50 μM; #8070.4; CARL ROTH, Karlsruhe, Germany) and mitochondrial depolarization with the protonophore carbonyl cyanide 4-(trifluoromethoxy)phenylhydrazone (FCCP; 0.5 μM) provided in the TMRM assay kit. H_2O_2 and FCCP were added 4 h and 10 min before adding fluorogenic probes, respectively. Note that FCCP, but not H_2O_2, was present throughout the incubation in PBS–glucose.

For each culture condition, fluorescent images from at least five randomly chosen fields were acquired with a 40× fluorescence objective using a Nikon Eclipse Ti-U fluorescence inverted microscope equipped with an ORCA Flash digital camera. The excitation and emission wavelengths for DHR-123 were 490 nm and 525 nm, respectively, whereas the corresponding wavelengths for TMRM were 548 nm and 575 nm, respectively. The open-source FIJI software was used for quantifying fluorescent signal intensities [38] over the surface area of each individual cell body morphologically identifiable by phase contrast (Phaco) optics [28]. The results were expressed in changes in fluorescence intensity relative to non-treated cultures.

2.10. Tau Aggregation Assay

For in vitro aggregation assays, we utilized recombinantly expressed full-length human tau (2N4R) obtained from BrinDx (www.brindx.com (accessed on 22 October 2024); #BRX-2002). Aggregation assays were based on protocols reported previously [29,41]. Briefly, samples containing tau monomers (tau_m; 22 μM) were resuspended in PBS and mixed with the polyanionic cofactor heparin (0.2 mg/mL) along with ThT (10 μM), a fluorescent reporter molecule used to monitor the formation of β-sheet-rich fibril structures [42].

Heparin-assembled tau fibrils (tau$_f$) (2.2 μM monomer equivalent) were also used as seeds instead of heparin.

Fibrillar aggregates of tau were generated by incubating test samples in an orbital shaker (Thermomixer Comfort; Eppendorf, Montesson, France) at a speed of 600 rpm for 72 h. After this incubation, end-stage products were measured by fluorescence emission with a Fluoromax-4 spectrofluorometer, setting the excitation wavelength at 450 nm. Prior to these measurements, each test molecule was subjected to standard biophysical assessments—specifically absorbance and fluorescence analyses—to confirm that it did not emit fluorescence within the spectral range of ThT. This precaution ensured that any detected ThT signal variations could be confidently attributed to conformational changes in tau and not to spectral interference from the test compound.

2.11. Statistical Analysis

Data are presented as the mean ± SEM. Statistical outliers were removed using the ROUT method (Q = 5%) [43]. The normality or near-normality of the datasets was checked by Shapiro–Wilk testing or QQ plot visualization, respectively [44]. When normality was assumed, we conducted a one-way ANOVA followed by a Tukey's test for all pairwise multiple comparisons or a Dunnett's test for multiple comparisons against a single control group. In cases where normality could not be assumed, data were analyzed with a Kruskal–Wallis ANOVA on ranks, followed by Dunn's multiple comparison test. A p-value of <0.05 was considered statistically significant. Statistical evaluation of the data is presented in the figure legend.

3. Results

3.1. Induction of P-Tau Abnormalities by Hemin

We aimed to establish an experimental cell culture model reproducing the accumulation of hyperphosphorylated tau, which is characteristic of tauopathies [14,15]. To achieve this, we utilized cortical cultures that were chronically exposed to redox-active iron from Hm. Specifically, cortical cultures maintained in ACM were treated repeatedly at div7, 8, and 9 with concentrations of Hm ranging from 3 to 30 μM with no medium change between treatments, allowing for final cumulative concentrations of 9 to 90 μM by div10, when the cultures were processed for analyses (Figure 1).

The induction of p-tau by Hm is described in Figure 2a–c. Precisely, we show that the overall number of neuronal somas accumulating tau phosphorylated at the AT8 epitope is relatively low (though not null) in control cortical cultures (Figure 2a). A faint and punctuated AT8 immunosignal was also detectable in the neurite network of the control cultures (Figure 2c), which is indicative of basal levels of AT8 tau phosphorylation in neurites under these conditions. There was, however, a sharp increase in the number of cell bodies accumulating hyperphosphorylated tau when the cultures were exposed to 3 to 10 μM of Hm for three consecutive days. We found that the pathological accumulation of p-tau in neuronal somas was significant at 3 μM and reached its peak at 10 μM of Hm, before decreasing at higher concentrations (Figure 2a). Neuronal death was not significant at 3 μM of Hm (Figure 2b). However, at concentrations of 10 μM Hm and higher, we observed a gradual decline in neuronal survival, reaching a peak at 30 μM. Note that at 3 and 10 μM of Hm, a strong AT8 immunosignal was also observed in proximal neurites originating from cell bodies, accumulating p-tau. The impact of repeated treatment regimens with Hm (3 and 10 μM) on p-tau accumulation and neuronal survival is illustrated by microphotographs from div10 cortical cultures (Figure 2c). In subsequent experiments, we applied a treatment regimen using Hm at a concentration of 3 μM to simulate p-tau accumulation without significantly affecting neuronal survival.

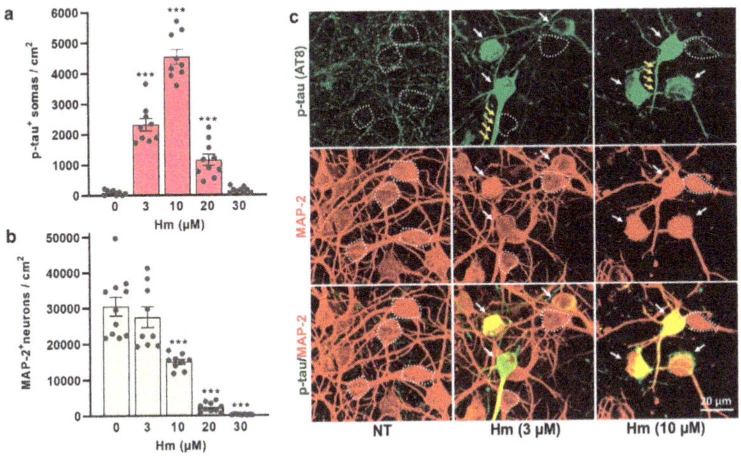

Figure 2. Induction of p-tau-related neurodegenerative changes by Hm. (**a**,**b**) Number of p-tau$^+$ somas (AT8 immunosignal) (**a**) and MAP-2$^+$ neurons (**b**) in div10 cortical cultures treated repeatedly at div7, 8, and 9 with 3–30 μM of Hm. (**a**,**b**) Data expressed in numbers of p-tau$^+$ somas/cm^2 or neuronal cells/cm^2 in div10 cultures are presented as the mean ± SEM. One-way ANOVA followed by Tukey's test, *** $p < 0.001$ vs. NT. (**c**) Single-color and merged images from div10 cortical cultures repeatedly exposed to Hm (3 and 10 μM) and then immunolabelled for p-tau (green) and MAP-2 (red). White arrows indicate cortical neurons accumulating p-tau in their somas. Small yellow arrows (upper panel) point to neuritic extensions where the p-tau immunosignal is strongly increased. The white dotted line shows the boundaries of neuronal cell bodies with a near absence of p-tau immunostaining. NT: non-treated.

3.2. Hm-Induced P-Tau Abnormalities Are Curtailed by the Purine Metabolic End-Product UA

We initially aimed to investigate whether the end-product of purine metabolism in humans, UA, could counteract p-tau accumulation after Hm exposure. To achieve this, cortical cultures that were repeatedly exposed to 3 μM Hm were concomitantly treated with varying concentrations of UA (0.01–300 μM). As shown in Figure 3a, a concentration of 0.1 μM UA reduced the number of cortical neuronal cell bodies accumulating p-tau by about 70%. UA demonstrated optimal anti-tau effects between 1 and 200 μM, reducing the number of p-tau$^+$ somas by over 95% within this range of concentrations. At 300 μM, UA appeared to lose some of its efficacy (Figure 3a). Importantly, we did not observe any neuronal loss, regardless of the treatment applied to cultures (Figure 3b). Figure 3c illustrates the inhibitory effects of 30 μM UA against Hm-induced p-tau accumulation.

3.3. UA Exerts Anti-Tau Effects by Curtailing Hm-Mediated Oxidative Insults

Because UA is known to operate as a potent antioxidant [45,46], we wished to determine whether the anti-tau effects of UA could result from the inhibition of ROS emission. To this aim, we used the ROS-sensitive dye DHR-123 for monitoring intracellular ROS production in Hm (3 μM)-treated cultures exposed or not to UA (30 μM), compared to NT cultures (Figure 4a). Precisely, we show that a treatment with 3 μM Hm is associated with an increase in ROS production and that a concomitant treatment with 30 μM UA totally prevents this effect. We also demonstrate that ROS emission is stimulated even more vigorously when div10 cortical cultures are acutely challenged with 50 μM H_2O_2 for 4 hrs. In that case, however, UA (30 μM) could not reduce ROS production (Figure 4b).

Concomitant to ROS quantification with DHR-123, we monitored changes in ΔΨm that could occur in ROS-emitting neurons [35,40]. Using the mitoprobe TMRM, we show that the ROS signal induced by 3 μM Hm is not associated with significant changes in

ΔΨm (Figure 4c). ΔΨm was similarly preserved in UA-treated cultures. There was a drop in ΔΨm, however, when cortical cultures were challenged with H_2O_2, regardless of the presence of UA (Figure 4d). An acute challenge with the protonophore FCCP (0.5 μM), which operates as an uncoupler of mitochondrial oxidative phosphorylation, led to an expected drop in ΔΨm (Figure 4d) but failed to stimulate ROS production (Figure 4b). Figure 4e provides representative illustrations showing the DHR-123 (upper panel) and TMRM (mid panel) fluorescent signals in div10 cortical cultures treated repeatedly with 3 μM Hm or acutely with H_2O_2 in the presence or not of 30 μM UA. The acute challenge with FCCP is also illustrated in comparison. The lower panel represents Phaco images merged with the DHR-123 and TMRM fluorescent signals.

As expected, treatment with 30 μM UA prevented p-tau build-up induced by 3 μM of Hm (Figure 4f). However, UA failed to curtail p-tau accumulation induced by acute exposure to H_2O_2. The treatment with FCCP, causing a reduction in ΔΨm and no increase in ROS production, failed to promote p-tau accumulation (Figure 4f). This indicates that the transient dissipation of ΔΨm cannot lead per se to p-tau accumulation.

Cell counting of MAP-2$^+$ cell bodies confirmed that neuronal survival was not affected by Hm (3 μM) exposure, whether UA (30 μM) was present or not in these cultures (Figure 4g). On the contrary, a 4 h challenge with 50 μM H_2O_2 led to a 25% decrease in neuronal survival that was not compensated by UA. Finally, the acute challenge with 0.5 μM FCCP did not result in significant neuronal loss in the present experimental time frame. Photomicrographs illustrate the impact of various test treatments on p-tau accumulation in div10 cortical cultures (Figure 4h).

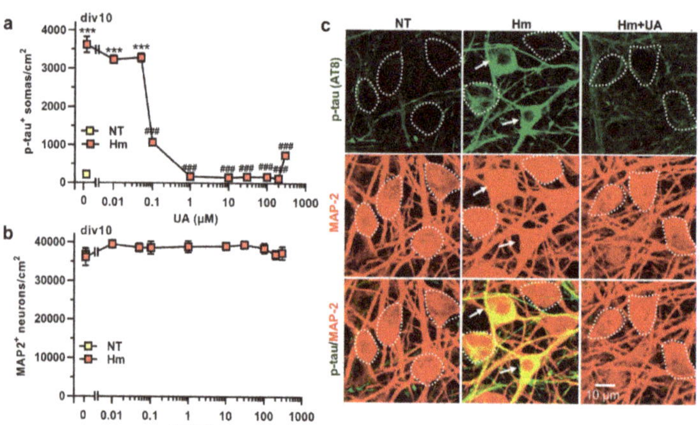

Figure 3. p-tau build-up induced by Hm is curtailed by uric acid. (**a**) Number of p-tau$^+$ somas (AT8 immunosignal) in div10 cortical cultures treated repeatedly at div7, 8, and 9 with 3 μM of Hm, together or not with 0.01–300 μM of UA. (**b**) Survival of MAP-2$^+$ neurons in div10 cortical cultures exposed to the same treatments as in (**a**). (**a**,**b**) Data expressed in numbers of p-tau$^+$ somas/cm^2 or MAP-2$^+$ neurons/cm^2 in div10 cultures are presented as the mean ± SEM (n = 6–9). One-way ANOVA followed by Tukey's test: *** $p < 0.001$ vs. NT. ### $p < 0.001$ vs. Hm. (**c**) Representative microphotographs illustrating the inhibitory effects of 30 μM UA on the accumulation of p-tau in the soma of MAP-2$^+$ neurons treated with 3 μM Hm. Note that neuronal survival is not impacted by any of the test treatments. White arrows point to cortical neurons accumulating p-tau in their somas. The white dotted line shows the boundaries of neuronal cell bodies with a near absence of p-tau immunostaining. NT: non-treated.

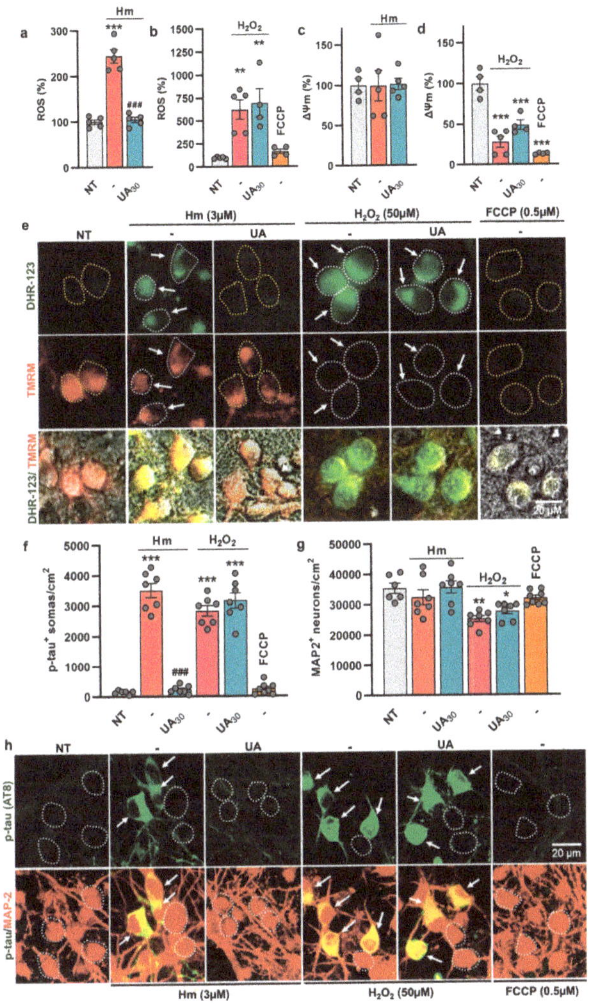

Figure 4. UA represses Hm-induced p-tau build-up by curtailing ROS-mediated insults. (**a**) ROS emission in div10 cortical cultures treated repeatedly with 3 μM Hm in the presence or not of 30 μM UA. (**b**) Comparison with cultures acutely exposed for 4 h to H_2O_2 (50 μM) in the presence or the absence of 30 μM UA. The impact of an acute challenge with 0.5 μM of the protonophore FCCP is also illustrated. (**c**) Estimation of ΔΨm in div10 cortical cultures treated as in (**a**). (**d**) Estimation of ΔΨm in div10 cortical cultures treated as in (**b**). (**a–d**) Data expressed in % of NT-cultures are presented as the mean ± SEM. One-way ANOVA followed by Tukey's test: ** $p < 0.01$ and *** $p < 0.001$ vs. NT; ### $p < 0.001$ vs. Hm. (**e**) Representative images showing DHR-123 (upper panel) and TMRM (mid panel) fluorescent signals in div10 cortical cultures receiving the same treatments as in (**a–d**). The lower panel represents Phaco images merged with the DHR-123 and TMRM fluorescent signals. Neuronal cell bodies exhibiting increased ROS levels are delineated by a white dotted line and pointed out by a white arrow. Neuronal cell bodies exhibiting basal ROS levels are delineated by a yellow dotted line. (**f**) Estimation of p-tau build-up in div10 cortical cultures receiving the same treatments as in (**a–d**). (**g**) Estimation of neuronal survival in div10 cortical cultures receiving the same treatments as in (**a–d**). (**f**,**g**) Data expressed in numbers of p-tau$^+$ somas/cm^2 or MAP-2$^+$ neurons /cm^2 in div10 cultures are presented as the mean ± SEM. One-way ANOVA followed by Tukey's test: * $p < 0.05$, ** $p < 0.01$, and *** $p < 0.001$ vs. NT; ### $p < 0.01$ vs. Hm. (**h**) Representative photomicrographs of div10

cortical cultures showing that UA prevents p-tau build-up induced by Hm but not H_2O_2. White arrows point to cortical neurons accumulating p-tau in their somas. The white dotted line shows the boundaries of neuronal cell bodies with a near absence of p-tau immunostaining.

3.4. The Suppression of P-Tau Build-Up by UA Is Reproduced by Compounds Inhibiting Iron-Mediated Lipid Peroxidation

To better understand the nature of the inhibitory effects of UA on p-tau accumulation, we performed a series of experiments comparing its suppressive action to that of other treatments susceptible of mimicking the effects of the purine compound. These treatments comprise the iron chelator DESF (50 μM), the inhibitors of lipid peroxidation and ferroptosis TROL (20 μM) and LIP (0.3 μM), the water-soluble vitamin VitC (25 μM), and APO (100 μg/mL) an iron-carrying glycoprotein reported to be protective against iron-mediated neurodegeneration [35,47]. We showed that DESF, TROL, and LIP mimicked the anti-tau effects of UA in cortical cultures treated repeatedly with 3 μM Hm. The efficacy of DES, TROL, and LIP was comparable to that of 30 μM UA (Figure 5a). VitC and APO, however, did not exert anti-tau effects in this setting. No neuronal cell loss was observed regardless of the treatments applied (Figure 5b). Photomicrographs from Figure 5c illustrate the impact of various test treatments on p-tau accumulation in div10 cortical cultures exposed to 3 μM Hm.

Then, we monitored ROS production and changes in $\Delta\Psi m$ in cultures receiving the same treatments as before. As expected, intracellular oxidative stress was significantly elevated in cortical cultures receiving a treatment regimen with 3 μM Hm. Conversely, ROS returned to basal levels when DESF, TROL, and LIP were added to Hm-treated cultures instead of UA (Figure 5d). However, APO and VitC were ineffective in reducing ROS production induced by Hm. $\Delta\Psi m$ was preserved whatever the test treatments applied to the cultures (Figure 5e). As expected, the oxidizer H_2O_2 (50 μM) caused a large increase in ROS emission and the protonophore FCCP (0.5 μM) a decrease in $\Delta\Psi$ (Figure 5d,e).

Figure 5f provides representative illustrations showing DHR-123 (upper panel) and TMRM (mid panel) fluorescent signals in div10 cortical cultures treated repeatedly with 3 μM Hm in the presence or not of UA and the other test treatments. The lower panel represents Phaco images merged with DHR-123 and TMRM fluorescent signals.

3.5. The Anti-Tau Effect of UA Is Not Reproduced Either by Its Immediate Precursor Xanthine or Its Oxidized Metabolite, Allantoin

We also tested whether the anti-tau and antioxidant effects of UA could be reproduced by its immediate precursor XANT (30 μM) or its oxidation product ALTN (30 μM) (Figure 6a). Neither of these two compounds could reproduce the anti-tau effects of UA (Figure 6b). In contrast, the synthetic 1,7-dimethyl derivative of UA DMUA (30 μM) retained the anti-tau effects of its natural parent compound. We also found that DMUA (30 μM) efficiently curtailed ROS production in Hm (3 μM)-treated cortical cultures. Its efficacy was similar to that of UA (30 μM). In contrast, XANT (30 μM) and ALTN (30 μM) were ineffective (Figure 6c). Figure 6d describes the impact of UA (30 μM), XANT (30 μM), ALTN (30 μM), and DMUA (30 μM) on p-tau accumulation induced by 3 μM Hm in div10 cortical cultures.

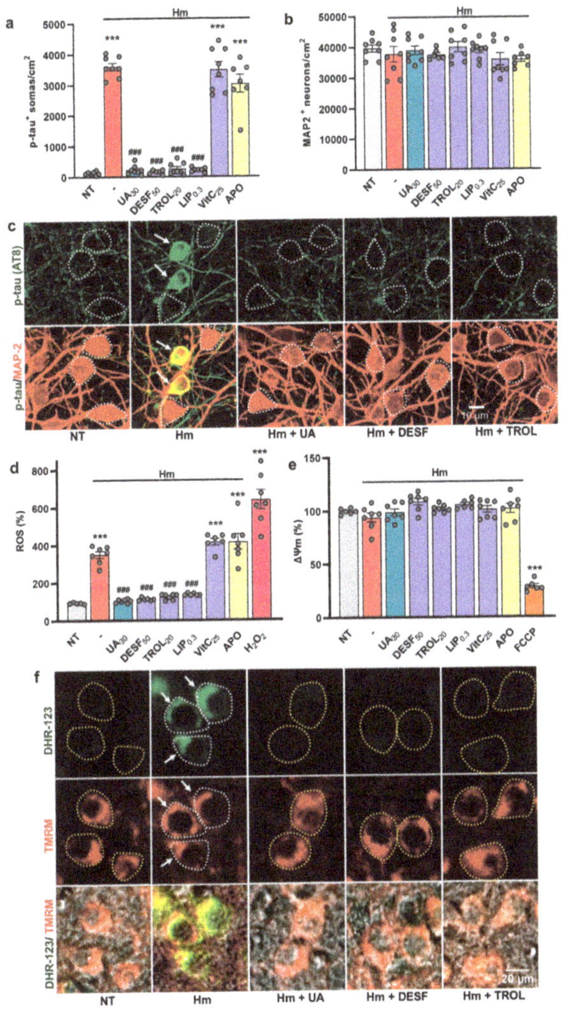

Figure 5. Prevention of Hm-induced p-tau build-up by UA is reproduced by compounds inhibiting iron-mediated lipid peroxidation. (**a**) Estimation of p-tau accumulation (AT8) in div10 cortical cultures treated repeatedly at div7, 8, and 9 with 3 μM Hm in the presence or not of UA (30 μM), the iron chelator DESF (50 μM), the inhibitors of lipid peroxidation and ferroptosis TROL (20 μM) and LIP (0.3 μM), the water-soluble vitamin VitC (25 μM), or the glycoprotein APO (100 μg/mL). (**b**) Estimation of neuronal survival in div10 cortical cultures undergoing the same treatments as in (**a**). (**a**,**b**) Data expressed in numbers of p-tau$^+$ somas/cm^2 or MAP-2$^+$ neurons/cm^2 in div10 cultures are presented as the mean ± SEM. One-way ANOVA followed by post hoc Tukey's test: *** $p < 0.001$ vs. NT, ### $p < 0.001$ vs. Hm. (**c**) Upper panel: Representative photomicrographs showing the impact of UA (30 μM), DESF (50 μM), and TROL (20 μM) on Hm (3 μM)-induced p-tau accumulation in div10 cortical cultures. Lower panel: Same cell culture field in which the p-tau immunosignal is combined with MAP-2 immunolabeling. White arrows point to cortical neurons accumulating p-tau in their somas. The white dotted line shows the boundaries of neuronal cell bodies with a near absence of p-tau immunostaining. (**d**) Estimation of ROS emission in div10 cortical cultures treated repeatedly at div7, 8, and 9 with 3 μM Hm in the presence or not of UA (30 μM) and the other test treatments described in (**a**). The oxidizer H_2O_2 (50 μM), acutely applied to the cultures for 4 h, is used as a positive control for ROS emission. (**e**) Estimation of ΔΨm in div10 cortical cultures treated repeatedly at div7, 8, and 9 with 3 μM Hm in the presence or not of UA (30 μM) or the other test treatments described in (**a**).

The protonophore FCCP (0.5 µM) applied acutely to the cultures is used as a reference compound to induce ΔΨm dissipation. (**d**,**e**) Data expressed in % of NT-cultures are presented as the mean ± SEM. One-way ANOVA followed by post hoc Tukey's test: *** $p < 0.001$ vs. NT, ### $p < 0.001$ vs. Hm. (**f**) Representative images showing DHR-123 (upper panel) and TMRM (mid panel) fluorescent signals in div10 cortical cultures treated with 3 µM Hm in the presence or not of UA (30 µM), DESF (50 µM), or TROL (20 µM). The lower panel represents Phaco images merged with DHR-123 and TMRM fluorescent signals. Neuronal cell bodies exhibiting increased ROS levels are delineated by a white dotted line and pointed out by a white arrow. Neuronal cell bodies exhibiting basal ROS levels are delineated by a yellow dotted line.

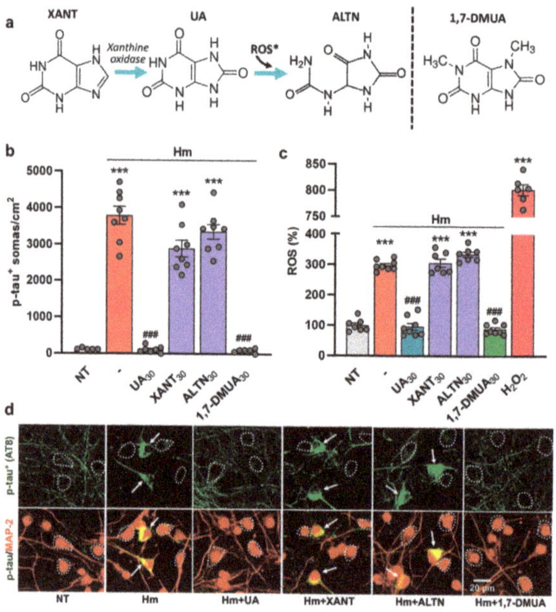

Figure 6. Inhibition of Hm-mediated p-tau accumulation by UA is not reproduced by xanthine or allantoin. (**a**) Simplified metabolic pathway of purine degradation with the chemical structures of UA, XANT, and ALTN. DMUA is a synthetic analog of UA. * In humans, the presence of ALTN reflects the non-enzymatic oxidative catabolism of UA. ROS: reactive oxygen species. (**b**) Estimation of p-tau accumulation (AT8) in div10 cortical cultures treated repeatedly at div7, 8, and 9 with Hm (3 µM) in the presence or not of UA (30 µM), XANT (30 µM), ALTN (30 µM), or DMUA (30 µM). (**c**) ROS emission in div10 cortical cultures undergoing the same treatments as in (**b**). The oxidizer H_2O_2 (50 µM) applied acutely for 4 h to the cultures is used as a positive control for ROS emission. (**b**,**c**) Data expressed in numbers of p-tau$^+$ somas/cm^2 (**b**) or ROS levels (**c**) in div10 cultures are presented as the mean ± SEM. One-way ANOVA followed by post hoc Tukey's test: *** $p < 0.001$ vs. NT, ### $p < 0.001$ vs. Hm. (**d**) Upper panel: Representative microphotographs showing the impact of test treatments on p-tau accumulation in div10 cortical cultures exposed to 3 µM Hm in the presence or not of the different purine derivatives. Lower panel: Same cell culture field where the p-tau immunosignal is combined with MAP-2 immunolabeling. White arrows point to cortical neurons accumulating p-tau in their somas. The white dotted line shows the boundaries of neuronal cell bodies with a near absence of p-tau immunostaining.

3.6. The Repressive Action of Uric Acid Against Hm-Induced P-Tau Accumulation Is Mimicked by DOT, a Non-Antibiotic Oxytetracycline Derivative

We tested whether a non-antibiotic TC compound that chemically derives from oxytetracycline could mimic the anti-tau effects of UA. This compound is of particular interest in the context of this study, as it demonstrated promising neuroprotective/antioxidant properties in another culture setting that mimics dopamine cell death in PD [28]. Specifically, we used a concentration of DOT of 3 µM reported to provide optimal rescue to dopamine neurons and compared its efficacy to that of UA, at 30 µM. Figure 7a shows that 3 µM of DOT efficiently mimicked the anti-tau effects of 30 µM UA. None of the test treatments had a significant impact on neuronal survival under the present experimental conditions (Figure 7b). The impact of the previous treatments on Hm-induced p-tau accumulation is illustrated by microphotographs in Figure 7c. In line with these observations, we found that DOT (3 µM) was as effective as UA (30 µM) in reducing ROS emission induced by Hm (3 µM) (Figure 7d). As expected, H_2O_2 (50 µM) used as a reference prooxidant treatment led to a strong increase in ROS emission. $\Delta\Psi m$ remained unchanged when Hm was applied to the cultures in the presence or not of UA or DOT (Figure 7e). As expected, there was a drop in $\Delta\Psi m$ when FCCP (0.5 µM) was used as a reference treatment to induce mitochondrial membrane depolarization. The results are illustrated by representative images in Figure 7f.

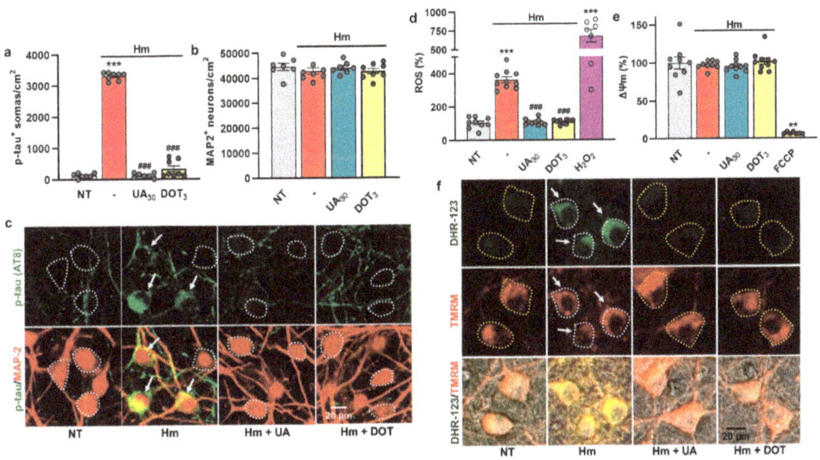

Figure 7. The inhibitory effect of UA against Hm-induced p-tau build-up is mimicked by DOT, a non-antibiotic oxytetracycline derivative. (**a**) Estimation of p-tau accumulation (AT8 immunosignal) in div10 cortical cultures treated repeatedly at div7, 8, and 9 with 3 µM Hm in the presence or not of UA (30 µM) or DOT (3 µM). (**b**) Estimation of neuronal survival in div10 cortical cultures undergoing the same treatments as in (**a**). (**a**,**b**) Data expressed in numbers of p-tau$^+$ somas/cm^2 or neuronal cells/cm^2 in div10 cultures are presented as the mean ± SEM. One-way ANOVA followed by Tukey's test. *** $p < 0.001$ vs. NT, ### $p < 0.001$ vs. Hm. (**c**) Upper panel: Representative microphotographs showing the impact of test treatments on p-tau accumulation in div10 cortical neurons in culture exposed repeatedly to 3 µM Hm in the presence or not of the treatments described in (**a**). Lower panel: Same cell culture field where the p-tau immunosignal is combined with MAP-2 immunolabeling. White arrows point to cortical neurons accumulating p-tau in their somas. The white dotted line shows the boundaries of neuronal cell bodies with a near absence of p-tau immunostaining. (**d**) ROS emission in div10 cortical cultures treated repeatedly at div7, 8, and 9 with 3 µM Hm in the presence or not of 3 µM of DOT or 30 µM of UA. Acute exposure to H_2O_2 (50 µM) is used as a positive control for ROS emission. Data expressed in % of NT cultures are presented as the mean ± SEM. One-way ANOVA followed by Tukey's test. *** $p < 0.001$ vs. NT and ### $p < 0.001$ vs. Hm. (**e**) Estimation of $\Delta\Psi m$ in div10 cortical cultures treated repeatedly at div7, 8, and 9 with 3 µM Hm in the presence or not of the treatments described in (**d**). An acute exposure to FCCP (0.5 µM) is used as a reference treatment to promote $\Delta\Psi m$ dissipation. Data expressed in % of NT cultures are presented as the mean ± SEM. Kruskal–Wallis followed by Dunn's test: ** $p < 0.01$ vs. NT. (**f**) Representative images showing DHR-123 (upper panel) and TMRM (mid panel) fluorescent signals in div10 cortical cultures

treated repeatedly with 3 μM Hm in the presence or not of UA (30 μM) or DOT (3 μM). The lower panel represents Phaco images merged with the DHR-123 and TMRM fluorescent signals. Neuronal cell bodies exhibiting increased ROS levels are delineated by a white dotted line and pointed out by a white arrow. Neuronal cell bodies exhibiting basal ROS levels are delineated by a yellow dotted line.

3.7. Glutamate-Mediated Tau Neurodegenerative Events Are Similarly Preventable by UA and DOT

We aimed to further investigate whether the inhibitory effects of UA and DOT on p-tau build-up could be observed in another context relevant to neurodegenerative tauopathies. To do this, we tested the impact of these compounds on div14 cortical cultures exposed to an excitotoxic stimulus of moderate intensity [48]. Precisely, we found that a 24 h challenge with 10 μM glutamate caused a robust increase in the number of p-tau$^+$ neurons in cortical neurons surviving the excitotoxic insult (Figure 8a). UA (10 μM) totally prevented this increase as well as neuronal loss resulting from excitotoxic stress (Figure 8a,b). The non-antibiotic TC DOT (3 μM) reproduced the anti-tau and neuroprotective effects of UA (Figure 8a,b). The NMDA glutamate receptor blocker, MK-801 (2 μM), the lipid peroxidation inhibitor TROL (20 μM), and the inhibitor of NADPH oxidase APOc (300 μM) also efficiently prevented glutamate-mediated tau neurodegenerative changes.

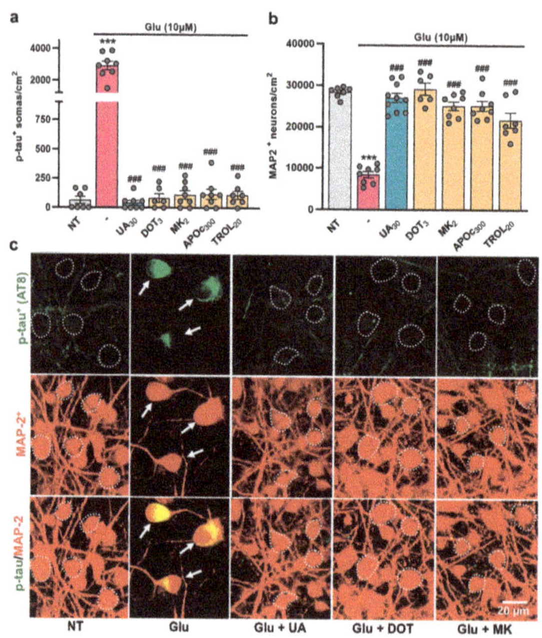

Figure 8. Glutamate-mediated tau neurodegenerative events are preventable by uric acid and the non-antibiotic tetracycline DOT. (**a**) Number of p-tau$^+$ cell bodies in div15 cortical cultures that had been previously exposed for 24 h to 10 μM glutamate in the presence or the absence of UA (30 μM), DOT (3 μM), MK-801 (MK, 2 μM), APOc (300 μM), or TROL (20 μM). (**b**) Estimation of neuronal survival in div15 cortical cultures undergoing the same treatments as in (**a**). (**a**,**b**) Data expressed in numbers of p-tau$^+$ somas/cm^2 or neuronal cells/cm^2 are presented as the mean ± SEM. One-way ANOVA followed by Tukey's test. *** $p < 0.001$ vs. NT and ### $p < 0.001$ vs. glutamate, only. (**c**) Representative microphotographs showing the impact of test treatments on p-tau accumulation in div15 cortical neuronal cultures challenged previously for 24 h with 10 μM glutamate in the presence or the absence of the treatments described in (**a**). White arrows point to cortical neurons accumulating p-tau in their somas. The white dotted line shows the boundaries of neuronal cell bodies with a near absence of p-tau immunostaining.

3.8. Potential of UA and DOT to Prevent Tau Amyloid Aggregation

The efficacy of UA and DOT in limiting p-tau build-up led us to determine whether these two molecules could also operate as inhibitors of tau aggregation (Figure 9). For that, we established a ThT fluorescence assay in which the aggregation of 2N4R tau_m is seeded by heparin (0.2 mg/mL), a polyanion commonly used as a cofactor to seed tau aggregation [24,30,49]. The estimation of steady-state ThT fluorescence levels after 72 h of continuous orbital agitation of tau samples revealed that UA and DOT significantly reduced tau aggregation (Figure 9a). The inhibitory effect of UA was significant at 100 μM and optimal at 200 μM. For DOT, we noted a substantial reduction in tau aggregation at 1 μM, with optimal inhibitory effects at 20 μM. To complete this characterization, we investigated the capacity of UA and DOT to reduce tau aggregation induced by 2.2 μM (monomer equivalent) heparin-assembled tau fibrils (tau_f), i.e., experimental conditions where tau aggregation does not directly rely on heparin as a cofactor (Figure 9b). In this context, 10 μM DOT was able to significantly reduce tau aggregation, while UA did not show inhibitory effects at 200 μM. Note that tau aggregation was minimal when tau_m were mixed with residual levels of heparin (0.02 mg/mL) present in test samples containing heparin–tau fibrils. As expected, when tau_m were agitated in the absence of heparin or tau_f, fibrillation did not occur.

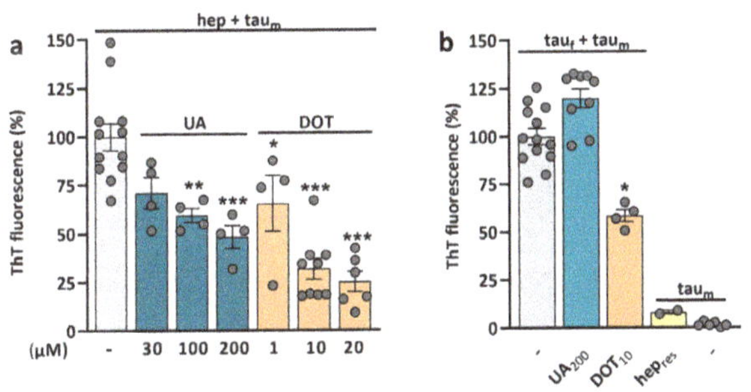

Figure 9. Comparison of the efficacy of uric acid and DOT in limiting tau amyloid aggregation induced by heparin or heparin-assembled tau fibrils. (**a**) Quantitation of heparin-induced tau aggregation in the presence or the absence of UA or DOT. Samples containing 22 μM of monomeric 2N4R tau together with 0.2 mg/mL heparin and 10 μM ThT are incubated for 72 h in the presence or the absence of UA (30, 100, and 200 μM) or DOT (1, 10, and 20 μM) using constant orbital agitation at 600 rpm. Data expressed in % of maximal ThT fluorescence at the endpoint stage are presented as mean values ± SEM. One-way ANOVA followed by post hoc Dunnett's test. * $p < 0.05$, ** $p < 0.01$, *** $p < 0.001$ vs. hep + tau_m. (**b**) Tau aggregation induced by 2.2 μM (monomer equivalent) heparin-assembled tau fibrils monitored in the presence or the absence of UA (200 μM) or DOT (10 μM) using the same incubation conditions as before. hep_{res}: residual concentration of heparin (0.02 mg/mL) estimated to be present in tau samples when tau fibrils are used as seeds. Data expressed in % of maximal ThT fluorescence at the endpoint stage are presented as mean values ± SEM. Kruskal–Wallis with a post hoc Dunn's test. * $p < 0.05$ vs. tau_f + tau_m.

4. Discussion

To replicate early tau neuropathological changes occurring in AD and other tauopathies, we utilized mouse primary cortical cultures, which were exposed to redox-active iron from Hm, a breakdown product of Hb. We show that Hm significantly promoted neuronal accumulation of pathological hyperphosphorylated tau. We established that UA, the end-product of human purine metabolism, effectively prevented this process by suppressing iron-mediated ROS emission. In contrast, UA's immediate precursor, XANT, and its oxidation product, ALTN, were ineffective. The anti-tau and antioxidant effects of UA were successfully reproduced by DOT, a non-antibiotic TC derivative of oxytetracycline that is structurally unrelated to purine compounds. Both UA and DOT were also effective when p-tau build-up was induced by moderate excitotoxic stimulation with glutamate. Using a ThT aggregation assay to mimic tau aggregation at more advanced stages of tauopathies, we demonstrate that UA and DOT could limit tau aggregation seeded by heparin. However, only DOT was effective when the aggregation process was induced by heparin-assembled tau fibrils.

4.1. Hemin Promotes Pathological P-Tau Accumulation in Cultured Cortical Neurons

Our initial objective was to model early cellular tau abnormalities that may occur before the formation of tau NFTs [1,50,51]. To achieve this, we established a culture system of mouse cortical neurons in which pathological p-tau accumulation is induced by redox-active iron from Hm, a degradation product of Hb. We chose this approach because several studies suggest that iron dyshomeostasis might actively contribute to the progression of tau lesions in AD pathology and other tauopathies [52–54], including chronic traumatic encephalopathy, a type of tauopathy arising from repeated traumatic head injuries and blood leakage around small vessels [8,55]. Brain imaging analyses provide the most compelling evidence for this hypothesis, showing that iron deposition is correlated with tau aggregates and neuronal loss in the brains of individuals diagnosed with AD [20]. A study by Yamamoto and colleagues further supports this hypothesis by indicating that iron (III) binds to hyperphosphorylated tau aggregates in tissue extracts from AD-affected brains [21].

Specifically, we monitored p-tau accumulation using the AT8 antibody, which detects a key phosphorylated epitope (Ser202/Thr205) associated with AD and other tauopathies [16,24]. The relevance of the AT8 antibody lies in its use for tracking the progression of the tau pathology both before and after the formation of tau NFTs [1,50,51].

We established that treatment with Hm over three consecutive days resulted in a concentration-dependent increase in the AT8 immunosignal within a significant portion of cortical neurons in culture. The accumulation of p-tau was generally detectable in both the soma and neuritic extensions, aligning with neuropathological findings in AD and other tauopathies [56,57]. The present observations are also consistent with data reported by Wan and colleagues [58], who demonstrated that hyperphosphorylated tau accumulates in cultured cortical neurons maintained in a medium enriched with soluble iron. Interestingly, the same authors found that mice on a high-iron diet exhibited pathological tau accumulation in the cortex in conjunction with cognitive deficits. The fact that only a fraction of cortical cell bodies accumulated p-tau upon treatment with Hm is consistent with neuropathological reports indicating that tau lesions affect only specific subsets of vulnerable neurons in tauopathies [59–61]. Interestingly, we found that Hm was capable of promoting p-tau accumulation in neuronal cell bodies without causing concurrent neuronal loss under specific conditions of treatment. Precisely, at a concentration of 3 μM of Hm, which did not lead to neuronal death, the accumulation of p-tau was observable in about 8–10% of neuronal somas. Since early tau neuropathological changes can cause neuronal dysfunction without leading to significant neurodegeneration [60,62,63], we decided to implement a treatment regimen with 3 μM of Hm to study the effects of tau-targeted therapies.

4.2. UA, the End Metabolic Product of Purines in Humans, Prevents Hm-Induced P-Tau Build-Up

Elevated circulating levels of UA—the final breakdown product of purine nucleotides in humans—play a significant role in gout and kidney stone formation. A high UA level in the blood is also often regarded as a potential cardiovascular risk factor [64]. However, several studies have reported that a reduction in serum UA is associated with an increased risk of AD dementia and other tauopathies [26,27,65]. This prompted us to investigate whether UA could influence the accumulation of pathological p-tau in our treatment paradigm with Hm. Our results demonstrate that UA very efficiently inhibited Hm-induced p-tau accumulation across a wide range of concentrations (0.1–300 μM), suggesting that restoring UA serum levels to baseline may be of therapeutic value for tauopathies. To further investigate the nature of UA's anti-tau effects, we used a working concentration of 30 μM, which reflects the levels of UA typically found in human cerebrospinal fluid, while plasma levels are generally more than 10-times higher [66,67].

4.3. UA Exerts Anti-Tau Effects by Curtailing Hm-Mediated Oxidative Stress

UA constitutes approximately 30% to 50% of the body's normal antioxidant capacity, making it a crucial physiological antioxidant [45,46,67,68]. This prompted us to investigate whether UA's suppressive effects on p-tau accumulation were due to its antioxidant properties. The use of the fluorogenic probe DHR-123 revealed that ROS levels increased in neurons treated with Hm and that UA prevented this effect, confirming the view that oxidative stress was responsible for p-tau accumulation in this setting. This conclusion aligns with data from other studies [58,69,70], although ROS have also been reported to promote tau dephosphorylation under certain conditions [71].

The essential role of ROS in pathological p-tau accumulation in our model system was further supported by the fact that an acute challenge with H_2O_2 led to the accumulation of the AT8 immunosignal in a subpopulation of cortical neurons. Unlike what we observed with Hm, UA was unable to prevent p-tau build-up or ROS production triggered by H_2O_2, despite its known ability to scavenge this oxidizing agent in in vitro studies [72]. One might assume that in neuron-enriched cultures, the antioxidant properties of UA are likely to be overwhelmed by the severity and acute nature of the insult elicited by H_2O_2. Our finding aligns with a previous report indicating that differentiated NSC-34 motor neuron-like cells exposed to H_2O_2 are not protected from degeneration when receiving urate as concomitant treatment [73]. Indeed, only a conditioned medium from UA-treated astrocytes was found to be protective against damage caused by H_2O_2 [73].

The concurrent use of the mitoprobe TMRM with the ROS indicator DHR-123 revealed that $\Delta\Psi m$ remained unaffected by Hm treatment, whether UA was present or not in the cultures. This suggests that the generation of ROS and the accumulation of p-tau induced by Hm were not directly linked to mitochondrial dysfunction. This stands in apparent contradiction to previous reports suggesting that a p-tau build-up could result from mitochondrial disturbances [74–76]. However, this may simply suggest that ROS from mitochondrial and non-mitochondrial origins may equally generate tau lesions.

4.4. UA Exerts Anti-Tau Effects by Preventing Hm-Mediated Lipid Peroxidation

Interestingly, DESF, which has iron-chelating properties [77], along with TROL and LIP—both of which inhibit lipid peroxidation and ferroptosis [78]—were all able to replicate the anti-tau and ROS inhibitory effects of UA. This suggests that UA exerts its anti-tau effects by disrupting a sequence of events in which the reaction between Fe(II) and H_2O_2—the Fenton reaction—produces hydroxyl radicals, subsequently leading to lipid peroxidation [79,80]. In line with this hypothesis, Hm has been shown to act as a catalyst for lipid peroxidation [81,82], whereas UA was reported to operate as an inhibitor of this process [72,83,84], possibly by scavenging hydroxyl radicals [45,85]. Additionally, the antioxidant capacity of UA may depend on its ability to reduce redox iron levels by forming coordination complexes with iron ions [86]. Due to limited access of Hm to cells [87], UA may primarily exert its ROS suppressive effect at the outer leaflet of the plasma membrane. This action may prevent ROS propagation into the intracellular compartment, thereby inhibiting p-tau accumulation. Note that this proposed scenario does not totally exclude that UA could also operate by stimulating neuronal iron efflux through the amyloid protein precursor/ferroportin complex as described before in another experimental setting [88].

It is important to note that the glycoprotein APO, which has a high capacity for iron chelation, and VitC, known for its strong antioxidant properties, failed to prevent p-tau accumulation in this context. The lack of inhibitory effects of vitamin C on Hm-induced tau accumulation may be linked to the fact that oxidative stress originates in the lipid environment of the plasma membrane. Indeed, VitC is known to be more effective as an antioxidant in aqueous environments than in lipid environments [89]. The lack of anti-tau effects observed with APO is somehow surprising, as we established previously that this glycoprotein is highly effective in preventing iron-mediated lipid peroxidation in other experimental settings [28,35,46]. We can assume that while ferric iron in the heme pocket of Hm remains accessible to low-molecular-weight molecules like DESF, it is not accessible, however, to large proteins with high molecular mass, such as APO [90].

4.5. The Anti-Tau Effect of UA Is Not Reproduced by Either Its Immediate Precursor XANT or Its Oxidative Product ALTN

To identify the structural features of UA that contribute to its antioxidant/anti-tau properties, we compared its effects with those of its immediate precursor XANT, which is produced through the degradation pathway of adenosine or guanosine [91]. Consistent with findings reported by Muraoka and Miura [83], our results indicate that XANT lacks intrinsic antioxidant properties. This likely explains why, in contrast to UA, XANT does not exert anti-tau effects in our model system. UA is chemically defined as 2,6,8-trioxy-purine and differs from XANT only by the presence of an 8-oxo group on the imidazole ring of the purine structure [92]. This suggests that this functional group contributes to UA's antioxidant and anti-tau activities. Supporting this observation, the 1,7-dimethyl derivative analogue of UA, which retains an 8-oxo group on its purine structure, was equally as effective as UA in preventing tau accumulation and oxidative stress caused by Hm exposure. Note that the oxidative product of UA ALTN—a pyrimidine ring-opened derivative—did not reduce p-tau burden, indicating that only an intact 2,6,8-trioxy-purine structure can provide anti-tau and antioxidant effects.

4.6. DOT, a Non-Antibiotic Oxytetracycline Derivative, Mimics the Anti-Tau Effects of UA

Compounds that can mimic the antioxidant effects of UA and penetrate the brain may offer a promising therapeutic option for reducing the pathological p-tau load. Non-antibiotic TC compounds, which have demonstrated neuroprotective and antioxidant properties in other studies, meet these criteria [28,35,93,94]. Here, we aimed to compare the anti-tau activity of UA with that of DOT, a non-antibiotic TC derived from oxytetracycline, which demonstrated potent antioxidant and neuroprotective properties in a culture system that models dopamine cell death in PD [28]. We found that DOT prevented p-tau accumulation induced by Hm with a similar efficacy as UA. As expected, the anti-tau effects of DOT were correlated with its ability to counteract oxidative stress induced by Hm.

4.7. P-Tau Build-Up Induced by Glutamate Is Preventable by UA and DOT

We wanted to investigate whether the suppressive action of UA and DOT against Hm-induced p-tau build-up could be observed in another experimental setting similarly relevant for neurodegenerative tauopathies. In addition to oxidative stress, neuronal hyperexcitability is suspected to contribute to tau pathological events in AD and other tauopathies [22,23,95]. This led us to implement a paradigm in which pathological p-tau build-up is triggered by an excitotoxic stimulation with the neurotransmitter glutamate [48,96].

After a 24 h challenge with 10 μM glutamate, we found that the AT8 immunosignal was robustly increased in the soma of a subpopulation of cortical neurons that survived the excitotoxic insult. Most interestingly, UA and DOT effectively suppressed p-tau build-up and neurodegenerative events associated with glutamate exposure. These results are consistent with earlier studies showing that UA [97,98] and certain TCs [35,99] can modulate glutamate-mediated neurotoxicity.

The NMDA receptor blocker MK-801, the inhibitor of the superoxide-generating enzyme NADPH oxidase APOc, and the inhibitor of lipid peroxidation TROL also prevented tau abnormalities caused by glutamate. This indicates that UA and DOT exerted their anti-tau effects by primarily blocking a mechanism conveyed by NMDA receptors, and secondarily by lipid peroxidation by-products induced through NADPH oxidase activation. These observations align with previous research studies demonstrating that NADPH oxidase is a major source of NMDA-induced superoxide formation in neurons [100–102]. This also indirectly confirms that ROS can cause tau lesions, regardless of the mechanisms that produce them. Whether UA and DOT modulated NMDA receptor activity or limited intracellular ROS production in response to receptor activation remains to be determined. Note that the inhibition of glutamate-mediated p-tau build-up by UA, DOT, and the other test compounds was closely linked to their capacity to prevent neuronal loss. This differs from what we observed in the Hm paradigm, where p-tau accumulation was preventable under conditions that do not result in neuronal death.

4.8. UA and DOT Have the Potential to Restrain Tau Aggregation

Following exposure to Hm or glutamate, we could not visually detect the presence of aggregates in cell bodies accumulating p-tau. This indicates that while the mouse tau protein can form amyloid fibrillar aggregates similar to the human tau protein [103], the post-translational modifications occurring in $AT8^+$ mouse tau species are not sufficient to promote aggregation under the present culture conditions.

Therefore, to investigate whether UA and DOT could also act as inhibitors of tau amyloid aggregation, we implemented a ThT-based aggregation assay using human recombinant 2N4R tau_m with the polyanion heparin operating as a cofactor for aggregation [30,104]. While this is an artificial model system designed to mimic tau aggregation in vitro, it may be valuable for identifying tau anti-aggregant molecules [30].

We found that both UA and DOT effectively reduced heparin-induced tau aggregation. However, DOT was significantly more potent than UA, with the lowest effective concentrations being estimated at 1 μM for DOT and 100 μM for UA. In addition, we investigated whether UA and DOT could reduce the tau aggregation induced by heparin-induced tau fibrils, i.e., conditions where heparin does not directly operate as a cofactor for tau fibrillation. While DOT remained effective in this context, UA did not. This suggests that DOT and UA reduced tau amyloidogenesis through distinct mechanisms.

The absence of anti-aggregant effects of UA when tau fibrillation was seeded with tau fibrils instead of heparin suggests that UA may reduce tau fibrillation by direct interaction with the cofactor heparin. This is notably supported by past studies showing the adsorption of heparin on sodium urate [105]. It has been proposed that the TC antibiotic doxycycline, which is closely related to DOT, could prevent tau aggregation by interacting with the 4R-repeat domain of the protein [29]. Therefore, we can assume that DOT could operate similarly; however, further research is necessary to confirm this point.

5. Conclusions

In conclusion, we have established a model system of neuronal cortical cultures in which low-intensity stressors serve to simulate pathological tau phosphorylation in tauopathies through a mechanism involving oxidative stress. Our findings show that UA, the end-product of purine catabolism in humans, and DOT, a non-antibiotic derivative of oxytetracycline chemically unrelated to purine compounds, can efficiently suppress p-tau build-up through their antioxidant potential. Additionally, we found that DOT and, to a lesser extent, UA, could reduce tau amyloid aggregation in a ThT aggregation assay. Overall, our results indicate that treatments involving urate-elevating agents might be therapeutically beneficial in preventing tau abnormalities associated with tauopathies. The non-purine TC compound DOT may represent an interesting alternative to UA-related therapies.

Author Contributions: Conceptualization: P.P.M., E.D.B., R.R.-V. and B.F.; Data curation: B.A.d.M., L.F., P.P.M. and R.H.T.-G.; Investigation: A.T., B.A.d.M., R.H.T.-G., T.A.A.F. (biology), I.C. (medicinal chemistry) and J.-M.B. (microbiology); Funding acquisition: B.F., E.D.B., R.R.-V., R.H.T.-G. and P.P.M.; Methodology: A.T., B.F., F.G.-L., I.C., J.-M.B., L.F., R.C., R.H.T.-G. and P.P.M.; Resources: B.F., L.F., R.C. and P.P.M.; Project administration: P.P.M.; Visualization: B.A.d.M. and P.P.M.; Writing original draft: P.P.M.; Review and editing: all authors. All authors have read and agreed to the published version of the manuscript.

Funding: This study received funding from Fondation Recherche Alzheimer (FRA 2021) and Institut Carnot (CM157). It was carried out in the framework of a Capes-Cofecub program between the French and Brazilian research institutions (88887.710933/2022-00; Me928/19). Young Brazilian researchers were supported by the Coordination for the Improvement of Higher Education Personnel Foundation (CAPES, 88887595435/2020-00). Researchers from Argentina were supported by the Argentinian National Scientific and Technological Promotion Agency (MINCyT, PICT-2020-SERIEA-02706) and the National University of Tucumán (UNT, grant PIUNT D759). This work also benefited from the Program Investissements d'Avenir [ANR-10-IAIHU-06], Translational Research Infrastructure for Biotherapies in Neurosciences [ANR-11-INBS-0011-NeurATRIS], and European Union together with the Guadeloupe Region through REG-MND Registre Guadeloupéen des Maladies Neurodégénératives [2019-FED-118].

Institutional Review Board Statement: The Committee on the Ethics of Animal Experiments Charles Darwin no. 5 approved the experimental protocols under authorization number Ce5/2017/005, approval date 30 May 2024.

Informed Consent Statement: Not applicable.

Data Availability Statement: Data is contained within the article. Further inquiries can be directed to the corresponding author.

Acknowledgments: This research work benefited from the equipment and services provided by the Cell Culture (ICV), Preclinical Functional Exploration (PhenoParc), and Cellular and Molecular Imaging (ICMQuant; RRID:SCR_026393) core facilities, at ICM. We are particularly grateful to David Akbar for technical assistance with confocal imaging. The TechMedILL facility (Illkirch, France) is acknowledged for DOT's PK data. The illustration in Figure 1 was created with Biorender.com.

Conflicts of Interest: P.P.M., R.R.-V., E.D.-B., L.F. and B.F. have filed patent applications for the use of non-antibiotic tetracycline derivatives for the treatment of Parkinson's disease and related neurodegenerative disorders.

Abbreviations

AD	Alzheimer's disease
ACM	astrocyte-conditioned medium
ALTN	allantoin
cNb	complete neurobasal medium
$\Delta\Psi m$	mitochondrial membrane potential
DESF	desferioxamine
DHR-123	dihydrorhodamine-123
DMEM	Dulbecco's Modified Eagle's Medium
div	day(s) in vitro

DMUA	1,7-dimethyluric acid
DOT	12a-deoxy-dedimethylamino-oxytetracycline
FCCP	carbonyl cyanide 4-(trifluoromethoxy) phenylhydrazone
Hb	hemoglobin
Hm	hemin
LIP	Liproxstatin-1
NFTs	neurofibrillary tangles
PBS	Dulbecco's phosphate-buffered saline
Phaco	phase contrast
ROS	reactive oxygen species
tau_f	tau fibrils
tau_m	tau monomers
TC	tetracycline
TMRM	tetramethylrhodamine methyl ester
TROL	trolox
UA	uric acid
XANT	xanthine

References

1. Braak, H.; Braak, E. Neuropathological Stageing of Alzheimer-Related Changes. *Acta Neuropathol.* **1991**, *82*, 239–259. [CrossRef] [PubMed]
2. Spillantini, M.G.; Goedert, M. Tau Pathology and Neurodegeneration. *Lancet. Neurol.* **2013**, *12*, 609–622. [CrossRef]
3. Ganguly, J.; Jog, M. Tauopathy and Movement Disorders-Unveiling the Chameleons and Mimics. *Front. Neurol.* **2020**, *11*, 599384. [CrossRef] [PubMed]
4. Cummings, J.L.; Gonzalez, M.I.; Pritchard, M.C.; May, P.C.; Toledo-Sherman, L.M.; Harris, G.A. The Therapeutic Landscape of Tauopathies: Challenges and Prospects. *Alzheimer's Res. Ther.* **2023**, *15*, 168. [CrossRef]
5. Lee, G.; Leugers, C.J. Tau and Tauopathies. *Prog. Mol. Biol. Transl. Sci.* **2012**, *107*, 263–293. [CrossRef]
6. Chung, D.C.; Roemer, S.; Petrucelli, L.; Dickson, D.W. Cellular and Pathological Heterogeneity of Primary Tauopathies. *Mol. Neurodegener.* **2021**, *16*, 57. [CrossRef]
7. Vaquer-Alicea, J.; Diamond, M.I.; Joachimiak, L.A. Tau Strains Shape Disease. *Acta Neuropathol.* **2021**, *142*, 57–71. [CrossRef]
8. Tagge, C.A.; Fisher, A.M.; Minaeva, O.V.; Gaudreau-Balderrama, J.A.; Zhang, X.L.; Mark, W.; Wojnarowicz, N.C.; Lu, H.; Kokiko-Cochran, O.N.; Saman, S.; et al. Concussion, Microvascular Injury, and Early Tauopathy in Young Athletes after Impact Head Injury and an Impact Concussion Mouse Model. *Brain* **2018**, *141*, 422–458. [CrossRef] [PubMed]
9. Cleret de Langavant, L.; Roze, E.; Petit, A.; Tressières, B.; Gharbi-Meliani, A.; Chaumont, H.; Michel, P.P.; Bachoud-Lévi, A.C.; Remy, P.; Edragas, R.; et al. Annonaceae Consumption Worsens Disease Severity and Cognitive Deficits in Degenerative Parkinsonism. Movement Disorders: Official Journal of the Movement Disorder Society. *Mov. Disord.* **2022**, *37*, 2355–2366. [CrossRef]
10. González-Lizárraga, F.; Boluda, S.; Ruiz Hidalgo, J.; Avila, C.L.; Dos Santos, C.; Socias, B.; Medina, L.; Chaumont, H.; Akbar, D.; Roze, E.; et al. Protein coaggregation in Caribbean atypical parkinsonism: The contribution of annonacin. *Neuropathol. Appl. Neurobiol.* 2024, in press.
11. Götz, J.; Probst, A.; Spillantini, M.G.; Schäfer, T.; Jakes, R.; Bürki, K.; Goedert, M. Somatodendritic Localization and Hyperphosphorylation of Tau Protein in Transgenic Mice Expressing the Longest Human Brain Tau Isoform. *EMBO J.* **1995**, *14*, 1304–1313. [CrossRef] [PubMed]
12. Yilmazer-Hanke, D.M. Pathogenesis of Alzheimer-Related Neuritic Plaques: AT8 Immunoreactive Dystrophic Neurites Precede Argyrophilic Neurites in Plaques of the Entorhinal Region, Hippocampal Formation, and Amygdala. *Clin. Neuropathol.* **1998**, *17*, 194–198. [PubMed]
13. Uchihara, T.; Nakamura, A.; Yamazaki, M.; Mori, O. Evolution from Pretangle Neurons to Neurofibrillary Tangles Monitored by Thiazin Red Combined with Gallyas Method and Double Immunofluorescence. *Acta Neuropathol.* **2001**, *101*, 535–539. [CrossRef]
14. Despres, C.; Byrne, C.; Qi, H.; Cantrelle, F.X.; Huvent, I.; Chambraud, B.; Baulieu, E.E.; Jacquot, Y.; Landrieu, I.; Lippens, G.; et al. Identification of the Tau Phosphorylation Pattern That Drives Its Aggregation. *Proc. Natl. Acad. Sci. USA* **2017**, *114*, 9080–9085. [CrossRef]
15. Xia, Y.; Prokop, S.; Gorion, K.M.; Kim, J.D.; Sorrentino, Z.A.; Bell, B.M.; Manaois, A.N.; Chakrabarty, P.; Davies, P.; Giasson, B.I. Tau Ser208 Phosphorylation Promotes Aggregation and Reveals Neuropathologic Diversity in Alzheimer's Disease and Other Tauopathies. *Acta Neuropathol. Commun.* **2020**, *8*, 88. [CrossRef] [PubMed]

16. Malia, T.J.; Teplyakov, A.; Ernst, R.; Wu, S.J.; Lacy, E.R.; Liu, X.; Vandermeeren, M.; Mercken, M.; Luo, J.; Sweet, R.W.; et al. Epitope Mapping and Structural Basis for the Recognition of Phosphorylated Tau by the Anti-Tau Antibody AT8. *Proteins* **2016**, *84*, 427–434. [CrossRef]
17. Goldstein, L.; Teng, Z.P.; Zeserson, E.; Patel, M.; Regan, R.F. Hemin Induces an Iron-Dependent, Oxidative Injury to Human Neuron-like Cells. *J. Neurosci. Res.* **2003**, *73*, 113–121. [CrossRef]
18. Zille, M.; Oses-Prieto, J.A.; Savage, S.R.; Karuppagounder, S.S.; Chen, Y.; Kumar, A.; Morris, J.H.; Scheidt, K.A.; Burlingame, A.L.; Ratan, R.R. Hemin-Induced Death Models Hemorrhagic Stroke and Is a Variant of Classical Neuronal Ferroptosis. *J. Neurosci. Off. J. Soc. Neurosci.* **2022**, *42*, 2065–2079. [CrossRef] [PubMed]
19. Schubert, D.; Piasecki, D. Oxidative Glutamate Toxicity Can Be a Component of the Excitotoxicity Cascade. *J. Neurosci. Off. J. Soc. Neurosci.* **2001**, *21*, 7455–7462. [CrossRef]
20. Spotorno, N.; Acosta-Cabronero, J.; Stomrud, E.; Lampinen, B.; Strandberg, O.T.; van Westen, D.; Hansson, O. Relationship between cortical iron and tau aggregation in Alzheimer's disease. *Brain A J. Neurol.* **2020**, *143*, 1341–1349. [CrossRef]
21. Yamamoto, A.; Shin, R.W.; Hasegawa, K.; Naiki, H.; Sato, H.; Yoshimasu, F.; Kitamoto, T. Iron (III) induces aggregation of hyperphosphorylated tau and its reduction to iron (II) reverses the aggregation: Implications in the formation of neurofibrillary tangles of Alzheimer's disease. *J. Neurochem.* **2002**, *82*, 1137–1147. [CrossRef]
22. Kazim, S.F.; Seo, J.H.; Bianchi, R.; Larson, C.S.; Sharma, A.; Wong, R.K.S.; Gorbachev, K.Y.; Pereira, A.C. Neuronal Network Excitability in Alzheimer's Disease: The Puzzle of Similar versus Divergent Roles of Amyloid β and Tau. *eNeuro* **2021**, *8*, ENEURO.0418-20.2020. [CrossRef]
23. Targa Dias Anastacio, H.; Matosin, N.; Ooi, L. Neuronal Hyperexcitability in Alzheimer's Disease: What Are the Drivers behind This Aberrant Phenotype? *Transl. Psychiatry* **2022**, *12*, 257. [CrossRef] [PubMed]
24. Goedert, M.; Jakes, R.; Vanmechelen, E. Monoclonal Antibody AT8 Recognises Tau Protein Phosphorylated at Both Serine 202 and Threonine 205. *Neurosci. Lett.* **1995**, *189*, 167–169. [CrossRef] [PubMed]
25. Morris, S.L.; Tsai, M.Y.; Aloe, S.; Bechberger, K.; König, S.; Morfini, G.; Brady, S.T. Defined Tau Phosphospecies Differentially Inhibit Fast Axonal Transport Through Activation of Two Independent Signaling Pathways. *Front. Mol. Neurosci.* **2021**, *13*, 610037. [CrossRef]
26. Schirinzi, T.; Di Lazzaro, G.; Colona, V.L.; Imbriani, P.; Alwardat, M.; Sancesario, G.M.; Martorana, A.; Pisani, A. Assessment of Serum Uric Acid as Risk Factor for Tauopathies. *J. Neural Transm.* **2017**, *124*, 1105–1108. [CrossRef]
27. Zhou, Z.; Zhong, S.; Liang, Y.; Zhang, X.; Zhang, R.; Kang, K.; Qu, H.; Xu, Y.; Zhao, C.; Zhao, M. Serum Uric Acid and the Risk of Dementia: A Systematic Review and Meta-Analysis. *Front. Aging Neurosci.* **2021**, *13*, 625690. [CrossRef] [PubMed]
28. Alves Fernandes, T.A.; Tourville, A.; Ciss, I.; Ribeiro Silva, R.; Andretto de Mattos, B.; Dos Santos Pereira, M.; Oblaza, M.; Brunel, J.M.; Ferrié, L.; Raisman-Vozari, R.; et al. Oxytetracycline and Its Non-Antibiotic Derivative DOT Protect Midbrain Dopamine Neurons from Iron-Driven Oxidative Damage. *Neurotox. Res.* **2025**, *43*, 16. [CrossRef]
29. Medina, L.; González-Lizárraga, F.; Dominguez-Meijide, A.; Ploper, D.; Parrales, V.; Sequeira, S.; Cima-Omori, M.S.; Zweckstetter, M.; Del Bel, E.; Michel, P.P.; et al. Doxycycline Interferes with Tau Aggregation and Reduces Its Neuronal Toxicity. *Front. Aging Neurosci.* **2021**, *13*, 635760. [CrossRef]
30. Montgomery, K.M.; Carroll, E.C.; Thwin, A.C.; Quddus, A.Y.; Hodges, P.; Southworth, D.R.; Gestwicki, J.E. Chemical Features of Polyanions Modulate Tau Aggregation and Conformational States. *J. Am. Chem. Soc.* **2023**, *145*, 3926–3936. [CrossRef]
31. Tourville, A.; Akbar, D.; Corti, O.; Prehn, J.H.M.; Melki, R.; Hunot, S.; Michel, P.P. Modelling α-Synuclein Aggregation and Neurodegeneration with Fibril Seeds in Primary Cultures of Mouse Dopaminergic Neurons. *Cells* **2022**, *11*, 1640. [CrossRef] [PubMed]
32. Ferreira Junior, N.C.; dos Santos Pereira, M.; Francis, N.; Ramirez, P.; Martorell, P.; González-Lizarraga, F.; Figadère, B.; Chehin, R.; Del Bel, E.; Raisman-Vozari, R.; et al. The Chemically-Modified Tetracycline COL-3 and Its Parent Compound Doxycycline Prevent Microglial Inflammatory Responses by Reducing Glucose-Mediated Oxidative Stress. *Cells* **2021**, *10*, 2163. [CrossRef] [PubMed]
33. Dos Santos Pereira, M.; Dias de Abreu, G.H.; Vanderlei, L.C.A.; Raisman-Vozari, R.; Guimarães, F.S.; Lu, H.C.; Michel, P.P.; Del Bel, E. 4′-Fluorocannabidiol Associated with Capsazepine Restrains L-DOPA-Induced Dyskinesia in Hemiparkinsonian Mice: Contribution of Anti-Inflammatory and Anti-Glutamatergic Mechanisms. *Neuropharmacology* **2024**, *251*, 109926. [CrossRef] [PubMed]
34. Michel, P.P.; Ruberg, M.; Agid, Y. Rescue of Mesencephalic Dopamine Neurons by Anticancer Drug Cytosine Arabinoside. *J. Neurochem.* **1997**, *69*, 1499–1507. [CrossRef]
35. Tourville, A.; Viguier, S.; González-Lizárraga, F.; Tomas-Grau, R.H.; Ramirez, P.; Brunel, J.M.; Dos Santos Pereira, M.; Del-Bel, E.; Chehin, R.; Ferrié, L.; et al. Rescue of Dopamine Neurons from Iron-Dependent Ferroptosis by Doxycycline and Demeclocycline and Their Non-Antibiotic Derivatives. *Antioxidants* **2023**, *12*, 575. [CrossRef]

36. dos Santos Pereira, M.; Abreu, G.H.D.; Rocca, J.; Hamadat, S.; Raisman-Vozari, R.; Michel, P.P.; Del Bel, E. Contributive Role of TNF-α to L-DOPA-Induced Dyskinesia in a Unilateral 6-OHDA Lesion Model of Parkinson's Disease. *Front. Pharmacol.* **2021**, *11*, 617085. [CrossRef]
37. Golub, L.M.; Ramamurthy, N.S.; McNamara, T.F.; Greenwald, R.A.; Rifkin, B.R. Tetracyclines Inhibit Connective Tissue Breakdown: New Therapeutic Implications for an Old Family of Drugs. *Crit. Rev. Oral Biol. Med.* **1991**, *2*, 297–321. [CrossRef]
38. Schindelin, J.; Arganda-Carreras, I.; Frise, E.; Kaynig, V.; Longair, M.; Pietzsch, T.; Preibisch, S.; Rueden, C.; Saalfeld, S.; Schmid, B.; et al. Fiji: An Open-Source Platform for Biological-Image Analysis. *Nat. Methods* **2012**, *9*, 676–682. [CrossRef]
39. Joshi, D.C.; Bakowska, J.C. Determination of Mitochondrial Membrane Potential and Reactive Oxygen Species in Live Rat Cortical Neurons. *J. Vis. Exp. JoVE* **2011**, 2704. [CrossRef]
40. Parrales-Macias, V.; Michel, P.P.; Tourville, A.; Raisman-Vozari, R.; Haïk, S.; Hunot, S.; Bizat, N.; Lannuzel, A. The Pesticide Chlordecone Promotes Parkinsonism-like Neurodegeneration with Tau Lesions in Midbrain Cultures and *C. elegans* Worms. *Cells* **2023**, *12*, 1336. [CrossRef]
41. Barghorn, S.; Biernat, J.; Mandelkow, E. Purification of Recombinant Tau Protein and Preparation of Alzheimer-Paired Helical Filaments in Vitro. *Methods Mol. Biol.* **2005**, *299*, 35–51. [CrossRef] [PubMed]
42. LeVine, H., 3rd. Quantification of Beta-Sheet Amyloid Fibril Structures with Thioflavin, T. *Methods Enzymol.* **1999**, *309*, 274–284. [CrossRef]
43. Upchurch, C.M.; Yeudall, S.; Pavelec, C.M.; Merk, D.; Greulich, J.; Manjegowda, M.; Raghavan, S.S.; Bochkis, I.M.; Scott, M.M.; Perez-Reyes, E.; et al. Targeting Oxidized Phospholipids by AAV-Based Gene Therapy in Mice with Established Hepatic Steatosis Prevents Progression to Fibrosis. *Sci. Adv.* **2022**, *8*, Eabn0050. [CrossRef] [PubMed]
44. Mishra, P.; Pandey, C.M.; Singh, U.; Gupta, A.; Sahu, C.; Keshri, A. Descriptive Statistics and Normality Tests for Statistical Data. *Ann. Card. Anaesth.* **2019**, *22*, 67–72. [CrossRef]
45. Ames, B.N.; Cathcart, R.; Schwiers, E.; Hochstein, P. Uric Acid Provides an Antioxidant Defense in Humans against Oxidant- and Radical-Caused Aging and Cancer: A Hypothesis. *Proc. Natl. Acad. Sci. USA* **1981**, *78*, 6858–6862. [CrossRef]
46. Guerreiro, S.; Ponceau, A.; Toulorge, D.; Martin, E.; Alvarez-Fischer, D.; Hirsch, E.C.; Michel, P.P. Protection of Midbrain Dopaminergic Neurons by the End-Product of Purine Metabolism Uric Acid: Potentiation by Low-Level Depolarization. *J. Neurochem.* **2009**, *109*, 1118–1128. [CrossRef] [PubMed]
47. Michel, P.P.; Vyas, S.; Agid, Y. Toxic Effects of Iron for Cultured Mesencephalic Dopaminergic Neurons Derived from Rat Embryonic Brains. *J. Neurochem.* **1992**, *59*, 118–127. [CrossRef]
48. Kobayashi, S.; Tanaka, T.; Soeda, Y.; Almeida, O.F.X.; Takashima, A. Local Somatodendritic Translation and Hyperphosphorylation of Tau Protein Triggered by AMPA and NMDA Receptor Stimulation. *EBioMedicine* **2017**, *20*, 120–126. [CrossRef]
49. Huynh, M.B.; Rebergue, N.; Merrick, H.; Gomez-Henao, W.; Jospin, E.; Biard, D.S.F.; Papy-Garcia, D. HS3ST2 Expression Induces the Cell Autonomous Aggregation of Tau. *Sci. Rep.* **2022**, *12*, 10850. [CrossRef]
50. Braak, H.; Thal, D.R.; Ghebremedhin, E.; Del Tredici, K. Stages of the Pathologic Process in Alzheimer Disease: Age Categories from 1 to 100 Years. *J. Neuropathol. Exp. Neurol.* **2011**, *70*, 960–969. [CrossRef]
51. Attems, J.; Thal, D.R.; Jellinger, K.A. The Relationship between Subcortical Tau Pathology and Alzheimer's Disease. *Biochem. Soc. Trans.* **2012**, *40*, 711–715. [CrossRef]
52. Connor, J.R.; Snyder, B.S.; Beard, J.L.; Fine, R.E.; Mufson, E.J. Regional Distribution of Iron and Iron-Regulatory Proteins in the Brain in Aging and Alzheimer's Disease. *J. Neurosci. Res.* **1992**, *31*, 327–335. [CrossRef] [PubMed]
53. Ding, B.; Chen, K.M.; Ling, H.W.; Sun, F.; Li, X.; Wan, T.; Chai, W.M.; Zhang, H.; Zhan, Y.; Guan, Y.J. Correlation of Iron in the Hippocampus with MMSE in Patients with Alzheimer's Disease. *J. Magn. Reson. Imaging JMRI* **2009**, *29*, 793–798. [CrossRef] [PubMed]
54. Tao, Y.; Wang, Y.; Rogers, J.T.; Wang, F. Perturbed Iron Distribution in Alzheimer's Disease Serum, Cerebrospinal Fluid, and Selected Brain Regions: A Systematic Review and Meta-Analysis. *J. Alzheimer's Dis. JAD* **2014**, *42*, 679–690. [CrossRef] [PubMed]
55. Alosco, M.L.; Su, Y.; Stein, T.D.; Protas, H.; Cherry, J.D.; Adler, C.H.; Balcer, L.J.; Bernick, C.; Pulukuri, S.V.; Abdolmohammadi, B.; et al. Associations between near End-of-Life Flortaucipir PET and Postmortem CTE-Related Tau Neuropathology in Six Former American Football Players. *Eur. J. Nucl. Med. Mol. Imaging* **2023**, *50*, 435–452. [CrossRef]
56. Paterno, G.; Bell, B.M.; Gorion, K.M.; Prokop, S.; Giasson, B.I. Reassessment of Neuronal Tau Distribution in Adult Human Brain and Implications for Tau Pathobiology. *Acta Neuropathol. Commun.* **2022**, *10*, 94. [CrossRef]
57. Combs, B.; Mueller, R.L.; Morfini, G.; Brady, S.T.; Kanaan, N.M. Tau and Axonal Transport Misregulation in Tauopathies. *Adv. Exp. Med. Biol.* **2019**, *1184*, 81–95. [CrossRef]
58. Wan, W.; Cao, L.; Kalionis, B.; Murthi, P.; Xia, S.; Guan, Y. Iron Deposition Leads to Hyperphosphorylation of Tau and Disruption of Insulin Signaling. *Front. Neurol.* **2019**, *10*, 607. [CrossRef]
59. Fu, H.; Possenti, A.; Freer, R.; Nakano, Y.; Hernandez Villegas, N.C.; Tang, M.; Cauhy, P.V.; Lassus, B.A.; Chen, S.; Fowler, S.L.; et al. A Tau Homeostasis Signature Is Linked with the Cellular and Regional Vulnerability of Excitatory Neurons to Tau Pathology. *Nat. Neurosci.* **2019**, *22*, 47–56. [CrossRef]

60. Carroll, T.; Guha, S.; Nehrke, K.; Johnson, G.V.W. Tau Post-Translational Modifications: Potentiators of Selective Vulnerability in Sporadic Alzheimer's Disease. *Biology* **2021**, *10*, 1047. [CrossRef]
61. Otero-Garcia, M.; Mahajani, S.U.; Wakhloo, D.; Tang, W.; Xue, Y.Q.; Morabito, S.; Pan, J.; Oberhauser, J.; Madira, A.E.; Shakouri, T.; et al. Molecular Signatures Underlying Neurofibrillary Tangle Susceptibility in Alzheimer's Disease. *Neuron* **2022**, *110*, 2929–2948.e8. [CrossRef]
62. Chiasseu, M.; Alarcon-Martinez, L.; Belforte, N.; Quintero, H.; Dotigny, F.; Destroismaisons, L.; Vande Velde, C.; Panayi, F.; Louis, C.; Di Polo, A. Tau Accumulation in the Retina Promotes Early Neuronal Dysfunction and Precedes Brain Pathology in a Mouse Model of Alzheimer's Disease. *Mol. Neurodegener.* **2017**, *12*, 58. [CrossRef]
63. Buée, L.; Troquier, L.; Burnouf, S.; Belarbi, K.; Van Der Jeugd, A.; Ahmed, T.; Fernandez-Gomez, F.; Caillierez, R.; Grosjean, M.E.; Begard, S.; et al. From Tau Phosphorylation to Tau Aggregation: What about Neuronal Death? *Biochem. Soc. Trans.* **2010**, *38*, 967–972. [CrossRef]
64. Hediger, M.A.; Johnson, R.J.; Miyazaki, H.; Endou, H. Molecular Physiology of Urate Transport. *Physiology* **2005**, *20*, 125–133. [CrossRef] [PubMed]
65. Ye, B.S.; Lee, W.W.; Ham, J.H.; Lee, J.J.; Lee, P.H.; Sohn, Y.H. Alzheimer's Disease Neuroimaging Initiative Does Serum Uric Acid Act as a Modulator of Cerebrospinal Fluid Alzheimer's Disease Biomarker Related Cognitive Decline? *Eur. J. Neurol.* **2016**, *23*, 948–957. [CrossRef]
66. Cao, B.; Li, Q.; Xiong, L.; Ruan, H.; Lu, Y.; Peng, F.; Li, H.; Huang, Y.; Luo, M.; Shu, Y. Cerebrospinal Fluid Uric Acid Levels Associated with Disease Severity in Patients with Anti-N-Methyl-d-Aspartate Receptor Encephalitis. *J. Neuroimmunol.* **2023**, *384*, 578221. [CrossRef] [PubMed]
67. Hasíková, L.; Závada, J.; Serranová, T.; Kozlík, P.; Kalíková, K.; Kotačková, L.; Trnka, J.; Zogala, D.; Šonka, K.; Růžička, E.; et al. Serum but Not Cerebrospinal Fluid Levels of Allantoin Are Increased in de Novo Parkinson's Disease. *NPJ Park. Dis.* **2023**, *9*, 60. [CrossRef]
68. Crawley, W.T.; Jungels, C.G.; Stenmark, K.R.; Fini, M.A. U-Shaped Association of Uric Acid to Overall-Cause Mortality and Its Impact on Clinical Management of Hyperuricemia. *Redox Biol.* **2022**, *51*, 102271. [CrossRef]
69. Su, B.; Wang, X.; Lee, H.G.; Tabaton, M.; Perry, G.; Smith, M.A.; Zhu, X. Chronic Oxidative Stress Causes Increased Tau Phosphorylation in M17 Neuroblastoma Cells. *Neurosci. Lett.* **2010**, *468*, 267–271. [CrossRef] [PubMed]
70. Wang, C.; Shao, C.; Zhang, L.; Siedlak, S.L.; Meabon, J.S.; Peskind, E.R.; Lu, Y.; Wang, W.; Perry, G.; Cook, D.G.; et al. Oxidative Stress Signaling in Blast TBI-Induced Tau Phosphorylation. *Antioxidants* **2021**, *10*, 955. [CrossRef]
71. Zambrano, C.A.; Egaña, J.T.; Núñez, M.T.; Maccioni, R.B.; González-Billault, C. Oxidative Stress Promotes Tau Dephosphorylation in Neuronal Cells: The Roles of Cdk5 and PP1. *Free Radic. Biol. Med.* **2004**, *36*, 1393–1402. [CrossRef]
72. Gulcin, I.; Oktay, M.; Koksal, E.; Serbetci, H.; Beydemir, S.; Kufrevioglu, O.I. Antioxidant and Radical Scavenging Activities of Uric Acid. *Asian J. Chem.* **2008**, *20*, 2079–2090.
73. Bakshi, R.; Zhang, H.; Logan, R.; Joshi, I.; Xu, Y.; Chen, X.; Schwarzschild, M.A. Neuroprotective effects of urate are mediated by augmenting astrocytic glutathione synthesis and release. *Neurobiol. Dis.* **2015**, *82*, 574–579. [CrossRef]
74. Escobar-Khondiker, M.; Höllerhage, M.; Muriel, M.P.; Champy, P.; Bach, A.; Depienne, C.; Respondek, G.; Yamada, E.S.; Lannuzel, A.; Yagi, T.; et al. Annonacin, a natural mitochondrial complex I inhibitor, causes tau pathology in cultured neurons. *J. Neurosci. Off. J. Soc. Neurosci.* **2007**, *27*, 7827–7837. [CrossRef]
75. Eckert, A.; Nisbet, R.; Grimm, A.; Götz, J. March Separate, Strike Together--Role of Phosphorylated TAU in Mitochondrial Dysfunction in Alzheimer's Disease. *Biochim. Et Biophys. Acta* **2014**, *1842*, 1258–1266. [CrossRef]
76. Samluk, L.; Ostapczuk, P.; Dziembowska, M. Long-Term Mitochondrial Stress Induces Early Steps of Tau Aggregation by Increasing Reactive Oxygen Species Levels and Affecting Cellular Proteostasis. *Mol. Biol. Cell* **2022**, *33*, Ar67. [CrossRef]
77. Hershko, C.; Abrahamov, A.; Konijn, A.M.; Breuer, W.; Cabantchik, I.Z.; Pootrakul, P.; Link, G. Objectives and Methods of Iron Chelation Therapy. *Bioinorg. Chem. Appl.* **2003**, *1*, 151–168. [CrossRef]
78. Zilka, O.; Shah, R.; Li, B.; Friedmann Angeli, J.P.; Griesser, M.; Conrad, M.; Pratt, D.A. On the Mechanism of Cytoprotection by Ferrostatin-1 and Liproxstatin-1 and the Role of Lipid Peroxidation in Ferroptotic Cell Death. *ACS Cent. Sci.* **2017**, *3*, 232–243. [CrossRef]
79. Li, J.; Cao, F.; Yin, H.L.; Huang, Z.J.; Lin, Z.T.; Mao, N.; Sun, B.; Wang, G. Ferroptosis: Past, Present and Future. *Cell Death Dis.* **2020**, *11*, 88. [CrossRef] [PubMed]
80. Von Krusenstiern, A.N.; Robson, R.N.; Qian, N.; Qiu, B.; Hu, F.; Reznik, E.; Smith, N.; Zandkarimi, F.; Estes, V.M.; Dupont, M.; et al. Identification of Essential Sites of Lipid Peroxidation in Ferroptosis. *Nat. Chem. Biol.* **2023**, *19*, 719–730. [CrossRef] [PubMed]
81. Schmitt, T.H.; Frezzatti, W.A., Jr.; Schreier, S. Hemin-Induced Lipid Membrane Disorder and Increased Permeability: A Molecular Model for the Mechanism of Cell Lysis. *Arch. Biochem. Biophys.* **1993**, *307*, 96–103. [CrossRef] [PubMed]
82. Balla, G.; Jacob, H.S.; Eaton, J.W.; Belcher, J.D.; Vercellotti, G.M. Hemin: A Possible Physiological Mediator of Low Density Lipoprotein Oxidation and Endothelial Injury. *Arterioscler. Thromb. A J. Vasc. Biol.* **1991**, *11*, 1700–1711. [CrossRef]

83. Muraoka, S.; Miura, T. Inhibition by Uric Acid of Free Radicals That Damage Biological Molecules. *Pharmacol. Toxicol.* **2003**, *93*, 284–289. [CrossRef]
84. Howell, R.R.; Wyngaarden, J.B. On the Mechanism of Peroxidation of Uric Acids by Hemoproteins. *J. Biol. Chem.* **1960**, *235*, 3544–3550. [CrossRef] [PubMed]
85. Regoli, F.; Winston, G.W. Quantification of Total Oxidant Scavenging Capacity of Antioxidants for Peroxynitrite, Peroxyl Radicals, and Hydroxyl Radicals. *Toxicol. Appl. Pharmacol.* **1999**, *156*, 96–105. [CrossRef] [PubMed]
86. Davies, K.J.; Sevanian, A.; Muakkassah-Kelly, S.F.; Hochstein, P. Uric acid-iron ion complexes. A new aspect of the antioxidant functions of uric acid. *Biochem. J.* **1986**, *235*, 747–754. [CrossRef]
87. Sarkar, A.R.; Pal, S.; Sarkar, A.K.; Jana, N.R. Hemin-Based Cell Therapy via Nanoparticle-Assisted Uptake, Intracellular Reactive Oxygen Species Generation and Autophagy Induction. *New J. Chem.* **2022**, *46*, 21746–21755. [CrossRef]
88. Cahill, C.M.; Sarang, S.S.; Bakshi, R.; Xia, N.; Lahiri, D.K.; Rogers, J.T. Neuroprotective Strategies and Cell-Based Biomarkers for Manganese-Induced Toxicity in Human Neuroblastoma (SH-SY5Y) Cells. *Biomolecules* **2024**, *14*, 647. [CrossRef]
89. Pandithavidana, D.R.; Jayawardana, S.B. Comparative Study of Antioxidant Potential of Selected Dietary Vitamins; Computational Insights. *Molecules* **2019**, *24*, 1646. [CrossRef]
90. Pongratz, M.; Schluga, P.; Jakupec, M.A.; Arion, V.B.; Hartinger, C.G.; Allmaier, G.; Keppler, B.K. Transferrin Binding and Transferrin-Mediated Cellular Uptake of the Ruthenium Coordination Compound KP1019, Studied by Means of AAS, ESI-MS and CD Spectroscopy. *J. Anal. At. Spectrom.* **2004**, *19*, 46–51. [CrossRef]
91. Maiuolo, J.; Oppedisano, F.; Gratteri, S.; Muscoli, C.; Mollace, V. Regulation of Uric Acid Metabolism and Excretion. *Int. J. Cardiol.* **2016**, *213*, 8–14. [CrossRef]
92. Smith, R.C.; Reeves, J.C.; Dage, R.C.; Schnettler, R.A. Antioxidant Properties of 2-Imidazolones and 2-Imidazolthiones. *Biochem. Pharmacol.* **1987**, *36*, 1457–1460. [CrossRef]
93. Tomas-Grau, R.; González-Lizárraga, F.; Ploper, D.; Avila, C.L.; Socías, S.B.; Besnault, P.; Tourville, A.; Mella, R.M.; Villacé, P.; Salado, C.; et al. Neuroprotective Effects of a Novel Demeclocycline Derivative Lacking Antibiotic Activity: From a Hit to a Promising Lead Compound. *Cells* **2022**, *11*, 2759. [CrossRef]
94. Rose, C.; Tomas-Grau, R.H.; Zabala, B.; Michel, P.P.; Brunel, J.M.; Chehín, R.; Raisman-Vozari, R.; Ferrié, L.; Figadère, B. C9-Functionalized Doxycycline Analogs as Drug Candidates to Prevent Pathological α-Synuclein Aggregation and Neuroinflammation in Parkinson's Disease Degeneration. *ChemMedChem* **2024**, *19*, E202300597. [CrossRef]
95. Hwang, K.; Vaknalli, R.N.; Addo-Osafo, K.; Vicente, M.; Vossel, K. Tauopathy and Epilepsy Comorbidities and Underlying Mechanisms. *Front. Aging Neurosci.* **2022**, *14*, 903973. [CrossRef]
96. Couratier, P.; Lesort, M.; Sindou, P.; Esclaire, F.; Yardin, C.; Hugon, J. Modifications of Neuronal Phosphorylated Tau Immunoreactivity Induced by NMDA Toxicity. *Mol. Chem. Neuropathol.* **1996**, *27*, 259–273. [CrossRef]
97. Yu, Z.F.; Bruce-Keller, A.J.; Goodman, Y.; Mattson, M.P. Uric Acid Protects Neurons against Excitotoxic and Metabolic Insults in Cell Culture, and against Focal Ischemic Brain Injury in Vivo. *J. Neurosci. Res.* **1998**, *53*, 613–625. [CrossRef]
98. Du, Y.; Chen, C.P.; Tseng, C.Y.; Eisenberg, Y.; Firestein, B.L. Astroglia-Mediated Effects of Uric Acid to Protect Spinal Cord Neurons from Glutamate Toxicity. *Glia* **2007**, *55*, 463–472. [CrossRef]
99. Lu, Y.; Yang, Y.; Chen, W.; Du, N.; Du, Y.; Gu, H.; Liu, Q. Minocycline, but Not Doxycycline Attenuates NMDA-Induced [Ca2+]i and Excitotoxicity. *NeuroReport* **2021**, *32*, 38–43. [CrossRef]
100. Brennan, A.M.; Suh, S.W.; Won, S.J.; Narasimhan, P.; Kauppinen, T.M.; Lee, H.; Edling, Y.; Chan, P.H.; Swanson, R.A. NADPH Oxidase Is the Primary Source of Superoxide Induced by NMDA Receptor Activation. *Nat. Neurosci.* **2009**, *12*, 857–863. [CrossRef]
101. Lam, T.I.; Brennan-Minnella, A.M.; Won, S.J.; Shen, Y.; Hefner, C.; Shi, Y.; Sun, D.; Swanson, R.A. Intracellular pH Reduction Prevents Excitotoxic and Ischemic Neuronal Death by Inhibiting NADPH Oxidase. *Proc. Natl. Acad. Sci. USA* **2013**, *110*, E4362–E4368. [CrossRef] [PubMed]
102. Lavaur, J.; Le Nogue, D.; Lemaire, M.; Pype, J.; Farjot, G.; Hirsch, E.C.; Michel, P.P. The Noble Gas Xenon Provides Protection and Trophic Stimulation to Midbrain Dopamine Neurons. *J. Neurochem.* **2017**, *142*, 14–28. [CrossRef] [PubMed]
103. Kampers, T.; Pangalos, M.; Geerts, H.; Wiech, H.; Mandelkow, E. Assembly of Paired Helical Filaments from Mouse Tau: Implications for the Neurofibrillary Pathology in Transgenic Mouse Models for Alzheimer's Disease. *FEBS Lett.* **1999**, *451*, 39–44. [CrossRef] [PubMed]
104. Goedert, M.; Jakes, R.; Spillantini, M.G.; Hasegawa, M.; Smith, M.J.; Crowther, R.A. Assembly of Microtubule-Associated Protein Tau into Alzheimer-like Filaments Induced by Sulphated Glycosaminoglycans. *Nature* **1996**, *383*, 550–553. [CrossRef]
105. Finlayson, B.; Du Bois, L. Adsorption of Heparin on Sodium Acid Urate. *Clin. Chim. Acta Int. J. Clin. Chem.* **1978**, *84*, 203–206. [CrossRef]

biomolecules

MDPI

Article

Pharmacological Characterization of the Novel CRF1 Receptor Antagonist, Thiazolo[4,5-d] Pyrimidine Analog, M43

Spyridon Marios Giatro [1,†], George Komontachakis [2,†], Aikaterini Kalantidou [3], Nastazia Lesgidou [1], Vlasios Karageorgos [2], Mohamed Teleb [4,5], Md Rabiul Islam [6], Thomas Mavromoustakos [7], Hesham Fahmy [6], Maria Venihaki [3], Minos-Timotheos Matsoukas [1,*] and George Liapakis [2,*]

1 Department of Biomedical Engineering, University of West Attica, 12210 Egaleo, Greece; sgiatro@uniwa.gr (S.M.G.); nlesgidou@uniwa.gr (N.L.)
2 Department of Pharmacology, School of Medicine, University of Crete, 71003 Heraklion, Greece; geokomontas@gmail.com (G.K.); bkarageorgos@hotmail.com (V.K.)
3 Department of Clinical Chemistry, School of Medicine, University of Crete, 71003 Heraklion, Greece; katekld7@gmail.com (A.K.); venycham@uoc.gr (M.V.)
4 Department of Pharmaceutical Chemistry, Faculty of Pharmacy, Alexandria University, Alexandria 21521, Egypt; mohamed.t.ismail@alexu.edu.eg
5 Department of Medicinal Chemistry, Faculty of Pharmacy, Alamein International University, Alamein 51718, Egypt
6 Department of Pharmaceutical Science, College of Pharmacy & Allied Health Professions, South Dakota State University, Brookings, SD 57007, USA; mdrabiul.islam925@jacks.sdstate.edu (M.R.I.); hesham.fahmy@sdstate.edu (H.F.)
7 Department of Chemistry, National and Kapodistrian University of Athens, 15771 Athens, Greece; tmavrom@chem.uoa.gr
* Correspondence: mmatsoukas@uniwa.gr (M.-T.M.); liapakig@uoc.gr (G.L.)
† These authors contributed equally to this work.

Abstract: The corticotropin-releasing factor (CRF) and its type 1 receptor (CRF_1R) play a key role in the regulation of the hypothalamic–pituitary–adrenal (HPA) axis. Dysregulation of the HPA axis is associated with congenital adrenal hyperplasia (CAH) and depression. Non-peptide CRF_1R-selective antagonists displayed antidepressant effects on animal models and are used for the management of CAH. To develop novel non-peptide CRF_1R antagonists, we have previously designed and synthesized a series of substituted pyrimidines. Among these analogs, molecule 43 (M43) binds to CRF_1R with the highest affinity. Based on this finding, we selected M43 for further pharmacological characterization in the present study. The results suggest that M43 is a potent CRF_1R antagonist, blocking the ability of the CRF-related agonist, Tyr^0-sauvagine, to stimulate (1) cAMP accumulation in HEK 293 cells expressing CRF_1R and (2) the proliferation rate of RAW 264.7 macrophages. Computational studies suggest that the antagonist properties of M43 are mostly attributed to its ability to interact with residues in the allosteric pocket of CRF_1R, comprised of the third, fifth, and sixth transmembrane domain residues, which block activation-associated structural rearrangements of the receptor. Our data will be used to design novel non-peptide CRF_1R antagonists for clinical use.

Keywords: non-peptide antagonists; corticotropin-releasing factor; receptor; signaling; structure

1. Introduction

The hypothalamic corticotropin-releasing factor (CRF or CRH) is a 41-amino acid peptide that exerts multiple biological actions through its interaction with the type 1 (CRF_1R).

The CRF receptors belong to family B of G-protein-coupled receptors (GPCRs) [1,2]. Activation of CRF receptors by the CRF and related agonists, including Tyr^0-sauvagine, leads to a biological effect by stimulating, through the Gs-proteins, the intracellular cAMP accumulation [1,2].

CRF and its receptors are essential for the maintenance of homeostasis because they regulate the hypothalamic–pituitary–adrenal axis (HPA) [3]. Specifically, CRF is secreted from the hypothalamus and stimulates the secretion of corticotropin (ACTH) from the pituitary through its interaction with the CRF_1R [2,4]. The ACTH subsequently stimulates the release of glucocorticoids from the adrenals [2,3]. The HPA axis undergoes negative feedback regulation by the glucocorticoids, which suppress the release of CRF and ACTH [5]. In addition to the regulation of the HPA axis, the CRF and CRF_1R play an important role in the function of the central nervous system (CNS) [6–9].

Hyperactivity of the HPA axis and dysregulation of the CRF/CRF_1R circuits of the CNS caused by various factors, such as chronic stress, are associated with maladaptation to stress and the appearance of stress-related disorders, such as anxiety and depression [5,7,10–14]. Injection of CRF in the CNS of model animals or overexpression of CRF in transgenic mice is associated with anxiety and depressive-like symptoms [15–19]. These actions of CRF in the CNS are mostly mediated through its interaction with the CRF_1R, as suggested in previous studies involving CRF_1R antisense oligonucleotides or CRF_1R-deficient animals [9,20]. Accordingly, non-peptide CRF_1R antagonists appeared to have antidepressant and anxiolytic properties in experimental animals, with several of them entering clinical trials [21,22].

Dysregulation of the HPA axis is also observed in several pathophysiological conditions, such as congenital adrenal hyperplasia (CAH), which is characterized by decreased glucocorticoid secretion from the adrenals and impaired glucocorticoid feedback inhibition of the HPA axis [23,24]. This results in augmented secretion of CRF and ACTH, leading to the excessive production of adrenal androgens. The CAH is an autosomal recessive disorder caused by 21-hydroxylase deficiency [23,24]. The severity of CAH is closely associated with the degree of 21-hydroxylase deficiency, with the most severe type being the classic salt-wasting one, which is characterized by mineralocorticoid and cortisol deficiency that leads to salt loss and varying degrees of prenatal virilization of external genitalia [23,24]. Kyritsi et al. have proposed that CAH carriers could be more prone to the development of anxiety disorders, in agreement with their findings, which indicated a chronic hyperactivity of the HPA axis due to its impaired feedback inhibition by glucocorticoids [25]. Similarly, Harasymiw et al. have shown that children, adolescents, and young adults with CAH were more likely to be diagnosed with a depressive or anxiety disorder compared to their healthy counterparts [26]. Treatment of CAH aims to replace cortisol and reduce the excess of androgens [23,24]. Administration of cortisol restores its decreased levels and decreases the excessive production of adrenal androgens by re-establishing the negative feedback of the HPA axis [23,24,27]. However, glucocorticoid replacement therapy often requires supraphysiological doses to suppress androgen excess, thus putting CAH patients at the risk of glucocorticoid overexposure, leading to conditions such as growth suppression, metabolic disorders, insulin resistance, and hypertension [23,24,27,28]. An approach to reducing the overproduction of adrenal androgens while administering glucocorticoids at physiological doses is to decrease the ACTH secretion by administering CRF_1R antagonists [27–30]. Based on this concept, two CRF_1R non-peptide antagonists, tildacerfont and crinecerfont, have been developed and tested for the management of CAH [29,31].

The development of tildacerfont for the treatment of CAH was terminated because it failed to meet its primary endpoint of glucocorticoid dose reduction in a clinical phase 2b trial [29]. Similarly, although several CRF_1R non-peptide antagonists are currently in clinical trials for the treatment of depression and anxiety, none is in clinical use [21]. In contrast, crinecerfont, after successfully passing phases 2 and 3 of clinical trials, was approved in December 2024 for the treatment of pediatric and adult patients with classic CAH [28,29,31]. However, despite its high tolerability, crinecerfont has several potentially severe side effects, although less common, including hypersensitivity reactions and suicidal ideation and behaviors [32].

The drawbacks of using CRF_1R non-peptide antagonists as drugs for the treatment of CAH and stress-related disorders highlight the need for the development of novel molecules with better pharmacological and safety profiles. In an effort to develop novel CRF_1R antagonists, we have designed, synthesized, and tested the binding properties of 45 non-peptide molecules [33]. Among these compounds, the analog 43 (or M43) has been shown to bind to CRF_1R with the highest affinity. In the present study, we rationalized the relationship between M43/CRF_1R interactions and the binding affinity of M43 by constructing several molecular models and experimentally determined the antagonistic properties of M43.

2. Materials and Methods

2.1. Cell Culture

Human embryonic kidney (HEK) 293 cells stably expressing the human CRF1R were generated as previously described [34] and were grown in Dulbecco's modified Eagle's medium (DMEM)/F-12 (1:1) containing 3.15 g/L of glucose and 10% bovine calf serum at 37 °C and 5% CO_2. One day before the experiment, the cells were plated in 96-well cell culture plates (pretreated with 0.1 mg/mL poly-L-lysine) at a density that resulted in 95–100% confluency within one day.

Mouse macrophage cell line RAW 264.7 was cultured using DMEM-high glucose medium (Gibco, Jenks, OK, USA, 41966-029) supplemented with 10% heat-inactivated Fetal Bovine Serum (Gibco, 10270-106) and 1% penicillin streptomycin (Gibco, 15140-122) at 37 °C in the presence of 5% CO_2. Macrophage activation was induced using 100 ng/mL LPS (L2630, Sigma, St. Louis, MO, USA). For proliferation assays, RAW 264.7 cells were plated in 96-well plates at a concentration of 4.000 cells/well for 24 h. Then, cells were treated with analog M43 at 1 μM and/or Tyr^0-Sauvagine at 10 nM for 72 h.

2.2. cAMP Accumulation Assays

HEK 293 cells stably expressing CRF_1R were initially incubated in 100 μL of assay buffer at 37 °C. After 1 h, additional buffer—with or without ligands (to assess basal activity)—was added to a final volume of 200 μL, and incubation continued for another 30 min at 37 °C. Varying concentrations of Tyr^0-sauvagine were applied, both in the presence and absence of 1 μM analog M43, to assess the ligand's potency. Parallel assays were conducted with a fixed 10 nM concentration of Tyr^0-sauvagine, with increasing concentrations of analog M43 to determine its half-maximal inhibitory concentration (IC_{50}). Following incubation, the buffer was removed, and the cells were placed on ice and lysed using 3% trichloroacetic acid (TCA). Lysates were kept on ice for 30–60 min and then stored at −20 °C. After storage for 1–5 days, samples were thawed and centrifuged at 1800× *g* for 10 min at 4 °C, and supernatants were neutralized using 2 N NaOH. Intracellular cAMP levels were quantified via a competitive binding assay. In brief, 20 μL of each neutralized sample was transferred to 3 mL polypropylene tubes containing buffer A (100 mM Tris-HCl, pH 7.4, 100 mM NaCl, and 5 mM EDTA) and 1 nM of [2,8-^{3}H] cAMP. Approximately 100 mg of bovine adrenal cortex extract (serving as the cAMP-binding protein) diluted in 500 mL of buffer A was then added. After 3 h of incubation on ice, samples were filtered through

Whatman 934AH glass fiber filters using cold buffer C (120 mM NaCl and 10 mM Tris-HCl, pH 7.4) as the wash solution and a Brandel filtering device.

The amount of cAMP was measured against a logarithmic standard curve (1–100 pmol/tube) using known concentrations of unlabeled cAMP. IC_{50} values for analog M43 (expressed as pIC_{50}) were derived using a one-site competition model via nonlinear regression (Prism 8.0). The IC_{50} represents the concentration of analog M43 needed to inhibit 50% of cAMP production induced by 10 nM sauvagine. Potency values for Tyr^0-sauvagine (expressed as $-LogEC_{50}$), in the presence and absence of analog M43, were determined using a one-site sigmoidal fit model.

2.3. *Proliferation Assays*

Cell proliferation in RAW 264.7 macrophages was assessed following treatment with 10 nM of sauvagine, 1 nM of analog M43, or their combination, using the MTT assay (Sigma). Cells were seeded into flat-bottomed 96-well plates and, after 24 h, were incubated with 100 ng/mL of LPS (L2630, Sigma), analog M43 at 1 μM, and/or Tyr^0-sauvagine at 10 nM at 37 °C in the presence of 5% CO_2. After 72 h, MTT (3-(4,5-dimethylthiazol-2-yl)-2,5-diphenyltetrazolium bromide) was added to each well at a final concentration of 0.5 mg/mL, and cells were incubated for 4 h at 37 °C. Mitochondrial dehydrogenases in viable cells converted MTT into insoluble formazan crystals. These were subsequently dissolved in DMSO, and absorbance was measured at 595 nm using a Dynatech MicroElisa plate reader (Chantilly, VA, USA).

2.4. *Model Generation of Receptor–Ligand Complexes*

We utilized AlphaFold3 to model the binding of ligands to the CRF_1R [35]. The sequence used to generate the model was derived from UniProt (P34998) (post-translational modification sequence ID: PRO_0000012814), encompassing amino acids 24-415. We generated the apo receptor model using the abovementioned sequence (protein only) and complex models (protein with each of the nine compounds). Compounds were added to the JSON input file as SMILES strings. Additionally, 50 phospholipids (ccd code: HWP) were added to each model to better simulate the lipid bilayer environment. The generated models were compared with known crystallographic CRF_1R structures, specifically 8GTI (inhibitor-bound) and 6P9X (activated complex) [36,37]. We assessed model confidence and relative positioning using AlphaFold's provided metrics: the predicted local distance difference test (pLDDT) for local confidence and the predicted alignment error (PAE) for global confidence and relative positioning.

2.5. *Molecular Dynamics Simulations*

Molecular dynamics simulations of the CRF_1R in complex with the small-molecule antagonist M43 were performed using GROMACS 2023.4 [38]. The initial structure used was the AlphaFold-predicted model of the M43–CRF_1R complex. Parameterization of the protein–ligand complex was performed using CHARMM-GUI [39]. The ligand topology was generated using the CHARMM General Force Field (CGenFF), and the protein was embedded into a pre-equilibrated 1-palmitoyl-2-oleoyl-sn-glycero-3-phosphocholine (POPC) lipid bilayer of dimensions 120 Å × 120 Å, resulting in a membrane patch containing 387 POPC molecules [40]. The system was solvated with TIP3P water and neutralized with Na^+ and Cl^- ions to achieve an approximate ionic strength of 0.15 M. All simulations were conducted using the CHARMM36 force field [41]. The system was energy-minimized using the steepest descent algorithm. This was performed under positional restraints on the protein backbone, side chains, and lipid head groups with force constants of 4000, 2000, and 1000 $kJ \cdot mol^{-1} \cdot nm^{-2}$, respectively. Additionally, dihedral restraints were applied to backbone φ/ψ angles (force constant: 1000 $kJ \cdot mol^{-1} \cdot rad^{-2}$). The minimization continued until the maximum force on any atom was below 1000 kJ·mol^{-1}·nm^{-1}, or for a maximum of 5000 steps. A six-step equilibration scheme was employed to gradually relax the system toward physiologically relevant dynamics while preserving the overall structural integrity of the receptor–ligand complex: first, 1.5 ns of NVT simulation at 310 K with strong positional and dihedral restraints on protein and lipids (randomized velocities) to allow solvent and lipid tails to settle; next, 1.5 ns NVT with all restraints halved to increase solvent and side-chain mobility; then, 1.5 ns NPT with semi-isotropic C-rescale barostat (1 bar, τ = 5 ps, κ = 4.5 $\times$ 10^{-5} bar^{-1}) and further reduced restraints to equilibrate the membrane; a 5 ns NPT run (2 fs timestep) with backbone restraints lowered to 500 kJ·mol^{-1}·nm^{-2} for initial protein flexibility; a 10 ns NPT interval with minimal restraints (backbone 200, side chains 50, lipids 40 kJ·mol^{-1}·nm^{-2}) to approach near-native sampling; and finally, another 10 ns NPT under the same light restraints to fully stabilize the system before production. Throughout all equilibration steps, hydrogen bond constraints were maintained using the LINCS algorithm, and short-range van der Waals and Coulomb interactions were treated with a 1.2 nm cutoff, using force-switching from 1.0 to 1.2 nm. Long-range electrostatics were computed using the Particle-Mesh Ewald (PME) method. Following equilibration, a 500 ns production simulation was carried out at 310 K and 1 bar. All restraints were removed, allowing the receptor–ligand complex and lipid bilayer to evolve freely. A 2 fs integration timestep was used, with all bonds to hydrogen atoms constrained. Temperature coupling was applied separately to solute (protein + ligand), membrane, and solvent groups using the velocity-rescale thermostat. Semi-isotropic pressure coupling was maintained using the C-rescale barostat.

Following molecular dynamics (MD) simulations, root mean square deviation (RMSD) analyses were conducted using only the Cα atoms of residues embedded within the membrane. The highly flexible extracellular and intracellular regions were excluded to focus on assessing the structural stability of the membrane-embedded portion of the receptor.

3. Results

3.1. Computational Models of the Interactions Between CRF1R Non-Peptide Antagonists and CRF$_1$R

We have previously shown that out of 45 novel non-peptide molecules we designed and synthesized, eight of them bound to CRF$_1$R with relatively high affinities (Figure 1) [33]. The synthesized compounds were inspired by existing CRF$_1$R antagonists in the literature, but with a different scaffold (thiazolo[4,5-d] pyrimidine) in an effort to increase affinity, provide novel chemical entities, and further advance research on small molecules in the field. All the selected compounds that displayed promising activities are thiazolo[4,5]pyrimidines carrying a small alkylimino (*N*-methyl, *N*-propyl, or hydroxylamine) group at position 2, a secondary amine (diethylamine or ethyl, butylamine) at position 7, and a substituted phenyl group (2,4,6-trichlorophenyl or 2-bromo-4-isopropylphenyl) at position 3. MC43, which displayed the highest binding affinity of 19.2 nM, which is only two times lower than the prototype non-peptide antagonist antalarmin (Figure 1) [33], has a hydroxylimino functional group at position 2, a butyl, ethylamino functional group at position 7, and a 2,4,6-trichlorophenyl functional group at position 3.

To understand the molecular determinants of the identified non-peptide ligands (M6, M7, M8, M21, M22, M31, M42, M43, and the control antalarmin) that bind to CRF$_1$R, we produced AlphaFold models of the apo state and complexes with these nine compounds (Figure 2). According to AlphaFolds' PAE (Figure 2A), the secondary structures are accurately positioned relative to each other, except for the first 80 residues forming the extracellular domain of the receptor (ECD) in relation to the rest of the protein. This indicates flexibility in the ECD. The pLDDT scores were notably high for the core amino acids of the transmembrane domain in both the apo and ligand-bound models (Figure 2B,C). All ligand-bound models exhibited higher pLDDT scores, especially within the core region of the binding pocket and its surrounding helices. This suggests that the compounds stabilize the receptor in the inactive state by interacting with the allosteric pocket between the third (TM3), fifth (TM5), and sixth (TM6) transmembrane domains, already identified by others [42]. By comparing the AlphaFold3 models, it is evident that the ligand-free model adopts an intermediate state, while the protein–ligand complex is in an inactive state. This is highlighted by an approximate 30-degree outward shift of the lower part of TM 6 (Figure 2D).

No significant local sidechain shifts around the binding site were observed among the nine compounds (Figure 2E). A primary observation across all compounds is that the thienopyrimidine group forms a hydrogen bond with $N283^{5.50b}$, which is consistent with the previously reported literature [42]. The superscripts of receptor residues represent their positions in the TMs of the receptor, with the most conserved residue in each TM of subfamily B1 GPCRs to be assigned the position index 0.50, and this number is preceded by the TM number (TM1–TM7) [43]. Thus, $N283^{5.50b}$ denotes that Asn283 is the most conserved residue in the TM5 of subfamily B1 GPCRs.

Furthermore, the trichlorophenyl group, which extends deep into the cavity, appears to benefit from all three halogens, as exemplified by M43. These halogens interact with hydrophobic residues ($L287^{5.54b}$, $L320^{6.46b}$, $F284^{5.51b}$, $I290^{5.57b}$) and potentially form a weak halogen bond with $Y363^{7.57b}$. More importantly, the three chlorine atoms enhance the negative electron density on the phenyl ring, while the heavy bromine atom on M22 or M8 makes the interaction asymmetric and subsequently less stable.

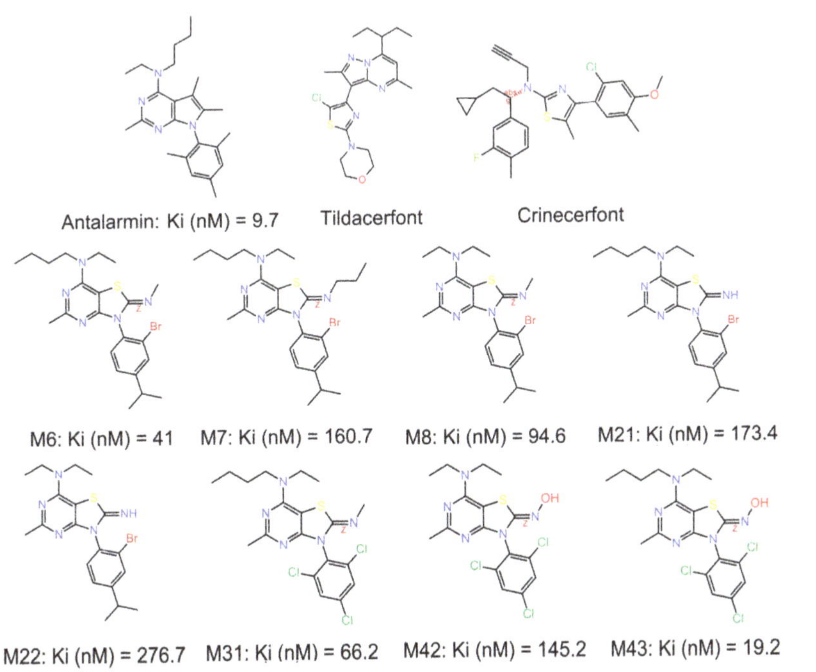

Figure 1. Chemical structures of eight novel non-peptide CRF antagonists and known antagonists, such as antalarmin, tildacerfont, and crinecerfontand, as well as their measured (where applicable) binding affinities for the CRF_1R [33].

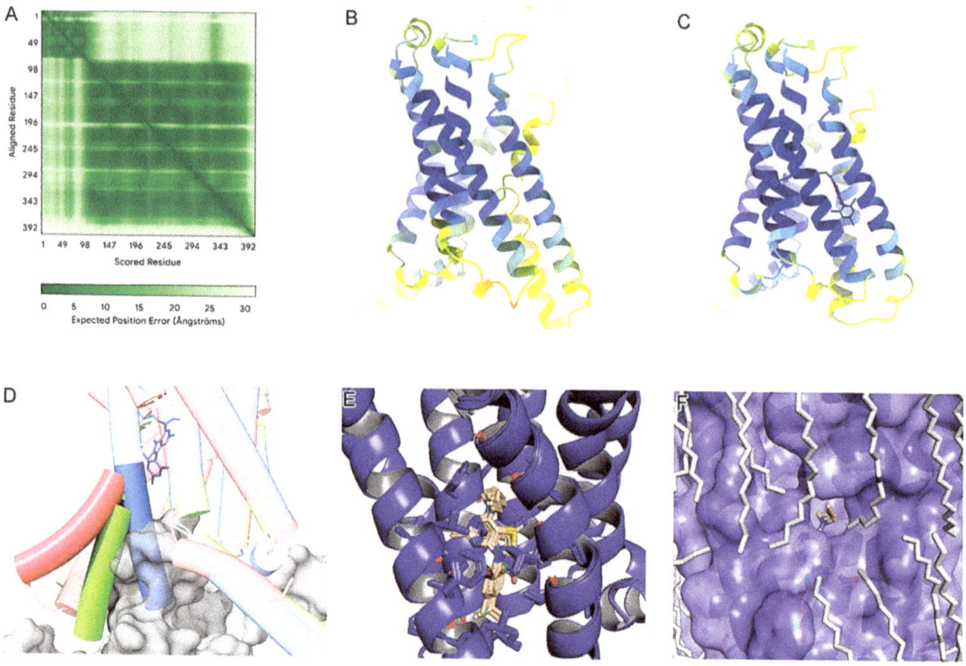

Figure 2. (**A**) PAE plot for the CRF_1R ligand-free model. (**B**) Ligand-free CRF_1R model, colored by pLDDT. (**C**) CRF_1R in complex with M43, colored by pLDDT. (**D**) Superimposition of the M43–CRF_1R complex (blue), CRF_1R ligand-free model (green), and CRF_1R active state experimental structure in a complex with CRF_1R peptide and guanine nucleotide binding protein G(s) (pdb id: 6p9x) (red). (**E**) Superimposition of 9 ligand-protein complexes generated using AlphaFold3. (**F**) Protein–ligand–membrane complex generated using AlphaFold, highlighting a potential interface between the ligand and the membrane.

Regarding the *N*-butyl-*N*-ethylamino group, consistent with K_i values, we propose that the addition of an even longer chain or the addition of other groups primarily on the butyl part might improve binding affinity. This is evident when comparing M43 with M42 K_i values, and it is supported by the AlphaFold3 model (Figure 2F), in which we observe an interface between the phospholipids and the *N*-butyl-*N*-ethylamino group of M43. This region is highly hydrophobic and positioned in the middle of the lipid bilayer, suggesting ample space for additional hydrophobic groups at this location. Potential additions in this region could include an aromatic moiety or halogens.

Finally, we observe that the oxime group of M43 and M42 forms a hydrogen bond with the carbonyl group of $T316^{6.42b}$ backbone. This is not the case for other molecules, such as M31, which, compared to M43, lacks the hydroxyl group of the oxime, but instead has a methyl group; therefore, it does not have the ability to form such a hydrogen bond and loses affinity. Also, by substituting the oxime hydroxyl group with hydrogen (as is the case for M22), although there is a potential hydrogen donor and acceptor, the angle does not favor the formation of a hydrogen bond.

Comparing the predicted binding mode of M43 with that of antalarmin (Figure S1), it is evident that both molecules have similar binding characteristics. The thiazole of M43 has one less substituent than the pyrrolopyrimidine of antalarmin, of which the three methyl substituents increase hydrophobicity. The main difference is that M43 seems to form an extra hydrogen bond with the oxime on the thiazole group. The fact that antalarmin had a slightly higher binding affinity than M43 may be explained by the increased hydrophobicity of the core scaffold provided by its methyl groups, which is partially compensated by the extra hydrogen bond of M43 with T316$^{6.42b}$.

3.2. M43 Forms Specific Hydrogen Bond Interactions with the Allosteric Pocket

To investigate the structural dynamics of the CRF_1R in complex with the top-performing small-molecule antagonist M43, unbiased molecular dynamics (MD) simulations were performed (see 1.5 Methods). The initial complex, based on the AlphaFold3-predicted model, was embedded in a POPC bilayer. The system, solvated with TIP3P water and 0.15 M NaCl, underwent energy minimization and multi-stage restrained equilibration before a 500 ns production run under physiological temperature and pressure conditions. As shown in Supplementary Figure S2, peripheral regions—particularly the extracellular domain (ECD)—exhibit pronounced conformational fluctuations and were excluded from global RMSD values. This finding is consistent with the AlphaFold3 PAE plot (Figure 2A). The resulting C_α RMSD remained below 3 Å throughout the simulation, indicating a stable transmembrane core (Figure 3A). Furthermore, when fitting on the C_α atoms of the membrane-inserted residues, the RMSD of the bound molecule M43 remained below 2 Å, demonstrating that the ligand maintains a well-defined binding pose relative to the stable membrane-embedded region of the protein (Figure 3B). These results support the structural integrity of the transmembrane domain and highlight the robustness of M43 binding during the simulation. To further validate the model and the simulation, RMSD calculations of the TMD were performed among the AlphaFold model, the most representative state from the MD simulation, and the complex of the receptor bound to BMK-C205 (pdb: 8gti) crystal structure (Figure S3A,B) [36]. BMK-C205 is an allosteric antagonist, structurally characterized in the same allosteric site of CRF_1R. Calculations show that the TMD of the AlphaFold starting model and the representative model from the MD have an RMSD of 1.73 and 1.78, respectively, to the resolved structure. RMSD calculations of frames from the M43-CRF_1R complex simulation, with the BMK-C205-CRF_1R crystal structure (Figure S3C), further validate and indicate a stable dynamics simulation for the M43-CRF_1R complex.

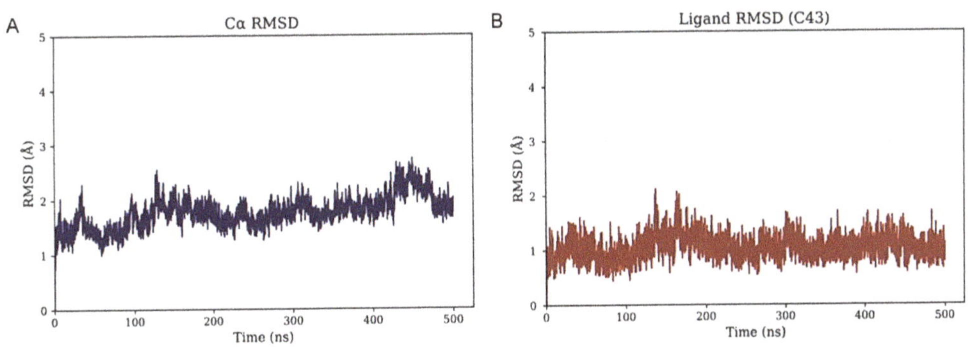

Figure 3. Root mean square deviation (RMSD) analysis of the protein–ligand complex. (**A**) C_α RMSD of the protein backbone over time, shown in blue. (**B**) Ligand RMSD after alignment on the protein C_α atoms, shown in red.

During the MD simulation, several hydrophobic interactions stabilize the ligand. In the co-folding AlphaFold models, the pocket is formed by L287$^{5.54b}$, L320$^{6.46b}$, F284$^{5.51b}$, and I290$^{5.57b}$, which accommodates the halogenated moiety of M43. One of the two terminal alkyl chains (M5–M6), linked via N1, is oriented toward F203$^{3.44b}$, supporting favorable hydrophobic packing. The second alkyl chain (M1–M4) extends toward the membrane interface, positioning itself in close proximity to the surrounding POPC lipids, suggesting membrane-assisted stabilization of the ligand conformation.

To further assess the persistence of key interactions stabilizing ligand M43, time-resolved distances of three putative hydrogen bonds were monitored throughout the trajectory: dist1: $N283^{5.50b}{}_{ND2}$–$M43_{N3}$, dist2: $T316^{6.42b}{}_{O}$–$M43_{O1X}$, and dist3: $T316^{6.42b}{}_{OG1}$–$M43_{N5}$ (Figure 4A). The $N283^{5.50b}$–$M43_{N3}$ interaction (dist1) proved to be highly stable, as known from previous studies, consistently maintaining a distance well below the hydrogen bond threshold throughout the simulation (Figure 4A) [42]. The $T316^{6.42b}{}_{O}$–$M43_{O1X}$ interaction monitored via dist2 exhibited moderate fluctuations around the cutoff value, occasionally breaking the hydrogen bond (Figure 4C). Notably, during these periods of dist2 disruption, an alternative hydrogen bond formed between $T316^{6.42b}{}_{OG1}$ and the N5 atom of M43 (dist3) (Figure 4D), indicating a dynamic switch between the two interactions. This complementary behavior suggests that $T316^{6.42b}$ maintains stabilizing contact with M43 by alternating between O1X and N5, depending on local structural fluctuations. Collectively, these observations highlight a cooperative hydrogen-bonding network, with the $N283^{5.50b}$ interaction serving as a persistent anchor and $T316^{6.42b}$ contributing adaptable support through transient interactions with multiple sites on M43. It also indicates that dist3 is important for compounds in the series where the N5 serves as a hydrogen bond acceptor that interacts with $T316^{6.42b}$. The fact that M31, M6, or M8 lack the hydroxyl group of the oxime, but instead have a methyl group, maintaining reasonable affinities, also indicates the important role of this hydrogen bond to the series in general.

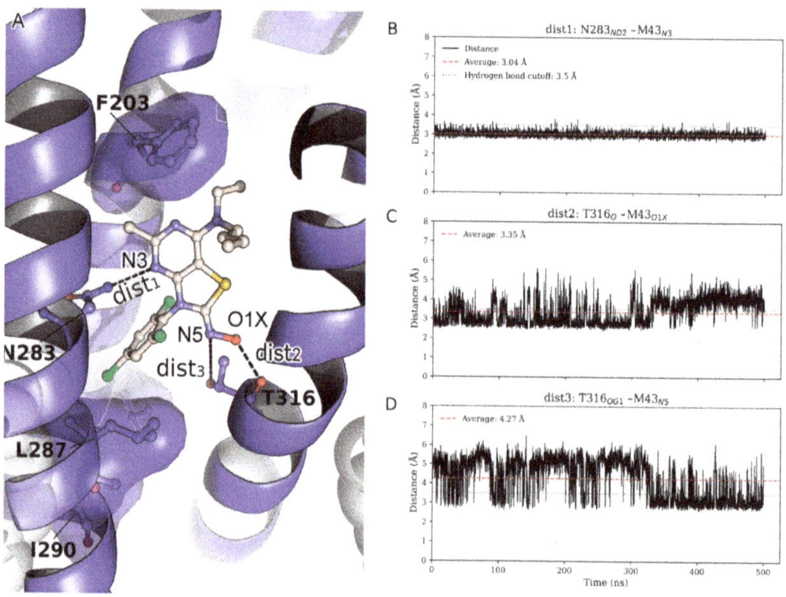

Figure 4. (**A**) Representative 3D structure from the simulation, showing M43 ligand (yellow) bound to CRF_1R (blue). Dashed lines indicate the measured distances between $N283^{5.50b}{}_{ND2}$–$M43_{N3}$, $T316^{6.42b}{}_{O}$–$M43_{O1X}$, and $T316^{6.42b}{}_{OG1}$–$M43_{N5}$, with corresponding labels. Key hydrophobic residues are highlighted with a transparent surface representation, and white spheres denote the membrane bilayer. (**B**) Time-resolved distance between $N283^{5.50b}{}_{ND2}$ and $M43_{N3}$ (dist1). (**C**) Time-resolved distance between $T316^{6.42b}{}_{O}$ and $M43_{O1X}$ (dist2). (**D**) Time-resolved distance between $T316^{6.42b}{}_{OG1}$ and $M43_{N5}$ (dist3). Dashed red lines represent the average distance over time, while dotted gray lines indicate the 3.5 Å hydrogen bond threshold.

3.3. M43 Antagonizes the Stimulation of cAMP Accumulation by Tyr^{0}*-Sauvagine*

The ability of M43 to bind to CRF_1R with the highest affinity compared to the other compounds tested in our previous study prompted us to test its antagonistic properties [33]. To accomplish this, we first determined its ability to decrease the potency of the CRF-related agonist Tyr^0-sauvagine in stimulating cAMP accumulation in HEK 293 cells expressing the CRF_1R. As shown in Figure 5, M43 at a concentration of 1 μM significantly decreased the potency of Tyr^0-sauvagine by 6.2-fold ($p < 0.005$, paired t-test). Specifically, in the absence of M43, Tyr^0-sauvagine's potency was 2.2 nM (or $-\text{LogEC}_{50} = 8.65 \pm 0.15$, n = 12), whereas in its presence, the potency of Tyr^0-sauvagine was 13.9 nM (or $-\text{logEC}_{50} = 7.86 \pm 0.22$, n = 12).

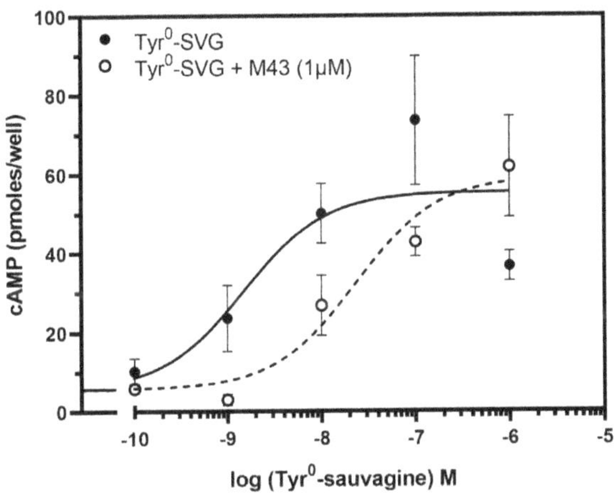

Figure 5. M43 decreases the potency of Tyr^0-sauvagine to stimulate cAMP accumulation. The stimulation of cAMP accumulation in HEK293 cells stably expressing the CRF_1R by increasing concentrations of Tyr^0-sauvagine (Tyr^0-SVG) in the absence or presence of 1 μM of M43 was determined as described in the "Materials and Methods". The means and S.E. (duplicate determination) are shown from a representative experiment performed 12 times. The potencies ($-\text{logEC}_{50}$) of SVG without or with M43, as determined from these experiments, were 8.65 ± 0.15 and 7.86 ± 0.22, respectively.

To further examine the antagonistic properties of M43, we determined its half-maximal inhibitory concentration (or antagonistic potency, pIC_{50}). The IC_{50} of M43 is determined as its concentration required to decrease the cAMP accumulation stimulated by 10 nM Tyr^0-sauvagine to half. As shown in Figure 6, M43 inhibited cAMP accumulation stimulated by 10 nM Tyr^0-sauvagine in a dose–response manner, with an antagonistic potency of 43.5 nM ($pIC_{50} = 7.36 \pm 0.18$, n = 6).

3.4. M43 Antagonizes the Effect of Tyr^{0}*-Sauvagine on the Proliferation of RAW 264.7 Macrophages*

To support the antagonistic properties of M43, we determined whether this molecule could block the effects of Tyr^0-sauvagine to induce the proliferation of RAW 264.7 macrophage cells. Cells were incubated with 10 nM of Tyr^0-sauvagine in the presence or absence of (control) of 1 μM of M43. As shown in Figure 7, the stimulation of the proliferation rate of RAW 264.7 cells by 10 nM of Tyr^0-sauvagine after 72 h was significantly inhibited by M43 at a concentration of 1 μM.

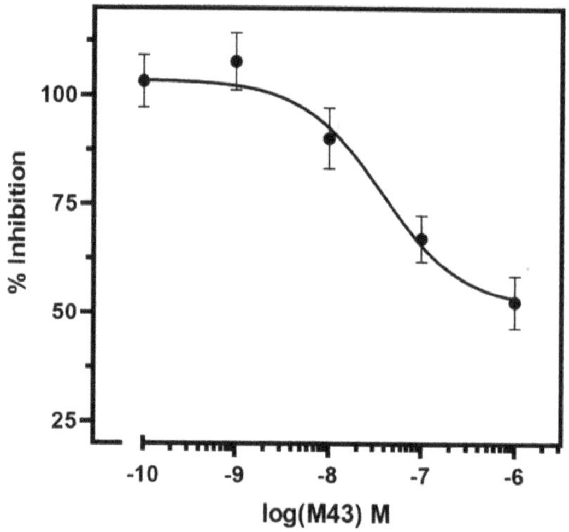

Figure 6. M43 inhibits cAMP accumulation stimulated by 10 nM of Tyr^0-sauvagine. The stimulation of cAMP accumulation in HEK 293 cells stably expressing the CRF$_1$R by 10 nM of Tyr^0-sauvagine in the absence or presence of increasing concentrations of M43 was determined as described in the "Materials and Methods". Data represent the mean ± S.E. from six independent experiments. The cAMP accumulation was normalized by assigning 100% to the maximal response after stimulation with Tyr^0-sauvagine in the absence of M43.

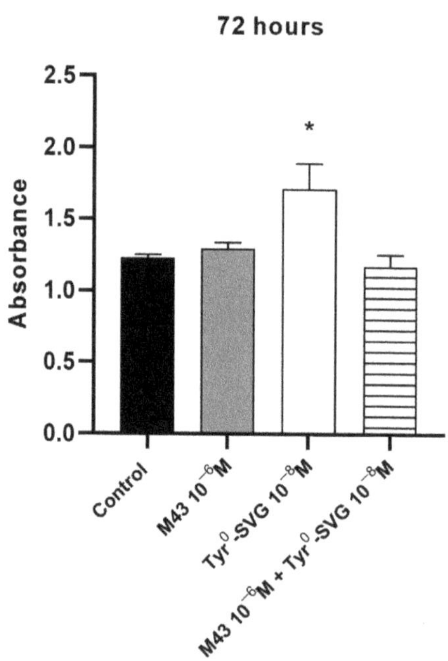

Figure 7. M43 inhibits the proliferation rate of RAW 264.7 macrophages stimulated by 10 nM of Tyr^0-sauvagine. The proliferation rate of RAW 264.7 macrophages following 72 h treatment with 10 nM of Tyr^0-sauvagine in the absence or presence of 1 μM of M43 was determined as described in the "Materials and Methods". The means and S.E. are shown from a representative experiment. (*) Represents a comparison between the 10 nM Tyr^0-sauvagine group and 10 nM Tyr^0-sauvagine+1 μM M43 group, n = 5. * $p < 0.01$. Statistical analysis was performed using one-way ANOVA, followed by Tukey's multiple comparison test.

4. Discussion

In our previous study, we have designed, synthesized, and tested the abilities of 45 novel non-peptide molecules to bind to CRF_1R [33]. Among these compounds, the analogs M6, M7, M8, M21, M22, M31, M42, and M43 bound to CRF_1R with high affinities (41, 160.7, 94.6, 173.4, 276.7, 66.2, 145.2, and 19.2 nM, respectively), with M43 being the best binder (Figure 1). Molecular modeling studies performed in the present study highlighted the essential role of the hydrogen bond between $N283^{5.50b}$ of CRF_1R and the common pyridine nitrogen of analogs M6, M7, M8, M21, M22, M31, M42, and M43, as well as all other compounds tested in our previous study [33]. In addition, the three chlorides of the phenyl group that exist in several compounds seem to have equally distributed charges, facilitating hydrophobic interactions, compared to the bromide and isopropyl groups of other compounds, thereby yielding better docking results. Comparison of the analogs M6 and M8, which have similar chemical structures but different groups at the tertiary amine position, suggests that the presence of ethyl and butyl groups (M6) at this position contributes to a higher affinity binding than the presence of two ethyl groups (M8), perhaps due to the tunnel formed toward the membrane (Figure 2). A similar relationship exists between the chemical groups at position 7 and the binding affinity of the analogs M43 (with ethyl and butyl groups) and M42 (with two ethyl groups). Molecular modeling data suggest that the contribution of the butyl group at the tertiary amine position of molecules to their high-affinity binding is mostly due to its interaction with $L320^{6.46b}$ of CRF_1R and the membrane. A comparison of M43 with other compounds featuring ethyl and butyl groups at the tertiary amine position, including M6, and analysis of molecular dynamics simulations suggest that the presence of the oxime hydroxyl group in M43 could explain its higher binding affinity compared with the other molecules. This hydroxyl group of M43 possibly forms an extra hydrogen bond with the main chain carbonyl oxygen or the side chain hydroxyl of $Thr316^{6.42b}$ (Figure 4).

Non-peptide CRF_1R antagonists—including antalarmin and more clinically advanced scaffolds, such as tildacerfont and crinecerfont—bind the intramembrane, allosteric site that is buried within the transmembrane (TM) bundle and distinct from the N-terminal peptide orthosteric site. This site is primarily formed by TM3, TM5, and TM6, and antagonist binding stabilizes inactive-state conformations that impede the outward displacement of TM6 required for G_s engagement and signaling. Our AlphaFold3 models and MD simulations place M43 in this canonical allosteric cleft and recapitulate the class-defining anchors: a persistent H-bond to $N283^{5.50b}$ and a complementary, switchable H-bonding interaction with $T316^{6.42b}$ (backbone carbonyl vs. side-chain hydroxyl). In contrast, the *N*-ethyl–*N*-butyl side chain extends toward a hydrophobic, membrane-facing subpocket (Figures 2 and 4). This pose and network are fully consistent with the allosteric antagonism established for CRF_1R [34,36,42]. Mechanistically, CRF_1R small-molecule antagonists commonly display insurmountable antagonism attributed to slow dissociation from the TM pocket; the persistent inhibition we observe for M43 is consonant with this kinetic trapping paradigm [21,42].

The ability of M43 to bind to CRF_1R with the highest binding affinity compared to the other tested compounds prompted us to further pharmacologically characterize it by determining its ability to antagonize the CRF-related agonist, Tyr^0-sauvagine. M43 is a CRF potent antagonist, blocking the ability of 10 nM Tyr^0-sauvagine to stimulate cAMP accumulation in HEK 293 cells that stably express the CRF_1R, displaying a half-maximal inhibitory concentration of 43.5 nM. As a potent antagonist, M43, at a concentration of 1 μM, was able to reduce the potency of Tyr^0-sauvagine to stimulate the intracellular cAMP accumulation by 6.2-fold. Similarly, the prototype CRF_1R selective antagonist antalarmin decreased the potency of Tyr^0-sauvagine by 33-fold to stimulate cAMP accumulation in

HEK 293 expressing the CRF_1R [34]. The lower antagonistic ability of M43 compared to that of antalarmin could be attributed to its lower binding affinity (19.2 nM) compared with antalarmin (9.7 nM) [33]. To provide supportive evidence for the antagonistic properties of M43, we tested its ability to block the effect of Tyr^0-sauvagine on the proliferation of the RAW 264.7 macrophage cells [44]. Our data suggested that M43 antagonized Tyr^0-sauvagine, thereby stimulating the proliferation of RAW 264.7 cells.

The activation of CRF_1R is associated with the disruption of interactions between receptor amino acids, and the formation of a new comparison of the high-resolution inactive and active structures of CRF_1R suggests that its activation is closely related to structural changes, involving an upward movement of the fourth (TM4) and fifth (TM5) transmembrane domains that repositions the second extracellular (EL2) and intracellular (IL2) loops [45]. Receptor activation is also associated with other orchestrated movements of the transmembrane domains, including the third (TM3), fifth, and sixth (TM6) ones [34,45]. These movements are responsible for the high-affinity binding of agonists to the receptor and the interaction of the latter with the G-proteins. M43 most likely impedes the activation-associated movements of CRF_1R by interacting with the described allosteric pocket formed by TM3, TM5, and TM6, as shown in Figure 4. This conformational trapping provides a coherent structural basis for the insurmountable-like antagonism observed across this class of molecules [21,42]. While our modeling and cellular pharmacology place M43 firmly within the CRF_1R allosteric antagonist class and align its behavior with antalarmin and clinically advanced scaffolds, additional experiments would sharpen differentiation. Such data will enable a more granular, mechanism-focused comparison of M43 with antalarmin, tildacerfont, and crinecerfont.

In future studies, we will evaluate the in vivo pharmacological properties and toxicities of the thiazolopyrimidine analog M43, as well as its ability to act on different GPCRs such as serotonin or adenosine receptors, which play crucial roles in depression [14,46,47]. Previous studies have shown that thiazolopyrimidine derivatives are antagonists for serotonin and adenosine receptors [48,49]. Evaluating the thiazolopyrimidine derivative M43 against other GPCRs could reveal an important pharmacological property, namely, a desirable non-selectivity for different GPCRs. As proposed by Roth et al., non-selective molecules could be used as "magic shotguns" to treat CNS disorders, including depression [50].

5. Conclusions

The results of this study suggest that the analog M43 is a potent CRF_1R antagonist that could be used as a lead molecule for the development of novel non-peptide CRF_1R antagonists. The antagonistic properties of M43 are attributed to its interaction with residues in the TM3, TM5, and TM6 of CRF_1R. These interactions block activation-associated structural rearrangements of the CRF_1R. The pharmacological properties of M43 render it an optimal lead compound in the rational design of novel non-peptide CRF antagonists, which will enrich the pharmaceutical arsenal against CAH and stress-related disorders.

Supplementary Materials: The following supporting information can be downloaded at https://www.mdpi.com/article/10.3390/biom15091265/s1: Table S1: Codes and chemical names of the compounds in this study. Figure S1: AphaFold3 models of CRF_1R in a complex with M43 and antalarmin, as well as a comparison with the crystal structure of CRF_1R with BMK-C205; Figure S2: Three-dimensional representation of residue flexibility during the simulation; Figure S3: Model comparison with the MD representative state and the crystal structure complex of the receptor bound to BMK-C205, as well as RMSD calculations compared to the BMK-C205-CRF_1R structure.

Author Contributions: Conceptualization, M.-T.M., H.F., and G.L.; investigation M.-T.M., M.V., and G.L.; methodology, G.K., A.K., S.M.G., N.L., M.V., M.-T.M., and G.L.; supervision, M.-T.M.,

M.V., and G.L.; validation, V.K., M.-T.M., M.V., and G.L.; software, V.K., S.M.G., N.L., and M.-T.M.; writing—original draft preparation, G.K., A.K., V.K., S.M.G., M.R.I., N.L., T.M., M.-T.M., M.V., and G.L.; writing—review and editing, G.K., A.K., V.K., S.M.G., M.T., M.R.I., N.L., T.M., M.-T.M., M.V., and G.L.; funding acquisition, M.-T.M. and G.L. All authors have read and agreed to the published version of the manuscript.

Funding: This research was funded by a grant in the framework of the Hellenic Foundation for Research and Innovation (H.F.R.I.) call "3rd Call for H.F.R.I.'s Research Projects to Support Faculty Members & Researchers" (Project Number: 25255) to M.-T.M. and by a grant from the Special Account for Research Funds of the University of Crete (SARF) (funding number: KA 10674) to G.L.

Institutional Review Board Statement: Not applicable.

Informed Consent Statement: Not applicable.

Data Availability Statement: Data are contained within the article.

Conflicts of Interest: The authors declare no conflicts of interest.

Abbreviations

The following abbreviations are used in this manuscript:

CRF	corticotropin-releasing factor
CRF1R	type 1 receptor for the corticotropin-releasing factor
HPA	hypothalamic–pituitary–adrenal
CAH	congenital adrenal hyperplasia
ACTH	adrenocorticotropic hormone or corticotropin
HEK	human embryonic kidney
DMEM	Dulbecco's modified Eagle's medium
cAMP	cyclic adenosine monophosphate
MTT	(3-(4,5-dimethylthiazol-2-yl)-2,5-diphenyltetrazolium bromide
pLDDT	predicted local distance difference test
PAE	predicted alignment error
POPC	1-palmitoyl-2-oleoyl-sn-glycero-3-phosphocholine
CGenFF	CHARMM General Force Field
ECD	extracellular domain
TCA	trichloroacetic acid
PME	Particle-Mesh Ewald

References

1. Grammatopoulos, D.K.; Chrousos, G.P. Functional characteristics of CRH receptors and potential clinical applications of CRH-receptor antagonists. *Trends Endocrinol. Metab.* **2002**, *13*, 436–444. [CrossRef] [PubMed]
2. Liapakis, G.; Venihaki, M.; Margioris, A.; Grigoriadis, D.; Gkountelias, K. Members of CRF Family and their Receptors: From Past to Future. *Curr. Med. Chem.* **2011**, *18*, 2583–2600. [CrossRef] [PubMed]
3. Chrousos, G.P. The hypothalamic-pituitary-adrenal axis and immune-mediated inflammation. *N. Engl. J. Med.* **1995**, *332*, 1351–1362. [CrossRef]
4. Vale, W.; Spiess, J.; Rivier, C.; Rivier, J. Characterization of a 41-residue ovine hypothalamic peptide that stimulates secretion of corticotropin and beta-endorphin. *Science* **1981**, *213*, 1394–1397. [CrossRef]
5. Smith, S.M.; Vale, W.W. The role of the hypothalamic-pituitary-adrenal axis in neuroendocrine responses to stress. *Dialogues Clin. Neurosci.* **2006**, *8*, 383–395. [CrossRef]
6. Owens, M.J.; Nemeroff, C.B. Physiology and pharmacology of corticotropin-releasing factor. *Pharmacol. Rev.* **1991**, *43*, 425–473. [CrossRef]
7. Gold, P.W.; Chrousos, G.P. Organization of the stress system and its dysregulation in melancholic and atypical depression: High vs low CRH/NE states. *Mol. Psychiatry* **2002**, *7*, 254–275. [CrossRef]
8. Keck, M.E.; Holsboer, F. Hyperactivity of CRH neuronal circuits as a target for therapeutic interventions in affective disorders. *Peptides* **2001**, *22*, 835–844. [CrossRef]

9. Reul, J.M.; Holsboer, F. Corticotropin-releasing factor receptors 1 and 2 in anxiety and depression. *Curr. Opin. Pharmacol.* **2002**, *2*, 23–33. [CrossRef]
10. Chrousos, G.P. Stress and disorders of the stress system. *Nat. Rev. Endocrinol.* **2009**, *5*, 374–381. [CrossRef] [PubMed]
11. Henckens, M.J.; Deussing, J.M.; Chen, A. Region-specific roles of the corticotropin-releasing factor-urocortin system in stress. *Nat. Rev. Neurosci.* **2016**, *17*, 636–651. [CrossRef]
12. Mikulska, J.; Juszczyk, G.; Gawrońska-Grzywacz, M.; Herbet, M. HPA Axis in the Pathomechanism of Depression and Schizophrenia: New Therapeutic Strategies Based on Its Participation. *Brain Sci.* **2021**, *11*, 1298. [CrossRef]
13. Guerry, J.D.; Hastings, P.D. In Search of HPA Axis Dysregulation in Child and Adolescent Depression. *Clin. Child Fam. Psychol. Rev.* **2011**, *14*, 135–160. [CrossRef]
14. Stahl, S.M. *Stahl's Essential Psychopharmacology: Neuroscientific Basis and Practical Applications*, 5th ed.; Cambridge University Press: Cambridge, UK, 2021.
15. Muller, M.B.; Wurst, W. Getting closer to affective disorders: The role of CRH receptor systems. *Trends Mol. Med.* **2004**, *10*, 409–415. [CrossRef]
16. Stenzel-Poore, M.P.; Heinrichs, S.C.; Rivest, S.; Koob, G.F.; Vale, W.W. Overproduction of corticotropin-releasing factor in transgenic mice: A genetic model of anxiogenic behavior. *J. Neurosci.* **1994**, *14*, 2579–2584. [CrossRef] [PubMed]
17. Muller, M.B.; Zimmermann, S.; Sillaber, I.; Hagemeyer, T.P.; Deussing, J.M.; Timpl, P.; Kormann, M.S.; Droste, S.K.; Kuhn, R.; Reul, J.M.; et al. Limbic corticotropin-releasing hormone receptor 1 mediates anxiety-related behavior and hormonal adaptation to stress. *Nat. Neurosci.* **2003**, *6*, 1100–1107. [CrossRef] [PubMed]
18. Nguyen, N.K.; Keck, M.E.; Hetzenauer, A.; Thoeringer, C.K.; Wurst, W.; Deussing, J.M.; Holsboer, F.; Muller, M.B.; Singewald, N. Conditional CRF receptor 1 knockout mice show altered neuronal activation pattern to mild anxiogenic challenge. *Psychopharmacology* **2006**, *188*, 374–385. [CrossRef] [PubMed]
19. Refojo, D.; Echenique, C.; Muller, M.B.; Reul, J.M.; Deussing, J.M.; Wurst, W.; Sillaber, I.; Paez-Pereda, M.; Holsboer, F.; Arzt, E. Corticotropin-releasing hormone activates ERK1/2 MAPK in specific brain areas. *Proc. Natl. Acad. Sci. USA* **2005**, *102*, 6183–6188. [CrossRef]
20. Keck, M.E.; Ohl, F.; Holsboer, F.; Muller, M.B. Listening to mutant mice: A spotlight on the role of CRF/CRF receptor systems in affective disorders. *Neurosci. Biobehav. Rev.* **2005**, *29*, 867–889. [CrossRef]
21. Zorrilla, E.P.; Koob, G.F. Progress in corticotropin-releasing factor-1 antagonist development. *Drug Discov. Today* **2010**, *15*, 371–383. [CrossRef] [PubMed]
22. Fahmy, H.; Spyridaki, K.; Kuppast, B.; Liapakis, G. The "homeostasis hormone" and its CRF(1) receptor. From structure to function. *Hormones* **2012**, *11*, 254–271. [CrossRef] [PubMed]
23. Auer, M.K.; Nordenström, A.; Lajic, S.; Reisch, N. Congenital adrenal hyperplasia. *Lancet* **2023**, *401*, 227–244. [CrossRef] [PubMed]
24. Merke, D.P.; Bornstein, S.R. Congenital adrenal hyperplasia. *Lancet* **2005**, *365*, 2125–2136. [CrossRef] [PubMed]
25. Kyritsi, E.M.; Koltsida, G.; Farakla, I.; Papanikolaou, A.; Critselis, E.; Mantzou, E.; Zoumakis, E.; Kolaitis, G.; Chrousos, G.P.; Charmandari, E. Psychological vulnerability to stress in carriers of congenital adrenal hyperplasia due to 21-hydroxylase deficiency. *Hormones* **2017**, *16*, 42–53. [CrossRef]
26. Harasymiw, L.A.; Grosse, S.D.; Cullen, K.R.; Bitsko, R.H.; Perou, R.; Sarafoglou, K. Depressive and anxiety disorders and antidepressant prescriptions among insured children and young adults with congenital adrenal hyperplasia in the United States. *Front. Endocrinol.* **2023**, *14*, 1129584. [CrossRef]
27. Reisch, N. Block and Replace—A New Therapeutic Concept in Congenital Adrenal Hyperplasia? *J. Clin. Endocrinol. Metab.* **2021**, *107*, e423–e425. [CrossRef]
28. Auchus, R.J.; Hamidi, O.; Pivonello, R.; Bancos, I.; Russo, G.; Witchel, S.F.; Isidori, A.M.; Rodien, P.; Srirangalingam, U.; Kiefer, F.W.; et al. Phase 3 Trial of Crinecerfont in Adult Congenital Adrenal Hyperplasia. *N. Engl. J. Med.* **2024**, *391*, 504–514. [CrossRef]
29. Sarafoglou, K.; Auchus, R.J. Future Directions in the Management of Classic Congenital Adrenal Hyperplasia Due to 21-Hydroxylase Deficiency. *J. Clin. Endocrinol. Metab.* **2025**, *110*, S74–S87. [CrossRef]
30. Sarafoglou, K.; Barnes, C.N.; Huang, M.; Imel, E.A.; Madu, I.-J.; Merke, D.P.; Moriarty, D.; Nakhle, S.; Newfield, R.S.; Vogiatzi, M.G.; et al. Tildacerfont in Adults With Classic Congenital Adrenal Hyperplasia: Results from Two Phase 2 Studies. *J. Clin. Endocrinol. Metab.* **2021**, *106*, e4666–e4679. [CrossRef]
31. Chrousos, G.P. Crinecerfont in a First Clinical Application of a CRH Antagonist: Further Potential Uses Are Still an Open Chapter! *J. Clin. Endocrinol. Metab.* **2024**, *109*, e1365–e1366. [CrossRef]
32. LiverTox: Clinical and Research Information on Drug-Induced Liver Injury [Internet]. Available online: https://www.ncbi.nlm.nih.gov/books/NBK612555/ (accessed on 15 March 2025).
33. Islam, M.R.; Teleb, M.; Karageorgos, V.; Sakellaris, S.; Papadopoulos, M.; Pirmettis, I.; Fronczek, F.R.; Liapakis, G.; Fahmy, H. Design, synthesis, structural optimization, SAR, in silico prediction of physicochemical properties and pharmacological evaluation of novel & potent thiazolo [4,5-d]pyrimidine corticotropin releasing factor (CRF) receptor antagonists. *Eur. J. Pharm. Sci.* **2022**, *169*, 106084. [CrossRef] [PubMed]

34. Spyridaki, K.; Matsoukas, M.T.; Cordomi, A.; Gkountelias, K.; Papadokostaki, M.; Mavromoustakos, T.; Logothetis, D.E.; Margioris, A.N.; Pardo, L.; Liapakis, G. Structural-Functional Analysis of the Third Transmembrane Domain of the Corticotropin-releasing Factor Type 1 Receptor: ROLE IN ACTIVATION AND ALLOSTERIC ANTAGONISM. *J. Biol. Chem.* **2014**, *289*, 18966–18977. [CrossRef] [PubMed]
35. Abramson, J.; Adler, J.; Dunger, J.; Evans, R.; Green, T.; Pritzel, A.; Ronneberger, O.; Willmore, L.; Ballard, A.J.; Bambrick, J.; et al. Accurate structure prediction of biomolecular interactions with AlphaFold 3. *Nature* **2024**, *630*, 493–500. [CrossRef] [PubMed]
36. Kim, H.; Lim, T.; Ha, G.E.; Lee, J.-Y.; Kim, J.-W.; Chang, N.; Kim, S.H.; Kim, K.H.; Lee, J.; Cho, Y.; et al. Structure-based drug discovery of a corticotropin-releasing hormone receptor 1 antagonist using an X-ray free-electron laser. *Exp. Mol. Med.* **2023**, *55*, 2039–2050. [CrossRef] [PubMed]
37. Liang, Y.L.; Belousoff, M.J.; Zhao, P.; Koole, C.; Fletcher, M.M.; Truong, T.T.; Julita, V.; Christopoulos, G.; Xu, H.E.; Zhang, Y.; et al. Toward a Structural Understanding of Class B GPCR Peptide Binding and Activation. *Mol. Cell* **2020**, *77*, 656–668 e655. [CrossRef]
38. Abraham, M.J.; Murtola, T.; Schulz, R.; Páll, S.; Smith, J.C.; Hess, B.; Lindahl, E. GROMACS: High performance molecular simulations through multi-level parallelism from laptops to supercomputers. *SoftwareX* **2015**, *1–2*, 19–25. [CrossRef]
39. Jo, S.; Kim, T.; Iyer, V.G.; Im, W. CHARMM-GUI: A web-based graphical user interface for CHARMM. *J. Comput. Chem.* **2008**, *29*, 1859–1865. [CrossRef]
40. Vanommeslaeghe, K.; Hatcher, E.; Acharya, C.; Kundu, S.; Zhong, S.; Shim, J.; Darian, E.; Guvench, O.; Lopes, P.; Vorobyov, I.; et al. CHARMM general force field: A force field for drug-like molecules compatible with the CHARMM all-atom additive biological force fields. *J. Comput. Chem.* **2010**, *31*, 671–690. [CrossRef]
41. Best, R.B.; Zhu, X.; Shim, J.; Lopes, P.E.M.; Mittal, J.; Feig, M.; MacKerell, A.D., Jr. Optimization of the Additive CHARMM All-Atom Protein Force Field Targeting Improved Sampling of the Backbone ϕ, ψ and Side-Chain $\chi1$ and $\chi2$ Dihedral Angles. *J. Chem. Theory Comput.* **2012**, *8*, 3257–3273. [CrossRef]
42. Hollenstein, K.; Kean, J.; Bortolato, A.; Cheng, R.K.; Dore, A.S.; Jazayeri, A.; Cooke, R.M.; Weir, M.; Marshall, F.H. Structure of class B GPCR corticotropin-releasing factor receptor 1. *Nature* **2013**, *499*, 438–443. [CrossRef]
43. Wootten, D.; Simms, J.; Miller, L.J.; Christopoulos, A.; Sexton, P.M. Polar transmembrane interactions drive formation of ligand-specific and signal pathway-biased family B G protein-coupled receptor conformations. *Proc. Natl. Acad. Sci. USA* **2013**, *110*, 5211–5216.
44. Sakellaris, S.; Matsoukas, M.T.; Karageorgos, V.; Poulaki, S.; Kuppast, B.; Margioris, A.; Venihaki, M.; Fahmy, H.; Liapakis, G. Selective antagonism of CRF1 receptor by a substituted pyrimidine. *Hormones* **2019**, *18*, 215–221. [CrossRef]
45. Matsoukas, M.T.; Panagiotopoulos, V.; Karageorgos, V.; Chrousos, G.P.; Venihaki, M.; Liapakis, G. Structural and Functional Insights into CRF Peptides and Their Receptors. *Biology* **2024**, *13*, 120. [CrossRef]
46. López-Cruz, L.; Salamone, J.D.; Correa, M. Caffeine and Selective Adenosine Receptor Antagonists as New Therapeutic Tools for the Motivational Symptoms of Depression. *Front. Pharmacol.* **2018**, *9*, 526. [CrossRef] [PubMed]
47. Wood, M.D.; Thomas, D.R.; Watson, J.M. Therapeutic potential of serotonin antagonists in depressive disorders. *Expert. Opin. Investig. Drugs* **2002**, *11*, 457–467. [CrossRef]
48. Peng, Y.; McCorvy, J.D.; Harpsøe, K.; Lansu, K.; Yuan, S.; Popov, P.; Qu, L.; Pu, M.; Che, T.; Nikolajsen, L.F.; et al. 5-HT2C Receptor Structures Reveal the Structural Basis of GPCR Polypharmacology. *Cell* **2018**, *172*, 719–730.e714. [CrossRef] [PubMed]
49. Varano, F.; Catarzi, D.; Falsini, M.; Vincenzi, F.; Pasquini, S.; Varani, K.; Colotta, V. Identification of novel thiazolo[5,4-d]pyrimidine derivatives as human A(1) and A(2A) adenosine receptor antagonists/inverse agonists. *Bioorganic Med. Chem.* **2018**, *26*, 3688–3695. [CrossRef] [PubMed]
50. Roth, B.L.; Sheffler, D.J.; Kroeze, W.K. Magic shotguns versus magic bullets: Selectively non-selective drugs for mood disorders and schizophrenia. *Nat. Rev. Drug Discov.* **2004**, *3*, 353–359. [CrossRef]

biomolecules

MDPI

Article

Nature's Synergy: Cellular and Molecular Evaluation of Snail Slime and Its Principal Component, Glycolic Acid, on Keratinocytes, with Preliminary Evidence from Endothelial Cells

Muhammad Rashad [1,†], Alessia Ricci [1,†], Serena Pilato [1], Amelia Cataldi [1,2], Marwa Balaha [1,3,‡] and Susi Zara [1,*,‡]

1 Department of Pharmacy, "G. d'Annunzio" University of Chieti-Pescara, 66100 Chieti, Italy; muhammad.rashad@unich.it (M.R.); alessia.ricci@unich.it (A.R.); serena.pilato@unich.it (S.P.); amelia.cataldi@unich.it (A.C.); marwa.balaha@unich.it (M.B.)

2 UdA-Tech Lab Research Center, "G. d'Annunzio" University of Chieti-Pescara, 66100 Chieti, Italy

3 Department of Pharmaceutical Chemistry, Faculty of Pharmacy, Kafrelsheikh University, Kafrelsheikh P.O. Box 33516, Egypt

* Correspondence: susi.zara@unich.it

† These authors contributed equally to this work.

‡ These authors also contributed equally to this work.

Abstract: Snail slime (SS) is a natural secretion rich in bioactive components such as glycoproteins, hyaluronic acid, glycolic acid (GA), and antimicrobial peptides. GA, a key component of SS, is known for its exfoliative properties. This study investigates SS's effects on keratinocytes (HaCaT) and endothelial cells (ECs), comparing its properties to those of GA. HaCaT cell viability and cytotoxicity, ROS release, and inflammation-related signaling (PI3K/Akt/NF-κB and COX-2 gene expression) were assessed. Extracellular matrix (ECM) remodeling was evaluated by gene expression of MMPs. In ECs, a preliminary evaluation of SS's effect was conducted in terms of cell viability and migration. Results demonstrated that SS is well tolerated by keratinocytes whereas GA exhibits cytotoxicity, suggesting that SS's natural composition mitigates GA's adverse effects. SS induced a controlled, brief inflammatory response, via the PI3K/Akt/NF-κB pathway, unlike GA, responsible for stronger and sustained pro-inflammatory events. Additionally, SS, through the upregulation of MMPs, contributes to ECM remodeling. In ECs, SS preserves viability and also enhances migration, thus supporting wound healing. These findings highlight SS's ability to balance pro-inflammatory events, making it a promising candidate for advanced dermatological applications, underscoring SS's potential in modulating key cellular signaling pathways, and supporting its future therapeutic prospects in wound healing.

Keywords: Snail slime; wound healing; matrix metalloproteinases; inflammation; glycolic acid; ROS

1. Introduction

Snail slime (SS), also referred to as snail mucus or snail mucin, is a complex viscoelastic fluid secreted by gastropod mollusks (e.g., *Helix aspersa*) which has garnered significant scientific interest due to its unique composition and potential applications [1]. Although it is primarily composed of water, it contains a diverse array of biologically active compounds [2,3] that contribute to the slime's multifunctional role in gastropod biology, facilitating locomotion, adhesion, and protection against desiccation and predators, and it even plays a role in reproduction [4,5].

SS has a complex composition, including water, the main component of SS, comprising up to 90–99.7% of its total weight [6] while in dry form it includes proteins [7], glycoproteins (mucins, lectins), polysaccharides (glycosaminoglycans), lipids, metal ions, aromatic amino acids [8], allantoin, collagen, elastin, glycolic acid (GA), achacin, antioxidants, hyaluronic acid, and peptides. These components contribute to its well-documented anti-inflammatory, antimicrobial, anti-aging, and antioxidant activities [9].

Each component contributes to the slime's regenerative and protective properties. For instance, hyaluronic acid is a powerful humectant, drawing moisture to the skin and helping maintain hydration [10]. Allantoin is known for its anti-inflammatory, soothing, and healing properties [11], while proteins and peptides in the mucin may support collagen production and skin elasticity [12]. GA, the smallest molecule in the family of α-hydroxy acids, has a keratinolytic effect acting as a gentle exfoliant, promoting cell turnover and improving skin texture [13]. GA is a nontoxic water-soluble white crystalline powder with a pH of 1.7 at 5% concentration. When used below 10%, GA is considered safe and effective in promoting skin regeneration by upregulating the synthesis of glycosaminoglycans and by stimulating collagen production [14]. However, its effects vary based on contact time, concentration, and availability of free acid. Prolonged exposure exceeding 10 min has been associated with skin irritation and inflammation [15].

Thanks to this unique combination of components, SS offers a powerful synergy of moisturizing, exfoliating, and healing effects; its components, in fact, work synergistically to promote skin regeneration, boost collagen production, and provide powerful antioxidant effects [16].

The unique properties of SS have sparked interest in various fields, particularly biomedicine, pharmaceutical industries, and cosmetics, which are currently exploring and exploiting snail mucin for its moisturizing, anti-aging, and skin-repair properties [17,18] In particular, in biomedicine, components like allantoin and hyaluronic acid show promise in wound healing and tissue regeneration [19,20], while antimicrobial peptides present in SS may serve as candidates for novel antimicrobial agents [21].

Human skin is composed of epidermis, dermis, and subcutaneous tissue. The epidermis is the outermost layer, primarily consisting of keratinocytes, while the lower dermis, with loose connective tissue with fibroblasts, contains blood vessels in which endothelial cells (ECs) are also represented. The epidermis is involved in skin regeneration by replacing the outer dead keratinocytes with new cells and the dermis supports wound healing by continuously remodeling the ECM and promoting new blood vessel formation [22]. Moreover, keratinocytes are also involved in the initiation of skin inflammation by specialized keratinocyte-intrinsic mechanisms [23].

Inflammation, in fact, can be considered a defensive mechanism in response to any internal or external damage to the skin or body organ. Inflammation can be beneficial or harmful depending on the cause and duration. Chronic inflammation is harmful while acute or brief inflammation is a part of the immune response and can be helpful in many ways. For example, for injuries or wounds, the release of cytokines and recruitment of macrophages to clean the debris, followed by cell migration and re-epithelialization to the site of injury [24], represents a key passage for the whole wound healing and recovery process.

In our previous study, we tested SS on fibroblasts and the results evidenced SS's capability to promote fibroblast viability and to trigger recovery mechanisms by activating the Erk protein. Moreover, an appreciable anti-inflammatory effect due to the significant reduction in cyclooxygenase-2 expression (COX-2) and the positive modulation of new blood vessel formation demonstrated by increased Angiopoietin 1 gene expression and a higher matrix deposition (evidenced by the augmented amount of released collagen I) could be identified [20]. Thus, based on this knowledge, this study aimed at deepening the effect of SS on the outermost skin layer, i.e., keratinocytes, represented by epidermidis.

Considering that GA is one of the main components of SS, we compared, for the first time, the effect of SS to that obtained by administering GA alone, elucidating the main differences in triggering the pro-inflammatory event that could be essential to promote skin exfoliation and re-epithelization. Starting from the assumption that an appreciable reparative/regenerative effect within the epidermidis takes place only if a functioning blood vessel system in the dermis exists, a further and initial evaluation of SS biocompatibility on ECs was also performed, in order to estimate SS's possible capability to support ECs and their migration, thus indirectly sustaining angiogenic events.

2. Materials and Methods

2.1. SS Extraction

Helix aspersa's slime was extracted using the cruelty-free Cherasco Muller method at Lumacheria Italiana srl (Cherasco, Italy), as already reported elsewhere [2,6,20]. The Cherasco Muller method focuses on animal welfare, preserving animal life and minimizing the risk of endangered species. Additionally, this method allows multiple extractions (up to 3) using the same animals with a gap of minimum two weeks to achieve better SS composition. The method is approved by the relevant authorities in charge, which not only monitor animals' health but also provide them an optimum environment for reproduction. Briefly, adult and healthy snails were recruited, washed in running water at room temperature, and transferred to a Muller extractor. A Muller extractor is an automatic machine with an installed software set for antimicrobial washing and awakening of snails (30 min) using ozonated water followed by three cycles (10 min each) of 10% citric acid-stimulated slime extraction. The collected SS was then filtered through 0.2-micron filter and 0.1% of both sodium benzoate and potassium sorbate preservatives were mixed to filtered SS to enhance shelf life and sustain quality.

2.2. Cell Culture

A human keratinocyte (HaCaT) cell line (catalog # T0020001), purchased from AddexBio (AddexBio, San Diego, CA, USA), and somatic hybrid ECs from the human umbilical vein (EA.hy926), purchased from the American Type Culture Collection (ATCC), CRL-2922 (p10), were both cultured in high-glucose Dulbecco's Modified Eagle Medium (DMEM), supplemented with 1% penicillin/streptomycin and 10% Fetal Bovine Serum (FBS) at 37 °C with 5% CO_2.

2.3. Cell Treatments

HaCaT cells were seeded at 25,000 cells/cm^2 in tissue-culture-treated multiwell plates. EA.hy926 cells were seeded at 8000/well in a 96-well plate and 80,000/well in a 12-well plate to perform wound healing. Both cell types were cultured in DMEM.

After an overnight incubation, both HaCaT and EA.hy296 cells were treated with SS dilutions of 1:40 (0.508 mg/mL), 1:60 (0.31 mg/mL), and 1:80 (0.247 mg/mL) and GA dilutions of 4.1 mM (0.31 mg/mL), 2.73 mM (0.21 mg/mL), and 2.05 mM (0.15 mg/mL) in DMEM while the control was represented by cells only receiving potassium sorbate and sodium benzoate preservatives at 0.1% and sodium citrate at 10% in DMEM, to normalize the effect of stabilizers. These concentrations were chosen according to the results obtained in the previous paper [20].

2.4. MTT Assay

Cell viability in HaCaT cells was measured after 24 and 48 h of treatment in the presence of SS (1:40, 1:60, 1:80) and GA (4.1, 2.73, and 2.05 mM), while in EA.hy926 cells, it was measured after 24, 48, and 72 h of treatment with SS (1:40, 1:60, 1:80) and GA (4.1 mM), by an MTT (3-(4,5-dimethylthiazol-2-yl)-2,5-diphenyltetrazolium bromide) assay (Merck Life Science, Milan, Italy). After the prescribed time, the culture medium was removed and refed with 10% MTT (0.5 mg/mL) in DMEM (100 µL/well) and incubated for 4 h at 37 °C with 5% CO_2 in the dark. After that, the same amount of DMSO (100 µL/well) was added to dissolve the formazan crystals, and again, the mixture was incubated for 20 min at 37 °C. The colored solution obtained by the dissolution of formazan crystals was an indication of the presence of viable cells. The plate was read at 540 nm using a GO microplate spectrophotometer (Thermo Fisher Scientific, Waltham, MA, USA) and the percentage of metabolically active cells was obtained by comparing the sample with the control, whose viability was established at 100%.

2.5. Lactate Dehydrogenase (LDH) Cytotoxicity Assay

Cytotoxicity was measured after 14 and 24 h of SS (1:40, 1:60, 1:80) and GA (4.1 mM, amount present in most concentrated dilution of SS (1:40)) treatment of HaCaT cells by using the CytoTox 96 non-radioactive cytotoxicity assay (Promega, Madison, WI, USA). It worked by measuring the LDH leakage from cells after treatment with SS and GA. The experiment was performed following the manufacturer's instructions and optical density (OD) was measured at 490 nm with a background absorbance of 690 nm. Measured LDH leakage was normalized with MTT OD values, as previously reported [25].

2.6. Flow Cytometry Analysis of ROS Release

HaCaT cells were seeded in 6-well plates at a density of 180,000 cells per well and treated with SS (1:40, 1:60, 1:80) and GA (4.1, 2.73, and 2.05 mM) for 6 and 18 h. Following treatment, cells were incubated with 5 µM of 5-(and-6)-chloromethyl-2′, 7′-dichlorodihydrofluorescein diacetate and acetyl ester (CM-H_2DCFDA; Cat. C6827, Molecular Probes, Invitrogen, Life Sciences Division, Milan, Italy) for 1 h at 37 °C, as previously described [26]. CM-H_2DCFDA is a cell-permeable fluorescent probe that, upon oxidation by intracellular ROS, emits green fluorescence. Fluorescence intensity was measured using a CytoFLEX flow cytometer (Beckman Coulter, Miami, FL, USA) equipped with an FL1 detector operating in logarithmic mode. The median fluorescence intensity (MFI) was calculated using CytExpert software (v2.3) to quantify ROS production. Dead cells were excluded from analysis by staining with Propidium Iodide (PI; 5 µg/mL, Cat. P4864, Sigma-Aldrich, St. Louis, MO, USA). A minimum of 10,000 events per sample was acquired for analysis.

2.7. Western Blot Analysis

HaCaT protein expression was measured after 14 and 24 or 48 h of treatment with SS (1:40, 1:60, 1:80) and GA (4.1 mM). Detached cells were centrifuged at 1200 rpm for 10 min at 4 °C to obtain the cell pellet. RIPA buffer with freshly added protease inhibitors (sodium orthovanadate at 1 mM, leupeptin at 50 µg/mL, aprotinin at 10 µg/mL and PMSF (Phenylmethyl sulfonyl fluoride) at 100 µg/mL, Merck Life Science, Milan, Italy), was used to suspend the pellet. The suspended pellet was then centrifuged at 15,000× *g* for 15 min at 4 °C and supernatant was collected for protein quantification using the Bicinchoninic acid assay (QuantiPro™ BCA Assay kit for 0.5–30 µg/mL protein, Merck Life Science, Milan, Italy) following the manufacturer's instructions. A total of 25 µg of protein was separated on an 8% and 12% (sodium dodecyl sulfate (SDS))-polyacrylamide gel by electrophoresis and transferred to a nitrocellulose membrane. The membranes were saturated with 5% non-fat milk or 3% Bovine Serum Albumin (BSA). The membranes were incubated overnight at 4 °C under gentle shaking with specific primary antibodies, including anti-PI3K (rabbit, 1:1000), anti-p-Akt (rabbit, 1:1000), anti-Akt (rabbit, 1:1000), (Cell Signaling Technology, Danvers, MA, USA) anti-Nf-kβ (rabbit, 1:800) (Santa Cruz

Biotechnology, Inc., Dallas, TX, USA), anti-COX2 (mouse, 1:1000) (Cayman Chemical, Ann Arbor, MI, USA), and anti-Tubulin (mouse, 1:1000) (Merck Life Science, Milan, Italy). On the next day, membranes were washed thrice with PBS tween−20 and membranes were incubated with specific IgG horseradish peroxidase (HRP)-conjugated secondary antibodies (Calbiochem, Darmstadt, Germany). Immunoreactive bands were detected using LiteAblot Extend Chemiluminescent Substrate (EuroClone S.p.a., Milan, Italy). The density of each band was identified using the ChemiDoc™ XRS system, and the QuantityOne 1-D analysis software, v 4.6.6 (BIORAD, Richmond, CA, USA), was applied to conduct densitometric analysis. Densitometric values of each band were normalized with OD values of internal tubulin.

2.8. RNA Extraction

HaCaT cells were collected from 6-well plates after 6 and 24 h of treatment and centrifuged at 1200 rpm to obtain cell pellets. The cells were lysed by 1% 2-mercaptoethanol lysis buffer followed by the addition of same volume of 70% ethanol (v/v) prepared in RNase-free water. Samples were transferred to spin cartridges, provided by the PureLink® RNA Mini Kit (Life Technologies, Carlsbad, CA, USA); to avoid DNA contamination, before RNA extraction, samples were incubated for 15 min with 80 µL of DNase mixture (On-column PureLink® Dnase Treatment, Life Technologies, Carlsbad, CA, USA). RNA was then eluted in 30 µL of RNase free-water and concentrations were measured (ng/µL) using the Qubit® RNA BR Assay kit using a Qubit® 4.0 Fluorometer reader (ThermoFisher Scientific Waltham, MA, USA).

2.9. Reverse Transcription and Real-Time Polymerase Chain Reaction (RT-PCR)

A high-capacity cDNA reverse transcription kit (Life Technologies, Carlsbad, CA, USA) was used for reverse transcription. For each sample, 360 ng of RNA was reverse-transcribed by mixing them in a 2X-RT master mix and incubating them in a thermal cycler (25 °C for 10 min, 37 °C for 120 min, and 85 °C for 5 min) to obtain cDNA. For assessing gene expression by quantitative PCR, the PowerUp™ SYBR™ Green 2X Master Mix (Thermo Fisher Scientific, Waltham, MA, USA) was used. The reaction volume (10 ng of cDNA, 1 µM of each forward and reverse primer, 10 µL of SYBR Green 2X Master Mix, and RNase free water to make up final volume of 20 µL) was transferred to a MicroAmp® Optical 96-well plate (Life Technologies, Carlsbad, CA, USA) for amplification. The sequence of primers is listed in Table 1.

Table 1. Sequence of primers used for qPCR gene expression.

Gene	Sequence (5′ to 3′)	Source
GAPDH-Fr	GGGTGTGAACCATGAGAAGTA	Primer blast
GAPDH-Rv	ACTGTGGTCATGAGTCCTTC	Primer blast
COX2-Fr	CCCTTCTGCCTGACACCTTT	Primer blast
COX2-Rv	TTCTGTACTGCGGGTGGAAC	Primer blast
MMP-2-Fr	GCTACGATGGAGGCGCTAAT	Primer blast
MMP-2-Rv	GGGCAGCCATAGAAGGTGTT	Primer blast
MMP-9-Fr	CGACGTCTTCCAGTACCGAG	Primer blast
MMP-9-Rv	GTTGGTCCCAGTGGGGATTT	Primer blast

The reaction plate was then loaded in QuantStudio 3.0 (Thermo Fisher Scientific, Waltham, MA, USA) and the amplification method was set according to the instructions of the supplier: 50 °C and 95 °C each for 2 min, followed by the 40 cycles of amplification at 95 °C for 15 s, and 60 °C for 60 s. QuantStudio™ Design and Analysis Software v1.5.1 (Thermo Fisher Scientific, Waltham, MA, USA) was employed for gene expression data analysis. Melt curve analysis confirmed the authenticity of the PCR products. Values of gene expression were normalized using glyceraldehyde 3-phosphate dehydrogenase (GAPDH) levels. For each time point, the fold changes in target genes were calculated relative to GAPDH expression. Relative mRNA abundance was quantified using the comparative $2^{-\Delta\Delta Ct}$ method.

2.10. IL-6 Secretion ELISA Assay

After 48 h of treatment with SS or GA, the amount of IL-6 released in the supernatants for each condition was detected by using an IL-6 ELISA kit (Enzo Life Sciences, Farmingdale, NY, USA) following the manufacturer's instructions. The absorbance was spectrophotometrically measured at 450 nm (Varioskan™ LUX Multimode Microplate Reader, Thermo Scientific, Waltham, MA, USA). The concentration of IL-6, expressed as pg/mL, was calculated thanks to an interpolation on the standard curve and then normalized on the MTT data.

2.11. Wound Healing Assay

The wound healing assay was performed on EA.hy926 endothelial cells and HaCaT keratinocytes by the normal scratch method. A scratch was made in the center of each well using a 200 μL micropipette tip. Dead and scratched cells were washed with normal media (DMEM) and adherent cells were treated with SS (concentrations: 1:40, 1:60, 1:80) and GA (concentration: 4.1 mM) in DMEM while the control was obtained without SS. To specifically assess cell migration independently of proliferation, the scratch assay was performed in serum-free medium (FBS-free). Photos were taken using an inverted light microscope (Leica DMi1, Wetzlar, Germany) equipped with a camera, using Leica LAS EZ software (LAS EZ 3.4 DVD 272), at different time intervals (0 h and 18 h of treatment for endothelial cells and 0 h, 24 h, and 48 h for keratinocytes). The scratch distance and area were measured using Fiji imageJ 1.54f software. The percentage of cut reduction after treatment was then normalized to the cut at T0.

2.12. Statistical Analysis

Statistical analysis was performed using GraphPad Prism v.8.0 software (San Diego, CA, USA) using one-way and two-way ANOVA (Analysis of variance) followed by Dunnett's post hoc test. The data were presented as the mean ± SD (standard deviation), and $p < 0.05$ was considered statistically significant.

3. Results

3.1. Assessment of HaCat Viability After SS and GA Treatment

HaCaT viability was assessed using the MTT assay after 24 and 48 h of treatment with different concentrations of SS (1:40, 1:60, and 1:80) and of GA (4.1, 2.73, and 2.05 mM, the same concentrations as those present in SS) and compared to that of the untreated control. At both time points, SS-treated cells showed no significant differences in viability compared to the control. In contrast, GA-treated cells exhibited no significant change after 24 h; however, after 48 h, treatment with GA at 4.1 and 2.73 mM significantly reduced cell viability compared to that of the control, as illustrated in Figure 1B.

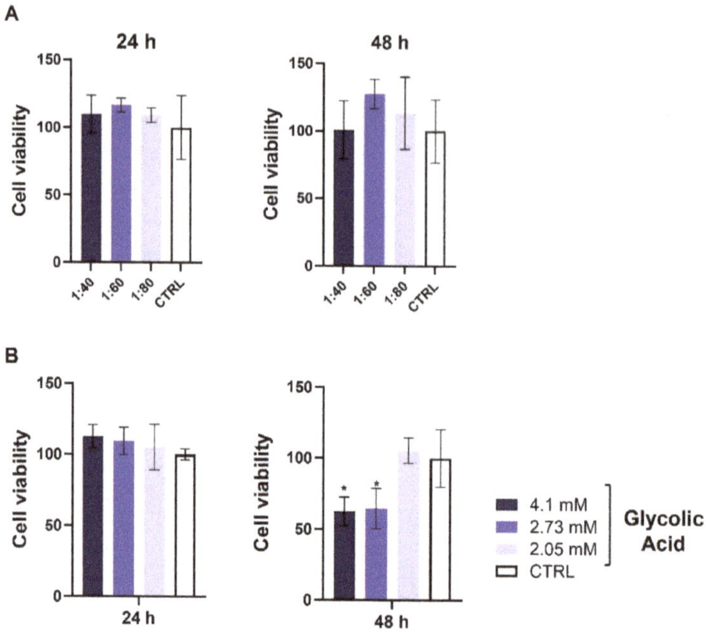

Figure 1. Cell viability test (MTT) was performed on HaCaT cells exposed to SS (1:40, 1:60, and 1:80 dilutions) (**A**) and GA (4.1, 2.73, and 2.05 mM) (**B**) for 24 and 48 h. Data are expressed as percentage of viable cells relative to control. Bars represent mean ± SD of three independent experiments. * $p < 0.05$ vs. CTRL.

3.2. Evaluation of GA's and SS's Effects on HaCaT Cytotoxicity

At both experimental time points (14 and 24 h), SS significantly reduced LDH release at all tested dilutions (1:40, 1:60, and 1:80) compared to both GA and the control (Figure 2). In contrast, GA provoked a significant increase in LDH release after 14 h compared to the control.

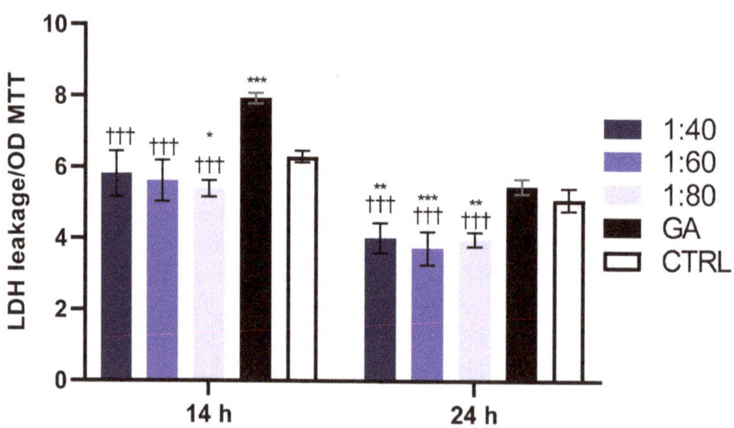

Figure 2. LDH release in HaCaT cells exposed to SS at 1:40, 1:60, and 1:80 dilutions and to GA (4.1 mM) for 14 and 24 h. LDH leakage values were normalized on MTT OD values. Data are presented as mean ± SD of three independent experiments; 14 h: * $p < 0.05$ vs. CTRL; *** $p < 0.01$ vs. CTRL; ††† $p < 0.001$ vs. GA; 24 h: ** $p < 0.01$ vs. CTRL; *** $p < 0.001$ vs. CTRL; ††† $p < 0.001$ vs. GA.

3.3. Measurement of ROS Level After SS and GA Treatment

To evaluate the effect of SS and GA treatments on ROS production, a flow cytometry analysis was performed using H2DCFDA-labeled HaCaT cells. After 6 h of treatment, 1:60 and 1:80 SS dilutions, as well as GA, significantly increased ROS levels compared to the control, while 1:40 dilution led to a reduction in ROS production. Interestingly, all SS-treated cells exhibited lower ROS levels compared to GA-treated cells (Figure 3A). After 18 h, a plateau in ROS levels was observed, with no significant changes detected in ROS levels between the SS- or GA-treated groups and the control, nor among cells treated with different SS dilutions compared to GA-treated cells (Figure 3B).

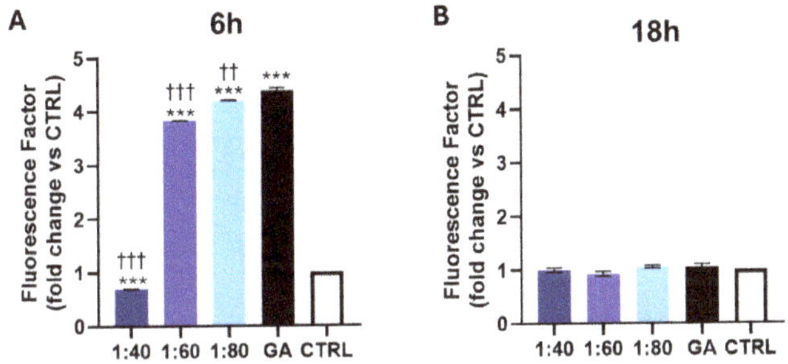

Figure 3. ROS levels measured by flow cytometry after 6 (**A**) and 18 h (**B**) of treatment with SS at 1:40, 1:60, and 1:80 dilutions and with GA (4.1 mM). Histograms represent median values ± SD of three independent experiments in which the Y axis reports mean fluorescence intensity (MFI) generated by the oxidation of H2DCFDA (generation of intracellular ROS) and is reported as fold change vs. CTRL. *** $p < 0.001$ vs. CTRL; ††† $p < 0.001$ vs. GA; †† $p = 0.01$ vs. GA.

3.4. Effect of SS and GA Treatment on PI3K/Akt/NF-κB Protein Levels

The PI3K/Akt/NF-κB pathway is a key cascade activated in response to ROS generation during mild inflammatory states. Therefore, an evaluation of their protein expression levels was performed after 14 and 24 h of treatment with SS and GA. After 14 h, PI3K expression level significantly increases in all SS-treated groups (1:40, 1:60, 1:80) and in a GA-treated group (4.1 mM), compared to the control. However, a significant reduction was observed in 1:40 and 1:60 SS-treated samples compared to that in GA-treated cells. After 24 h, PI3K expression levels were significantly decreased in all SS- and GA-treated samples compared to those in the control. Notably, the 1:80 SS-treated sample showed a significant reduction compared to GA-treated cells, as shown in Figure 4A. Regarding the p-Akt/Akt expression ratio, 14 h was sufficient to significantly increase the ratio in all SS- and GA-treated samples compared to that in the control. Specifically, 1:40 SS treatment resulted in a significant reduction in the p-Akt/Akt ratio compared to GA, while 1:80 SS treatment showed a significant increase compared to GA. After 24 h, both the 1:40 and 1:80 SS-treated samples disclosed a significant reduction compared to both the control and GA-treated cells (Figure 4B). For NF-κB, protein expression significantly increased after 14 h of treatment with 1:60 and 1:80 SS and GA compared to that in the control, while a significant reduction in 1:40 and 1:60 SS-treated samples was detected with respect to GA-treated cells. After 24 h, a significant reduction in NF-κB expression was recorded in all SS- and GA-treated samples, with respect to the control, with the 1:40 SS-treated group showing a further significant decrease relative to the GA-treated group (Figure 4C).

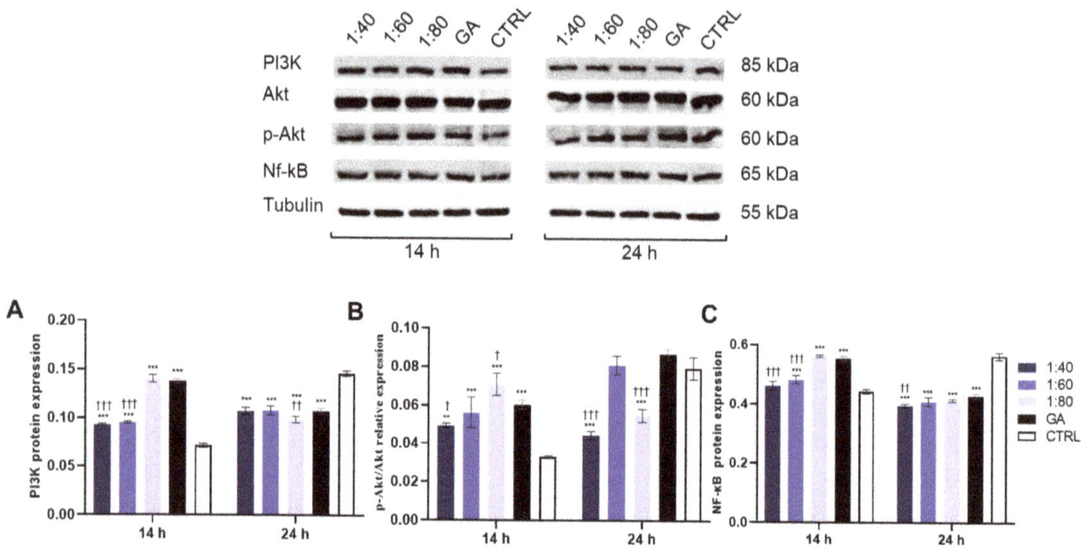

Figure 4. Western blotting analysis of PI3K (**A**), Phospho-Akt (p-Akt) and Akt (**B**), and NF-kB (**C**) expression levels in HaCaT cells treated with SS at 1:40, 1:60, and 1:80 dilutions and with GA (4.1 mM) for 14 and 24 h. The graph represents a densitometric analysis of protein expression levels normalized on tubulin expression levels. Data are presented as the mean ± SD of three independent experiments. A: 14 h: *** $p < 0.001$ vs. CTRL; ††† $p < 0.001$ vs. GA; 24 h: *** $p < 0.001$ vs. CTRL; †† $p < 0.01$ vs. GA. B: 14 h: ** $p < 0.01$ vs. CTRL; *** $p < 0.001$ vs. CTRL; † $p < 0.05$ vs. GA; 24 h: *** $p < 0.001$ vs. CTRL; ††† $p < 0.001$ vs. GA. C: 14 h: *** $p < 0.001$ vs. CTRL; ††† $p < 0.001$ vs. GA; 24 h: *** $p < 0.001$ vs. CTRL; †† $p < 0.01$ vs. GA, For the original data of Western blot in Figure 4, please refer to Supplementary Materials Figures S1–S5 and Tables S1–S3.

3.5. Assessment of Inflammatory Response

In order to evaluate the inflammatory response of HaCaT cells after treatment with SS and GA, COX2 gene and protein expression was assessed. COX2 gene expression was analyzed after 6 and 24 h of treatment. After 6 h, all SS-treated samples showed a statistically significant decrease in COX2 gene expression compared to the control; moreover, 1:40 and 1:80 SS-treated samples showed a significant reduction in COX2 gene expression, also compared to GA-treated cells. After 24 h, GA-treated sample revealed a significant increase in COX2 gene expression with respect to the control while all the SS-treated samples (1:40, 1:60, 1:80) disclosed a significant reduction compared to GA-treated cells (Figure 5A). COX2 protein expression was evaluated after 48 h of treatment. GA induced a statistically significant increase in COX2 protein expression compared to the control, while no statistically significant changes were recorded comparing all SS dilutions with the control. In addition, 1:60 and 1:80 SS showed a significant reduction in protein expression with respect to GA (Figure 5B).

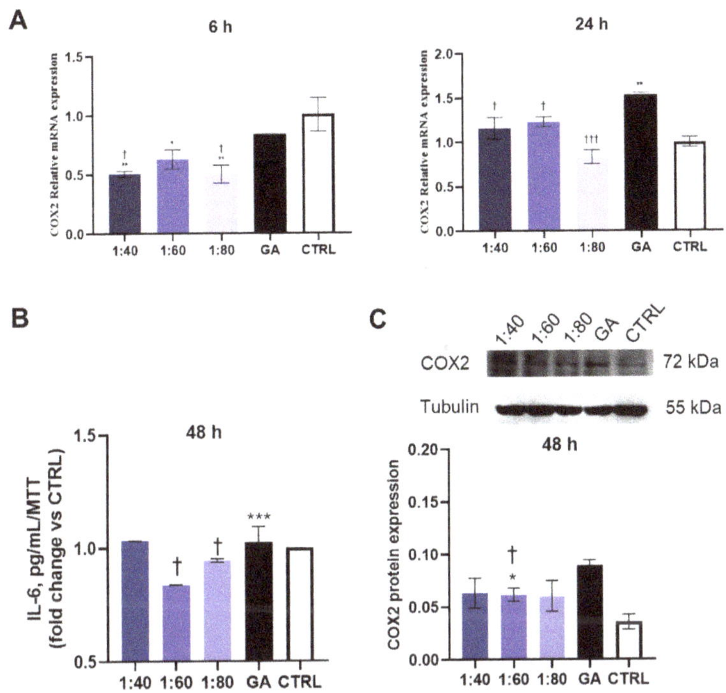

Figure 5. (**A**) COX2 gene expression in HaCaT cells treated with SS at 1:40, 1:60, and 1:80 dilutions and with GA (4.1 mM) for 6 and 24 h, analyzed by RT-PCR. COX2 gene expression was normalized with respect to GAPDH. Data are presented as mean ± SD; 6 h: ** $p < 0.01$ vs. CTRL; * $p < 0.05$ vs. CTRL; † $p < 0.05$ vs. GA; 24 h: ** $p < 0.01$ vs. CTRL; † $p < 0.05$ vs. GA; ††† $p < 0.001$ vs. GA. (**B**) COX2 protein expression levels in HaCaT cells treated with SS at 1:40, 1:60, and 1:80 dilutions and with GA (4.1 mM) for 48 h. The graph represents a densitometric analysis of protein expression levels normalized on tubulin. Data are presented as mean ± SD of three independent experiments. *** $p < 0.001$ vs. CTRL; † $p < 0.05$ vs. GA. (**C**) ELISA assay for IL-6 secretion after 48 h of SS and GA treatment. IL-6 levels are reported as pg/mL normalized on MTT and presented as fold change vs. CTRL. * $p < 0.05$ vs. CTRL; † $p < 0.05$ vs. GA. The original data of Western blot in Figure 5C, please refer to Supplementary Materials Figure S6 and Table S4.

In order to consolidate these findings, a measurement of the release of pro-inflammatory cytokine IL-6 within the supernatant, after 48 h of treatment with SS and GA, was carried out by means of ELISA. The 1:60 condition produced a reduction statically significant with respect to both the control and GA (Figure 5C).

3.6. Analysis of MMP Gene Expression in Response to SS and GA Treatment

In order to investigate the effect of SS and GA on ECM remodeling, an evaluation of MMPs was conducted in HaCaT cells after 6 and 24 h of treatment. In particular, MMP-2 and MMP-9 gene expression was measured. After 6 h, 1:40 and 1:80 SS-treated samples showed a statistically significant increase in MMP-2 gene expression, with respect to the control and GA-treated cells. After 24 h, only the 1:80 SS-treated sample maintained a significantly higher level of MMP-2 compared to the GA-treated cells and control (Figure 6A).

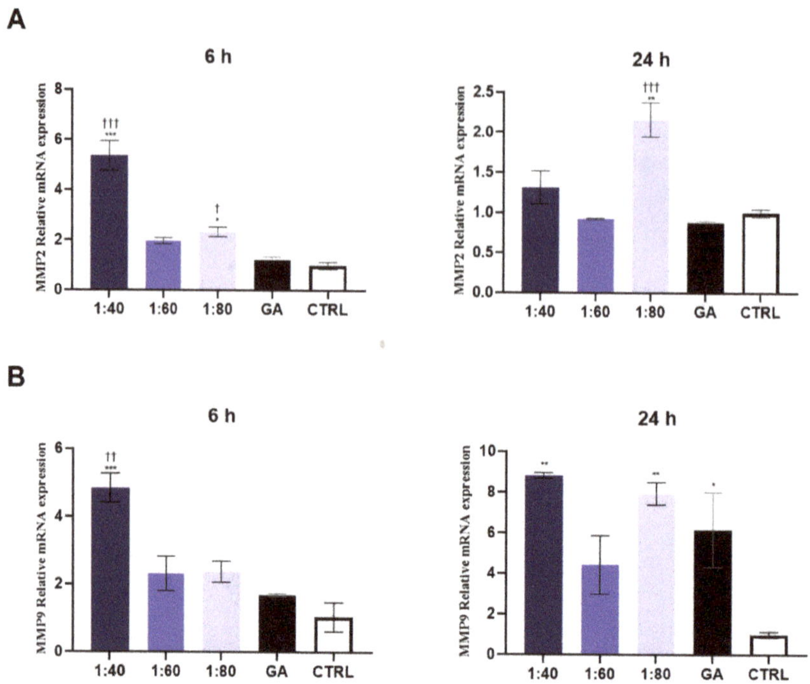

Figure 6. MMP-2 (**A**) and MMP-9 (**B**) gene expression in HaCaT cells treated with SS at 1:40, 1:60, and 1:80 dilutions and with GA (4.1 mM) for 6 and 24 h, analyzed by RT-PCR. Data are normalized on GAPDH and are expressed as mean ± SD of three independent experiments. (**A**) 6 h: *** $p < 0.001$ vs. CTRL; * $p < 0.05$ vs. CTRL; ††† $p < 0.001$ vs. GA; † $p < 0.05$ vs. GA; 24 h: ** $p < 0.01$ vs. CTRL; ††† $p < 0.001$ vs. GA. (**B**) 6 h: *** $p < 0.001$ vs. CTRL; †† $p < 0.01$ vs. GA; 24 h: ** $p < 0.01$ vs. CTRL; * $p = 0.05$ vs. CTRL.

After 6 h, the 1:40 SS-treated sample disclosed an increased MMP-9 gene expression compared to the control and GA-treated cells, while after 24 h, the 1:40 and 1:80 SS- and GA-treated samples revealed a significant increase compared to the control (Figure 6B).

3.7. Wound Healing Assessment in HaCaT Cells After SS and GA Treatment

To investigate the effects of SS and GA on keratinocyte migration—a key step in epidermal regeneration—a wound healing assay was performed using HaCaT cells. A scratch was introduced in each well to simulate a wound, and baseline images were captured at 0 h. Cells were then treated with SS (1:40, 1:60, 1:80) or GA (4.1 mM), and wound closure was assessed after 24 and 48h h by comparing the residual wound area to the baseline area. Results demonstrated that SS promoted HaCaT cell migration in a dilution-dependent manner. After 24 h and 48 h of treatment, both the 1:40 and 1:80 SS dilutions showed a significantly higher percentage of wound closure compared to the control. Although GA treatment also led to a significant increase in wound closure after 48 h, its effect remained lower than that observed with the 1:40 SS dilution at both time points. (Figure 7). These findings further support the regenerative potential of SS over isolated GA in promoting re-epithelialization.

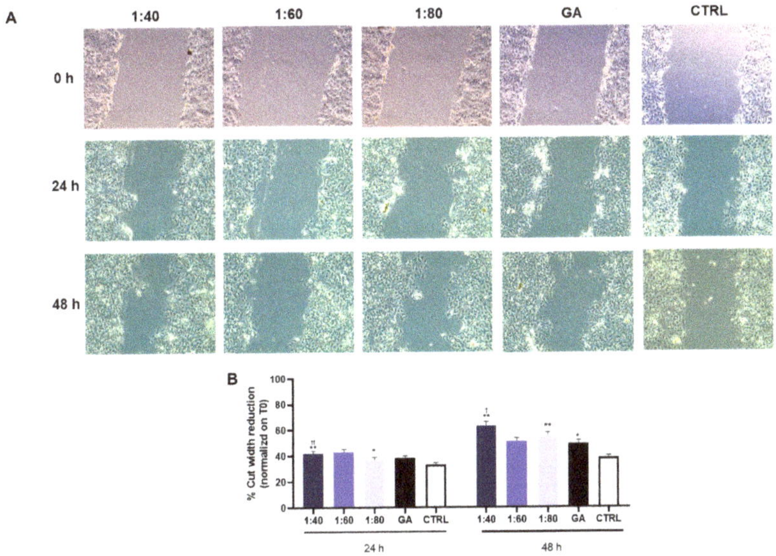

Figure 7. Wound healing assay in HaCaT cells after treatment with SS at different concentrations (1:40, 1:60, 1:80) and GA (4.1 mM). (**A**) Representative images are reported, acquired at 10X magnification. (**B**) Bar graph displays the percentage of cut width reduction after 24 h and 48 h of treatment, normalized on the cut measurement at T0. * $p < 0.05$ vs. CTRL; ** $p < 0.01$ vs. CTRL; † $p < 0.05$ vs. GA; †† $p < 0.01$ vs. GA.

3.8. Evaluation of Endothelial Cell Viability

The percentage of metabolically active EA.hy926 cells was assessed using the MTT assay after 24, 48, and 72 h of treatment with SS and GA. After 24 h, cells exposed to the 1:40 SS dilution recorded a significant viability reduction compared to GA-treated cells, while the 1:60 dilution showed a significant increase in cell viability compared to GA-treated cells and the control. At 48 h, 1:40 and 1:60 treated-cells showed a significant increase in cell viability compared to GA-treated cells and the control. Notably, after 72 h, all the SS dilutions, 1:40, 1:60, and 1:80, recorded a significant improvement in cell viability with respect to GA-treated cells and the control (Figure 8).

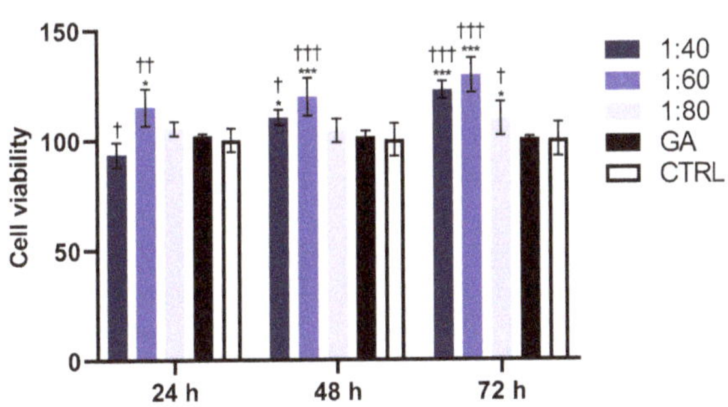

Figure 8. MTT assay performed on endothelial cell lineage (EA.hy926) after treatment with 1:40, 1:60, and 1:80 SS dilutions and with GA (4.1 mM) for 24, 48, and 72 h. Percentage of viable cells was normalized with control treated with DMEM and adjusted at 100%. Data are shown as mean ± SD of three independent experiments. For 24 h: * $p < 0.01$ vs. CTRL; † $p < 0.05$ vs. GA; †† $p < 0.01$ vs. GA; for 48 h: * $p < 0.05$ vs. CTRL; *** $p < 0.001$ vs. CTRL; † $p < 0.05$ vs. GA; ††† $p < 0.001$ vs. GA; for 72 h: * $p < 0.05$ vs. CTRL; *** $p < 0.001$ vs. CTRL; † $p < 0.05$ vs. GA; ††† $p < 0.001$ vs. GA.

3.9. Wound Healing Assessment in Endothelial Cells After SS and GA Treatments

The wound healing assay was performed on endothelial EA.hy926 cells in order to investigate the migratory capability of the cells after SS and GA exposure. A scratch was made in each well of cultured cells and photos of the cut were taken before starting the treatment (0 h). After 18 h of treatment with SS and GA, photos were retaken to analyze cell migration by measuring the dimensions of the cut. Data was analyzed to estimate the percentage of wound closure after 18 h; results showed that all the SS dilutions (1:40, 1:60, 1:80) significantly improved the cell migration and filled the gap compared to GA, while a statistically significant reduction in wound closure is seen in GA-treated cells compared to the control (Figure 9).

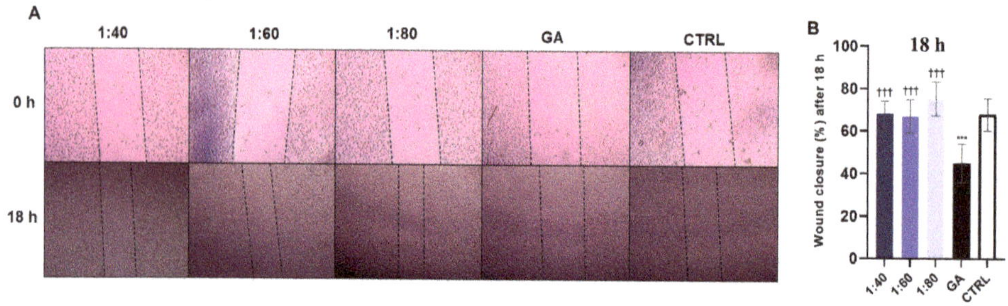

Figure 9. Wound healing assay in EA.hy926 cells after treatment with SS at different concentrations (1:40, 1:60, 1:80) and GA (4.1 mM). (**A**) Representative images are reported, acquired at 5× magnification. (**B**) Bar graph displays percentage of cut width reduction after 18 h of treatment, normalized on cut measurement at T0. *** $p < 0.001$ vs. CTRL; ††† $p < 0.001$ vs. GA.

4. Discussion

SS has extensively been used for treating inflammation, burns, and other skin disorders since ancient times. Its traditional therapeutic applications have triggered researchers to make contributions to advancing and exploring the beneficial effects of this unique secretion. Researchers have been working to delve into the use of SS in cosmetics and sunscreen products [27], as well as its antibacterial, anti-inflammatory and wound-healing activities across different dosage forms [28–31]. In contemporary research, particular attention has been given to the development of SS-loaded patches and nanoparticles for targeted drug delivery, aiming to treat a range of diseases and biological disorders [32–34].

Based on this knowledge, the present work aims at investigating three points: (I) an in-depth analysis of SS's role on the outermost skin layer, represented by keratinocytes; (II) a comparative evaluation of GA, one of the main components of SS, when administered alone versus in its natural SS matrix, focusing on modulation of the pro-inflammatory response and ECM remodeling; (III) a preliminary evaluation of SS's effect on EC viability and migration as a foundation for future in-depth investigation on deeper skin layers.

Firstly, an evaluation of SS's tolerance and cytotoxicity on keratinocytes, the primary skin cell population exposed to topical applications, was required. This evaluation was performed by administering SS and GA alone. The latter is widely recognized as an abundant component of SS, mainly responsible for peeling and gentle exfoliating effects, promoting cell turnover, when applied at low concentrations and for a short period of time [35], by upregulating the synthesis of glycosaminoglycans and stimulating collagen production. The viability analysis revealed that SS appeared appreciably tolerated by keratinocytes at dilutions ranging from 1:40 to 1:80, while when administering GA alone at the same concentrations as those present within SS, a biological tolerance was admitted only after short exposure. In fact, differently from SS, GA exerted a keratinolytic effect when the exposure time was prolonged up to 48 h. This preliminary result led us to hypothesize that the complex composition of SS is able to mitigate the marked keratinolytic effect of GA, even when GA content is higher (dilution 1:40), probably thanks to many biologically active compounds. This hypothesis is further and strongly supported by cytotoxicity evaluation aiming at measuring the prompt release of LDH enzyme when cells face toxic stimuli. Our results, indeed, underline completely different behaviors in keratinocytes in the presence of SS and GA: in fact, it could be admitted that the beneficial composition of SS allows this product to be positively tolerated by keratinocytes, considerably reducing LDH release. Diametrically opposite is the biological response in the presence of GA: a considerable peak in enzyme release leads us to assume that, when administered alone, GA could trigger cytotoxic responses, which are totally reversed when GA is included in SS composition. This hypothesis aligns with earlier findings by Van Scott and colleagues who reported that, when applied for more than 10 min, GA possesses an intrinsic cytotoxicity. This is attributed to a possible deeper penetration into the skin and a conversion into free acid, thus showing disadvantages like apoptosis, pruritus, swelling, and erythema [36]. These results highlight a key advantage of SS: it retains the beneficial exfoliating properties of GA while minimizing cytotoxic effects, thereby avoiding excessive stress on epidermidis cells.

Keratinocytes undergo continuous regeneration and exfoliation. This process is essential for maintaining skin hydration and a youthful appearance, as it involves the shedding of dead cells and their replacement with newly formed cells [37]. If the balance between the regeneration of new cells and the elimination of dead cells is disturbed, it leads to inflammatory reactions causing, in some cases, skin disorders such as acne and melasma [38]. Thus, a physical peeling of the dead skin cells, alongside anti-inflammatory and antibacterial treatment, is necessary [39]. SS is a unique blend of multifunctional macromolecules that not only gently exfoliates dead epidermidis layers but also possesses anti-inflammatory properties, as widely demonstrated [30,34]. However, inflammation is also considered a physiological process with beneficial effects when it occurs for brief periods, aiming at triggering positive events such as activation of the immune system to fight off foreign invaders, remove debris, and heal injuries [40]. For instance, formulations containing GA are known to induce an initial temporary inflammatory process able to stimulate exfoliation and regeneration of the outermost skin layer [41]. Our results support the hypothesis that, when GA is applied alone, a pro-inflammatory and marked event occurs, and it seems to require a prolonged time to be resolved respect to the administration of GA in the SS mixture. This hypothesis emerged by analyzing the involvement of the PI3K/Akt/NF-κB pathway. Indeed, emerging evidence suggests that PI3K/Akt pathway is involved in skin development and homeostasis [42] through the activation of several intracellular molecules, such as NF-κB nuclear transcription factor. The latter, in fact, by activating further intracellular mediators, is a crucial regulator of numerous cellular events, such as proliferation, differentiation, migration, angiogenesis, and metabolism [43]. In particular, NF-κB is often associated with inflammatory responses: its translocation into the nucleus after specific stimuli promotes the expression of pro-inflammatory genes, such as interleukins, chemokines, TNF-α and others [44]. When a specific pro-inflammatory stimulus occurs, activated PI3K can phosphorylate Akt which, in turn, activates NF-κB inducing its translocation into the nucleus [45]. Our results suggest that when HaCaT cells were exposed to GA alone, the pro-inflammatory response could be activated: indeed, after 14 h of treatment, upregulation, disclosed by an evident similar trend for PI3K, Akt, and NF-κB proteins, was detected, thus suggesting a possible cascade activation. It can be evidenced that SS was also able to upregulate the aforementioned pathway, even if to a minor extent, thus inducing a brief inflammatory event. This assumption is further supported by the evidence that the experimental conditions in which PI3K, Akt, and the NF-κB pathway were mainly downregulated were those in which the highest quantities of SS were administered, thus reinforcing and supporting the hypothesis that the rich and complex composition of SS was able to attenuate the marked response obtained with GA alone. Liu et al. [43] described that a feedback mechanism interlinks the upregulation and the downregulation of NF-κB factor. In fact, the initial upregulation of NF-κB not only triggers pro-inflammatory pathways but also directly regulates cell-mediated immunity, survival, proliferation, cell cycle, and anti-apoptotic signals by inhibiting the c-Myc gene [46], inducing, in turn, cell protection mechanisms by inhibiting the necrosis followed by downregulation of NF-κB [44]. Our results are coherent with the findings of Liu and colleagues: an initial upregulation of NF-κB, presumably due to PI3K/Akt pathway recruitment, is then followed by a downregulation of the pathway, suggesting a possible resolution of the pro-inflammatory event and the beginning of a cell protection mechanism involving further molecular cascades. One of the key pro-inflammatory genes, whose expression is regulated by NF-κB, is COX-2 [47]. It has been already reported that COX-2 can regulate chemokine production by reducing the recruitment of pro-inflammatory cells and promoting the local accumulation of regulatory T cells, thus stimulating cell protection mechanism through the activation of cell-mediated innate immunity [48]. In addition, it was reported that skin wound healing requires a transient and tightly regulated induction of COX-2, which is essential for initiating the tissue repair process. However, prolonged or excessive inflammation can disrupt this balance, ultimately impairing healing and compromising skin integrity [49]. To assess the inflammatory effects of SS and GA, we investigated the expression of COX-2, a key pro-inflammatory enzyme, at both the gene and protein levels in HaCaT keratinocytes. Additionally, interleukin-6 (IL-6) release was evaluated as a further marker of inflammatory activation. Our results let us argue that the activated PI3K/Akt/NF-κB pathway, in the presence of GA alone, resulted in increased COX-2 gene expression after 24 h of treatment, confirming pro-inflammatory stimulation, which could be responsible, after a longer exposure, for cell viability reduction, as reported in the MTT after 48 h of treatment. Furthermore, COX-2 gene expression data support our initial hypothesis: the activation of the PI3K/Akt/NF-κB pathway after 14 h in the presence of SS was not reflected in increased COX-2 gene expression after 24 h, differently from what was detected in the presence of GA alone, thus underlining the role of SS in the attenuation of inflammation. Considering the results of PI3K/Akt/NF-κB protein expression and COX-2 gene expression, it can be concluded that both GA and SS were able to trigger a brief inflammation in keratinocytes, presumably recruiting the PI3K/Akt/NF-κB pathway. However, in the presence of GA, the inflammatory response was triggered immediately after the administration of GA, as demonstrated by COX-2 gene expression after 6 h, while SS required prolonged exposure to activate this response and, even when activated, it resulted in positive modulation. COX-2 protein expression measured at 48 h further confirmed that SS markedly modulated the inflammatory response in a time-dependent manner. GA induced a significant increase in COX-2 protein levels, highlighting its prolonged inflammatory potential. Conversely, none of the SS-treated groups showed a significant increase in COX-2 protein expression compared to the control. Notably, the 1:60 and 1:80 SS dilutions reduced COX-2 protein expression compared to GA, suggesting that SS not only prevents excessive inflammatory signaling but may also counteract GA-induced upregulation when administered as part of a complex mixture. To further validate these findings, we assessed IL-6 release after 48 h of treatment. Interestingly, when compared to GA, both the 1:60 and 1:80 dilutions exhibited lower IL-6 levels, supporting the conclusion that SS exerted a dampening effect on the inflammatory milieu. Altogether, these findings suggest that SS offers a controlled and modulated inflammatory response, in contrast to the more aggressive and sustained activation observed with GA. The ability of SS to attenuate key inflammatory mediators, such as COX-2 and IL-6, reinforces its potential as a safer and more biocompatible alternative for topical applications, particularly in formulations intended for sensitive or damaged skin. Moreover, the observed anti-inflammatory profile of SS supports its multifunctional role in wound healing, where a balanced inflammatory response is critical for successful tissue regeneration without scarring or chronic irritation [50].

In healthy skin, keratinocytes coordinate with fibroblasts to regulate skin homeostasis, particularly regulating ECM remodeling. In a paracrine way, keratinocytes stimulate fibroblasts to release ECM remodeling factors; at the same time, they are able to drive ECM reshaping themselves, according to Pfisterer et al. [51] and Russo et al. [52]. In the field of ECM remodeling, an essential role is played by gelatinases which, by degrading the organic components of the ECM, create a physical space for cell migration from the deeper layers of the skin to the surface ones [53], thus confirming the activation of MMPs as an essential step for cell migration [54]. MMP-2 (gelatinase A) and MMP-9 (gelatinase B), involved in collagen proteolysis and ECM removal, essential for wound reepithelization [55], are also able to accelerate cell migration, angiogenesis, cell cycle regulation, and connective tissue remodeling [56]. In fact, MMP-2 and -9 have important roles in accelerating cell migration and are expressed by keratinocytes at the leading edge of wounds to promote re-epithelialization, as discussed by Krishnaswamy et al. [57]. Additionally, gelatinases are the predominant enzymes found in higher amounts compared to other MMPs in chronic wound tissues and have broader substrate preferences than collagenases (MMP-1 and -13), thus degrading a wide gamut of ECM molecules. However, considering the role of other MMPs, such as the 1, 3, and 13 isoforms, in remodeling events, further studies are required to explore SS's effect on the complete panel of MMPs [58,59]. This evidence is strongly supported by our experimental model in which the brief and initial inflammation was followed by subsequent MMP-2 and MMP-9 gene expression upregulation, thus pushing up ECM remodeling; this event appeared markedly more pronounced for SS than for GA alone, admitting that SS administration significantly sped up and sustained the ECM remodeling process.

Although ROS play a crucial role in signaling pathways and physiological processes, particularly in tissue repair, ROS function as a "double-edged sword", beneficial when controlled but harmful when sustained [60]. The evaluation of intracellular ROS levels provides valuable insights into the oxidative status induced by topical agents such as SS and GA. The early ROS elevation in the 1:60 and 1:80 samples suggests that, at these dilutions, SS transiently stimulated oxidative signaling, which could act as a trigger for cellular processes such as proliferation and differentiation. Interestingly, the 1:40 SS dilution led to a reduction in ROS levels, indicating a potential antioxidant or ROS-scavenging effect at higher SS concentrations. These findings imply that components of SS may exert dual effects depending on the concentration, promoting redox signaling at moderate levels and exerting antioxidant activity at higher concentrations. GA treatment, in contrast, caused a marked and statistically significant increase in ROS production after 6 h, more pronounced than that with any SS dilution. This is consistent with GA's known keratinolytic and exfoliative mechanisms, which are often accompanied by oxidative stress. The comparison between SS- and GA-treated cells supports the notion that the natural matrix of SS mitigates the oxidative stress associated with GA alone.

After 18 h of treatment, ROS levels in all groups (SS, GA, and control) converged; this plateau suggests the activation of cellular compensatory mechanisms and restoration of redox homeostasis over time. The absence of sustained ROS accumulation, especially in SS-treated cells, reinforces the idea that SS does not induce persistent oxidative stress, which is critical for avoiding chronic inflammation and cellular damage.

Overall, these findings highlight the redox-modulating properties of SS. Unlike GA, which triggers a stronger oxidative burst, SS induces a more controlled and transient ROS response that may support cellular activation without promoting prolonged oxidative damage. This behavior further emphasizes the potential of SS as a safer, well-tolerated alternative for skin regeneration therapies involving controlled oxidative stimuli.

Keratinocyte migration is a critical step in the re-epithelialization phase of wound healing and skin regeneration [61]. In the present study, a wound healing assay was employed to investigate the impact of SS and GA on keratinocyte migration. Our findings demonstrate that SS significantly enhanced HaCaT cell migration in a dose-dependent manner. Both the 1:40 and 1:80 dilutions of SS notably accelerated wound closure after 24 and 48 h of treatment. Although GA also significantly promoted wound closure at 48 h, its effect was considerably less pronounced than that of SS, particularly when compared to the 1:40 SS dilution, which maintained superior results at both 24 and 48 h. This discrepancy may be attributed to the limited functional scope of GA, which, while effective as a keratolytic agent, lacks the broader spectrum of bioactive molecules present in SS [62] that collectively support cell migration, proliferation, and ECM remodeling. The ability of SS to enhance keratinocyte migration more effectively than GA alone underscores its potential as a multifaceted regenerative agent in dermatological applications.

New blood vessel formation is a multistep and complex process, regulated by a tight balance between pro- and anti-angiogenic factors [63] and strictly connected to MMP activity, which is believed to promote degradation of the preexisting basement membrane and ECM components to control the release of angiogenic and growth factors, as well as to stimulate endogenous angiogenic inhibitors [64]. In addition, skin regeneration and reepithelization are processes requiring the activity of blood vessels, populating the dermis, similarly to the wound healing mechanism [65]. Based on this, an initial evaluation of SS's effect on ECs was carried out in order to estimate SS's capability to support an accurate skin regeneration process; thus, EC response, in terms of viability and migration, was assessed. Results showed that SS promoted EC viability throughout up to 72 h of treatment, compared to GA and the control. Moreover, the migratory capability of ECs, one of the main angiogenic factors in obtaining new blood network formation [66], was promoted when SS was administered. In fact, differently from GA, SS significantly improved cell migration and, interestingly, this effect appeared even more pronounced when SS was more diluted, achieving 75% wound closure compared to GA which achieved only 45%. This finding is in alignment with previous findings which already reported the valuable ability of SS in promoting angiogenesis during skin healing wounds [67].

5. Conclusions

This study underlines that SS, due to its unique blend of multiple beneficial macromolecules, appears well tolerated by skin cells. SS in comparison to GA does not impart cytotoxic burden to skin cells, while the latter does. Similarly, SS induces well-controlled and regulated brief inflammation, helpful for cellular recovery processes, while GA induces inflammation immediately after administration and requires a longer time to become restored, suggesting that the complex composition of SS is able not only to attenuate the strong effect of GA but also to counteract inflammation and to maintain normal skin homeostasis. In addition, SS promotes transient and controlled cellular activation rather than prolonged damage. This balanced response is evidenced by the downregulation of COX-2 and IL-6, alongside a temporary rise in intracellular ROS levels, which may play a role in initiating key regenerative processes such as keratinocyte migration. Moreover, this paper underlines SS's ability to effectively promote ECM remodeling by significantly raising MMP expression. A promising effect is also exerted on ECs populating the deeper layer of the skin, making SS suitable for tailored preparations. Therefore, considering the complexity of the wound healing process and the involvement of the different tissues, our results add one more piece of evidence in understanding the role of SS in skin regeneration, allowing us to admit that SS is a potent agent in regulating cellular signaling pathways, and opens up future avenues for its potential use in wound healing. While our work provides mechanistic insights through in vitro assays, future studies employing animal models will be essential to validate these findings and assess wound closure dynamics, granulation tissue formation, and collagen deposition, thereby strengthening translational relevance. This positions our findings as a foundation for forthcoming translational research.

Supplementary Materials: The following supporting information can be downloaded at https://www.mdpi.com/article/10.3390/biom15091302/s1, Figure S1: Original nitrocellulose membrane; Figure S2: Original blots of tubulin and PI3K protein expression; Figure S3: Original blots of tubulin and p-AKT protein expression; Figure S4: Original blots of tubulin and AKT protein expression; Figure S5: Original blots of tubulin and Nf-kB protein expression Figure S6: Original blots of tubulin and COX-2 protein expression; Table S1: Original densitometric data of tubulin and PI3K Western blot, Table S2: Original densitometric data of tubulin and pAKT/AKT Western blot, Table S3: Original densitometric data of tubulin and Nf-kB Western blot, Table S4: Original densitometric data of tubulin and COX-2 Western blot.

Author Contributions: Conceptualization M.B. and S.Z.; Methodology: M.R., A.R. and S.P.; Software: M.R., A.R. and S.P.; Validation: A.C. and S.Z.; Formal Analysis: A.R.; Investigation: A.R. and M.R.; Resources: A.C. and S.Z.; Data Curation: M.R. and A.R.; Writing—Original Draft Preparation: M.R., A.R. and M.B.; Writing—Review and Editing: M.B., S.Z. and S.P.; Visualization: A.C.; Supervision: S.Z.; Project Administration S.Z.; Funding Acquisition: A.C. and M.R. All authors have read and agreed to the published version of the manuscript.

Funding: This research was financed by Ministerial Decree no. 351 of 9 April 2022, based on the NRRP funded by the European Union—NextGenerationEU—Mission 4, "Education and Research", Component 1, "Enhancement of the offer of educational services: from nurseries to universities"—Investment 3.4, "Advanced teaching and university skills". The work was also funded by the European Union–NextGenerationEU, Mission 4, Component 1, under the Italian Ministry and Research (MUR) National Innovation Ecosystem grant ECS00000041-Vitality-CUP D73C22000840006, Spoke 4—WP4 (Amelia Cataldi).

Institutional Review Board Statement: Not applicable.

Informed Consent Statement: Not applicable.

Data Availability Statement: The raw data supporting the conclusions of this article will be made available by the authors on request.

Acknowledgments: This publication was produced during MR's attendance at the PhD program in Biomolecular and Pharmaceutical Sciences at "G. d'Annunzio" University of Chieti-Pescara, Italy, Cycle XXXVIII, with the support of a scholarship financed by the Ministerial Decree no. 351 of 9 April 2022, based on the National Recovery and Resilience Plan (NRRP) funded by the European Union—NextGenerationEU—Mission 4, "Education and Research", Component 1, "Enhancement of the offer of educational services: from nurseries to universities"—Investment 3.4, "Advanced teaching and university skills". This publication was made possible thanks to the post-doc fellowship attributed to PhD Alessia Ricci (National Innovation Ecosystem grant ECS00000041—Vitality). The results presented in the current paper were obtained thanks to the collaboration between University "G. d'Annunzio" Chieti-Pescara and Lumacheria Italiana S.r.l. (Cherasco, Italy) in accordance with the agreement signed on 4 February 2025.

Conflicts of Interest: The authors declare no conflicts of interest.

References

1. Cilia, G.; Fratini, F. Antimicrobial properties of terrestrial snail and slug mucus. *J. Complement. Integr. Med.* **2018**, *15*, 20170168. [CrossRef]
2. Rashad, M.; Sampò, S.; Cataldi, A.; Zara, S. Biological activities of gastropods secretions: Snail and slug slime. *Nat. Prod. Bioprospecting* **2023**, *13*, 42. [CrossRef]
3. Adikwu, M.; Nnamani, P. Some physiological and toxicological properties of snail mucin extracted from Archachatina marginata. *Bio-Res.* **2006**, *3*, 1–6. [CrossRef]
4. Barajas-Ledesma, E.; Holland, C. Probing the compositional and rheological properties of gastropod locomotive mucus. *Front. Soft Matter* **2023**, *3*, 1201511. [CrossRef]
5. Olaniyan, O.T.; Adetunji, C.O. *Biological, Biochemical, and Biodiversity of Biomolecules from Marine-Based Beneficial Microorganisms: Industrial Perspective*; Adetunji, C.O., Panpatte, D.G., Jhala, Y.K., Eds.; Springer: Singapore, 2021; Volume 27, pp. 57–81.
6. Rashad, M.; Sampò, S.; Cataldi, A.; Zara, S. From Nature to Nurture: The Science and Applications of Snail Slime in Health and Beauty. *J. Cosmet. Dermatol.* **2025**, *24*, e70002. [CrossRef] [PubMed]
7. Zhu, K.; Zhang, Z.; Li, G.; Sun, J.; Gu, T.; Ain, N.U.; Zhang, X.; Li, D. Extraction, structure, pharmacological activities and applications of polysaccharides and proteins isolated from snail mucus. *Int. J. Biol. Macromol.* **2023**, *258*, 128878. [CrossRef]
8. Liegertová, M.; Malý, J. Gastropod Mucus: Interdisciplinary Perspectives on Biological Activities, Applications, and Strategic Priorities. *ACS. Biomater. Sci. Eng.* **2023**, *9*, 5567–5579. [CrossRef]
9. Cabibbo, M.; Scialabba, C.; Drago, S.E.; Craparo, E.F.; Cavallaro, G. From nature to medicine: Snail slime-based functional excipients for oral dosage forms. *Int. J. Pharm.* **2025**, *682*, 125914. [CrossRef]
10. Rattanawiwatpong, P.; Wanitphakdeedecha, R.; Bumrungpert, A.; Maiprasert, M. Anti-aging and brightening effects of a topical treatment containing vitamin C, vitamin E, and raspberry leaf cell culture extract: A split-face, randomized controlled trial. *J. Cosmet. Dermatol.* **2020**, *19*, 671–676. [CrossRef]
11. Dinica, R.M.; Sandu, C.; Botezatu, A.V.D.; Busuioc, A.C.; Balanescu, F.; Mihaila, M.D.I.; Dumitru, C.N.; Furdui, B.; Iancu, A.V. Allantoin from Valuable Romanian Animal and Plant Sources with Promising Anti-Inflammatory Activity as a Nutricosmetic Ingredient. *Sustainability* **2021**, *13*, 10170. [CrossRef]
12. Kim, Y.; Sim, W.-J.; Lee, J.-S.; Lim, T.-G. Snail mucin is a functional food ingredient for skin. *J. Funct. Foods* **2022**, *92*, 105053. [CrossRef]
13. Kim, K.H.; Chung, C.B.; Kim, Y.H.; Kim, K.S.; Han, C.S.; Kim, C.H. Cosmeceutical properties of levan produced by Zy-momonas mobilis. *J. Cosmet. Sci.* **2005**, *56*, 395–406. [CrossRef]
14. Narda, M.; Trullas, C.; Brown, A.; Piquero-Casals, J.; Granger, C.; Fabbrocini, G. Glycolic acid adjusted to pH 4 stimulates collagen production and epidermal renewal without affecting levels of proinflammatory TNF-alpha in human skin explants. *J. Cosmet. Dermatol.* **2020**, *20*, 513–521. [CrossRef]
15. Van Scott, E.J.; Ditre, C.M.; Yu, R.J. Alpha-hydroxyacids in the treatment of signs of photoaging. *Clin. Dermatol.* **1996**, *14*, 217–226. [CrossRef]
16. Tsoutsos, D.; Kakagia, D.; Tamparopoulos, K. The efficacy of *Helix. aspersa* Müller extract in the healing of partial thickness burns: A novel treatment for open burn management protocols. *J. Dermatol. Treat.* **2009**, *20*, 219–222. [CrossRef]
17. Brieva, A.; Philips, N.; Tejedor, R.; Guerrero, A.; Pivel, J.P.; Alonso-Lebrero, J.L.; Gonzalez, S. Molecular basis for the re-generative properties of a secretion of the mollusk *Cryptomphalus aspersa*. *Skin Pharmacol. Physiol.* **2008**, *21*, 15–22. [CrossRef]
18. Dhiman, V.; Pant, D. Human health and snails. *J. Immunoass. Immunochem.* **2020**, *42*, 211–235. [CrossRef] [PubMed]
19. McDermott, M.; Cerullo, A.R.; Parziale, J.; Achrak, E.; Sultana, S.; Ferd, J.; Samad, S.; Deng, W.; Braunschweig, A.B.; Holford, M. Advancing Discovery of Snail Mucins Function and Application. *Front. Bioeng. Biotechnol.* **2021**, *9*, 734023. [CrossRef]
20. Ricci, A.; Gallorini, M.; Feghali, N.; Sampò, S.; Cataldi, A.; Zara, S. Snail Slime Extracted by a Cruelty Free Method Preserves Viability and Controls Inflammation Occurrence: A Focus on Fibroblasts. *Molecules* **2023**, *28*, 1222. [CrossRef] [PubMed]
21. Pitt, S.J.; Graham, M.A.; Dedi, C.G.; Taylor-Harris, P.M.; Gunn, A. Antimicrobial properties of mucus from the brown garden snail *Helix aspersa*. *Br. J. Biomed. Sci.* **2015**, *72*, 174–181. [CrossRef] [PubMed]
22. Yang, R.; Wang, J.; Chen, X.; Shi, Y.; Xie, J. Epidermal Stem Cells in Wound Healing and Regeneration. *Stem Cells Int.* **2020**, *2020*, 1–11. [CrossRef] [PubMed]
23. Pasparakis, M.; Haase, I.; Nestle, F.O. Mechanisms regulating skin immunity and inflammation. *Nat. Rev. Immunol.* **2014**, *14*, 289–301. [CrossRef] [PubMed]
24. Germolec, D.R.; Shipkowski, K.A.; Frawley, R.P.; Evans, E. Markers of Inflammation. *Methods Mol. Biol.* **2018**, *1803*, 57–79. [CrossRef]
25. De Colli, M.; Tortorella, P.; Marconi, G.D.; Agamennone, M.; Campestre, C.; Tauro, M.; Cataldi, A.; Zara, S. In vitro comparison of new bisphosphonic acids and zoledronate effects on human gingival fibroblasts viability, inflammation and matrix turnover. *Clin. Oral Investig.* **2015**, *20*, 2013–2021. [CrossRef]

26. di Giacomo, V.; Balaha, M.; Pinti, M.; Di Marcantonio, M.C.; Cela, I.; Acharya, T.R.; Kaushik, N.K.; Choi, E.H.; Mincione, G.; Sala, G.; et al. Cold atmospheric plasma activated media selectively affects human head and neck cancer cell lines. *Oral. Dis.* **2024**, *31*, 401–416. [CrossRef]
27. Rizzi, V.; Gubitosa, J.; Fini, P.; Nuzzo, S.; Agostiano, A.; Cosma, P. Snail slime-based gold nanoparticles: An interesting potential ingredient in cosmetics as an antioxidant, sunscreen, and tyrosinase inhibitor. *J. Photochem. Photobiol. B Biol.* **2021**, *224*, 112309. [CrossRef]
28. Gubitosa, J.; Rizzi, V.; Fini, P.; Fanelli, F.; Sibillano, T.; Corriero, N.; Cosma, P. Chitosan/snail slime films as multifunctional platforms for potential biomedical and cosmetic applications: Physical and chemical characterization. *J. Mater. Chem. B* **2022**, *11*, 2638–2649. [CrossRef]
29. Agada, D.E.; Sar, T.T.; Ujoh, J.A.; Ameh, L.O. Antibacterial susceptibility of staphylococcus aureus, salmonella typhi, bacillus subtilis and escherichia coli to snail slime. *Afr. Health Sci.* **2023**, *23*, 177–182. [CrossRef]
30. Aouji, M.; Rkhaila, A.; Bouhaddioui, B.; Zirari, M.; Harifi, H.; Taboz, Y.; Lrhorfi, L.A.; Bengueddour, R. Chemical composi-tion, mineral profile, anti-bacterial, and wound healing properties of snail slime of *Helix aspersa* Müller. *Biomedicine* **2023**, *13*, 10–19. [CrossRef]
31. Ellijimi, C.; Ben Hammouda, M.; Othman, H.; Moslah, W.; Jebali, J.; Ben Mabrouk, H.; Morjen, M.; Haoues, M.; Luis, J.; Marrakchi, N.; et al. *Helix aspersa maxima* mucus exhibits antimelanogenic and antitumoral effects against melanoma cells. *Biomed. Pharmacother.* **2018**, *101*, 871–880. [CrossRef] [PubMed]
32. Di Filippo, M.F.; Albertini, B.; Dolci, L.S.; Bonvicini, F.; Bigi, A.; Gentilomi, G.A.; Passerini, N.; Panzavolta, S. Novel drug-loaded film forming patch based on gelatin and snail slime. *Int. J. Pharm.* **2021**, *598*, 120408. [CrossRef]
33. Perpelek, M.; Tamburaci, S.; Aydemir, S.; Tihminlioglu, F.; Baykara, B.; Karakasli, A.; Havitcioglu, H. Bioactive snail mucus-slime extract loaded chitosan scaffolds for hard tissue regeneration: The effect of mucoadhesive and antibacterial extracts on physical characteristics and bioactivity of chitosan matrix. *Biomed. Mater.* **2021**, *16*, 065008. [CrossRef]
34. Gubitosa, J.; Rizzi, V.; Fini, P.; Laurenzana, A.; Fibbi, G.; Veiga-Villauriz, C.; Fanelli, F.; Fracassi, F.; Onzo, A.; Bianco, G.; et al. Biomolecules from snail mucus (*Helix. aspersa*) conjugated gold nanoparticles, exhibiting potential wound healing and anti-inflammatory activity. *Soft Matter* **2020**, *16*, 10876–10888. [CrossRef]
35. Houshmand, E.B. Effect of glycolic acid, phytic acid, soothing complex containing Emulsion on Hyperpigmentation and skin luminosity: A clinical evaluation. *J. Cosmet. Dermatol.* **2021**, *20*, 776–780. [CrossRef] [PubMed]
36. Prabakaran, L.; Yang, J.; Lee, B.; Kim, S.; Kim, J.W. Current trends and requirements in sensors for hydroxy acid-based skincare treatments: A mini-review. *J. Ind. Eng. Chem.* **2024**, *144*, 723–734. [CrossRef]
37. Fuchs, E.; Raghavan, S. Getting under the skin of epidermal morphogenesis. *Nat. Rev. Genet.* **2002**, *3*, 199–209. [CrossRef] [PubMed]
38. Knutsen-Larson, S.; Dawson, A.L.; Dunnick, C.A.; Dellavalle, R.P. Acne Vulgaris: Pathogenesis, Treatment, and Needs Assessment. *Dermatol. Clin.* **2012**, *30*, 99–106. [CrossRef]
39. Conforti, C.; Zalaudek, I.; Vezzoni, R.; Retrosi, C.; Fai, A.; Fadda, S.; Di Michele, E.; Dianzani, C. Chemical peeling for acne and melasma: Current knowledge and innovations. *Ital. J. Dermatol. Venereol.* **2020**, *155*, 280–285. [CrossRef]
40. Chen, L.; Deng, H.; Cui, H.; Fang, J.; Zuo, Z.; Deng, J.; Li, Y.; Wang, X.; Zhao, L. Inflammatory responses and inflamma-tion-associated diseases in organs. *Oncotarget* **2017**, *9*, 7204–7218. [CrossRef]
41. Denda, S.; Denda, M.; Inoue, K.; Hibino, T. Glycolic acid induces keratinocyte proliferation in a skin equivalent model via TRPV1 activation. *J. Dermatol. Sci.* **2010**, *57*, 108–113. [CrossRef]
42. Teng, Y.; Fan, Y.; Ma, J.; Lu, W.; Liu, N.; Chen, Y.; Pan, W.; Tao, X. The PI3K/Akt Pathway: Emerging Roles in Skin Homeostasis and a Group of Non-Malignant Skin Disorders. *Cells* **2021**, *10*, 1219. [CrossRef]
43. Bai, D.; Ueno, L.; Vogt, P.K. Akt-mediated regulation of NFkappaB and the essentialness of NFkappaB for the oncogenicity of PI3K and Akt. *Int. J. Cancer.* **2009**, *125*, 2863–2870. [CrossRef]
44. Liu, T.; Zhang, L.; Joo, D.; Sun, S.C. NF-κB signaling in inflammation. *Signal Transduct. Target Ther.* **2017**, *2*, 17023. [CrossRef]
45. Lee, D.H.; Lee, C.S. Flavonoid myricetin inhibits TNF-α-stimulated production of inflammatory mediators by suppressing the Akt, mTOR and NF-κB pathways in human keratinocytes. *Eur. J. Pharmacol.* **2016**, *784*, 164–172. [CrossRef] [PubMed]
46. Romashkova, J.A.; Makarov, S.S. NF-κB is a target of AKT in anti-apoptotic PDGF signalling. *Nature* **1999**, *401*, 86–90. [CrossRef] [PubMed]
47. Tak, P.P.; Firestein, G.S. NF-kappaB: A key role in inflammatory diseases. *J. Clin. Investig.* **2001**, *107*, 7–11. [CrossRef]
48. Kalinski, P. Regulation of Immune Responses by Prostaglandin E2. *J. Immunol.* **2012**, *188*, 21–28. [CrossRef]
49. Futagami, A.; Ishizaki, M.; Fukuda, Y.; Kawana, S.; Yamanaka, N. Wound Healing Involves Induction of Cyclooxygenase-2 Expression in Rat Skin. *Mod. Pathol.* **2002**, *82*, 1503–1513. [CrossRef] [PubMed]
50. Zhou, J.; Zhao, Y.; Simonenko, V.; Xu, J.J.; Liu, K.; Wang, D.; Shi, J.; Zhong, T.; Zhang, L.; Zeng, L.; et al. Simultaneous silencing of TGF-β1 and COX-2 reduces human skin hypertrophic scar through activation of fibroblast apoptosis. *Oncotarget* **2017**, *8*, 80651–80665. [CrossRef]

51. Pfisterer, K.; Shaw, L.E.; Symmank, D.; Weninger, W. The Extracellular Matrix in Skin Inflammation and Infection. *Front. Cell Dev. Biol.* **2021**, *9*, 682414. [CrossRef]
52. Russo, B.; Brembilla, N.C.; Chizzolini, C. Interplay Between Keratinocytes and Fibroblasts: A Systematic Review Providing a New Angle for Understanding Skin Fibrotic Disorders. *Front. Immunol.* **2020**, *11*, 648. [CrossRef]
53. Newby, A. Matrix metalloproteinases regulate migration, proliferation, and death of vascular smooth muscle cells by degrading matrix and non-matrix substrates. *Cardiovasc. Res.* **2006**, *69*, 614–624. [CrossRef]
54. Nabeshima, K.; Inoue, T.; Shimao, Y.; Sameshima, T. Matrix metalloproteinases in tumor invasion: Role for cell migration. *Pathol. Int.* **2002**, *52*, 255–264. [CrossRef]
55. Caley, M.P.; Martins, V.L.; O'Toole, E.A. Metalloproteinases and wound healing. *Adv. Wound Care* **2015**, *4*, 225–234. [CrossRef]
56. Kamarazaman, I.S.; Kiong, L.S.; Hasan, M.K.N.; Basherudin, N.; Kasim, N.A.M.; Ali, A.A.; Ramli, S.; Maniam, S.; James, R.J.; Rojsitthisak, P.; et al. *Baeckea frutescens* L. Promotes wound healing by upregulating expression of TGF-β, IL-1 β, VEGF and MMP-2. *Saudi Pharm. J.* **2024**, *32*, 102110. [CrossRef]
57. Krishnaswamy, V.R.; Mintz, D.; Sagi, I. Matrix metalloproteinases: The sculptors of chronic cutaneous wounds. *Biochim. Biophys. Acta (BBA)-Mol. Cell Res.* **2017**, *1864*, 2220–2227. [CrossRef]
58. Han, Y.P.; Tuan, T.L.; Wu, H.; Hughes, M.; Garner, W.L. TNF-alpha stimulates activation of pro-MMP2 in human skin through NF-(kappa)B mediated induction of MT1-MMP. *J. Cell Sci.* **2001**, *114*, 131–139. [CrossRef] [PubMed]
59. Shakibaei, M.; John, T.; Schulze-Tanzil, G.; Lehmann, I.; Mobasheri, A. Suppression of NF-kappaB activation by curcumin leads to inhibition of expression of cyclo-oxygenase-2 and matrix metalloproteinase-9 in human articular chondrocytes: Im-plications for the treatment of osteoarthritis. *Biochem. Pharmacol.* **2007**, *73*, 1434–1445. [CrossRef] [PubMed]
60. Ukaegbu, K.; Allen, E.; Svoboda, K.K.H. Reactive Oxygen Species and Antioxidants in Wound Healing: Mechanisms and Therapeutic Potential. *Int. Wound J.* **2025**, *22*, e70330. [CrossRef] [PubMed]
61. Ciulla, M.; Rashad, M.; Spano, M.; Mannina, L.; Garzoli, S.; Zengin, G.; Di Giacomo, S.; Bruni, P.; Di Profio, P.; Fontana, A.; et al. Characterization of Mucins, Glycosaminoglycans, and Bioactive Compounds in *Helix aspersa*'s slime by Spectroscopic and Biochemical Analysis. *J. Mol. Struct.* **2026**, *1349*, 143838. [CrossRef]
62. Balaha, M.; Cataldi, A.; Ammazzalorso, A.; Cacciatore, I.; De Filippis, B.; Di Stefano, A.; Maccallini, C.; Rapino, M.; Ko-rona-Glowniak, I.; Przekora, A.; et al. CAPE derivatives: Multifaceted agents for chronic wound healing. *Arch. Pharm.* **2024**, *357*, e2400165. [CrossRef]
63. Furtado, J.; Eichmann, A. Vascular development, remodeling and maturation. *Curr. Top. Dev. Biol.* **2024**, *159*, 344–370. [CrossRef] [PubMed]
64. Bajbouj, K.; Ramakrishnan, R.K.; Hamid, Q.; Pirozzi, C.J. Role of Matrix Metalloproteinases in Angiogenesis and Its Implications in Asthma. *J. Immunol. Res.* **2021**, *2021*, 1–12. [CrossRef] [PubMed]
65. Takeo, M.; Lee, W.; Ito, M. Wound healing and skin regeneration. *Cold Spring Harb. Perspect. Med.* **2015**, *5*, a023267. [CrossRef] [PubMed]
66. Petrillo, S.; Genova, T.; Chinigò, G.; Roato, I.; Scarpellino, G.; Kopecka, J.; Altruda, F.; Tolosano, E.; Riganti, C.; Mussano, F.; et al. Endothelial Cells Promote Osteogenesis by Establishing a Functional and Metabolic Coupling with Human Mesenchymal Stem Cells. *Front. Physiol.* **2022**, *12*, 813547. [CrossRef]
67. Rosanto, Y.B.; Hasan, C.Y.; Rahardjo, R.; Pangestiningsih, T.W. Effect of snail mucus on angiogenesis during wound healing. *F1000Research* **2021**, *10*, 181. [CrossRef]

biomolecules

MDPI

Article

Adamantyl-Substituted Chalcone CA13 Induces Cytoprotective Autophagy and JNK-Dependent Apoptosis in Lung Cancer Cells

Yuting Chen [1,†], Yaxin Liu [1,†], Jing Zhou [1], Tingting Bao [1], Jing Wang [2,*] and Mingtao Ao [1,3,*]

1 School of Pharmacy, Hubei University of Science and Technology, Xianning 437100, China
2 Institutes of Biomedical Science, Inner Mongolia University, Hohhot 010021, China
3 Xiantao First People's Hospital, The Affiliated Hospital of Hubei University of Science and Technology, Xiantao 433000, China
* Correspondence: jingwang@imu.edu.cn (J.W.); aomingtao@hbust.edu.cn (M.A.)
† These authors contributed equally to this work.

Abstract

Lung cancer remains a leading cause of cancer mortality worldwide, highlighting the need for novel therapeutics. Here, we designed and synthesized a series of adamantyl-substituted chalcones and identified CA13 as a lead compound with potent and selective antiproliferative activity against non–small cell lung cancer (NSCLC) cells. CA13 triggered both apoptosis and autophagy in H292 cells. Western blotting and confocal imaging confirmed the activation of complete autophagic flux, while inhibition of autophagy markedly enhanced CA13-induced apoptosis, suggesting a cytoprotective role of autophagy. Mechanistically, CA13 activated JNK phosphorylation in a dose- and time-dependent manner, and pharmacological blockade of JNK significantly attenuated apoptotic signaling. In vivo, CA13 effectively suppressed H292 xenograft tumor growth without apparent systemic toxicity. Collectively, these results demonstrate that CA13 exerts its antitumor effects through JNK-dependent apoptosis accompanied by cytoprotective autophagy, providing a promising structural framework for the development of chalcone-based anticancer agents targeting programmed cell death pathways.

Keywords: chalcone; apoptosis; autophagy; JNK; lung cancer

1. Introduction

Lung cancer remains one of the leading causes of cancer-related mortality worldwide [1]. Despite significant advances in targeted therapies and immunotherapies, therapeutic resistance and disease relapse remain major challenges, underscoring the urgent need for novel small-molecule agents with potent antitumor activity and high favorable selectivity [2,3]. Non-small cell lung cancer (NSCLC) represents approximately 85% of all lung cancer cases, and most patients are diagnosed at advanced or metastatic stages, where curative surgical options are limited [4]. Therefore, the discovery of new therapeutic compounds capable of targeting key molecular pathways that drive cancer progression remains a critical objective in lung cancer research.

Natural product-derived scaffolds continue to provide a valuable source of inspiration for anticancer drug discovery, among which chalcones (1,3-diphenyl-2- propen-1-one derivatives) have attracted particular interest due to their structural simplicity, ease of chemical modification, and broad range of biological activities [5–8]. Chalcones, characterized by their α,β-unsaturated carbonyl scaffold, are a class of naturally occurring flavonoids found in various edible plants [9]. They exhibit diverse pharmacological activities, including anti-inflammatory [10], antioxidant [11], and anticancer effects [12]. Increasing evidence suggests that the anticancer effects of chalcones are largely attributed to their ability to modulate various forms of programmed cell death, particularly apoptosis and autophagy [13–16]. Chalcone derivatives have been shown to provoke mitochondrial-mediated apoptosis through reactive oxygen species (ROS) accumulation, collapse of mitochondrial membrane potential, cytochrome c release, modulation of Bcl-2 family proteins, caspase activation and PARP cleavage [7,17–19]. Meanwhile, an increasing number of studies demonstrate that chalcones can also regulate autophagy: either by inducing autophagic flux or by inhibiting autophagosome degradation, thus influencing the delicate balance between cell survival and cell death [20–26]. For instance, 2′-Hydroxychalcone (2′-HC) (Figure 1) enhanced the autophagic levels and induced apoptosis in breast cancer cells [23]. Synthetic 2′-hydroxychalones induced apoptosis and autophagy in A549 lung cancer cells [24]. Similarly, Licochalcone A (Figure 1) and related chalcones triggered both apoptosis and autophagy while concurrently reducing cellular invasiveness, emphasizing the dual roles of these pathways in anticancer responses [25,26].

Autophagy, however, plays a dual role in cancer therapy. It may function as a cytoprotective mechanism, enabling tumor cells to survive metabolic stress and therapeutic insult, or conversely, act as a pro-death mechanism when excessively or aberrantly activated [27]. The biological outcome of autophagy modulation often depends on cancer type, treatment context, and dosage. Several studies have shown that chalcone derivatives can trigger autophagy as a survival mechanism, and that pharmacological or genetic inhibition of autophagy significantly enhances chalcone-induced apoptosis [26,28,29]. For example, blocking autophagy significantly increased flavokawain B-mediated cytotoxicity, indicating that flavokawain B induces protective autophagy in thyroid cancer cells [28].

In our continuing efforts to identify novel modulators of apoptosis and autophagy as potential anticancer candidates [30], we designed and synthesized a series of adamantyl-substituted chalcone derivatives. The introduction of the bulky hydrophobic adamantyl group was intended to enhance lipophilicity and facilitate stronger molecular interactions, while fine-tuning the electronic properties of aromatic substituents aimed to optimize biological activity. In this study, we characterized the anticancer effects of the most active compound, **CA13** (Figure 1), focusing on its ability to induce apoptosis and autophagy in lung cancer cells and evaluating its in vivo antitumor efficacy.

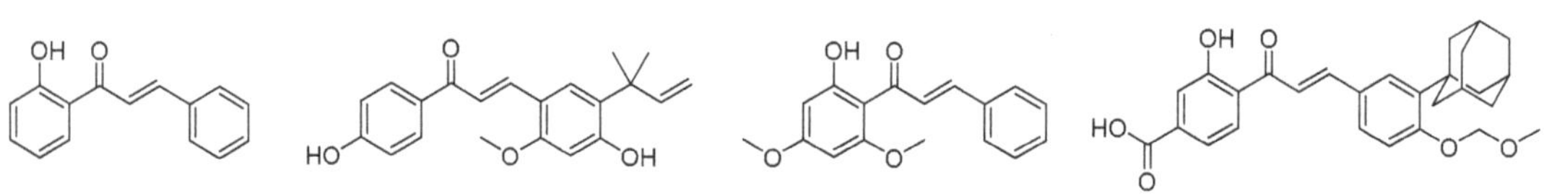

Figure 1. Chemical structures of representative chalcones.

2. Materials and Methods

2.1. Chemistry

All the reagents were obtained from Bide Pharmatech Ltd. (Shanghai, China). The progress of the reaction was monitored by thin-layer chromatography (TLC) on pre-coated silica gel 60 F254 plates, and a UV light was used to visualize agent. ^{1}H NMR (400 MHz) and ^{13}C NMR (100 MHz) are obtained by Bruker Avance III 400 spectrometer (Bruker Corporation, Billerica, MA, USA). Tetramethylsilane (TMS) is used as the internal standard, and ppm is used as the unit to record the Chemical shift (δ). The terms s, d, t, q, m refer to singlet, doublet, triplet, quartlet, multiplet signals. The J value (Coupling constant)

https://doi.org/10.3390/biom16010054

is expressed in Hertz (Hz). Mass spectra were obtained on a SHIMADZU LCMS-8040 spectrometer (Shimadzu, Kyoto, Japan) with ESI. Fourier transform infrared (FT-IR) spectra were recorded on a SHIMADZU IRAffinity-1 spectrometer. Melting points were determined using an SGW X-4 model apparatus (INESA, Shanghai, China).

2.1.1. The Synthesis of 3-(Adamant-1-yl)-4-(Methoxymethoxy)-Benzaldehyde (**3**)

Compound **3** was prepared according to our previously reported procedure [30] with necessary modifications. To a solution of 4-hydroxybenzaldehyde **1** (1.22 g, 10 mmol) and adamant-1-ol (1.69 g, 11.1 mmol) in glacial acetic acid (10 mL) was added concentrated H_2SO_4 (0.65 mL, 12 mmol). The reaction mixture was stirred for 6 h at 70 °C. Upon completion of the reaction, the reaction mixture was quenched with ice water, and the precipitated solid was filtered, washed with water and petroleum ether, and dried to get **2**. Then, compound **2** (1.28 g, 5 mmol) was dissolved in acetone (30 mL). Chloromethyl methyl ether (MOMCl) (1.14 mL, 15 mmol) and anhydrous potassium carbonate (2.07 g, 15 mmol) were added and the reaction mixture was stirred at 60 °C for 12 h. After completion, the reaction mixture was filtered and the filtrate was concentrated under reduced pressure. The crude was recrystallized with ethanol to give **3**. White solid, yield 55%.

2.1.2. General Procedure for the Preparation of Chalcones (**CA 5–13**)

A mixture of corresponding acetophenone (1 mmol) and benzaldehyde derivatives (1.1 mmol) in ethanol (10 mL) was added NaOH (10 eq.) and stirred at 70 °C for 12–24 h. Upon reaction completion, the reaction mixture was poured into ice water and neutralized with 1 N HCl. The solid was filtered and purified by column chromatography on silica gel to give the corresponding chalcones **CA 5–13**.

*(E)-3-hydroxy-4-(3-(4-hydroxyphenyl)acryloyl)benzoic acid (**CA-5**)*

A yellow solid, yield: 20.3%. mp 126−128 °C. ^{1}H NMR (400 MHz, DMSO-d_6) δ 8.55 (d, J = 2.3 Hz, 1H), 8.04 (dd, J = 2.1, 8.7 Hz, 1H), 7.75 (dd, J = 3.5, 9.3 Hz, 4H), 7.06 (d, J = 8.5 Hz, 1H), 6.85 (d, J = 8.8 Hz, 2H). ESI-MS: m/z calcd. for $C_{16}H_{11}O_5^-$ $[M-H]^-$ 283.06, found 283.06.

*(E)-3-(3-adamantan-1-yl)-4-(methoxymethoxy)phenyl)-1-(4-fluoro-2-hydroxyphenyl)prop-2-en-1-one (**CA-6**)*

A yellow solid, yield: 45.8%. mp 148−150 °C. IR (KBr, cm^{-1}): 2905, 2854 (aliphatic C–H, adamantyl); 1640 (C=O); 1572 (C=C, conjugated & Ar); 1492 (Ar C=C/C–H); 1234, 1134 (C–O–C); 991 (trans-CH=CH out-of-plane bending). ^{1}H NMR (400 MHz, DMSO-d_6) δ 12.83 (d, J = 7.5 Hz, 1H), 8.26 (t, J = 8.0 Hz, 1H), 7.81–7.85 (m, 2H), 7.75 (dd, J = 6.0, 7.5 Hz, 1H), 7.65 (d, J = 5.1 Hz, 1H), 7.11 (d, J = 5.7 Hz, 2H), 7.07 (dd, J = 6.0, 8.0 Hz, 1H), 5.34 (s, 2H), 3.47 (s, 3H), 2.13–2.08 (m, 9H), 1.80–1.75 (m, 6H). ^{13}C NMR (100 MHz, DMSO-d_6) δ 193.1, 162.8, 158.8, 146.5, 140.3, 138.8, 132.9, 129.3, 128.5, 127.8, 120.6, 119.8, 119.7, 117.8, 115.0, 94.3, 56.8, 40.5, 37.2, 37.0, 28.8. ESI-MS: m/z calcd. for $C_{27}H_{30}FO_4^+$ $[M+H]^+$ 437.21, found 437.30.

*(E)-3-(3-adamantan-1-yl)-4-(methoxymethoxy)phenyl)-1-(4-chloro-2-hydroxyphenyl)prop-2-en-1-one (**CA-7**)*

A yellow solid, yield: 57.4%. mp 130−132 °C. IR (KBr, cm^{-1}): 2910, 2852 (aliphatic C–H, adamantyl); 1640 (C=O); 1572 (C=C, conjugated & Ar); 1492 (Ar C=C/C–H); 1238, 1134 (C–O–C); 991 (trans-CH=CH out-of-plane bending). ^{1}H NMR (400 MHz, CHLOROFORM-d) δ 13.17 (s, 1H), 7.94 (d, J = 15.3 Hz, 1H), 7.86 (d, J = 8.6 Hz, 1H), 7.56 (d, J = 2.0 Hz, 1H), 7.50 (dd, J = 2.0, 8.6 Hz, 1H), 7.45 (d, J = 15.3 Hz, 1H), 7.15 (d, J = 8.4 Hz, 1H), 7.05 (d, J = 2.0 Hz, 1H), 6.93 (dd, J = 2.0, 8.6 Hz, 1 H), 5.30 (s, 2 H), 3.55 (s, 3 H), 2.17–2.12 (m, 9 H), 1.82–1.78 (m, 6 H). ^{13}C NMR (100 MHz, CHLOROFORM-d) δ 192.9, 164.3, 159.1, 146.8,

https://doi.org/10.3390/biom16010054

141.8, 139.3, 130.5, 128.1, 128.0, 127.7, 119.4, 118.7, 118.6, 117.1, 114.8, 94.1, 56.5, 40.5, 37.2, 37.0, 29.0. ESI-MS: m/z calcd. for $C_{27}H_{30}ClO_4^+$ $[M + H]^+$ 453.18, found 453.30.

*(E)-3-(3-adamantan-1-yl)-4-(methoxymethoxy)phenyl)-1-(4-bromo-2-hydroxyphenyl)prop-2-en-1-one (**CA-8**)*

A yellow solid, yield: 50.8%. mp 135−138 °C. IR (KBr, cm^{-1}): 2910, 2852 (aliphatic C–H, adamantyl); 1642 (C=O); 1572 (C=C, conjugated & Ar); 1492 (Ar C=C/C–H); 1236, 1131 (C–O–C); 991 (trans-CH=CH out-of-plane bending). ^1H NMR (400 MHz, DMSO-d_6) δ 12.78 (s, 1H), 8.21 (d, J = 8.4 Hz, 1H), 7.83–7.76 (m, 2H), 7.73 (dd, J = 1.7, 8.4 Hz, 1H), 7.62 (d, J = 1.5 Hz, 1H), 7.13 (d, J = 2.0 Hz, 1H), 7.10 (d, J = 8.4 Hz, 1H), 7.04 (dd, J = 1.8, 8.4 Hz, 1H), 5.33 (s, 2H), 3.46–3.44 (m, 3H), 2.11–2.07 (m, 9H), 1.78–1.73 (m, 6H). ^{13}C NMR (100 MHz, DMSO-d_6) δ 193.0, 162.5, 158.7, 146.3, 140.1, 138.8, 132.8, 129.2, 128.5, 127.8, 120.9, 120.0, 119.8, 117.8, 115.0, 94.3, 56.7, 40.4, 37.2, 37.0, 28.8. ESI-MS: m/z calcd. for $C_{27}H_{28}BrO_4^+$ $[M - H]^-$ 495.12, found 495.25.

*(E)-3-(3-adamantan-1-yl)-4-(methoxymethoxy)phenyl)-1-(2-hydroxy-4-methylphenyl)prop-2-en-1-one (**CA-9**)*

A yellow solid, yield: 47.6%. mp 125−127 °C. IR (KBr, cm^{-1}): 2910, 2846 (aliphatic C–H, adamantyl); 1648 (C=O); 1602 (C=C, conjugated & Ar); 1236, 1133 (C–O–C); 995 (trans-CH=CH out-of-plane bending). ^1H NMR (400MHz, DMSO-d_6) δ 12.98 (s, 1H), 8.23 (d, J = 8.3 Hz, 1H), 7.93–7.85 (m, 2H), 7.81–7.78 (m, 1H), 7.66 (d, J = 1.8 Hz, 1H), 7.14 (d, J = 8.6 Hz, 1H), 6.89–6.84 (m, 2H), 5.37 (s, 2H), 3.49 (s, 3H), 2.38 (s, 3H), 2.16–2.11 (m, 6H), 1.81–1.77 (m, 6 H). ^{13}C NMR (100 MHz, DMSO-d_6) δ 193.5, 163.0, 158.6, 148.0, 145.9, 138.7, 131.2, 129.0, 128.5, 127.9, 120.7, 119.1, 118.4, 118.2, 115.0, 94.2, 56.7, 40.4, 37.2, 37.0, 28.9, 21.9. ESI-MS: m/z calcd. for $C_{28}H_{33}O_4^+$ $[M + H]^+$ 433.24, found 433.30.

*(E)-3-(3-adamantan-1-yl)-4-(methoxymethoxy)phenyl)-1-(2-hydroxy-4-methoxyphenyl)prop-2-en-1-one (**CA-10**)*

A red solid, yield: 44.3%. mp 135−137 °C. IR (KBr, cm^{-1}): 2910, 2852 (aliphatic C–H, adamantyl); 1640 (C=O, chalcone); 1575 (C=C, conjugated & Ar); 1499 (Ar C=C/C–H); 1234, 1133 (C–O–C); 991 (trans-CH=CH out-of-plane bending). ^1H NMR (400MHz, DMSO-d_6) δ 13.58 (s, 1H), 8.31–8.26 (m, J = 9.2 Hz, 1H), 7.87–7.78 (m, 2H), 7.77–7.73 (m, 1H), 7.62 (s, 1H), 7.11–7.09 (m, J = 8.6 Hz, 1H), 6.56 (dd, J = 2.2, 9.0 Hz, 1H), 6.52 (d, J = 2.2 Hz, 1H), 5.33 (s, 2H), 3.85 (s, 3H), 3.45 (s, 3H), 2.12–2.07 (m, 9H), 1.78–1.74 (m, 6H). ^{13}C NMR (100 MHz, DMSO-d_6) δ 192.4, 166.3, 166.1, 158.5, 145.4, 138.7, 133.1, 129.0, 128.4, 128.0, 119.0, 115.0, 114.4, 107.8, 101.4, 94.3, 56.7, 56.2, 40.5, 37.2, 37.0, 28.8. ESI-MS: m/z calcd. for $C_{28}H_{33}O_5^+$ $[M + H]^+$ 449.23, found 449.35.

*(E)-3-(3-adamantan-1-yl)-4-(methoxymethoxy)phenyl)-1-(2-hydroxy-4,6-dimethoxyphenyl)prop-2-en-1-one (**CA-11**)*

A red solid, yield: 40.6%. mp 168−170 °C. IR (KBr, cm^{-1}): 2910, 2850 (aliphatic C–H, adamantyl); 1640 (C=O, chalcone); 1598 (C=C, conjugated & Ar); 1507 (Ar C=C/C–H); 1234, 1131 (C–O–C); 997 (trans-CH=CH out-of-plane bending). ^1H NMR (400 MHz, CHLOROFORM-d) δ 7.89–7.76 (m, 2H), 7.53 (d, J = 2.2 Hz, 1H), 7.44 (dd, J = 2.1, 8.4 Hz, 1H), 7.12 (d, J = 8.4 Hz, 1H), 6.12 (d, J = 2.3 Hz, 1H), 5.98 (d, J = 2.4 Hz, 1H), 5.28 (s, 2H), 3.92 (s, 3H), 3.85 (s, 3H), 3.54 (s, 3H), 2.15–2.11 (m, 9H), 1.83–1.77 (m, 6H). ^{13}C NMR (100 MHz, CHLOROFORM-d) δ 192.6, 168.4, 166.0, 162.5, 158.2, 143.2, 138.9, 128.8, 127.7, 127.3, 125.2, 114.7, 106.4, 94.1, 93.8, 91.3, 56.4, 55.8, 55.6, 40.6, 37.2, 37.1, 29.1. ESI-MS: m/z calcd. for $C_{29}H_{35}O_6^+$ $[M + H]^+$ 479.24, found 479.35.

*(E)-3-(3-((3r,5r,7r)-adamantan-1-yl)-4-(methoxymethoxy)phenyl)-1-(2,4,6-trimethoxyphenyl)prop-2-en-1-one (**CA-12**)*

A yellow solid, yield: 47.6%. mp 185−187 °C. IR (KBr, cm^{-1}): 2908, 2850 (aliphatic C–H, adamantyl); 1650 (C=O); 1598 (C=C, conjugated & Ar); 1507 (Ar C=C/C–H); 1235, 1133 (C–O–C); 997 (trans-CH=CH out-of-plane bending) ^{1}H NMR (400 MHz, DMSO-d_6) δ 7.50 (dd, *J* = 2.0, 8.6 Hz, 1H), 7.36 (d, *J* = 2.0 Hz, 1H), 7.15 (d, *J* = 16.0 Hz, 1H), 7.04 (d, *J* = 8.6 Hz, 1H), 6.79 (d, *J* = 16.0 Hz, 1H), 6.31 (s, 2H), 5.29 (s, 2H), 3.84 (s, 3H), 3.71 (s, 6H), 3.43 (s, 3H), 2.07–2.04 (m, 9H), 1.75–1.72 (m, 6H). ^{13}C NMR (100 MHz, DMSO-d_6) δ 193.9, 162.2, 158.4, 158.1, 145.0, 138.6, 127.9, 127.8, 127.7, 127.4, 115.1, 111.7, 94.2, 91.5, 56.6, 56.2, 55.9, 40.5, 37.1, 36.9, 28.8. ESI-MS: *m/z* calcd. for $C_{30}H_{37}O_6^+$ [M + H]$^+$ 493.26, found 493.35.

*4-((E)-3-(3-adamantan-1-yl)-4-(methoxymethoxy)phenyl)acryloyl)-3-hydroxybenzoic acid (**CA-13**)*

A yellow solid, yield: 30.4%. mp 196−198 °C. IR (KBr, cm^{-1}): 2910, 2852 (aliphatic C–H, adamantyl); 1640 (C=O); 1572 (C=C, conjugated & Ar); 1492 (Ar C=C/C–H); 1133 (C–O–C, ether); 983 (trans-CH=CH out-of-plane bending). ^{1}H NMR (400 MHz, CHLOROFORM-d) δ 11.26 (s, 1H), 8.52 (d, *J* = 2.3 Hz, 1H), 8.10 (dd, *J* = 2.2, 8.7 Hz, 1H), 7.73 (d, *J* = 15.7 Hz, 1H), 7.47 (d, *J* = 2.1 Hz, 1H), 7.41 (dd, *J* = 2.1, 8.6 Hz, 1H), 7.33 (d, *J* = 15.5 Hz, 1H), 7.07 (d, *J* = 8.4 Hz, 1H), 7.01 (d, *J* = 8.8 Hz, 1H), 5.21 (s, 2H), 3.46 (s, 3H), 2.09–2.02 (m, 9H), 1.75–1.70 (m, 6H). ^{13}C NMR (100 MHz, CHLOROFORM-d) δ 188.1, 169.9, 165.1, 158.6, 145.3, 139.1, 135.6, 131.4, 130.1, 128.1, 127.7, 127.5, 119.1, 117.9, 114.7, 112.5, 94.1, 56.4, 40.6, 37.2, 37.1, 29.0. ESI-MS: *m/z* calcd. for $C_{28}H_{29}O_6^-$ [M − H]$^-$ 461.20, found 461.20.

2.2. Cell Culture

Human lung cancer cell lines A549 (ATCC® CCL-185), H460 (ATCC® HTB-177), H292 (ATCC® CRL-1481), and the normal lung fibroblast line MRC-5 (ATCC® CCL-171) were obtained from the American Type Culture Collection (ATCC) (Manassas, VA, USA). A549, H460, and H292 cells were maintained in Dulbecco's Modified Eagle Medium (DMEM) supplemented with 10% fetal bovine serum (FBS) and 1% penicillin–streptomycin. MRC-5 cells were cultured in Minimum Essential Medium (MEM) containing 10% FBS. All cells were incubated at 37 °C in a humidified atmosphere with 5% CO_2.

2.3. Cell Viability Assay (MTT Assay)

Cell viability was determined using the MTT assay. Briefly, H460 and H292 cells were seeded into 96-well plates at a density of 1×10^4 cells per well and treated with various concentrations of the test compounds (2.5, 5, 10, 20, 30, 40, 50 and 80 μM) for 24 h. For mechanistic studies, cells were pretreated with 3-methyladenine (3-MA, 10 mM, 2 h), chloroquine (CQ, 3 μM, 1 h), or SP600125 (10 μM, 2 h) prior to CA13 exposure for an additional 24 h. Subsequently, 20 μL of MTT solution (5 mg/mL) was added to each well and incubated for 4 h at 37 °C. The supernatant was carefully removed, and the resulting formazan crystals were dissolved in 150 μL DMSO. Absorbance was measured at 490 nm using a microplate reader (PerkinElmer, Waltham, MA, USA). The density of formazan formed in the blank group was set as 100% of viability. Cell viability (%) = [sample (OD_{490})/blank (OD_{490})] × 100%. Blank group: cultured with fresh medium only. Sample group: treated with compounds. All experiments were performed in triplicate.

https://doi.org/10.3390/biom16010054

2.4. Western Blot Analysis

Western blotting was performed as previously described [30]. The following primary antibodies were used: anti-PARP (ab227244), LC3B (ab192890), ATG7 (ab52472), pJNK (ab124956), JNK (ab124956), β-actin (ab6276), and GAPDH (ab181602) (Abcam, Cambridge, UK). Total protein was extracted from cells using RIPA lysis buffer, and protein concentration was quantified with a BCA Protein Assay Kit (P0010, Beyotime Biotechnology, Shanghai, China). Protein samples were separated by SDS-PAGE and transferred onto PVDF membranes (Millipore, IPVH00010, Billerica, MA, USA). The membranes were blocked with 5% non-fat milk in TBST (pH 7.6) for 1.5 h at room temperature, followed by incubation with primary antibodies overnight at 4 °C. After three washes with TBST, the membranes were incubated with HRP-conjugated goat anti-mouse (1:5000, Abclonal, Hangzhou, China) or goat anti-rabbit (1:5000, Abclonal) secondary antibodies for 1 h at room temperature. Protein bands were visualized using an enhanced chemiluminescence kit (Advansta, Inc., San Jose, CA, USA) according to the manufacturer's instructions. Original figures can be found in Supplementary Materials.

2.5. Flow Cytometry Analysis of Apoptosis

Annexin V-FITC/PI apoptosis detection was performed as previously described [30]. Briefly, H292 cells were seeded in 6-well plates. After adherence, cells were treated with CA13 (0, 2.5, 5 and 10 μM) for 24 h, harvested using trypsin (without EDTA) at 37 °C, and washed twice with cold PBS. Cell pellets were resuspended in 250 μL of 1 × binding buffer, followed by staining with 5 μL Annexin V-FITC and 5 μL propidium iodide (PI) (Yeasen, Inc., Shanghai, China) for 15 min at room temperature in the dark. Apoptotic cell populations were analyzed using a flow cytometer (Beckman Coulter, CA, USA). Data analysis was conducted with FlowJo software (version 10.8.1), distinguishing early apoptotic (Annexin V+/PI−), late apoptotic (Annexin V+/PI+), and necrotic (Annexin V−/PI+) cells.

2.6. Autophagic Flux Analysis

mRFP-GFP-LC3 plasmids (Hanbio, Nanjing, China) were used to transfect H292 cells according to the manufacturer's instructions, and were then treated with CA13 (2.5 or 5 μM) for 24 h. The formation of autolysosomes was observed under a Leica TCS SP8 STED confocal laser scanning microscope (Leica Microsystems GmbH, Wetzlar, Germany), and images were captured at 400× magnification.

2.7. Xenograft Mouse Model

All animal experiments were approved by the Institutional Animal Care and Use Committee of Hubei University of Science and Technology. Five-week-old male BALB/c-nu/nu mice were maintained under specific pathogen-free conditions. H292 cells (1×10^6) suspended in 100 μL PBS were injected subcutaneously into the right flank of each mouse. When tumor volume reached approximately 100 mm^3, mice were randomized into three groups (n = 5 per group) and administered PBS (vehicle control), CA13 at 20 mg/kg, or CA13 at 40 mg/kg by intraperitoneal injection every other day. Tumor volume and body weight were measured every three days for 24 days.

https://doi.org/10.3390/biom16010054

2.8. Statistical Analysis

Data are expressed as means ± standard deviation (SD) from three independent experiments. Comparisons between two groups were assessed using Student's two-tailed unpaired *t*-test. Statistical analysis was performed using GraphPad Prism 8.0. A *p*-value < 0.05 was considered statistically significant.

3. Results

3.1. Chemistry

Chalcones **CA 1–13** were synthesized via aldol condensation of corresponding aldehydes (1 mmol) and various acetophenones (1 mmol) in methanolic basic medium as shown in Scheme 1. The characterization of **CA 1–4** was reported previously [31–33]. Friedel-Crafts alkylation of 4-hydroxybenzaldehyde yielded 3-adamanty-l-2- hydroxybenzaldehyde, which was then treated with chloromethyl methyl ether (MOMCl) to afford the adamantyl-substituted hydroxybenzaldehyde (**2**), following our reported procedure [30]. Aldol condensation of relevant acetophenones with the adamantyl-substituted hydroxybenzaldehyde, performed with NaOH in MeOH at 70 °C, provided new adamantyl-substituted chalcones **CA 6–13**. Structures were confirmed by melting points, NMR (^{1}H and ^{13}C NMR), and mass spectrometry.

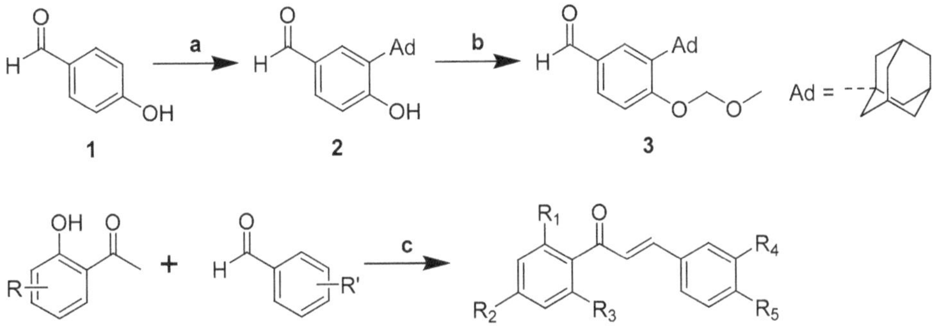

CA1-13

CA1: R_1=OH, R_2=R_3=R_4=H, R_5=OH
CA2: R_1=OH, R_2=Cl, R_3=R_4=H, R_5=OH
CA3: R_1=OH, R_2=CH_3, R_3=R_4=H, R_5=OH
CA4: R_1=OH, R_2=OCH_3, R_3=R_4=H, R_5=OH
CA5: R_1=OH, R_2=COOH, R_3=R_4=H, R_5=OH
CA6: R_1=OH, R_2=F, R_3=H, R_4=Ad, R_5=OCH_2OCH_3
CA7: R_1=OH, R_2=Cl, R_3=H, R_4=Ad, R_5=OCH_2OCH_3
CA8: R_1=OH, R_2=Br, R_3=H, R_4=Ad, R_5=OCH_2OCH_3
CA9: R_1=OH, R_2=CH_3, R_3=H, R_4=Ad, R_5=OCH_2OCH_3
CA10: R_1=OH, R_2=OCH_3, R_3=H, R_4=Ad, R_5=OCH_2OCH_3
CA11: R_1=OH, R_2=R_3=OCH_3, R_4=Ad, R_5=OCH_2OCH_3
CA12: R_1=R_2=R_3=OCH_3, R_4=Ad, R_5=OCH_2OCH_3
CA13: R_1=OH, R_2=COOH, R_3=H, R_4=Ad, R_5=OCH_2OCH_3

Scheme 1. Synthetic route of chalcones **CA 1–13**. Reagents and conditions: (a) H_2SO_4, acetic acid, 70 °C, 6 h; (b) K_2CO_3, MOMCl, acetone, 60 °C, 12 h; (c) NaOH, MeOH, 70 °C, 12–24 h.

https://doi.org/10.3390/biom16010054

3.2. Biological Activity

3.2.1. In Vitro Antiproliferative Activity of Adamantly-Substituted Chalcones on Lung Cancer Cell Lines

To explore the antitumor activity of these chalcone derivatives, all compounds were first assayed for in vitro antiproliferative activities against H292 and A549 cancer cell lines using an MTT assay with 5-fluorouracil (5-FU) included as a positive control. The concentrations of compounds required for 50% inhibition of cell viability (IC_{50}) were determined and summarized in Table 1. The results indicated that more than half of the synthesized chalcone derivatives exhibited antiproliferative activity against both H292 and A549 cells with IC_{50} values below 50 μM. Among them, five compounds (**CA7**, **CA8**, **CA10**, **CA11**, and **CA13**) were active toward H292 cells (IC_{50} < 35 μM), and three (**CA7**, **CA11**, and **CA13**) also exhibited good inhibitory activity against A549 cells (IC_{50} < 30 μM). Notably, compound **CA13** showed the most potent cytotoxic effect against both tested cell lines (IC_{50} = 10.15 μM for H292 and 12.24 μM for A549), significantly lower than those of the reference compounds 2′-HC (IC_{50} > 50 μM for both cell lines) and 5-FU (28.83 ± 2.31 μM for H292 and 21.56 ± 2.04 μM for A549). The results summarized in Table 1 suggest that different substituents on the phenyl rings (ring-A and ring-B) have remarkable influences on antiproliferative potency.

Table 1. In vitro antiproliferative activities of target compounds against cancer H292 cells.

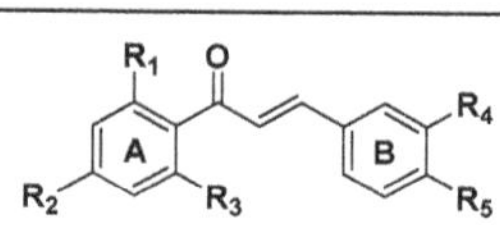

Comp.	R1	R2	R3	R4	R5	IC_{50} (μmol/L) [1]	
						H292	A549
2′-HC	OH	H	H	H	H	>50	>50
CA1	OH	H	H	H	OH	>50	>50
CA2	OH	Cl	H	H	OH	40.56 ± 2.16	48.04 ± 2.01
CA3	OH	CH_3	H	H	OH	>50	>50
CA4	OH	OCH_3	H	H	OH	45.25 ± 1.85	>50
CA5	OH	COOH	H	H	OH	38.17 ± 2.35	41.35 ± 2.13
CA6	OH	F	H	Ad	OCH_2OCH_3	>50	>50
CA7	OH	Cl	H	Ad	OCH_2OCH_3	22.24 ± 1.80	25.46 ± 1.21
CA8	OH	Br	H	Ad	OCH_2OCH_3	21.57 ± 2.17	27.32 ± 1.74
CA9	OH	CH_3	H	Ad	OCH_2OCH_3	>50	>50
CA10	OH	OCH_3	H	Ad	OCH_2OCH_3	34.18 ± 2.09	32.16 ± 2.28
CA11	OH	OCH_3	OCH_3	Ad	OCH_2OCH_3	21.06 ± 1.57	24.03 ± 1.82
CA12	OCH3	OCH_3	OCH_3	Ad	OCH_2OCH_3	40.31 ± 2.04	43.52 ± 2.24
CA13	OH	COOH	H	Ad	OCH_2OCH_3	10.15 ± 1.61	12.24 ± 1.28
5-FU	-	-	-	-	-	28.83 ± 2.31	21.56 ± 2.04

[1] Data are means ± SD of triplicate experiments.

Using OH substituents at R1 and R5, we examined the antiproliferative activities of **CA1–CA5** having different groups at R2 position (H, Cl, CH_3, OCH_3, and COOH, respectively). Among them, compound **CA5** with a COOH group at R2 exhibited the strongest antiproliferative activity against both cancer cell lines (IC_{50} = 38.17 μM for H292 and 41.35 μM for A549), followed by **CA2** with a Cl substituent. In contrast, compounds with electron-donating groups, such as CH_3 (**CA3**) or OCH_3 (**CA4**), showed weaker or no activity (IC_{50} > 45 μM). These results indicate that the introduction of electron-withdrawing substituents (COOH, Cl) at the R2 position of ring-A enhances the antiproliferative effect. Fur-

https://doi.org/10.3390/biom16010054

ther comparison between **CA2–CA5** (R4 = H) and **CA7–CA10** (R4 = Ad, R5 = OCH_2OCH_3) revealed that incorporation of the bulky adamantyl group at ring-B significantly improved antiproliferative potency. For example, **CA7** (R2 = Cl) and **CA8** (R2 = Br) displayed much lower IC_{50} values (22.24 and 21.57 μM against H292, respectively) than their non-adamantyl analogues **CA2** and **CA4**. This enhancement suggests that the adamantyl group might contribute to increased lipophilicity and better cellular uptake [30,34], thereby enhancing cytotoxic efficacy.

In addition, **CA11**, bearing dual methoxy groups at R2 and R3, exhibited improved activity (IC_{50} = 21.06 μM for H292 and 24.03 μM for A549) compared with its mono-methoxy analogue **CA10** (34.18 μM and 32.16 μM, respectively), indicating that multiple electron-donating groups can synergistically enhance antiproliferative potency when combined with the adamantyl moiety. Remarkably, **CA13**, which carries both a COOH group at R2 and an adamantyl substituent at R4, showed the best antiproliferative activity among all tested compounds (IC_{50} = 10.15 μM for H292 and 12.24 μM for A549), while its non-adamantyl analogue **CA5** was much less active. This observation highlights a synergistic effect between the hydrophilic carboxyl group and the bulky hydrophobic adamantyl moiety, which appears to be critical for high cytotoxic potency.

Next, the most active compound, **CA13**, was selected for further evaluation of its antiproliferative activity against the additional lung cancer cell line (H460) and a normal human lung fibroblast line (MRC-5). As shown in Table 2, **CA13** exhibited broad-spectrum cytotoxicity against A549, H460, and H292 cancer cells with IC_{50} values ranging from 10.15 to 12.24 μM. Notably, **CA13** showed negligible toxicity toward normal MRC-5 cells (IC_{50} > 50 μM), whereas the positive control 5-FU exhibited measurable toxicity (IC_{50} = 40.72 ± 2.26 μM) and the reference 2′-HC exhibited weak or no activity. These results highlight **CA13** as a promising lead compound with both potent and selective antiproliferative activity, warranting further investigation.

Table 2. In vitro antiproliferative activities of selected compounds against cancer and normal cells.

Comp.	IC_{50} (μmol/L) [1]			
	A549	**H460**	**H292**	**MRC-5**
CA13	12.24 ± 1.28	10.60 ± 0.67	10.15 ± 1.61	>50
2′-HC	>50	46.83 ± 2.15	>50	>50
5-FU	21.56 ± 2.04	27.57 ± 1.92	28.83 ± 2.31	40.72 ± 2.26

[1] Data are means ± SD of triplicate experiments.

3.2.2. Adamantly-Substituted Chalcone **CA13** Induced Autophagy in Lung Cancer Cells

Autophagy, a type II programmed cell death mechanism, plays a critical role in maintaining cellular homeostasis and modulating cancer cell fate under therapeutic stress. Among the core regulators of autophagy, microtubule-associated protein 1 light chain 3 (LC3) is essential for the biogenesis of autophagosomes and is widely recognized as a molecular hallmark of autophagic activation. During autophagy initiation, cytosolic LC3-I undergoes conjugation with phosphatidylethanolamine to form LC3-II through an Atg7/Atg3-mediated ubiquitin-like system, enabling autophagosomal membrane elongation and closure [27,35].

Previous studies have reported that 2′-hydroxychalcones are capable of inducing autophagy in lung cancer cells [24]. Based on these findings, we explored whether adamantyl-substituted chalcones exhibit similar autophagy-inducing properties. Western blot analysis revealed a concentration-dependent increase in LC3A/B-II expression following **CA13** treatment (Figure 2A), suggesting enhanced autophagosome formation. To further verify that **CA13** activates complete autophagic flux rather than impeding autophagosome

https://doi.org/10.3390/biom16010054

degradation, a tandem fluorescent mRFP-GFP-LC3 reporter assay was employed. As shown in Figure 2B, **CA13** treatment markedly increased the number of LC3 puncta in a concentration-dependent manner. The predominance of red-only puncta indicated efficient fusion between autophagosomes and lysosomes, demonstrating that **CA13** not only stimulates autophagosome formation but also facilitates their maturation into autolysosomes. Together, these findings indicate that **CA13** acts as a potent inducer of autophagy in H292 cells, facilitating both the initiation of autophagosome formation and the progression of autophagic flux.

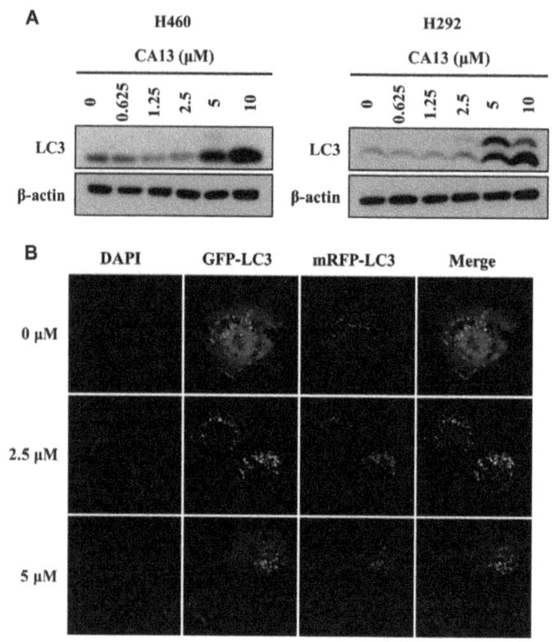

Figure 2. CA13 induced autophagy and promoted autophagy flux in lung cancer cells. (**A**) Western blot analyses of LC-3 in cells treated with **CA13** for 24 h. β-actin was used as an internal control. Original blots can be found in Supplementary Materials. (**B**) The H292 cells were transfected with a plasmid encoding mRFP-GFP-LC3 and treated with **CA13** for 24 h. Images were captured with confocal microscopy. Scale bar: 10 μm.

3.2.3. **CA13** Induced Apoptosis in Lung Cancer Cells

Apoptosis has been widely reported as a key mechanism of chalcone-induced cytotoxicity in various cancer cell lines [8]. To determine whether apoptosis contributes to the cytotoxic effects of **CA13** in lung cancer cells, we first examined the expression of apoptosis-related proteins. Western blot analysis (Figure 3A) demonstrated that **CA13** treatment increased the levels of cleaved PARP, a hallmark of apoptosis, in both a dose- and time-dependent manner. To further confirm the pro-apoptotic effect of **CA13**, Annexin V-FITC/PI double staining followed by flow cytometric analysis was performed in H292 cells. Annexin V binds to phosphatidylserine residues exposed on the outer leaflet of the plasma membrane, allowing discrimination between viable, early apoptotic, and late apoptotic cells when combined with PI staining. As shown in the representative flow cytometry plots (Figure 3B), **CA13** treatment significantly and dose-dependently increased the proportion of apoptotic cells. The percentage of late apoptotic cells rose from 11.1 ± 1.7% at 2.5 μM to 17.5 ± 2.4% at 10 μM after 12 h of treatment. These results indicate that **CA13** induces apoptotic cell death in H292 lung cancer cells.

https://doi.org/10.3390/biom16010054

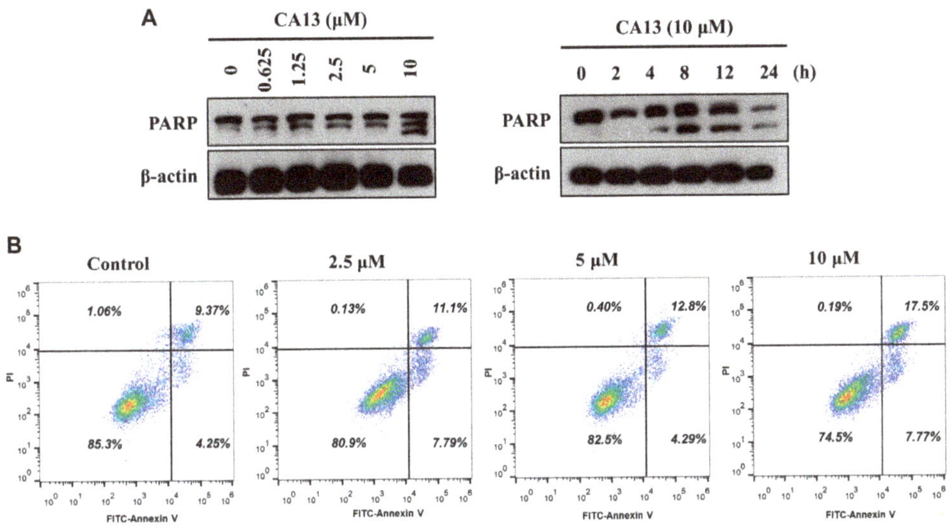

Figure 3. **CA13** induces apoptosis in H292 lung cancer cells. (**A**) Western blot analysis of cleaved PARP expression in H292 cells treated with various concentrations of **CA13** (0–10 μM) for 24 h or with 10 μM **CA13** for the indicated time periods. β-actin was used as a loading control. Original blots can be found in Supplementary Materials. (**B**) H292 cells were treated with **CA13** at the indicated concentrations for 24 h and analyzed by Annexin V FITC/PI staining. Apoptotic cell ratio was quantified by flow cytometry. Data are presented as mean ± standard deviation (SD) for three independent experiments.

3.2.4. **CA13**-Induced Autophagy Plays a Protective Role in Lung Cancer Cells

Accumulating evidence indicates that autophagy and apoptosis are closely interconnected in cancer cells, with autophagy exhibiting either cytoprotective or cytotoxic effects depending on the cellular context and tumor microenvironment [36,37]. To elucidate the role of **CA13**-induced autophagy in H292 cells, inhibitors targeting different stages of autophagy were employed. As shown in Figure 4, inhibition of autophagy by either 3-MA or CQ significantly enhanced CA13-induced cell death, as demonstrated by decreased cell viability in MTT assays. These results suggest that **CA13**-induced autophagy plays a protective role in H292 cells by attenuating apoptotic cell death.

3.2.5. JNK Activation Is Required for **CA13**-Induced Apoptosis in Lung Cancer Cells

Activation of the c-Jun N-terminal kinase (JNK) signaling pathway has been widely implicated in chalcone-mediated apoptosis across various cancer cell types [38–42]. To determine whether **CA13**-induced apoptosis depends on JNK activation, we first examined JNK phosphorylation in H292 cells. As shown in Figure 5A, **CA13** treatment markedly increased the phosphorylation level of JNK (p-JNK) in both dose- and time-dependent manners, while total JNK levels remained unchanged. To further validate the role of JNK activation in **CA13**-induced cytotoxicity, the specific JNK inhibitor SP600125 was applied. Pretreatment with SP600125 significantly restored cell viability compared with **CA13** treatment alone (Figure 5B). Western blot analysis confirmed that SP600125 effectively blocked JNK phosphorylation and attenuated **CA13**-induced cleavage of PARP (Figure 5C). These results demonstrate that JNK activation is essential for CA13-induced apoptosis in H292 cells, highlighting the critical role of the JNK signaling pathway in mediating the pro-apoptotic effects of **CA13**.

https://doi.org/10.3390/biom16010054

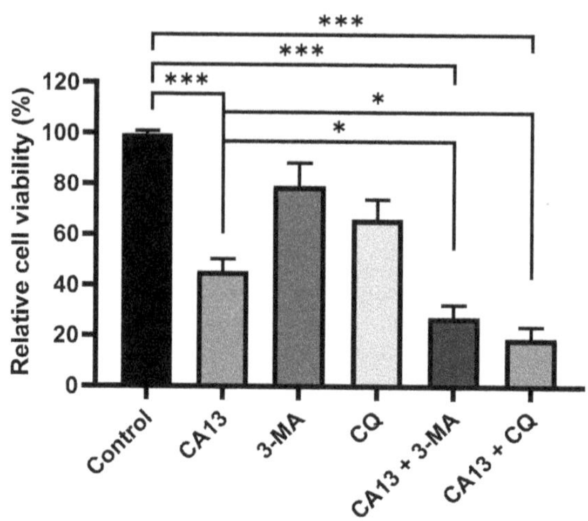

Figure 4. **WA13** induces cytoprotective autophagy in H292 cells. Cell viability of H292 cells was measured after treatment with **CA13** (10 μM), 3-methyladenine (3-MA, 10 mM), chloroquine (CQ, 3 μM), or their combination for 24h using MTT assays. Both early (3-MA) and late (CQ) autophagy inhibitors significantly reduced cell viability when combined with **CA13**, suggesting that inhibition of autophagy enhances **CA13**-induced cytotoxicity. Data are presented as mean ± standard deviation (SD) for three independent experiments. * $p < 0.05$, *** $p < 0.001$.

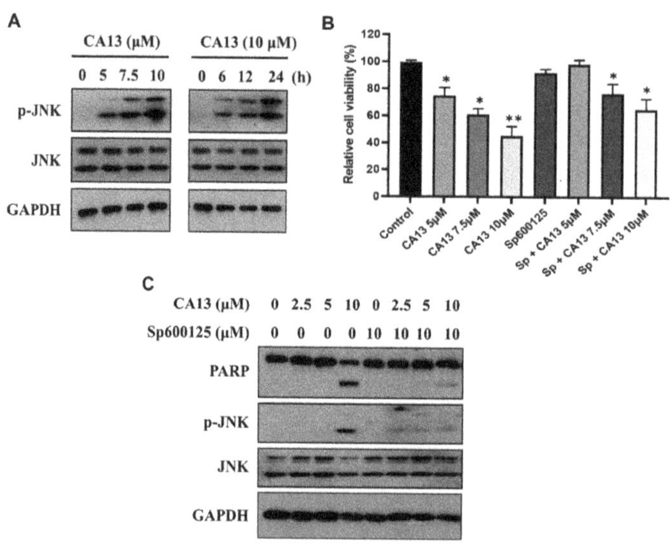

Figure 5. JNK activation is required for **CA13**-induced apoptosis in H292 cells. (**A**) H292 cells were treated with **CA13** (0–10 μM) for 24 h or with 10 μM **CA13** for the indicated time periods. The expression of JNK and p-JNK was analyzed by Western blotting, with GAPDH as a loading control. (**B**) Cell viability of H292 cells was measured using MTT assays after treatment with **CA13** (10 μM), SP600125 (10 μM), or their combination for 24 h. (**C**) H292 cells were pretreated with SP600125 (10 μM) for 6 h, followed by incubation with **CA13** (10 μM) for 24 h. The protein levels of PARP, JNK, and p-JNK were determined by Western blotting. Original blots can be found in Supplementary Materials. Data are presented as mean ± standard deviation (SD) for three independent experiments. * $p < 0.05$, ** $p < 0.01$.

https://doi.org/10.3390/biom16010054

3.2.6. **CA13** Suppresses Lung Cancer Growth In Vivo

To evaluate the antitumor efficacy of **CA13** in vivo, a xenograft model was established by subcutaneous implantation of H292 cells into Balb/c nude mice. Once tumors became palpable, mice were intraperitoneally administered **CA13** at doses of 20 or 40 mg/kg, or vehicle control, every three days. As shown in Figure 6A,B, **CA13** treatment significantly inhibited tumor growth in a dose-dependent manner compared with the control group ($p < 0.05$). Final tumor weight analysis confirmed the pronounced reduction in tumor burden in **CA13**-treated mice (Figure 6C). Importantly, no significant difference in body weight was observed among the groups throughout the treatment period (Figure 6D), suggesting that **CA13** did not cause overt systemic toxicity. These findings demonstrate that **CA13** exhibits potent antitumor activity against human H292 lung cancer xenografts in vivo, while maintaining a favorable safety profile.

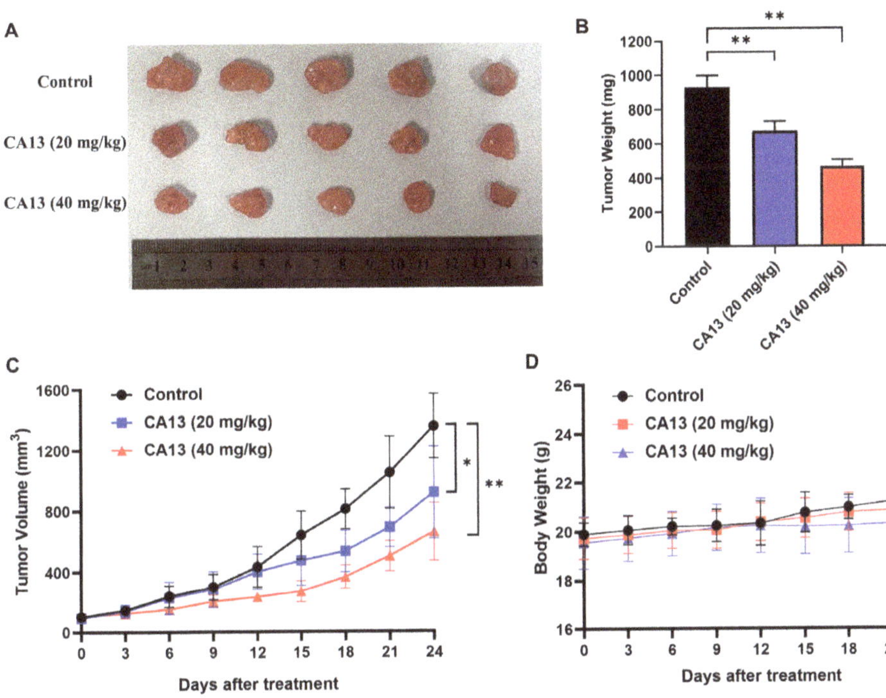

Figure 6. **CA13** suppresses tumor growth in the H292 xenograft mouse model. (**A**,**B**) Tumor-bearing Balb/c nude mice ($n = 5$ per group) were treated with vehicle or **CA13** (20 or 40 mg/kg) via intraperitoneal injection every three days. Tumor volume was measured at the indicated time points. (**C**) Tumors were excised and weighed after 24 days of treatment. (**D**) Body weight was monitored throughout the experimental period to evaluate potential systemic toxicity. Data are presented as mean ± standard deviation (SD) for three independent experiments. * $p < 0.05$, ** $p < 0.01$.

4. Discussion

In this study, a series of adamantyl-substituted chalcones were synthesized and evaluated for their anticancer properties. Among them, **CA13** exhibited the most potent antiproliferative activity against human lung cancer cells while showing minimal cytotoxicity toward normal lung fibroblasts. Structural–activity analysis suggested that the introduction of an adamantyl group and a carboxyl substituent synergistically enhanced cytotoxic potency, likely by improving molecular lipophilicity and π–π conjugation, consistent with prior studies reporting enhanced bioactivity through hydrophobic modification of chalcones [30,34]. Notably, our previous work identified WA15, a structurally analogous adamantyl-substituted chalcone that shares the core adamantyl and carboxyl pharmacophores with CA13 and exhibits comparable in vitro anti-NSCLC potency [30]. Nonetheless, the mechanistic focuses of the two studies are fundamentally distinct. The prior investigation of WA15 established its role as an RXRα modulator, demonstrating its ability to inhibit RXRα transactivation, induce proteasome-dependent RXRα degradation [30]. In contrast, the present study on CA13 centers on elucidating how CA13-induced autophagy regulates lung cancer cell apoptosis. Specifically, we confirm a cytoprotective role of autophagy against CA13-triggered apoptotic stress, a previously unexplored axis.

Mechanistic analyses revealed that **CA13** triggers both autophagy and apoptosis in H292 lung cancer cells. Western blot and fluorescence microscopy demonstrated significant LC3-II accumulation and formation of autolysosomes, confirming activation of complete autophagic flux. Interestingly, inhibition of autophagy potentiated **CA13**-induced apoptosis, suggesting that autophagy serves a cytoprotective role under **CA13**-induced stress. This duality of autophagy, functioning either as a survival or death mechanism, has been widely observed in tumor responses to chemotherapeutic chalcones [21–26].

Consistent with previous studies on chalcone analogues, **CA13** triggered apoptotic cell death in lung cancer cells and JNK activation is indispensable for CA13-induced apoptosis. Similar JNK-dependent apoptotic mechanisms have been described for other chalcones, such as xanthohumol and butein, where sustained JNK activation promotes mitochondrial dysfunction and caspase activation [39–41]. Our results reinforce that JNK signaling acts as a central mediator of **CA13**-induced apoptosis in lung cancer cells.

The interplay between autophagy and apoptosis represents a critical determinant of therapeutic response. While **CA13** concurrently activated both processes, functional studies revealed that autophagy mitigates **CA13**-induced apoptotic stress, functioning as a survival mechanism. Therefore, dual targeting of autophagy and apoptosis could represent a rational strategy to potentiate **CA13**'s anticancer activity.

5. Conclusions

In summary, **CA13** is a novel adamantyl chalcone derivative that exerts potent cytotoxic effects through JNK-dependent apoptosis and cytoprotective autophagy. These findings not only highlight **CA13** as a promising therapeutic lead against lung cancer but also provide mechanistic insights into how structural modification of chalcones can modulate the balance between survival and death pathways. Future studies will focus on identifying upstream molecular targets of **CA13** and evaluating combination therapies that inhibit autophagy to enhance its antitumor efficacy.

Supplementary Materials: The following supporting information can be downloaded at: https://www.mdpi.com/article/10.3390/biom16010054/s1, ^{1}H NMR and ^{13}C NMR data for compounds **CA6–CA13**. Figure S1. ^{1}H NMR Spectrum of compound CA 6. Figure S2. ^{13}C NMR Spectrum of compound CA 6. Figure S3. Mass Spectrum of compound CA 6. Figure S4. IR Spectrum of compound CA 6. Figure S5. ^{1}H NMR Spectrum of compound CA 7. Figure S6. ^{13}C NMR Spectrum of compound CA 7. Figure S7. Mass Spectrum of compound CA 7. Figure S8. IR Spectrum of compound CA 7. Figure S9. ^{1}H NMR Spectrum of compound CA 8. Figure S10. ^{13}C NMR Spectrum of compound CA 8. Figure S11. Mass Spectrum of compound CA 8. Figure S12. IR Spectrum of compound CA 8. Figure S13. ^{1}H NMR Spectrum of compound CA 9. Figure S14. ^{13}C NMR Spectrum of compound CA 9. Figure S15. Mass Spectrum of compound CA 9. Figure S16. IR Spectrum of compound CA 9. Figure S17. ^{1}H NMR Spectrum of compound CA 10. Figure S18. ^{13}C NMR Spectra of compound CA 10. Figure S19. Mass Spectrum of compound CA 10. Figure S20. IR Spectrum of compound CA 10. Figure S21. ^{1}H NMR Spectrum of compound CA 11. Figure S22. ^{13}C NMR Spectra of compound CA 11. Figure S23. Mass Spectrum of compound CA 11. Figure S24. IR Spectrum of compound

CA 11. Figure S25. ^{1}H NMR Spectra of compound CA 12. Figure S26. ^{13}C NMR Spectra of compound CA 12. Figure S27. Mass Spectrum of compound CA 12. Figure S28. IR Spectrum of compound CA 12. Figure S29. ^{1}H NMR Spectra of compound CA 13. Figure S30. ^{13}C NMR Spectra of compound CA 13. Figure S31. Mass Spectrum of compound CA 13. Figure S32. IR Spectrum of compound CA 13. Original Western blots.

Author Contributions: Conceptualization, Y.C. and M.A.; methodology, Y.L.; validation, Y.C., Y.L. and J.Z.; formal analysis, J.Z.; investigation, T.B.; resources, J.W.; data curation, J.W.; writing—original draft preparation, Y.C. and Y.L.; writing—review and editing, J.W. and M.A.; visualization, T.B.; supervision, M.A. and J.W.; project administration, M.A.; funding acquisition, M.A. and J.W. All authors have read and agreed to the published version of the manuscript.

Funding: This research was funded by the Health Commission of Hubei Province Scientific Research Project (No. WJ2025M099), the National Natural Science Foundation of China (No. 82202069), the Hubei Provincial Natural Science Foundation of China (No. 2022CFB347), and the Doctoral Research Fund of Hubei University of Science and Technology (No. BK202304).

Institutional Review Board Statement: The animal study protocol was approved by the Institutional Ethics Committee of Hubei University of Science and Technology (protocol code 2022-04-052, approval date 1 April 2022).

Informed Consent Statement: Not applicable.

Data Availability Statement: The data is available via the attached Supplementary Materials provided above.

Conflicts of Interest: The authors declare no conflicts of interest.

References

1. Zhou, J.; Xu, Y.; Liu, J.; Feng, L.; Yu, J.; Chen, D. Global burden of lung cancer in 2022 and projections to 2050: Incidence and mortality estimates from GLOBOCAN. *Cancer Epidemiol.* **2024**, *93*, 102693. [CrossRef]
2. Zhao, S.; Zhao, H.; Yang, W.; Zhang, L. The next generation of immunotherapies for lung cancers. *Nat. Rev. Clin. Oncol.* **2025**, *22*, 592–616. [CrossRef]
3. Li, S.; Wang, A.; Wu, Y.; He, S.; Shuai, W.; Zhao, M.; Zhu, Y.; Hu, X.; Luo, Y.; Wang, G. Targeted therapy for non-small-cell lung cancer: New insights into regulated cell death combined with immunotherapy. *Immunol. Rev.* **2024**, *321*, 300–334. [CrossRef]
4. Herbst, R.S.; Morgensztern, D.; Boshoff, C. The biology and management of non-small cell lung cancer. *Nature* **2018**, *553*, 446–454. [CrossRef] [PubMed]
5. Davison, E.K.; Brimble, M.A. Natural product derived privileged scaffolds in drug discovery. *Curr. Opin. Chem. Biol.* **2019**, *52*, 1–8. [CrossRef] [PubMed]
6. Mastachi-Loza, S.; Ramirez-Candelero, T.I.; Benitez-Puebla, L.J.; Fuentes-Benites, A.; Gonzalez-Romero, C.; Vazquez, M.A. Chalcones, a Privileged Scaffold: Highly Versatile Molecules in [4+2] Cycloadditions. *Chem. Asian J.* **2022**, *17*, e202200706. [CrossRef]
7. WalyEldeen, A.A.; Sabet, S.; El-Shorbagy, H.M.; Abdelhamid, I.A.; Ibrahim, S.A. Chalcones: Promising therapeutic agents targeting key players and signaling pathways regulating the hallmarks of cancer. *Chem. Biol. Interact.* **2023**, *369*, 110297. [CrossRef]
8. Ouyang, Y.; Li, J.; Chen, X.; Fu, X.; Sun, S.; Wu, Q. Chalcone Derivatives: Role in Anticancer Therapy. *Biomolecules* **2021**, *11*, 894. [CrossRef] [PubMed]
9. Rudrapal, M.; Khan, J.; Dukhyil, A.A.B.; Alarousy, R.; Attah, E.I.; Sharma, T.; Khairnar, S.J.; Bendale, A.R. Chalcone Scaffolds, Bioprecursors of Flavonoids: Chemistry, Bioactivities, and Pharmacokinetics. *Molecules* **2021**, *26*, 7177. [CrossRef]
10. Nowakowska, Z. A review of anti-infective and anti-inflammatory chalcones. *Eur. J. Med. Chem.* **2007**, *42*, 125–137. [CrossRef]
11. Mittal, A.; Vashistha, V.K.; Das, D.K. Recent advances in the antioxidant activity and mechanisms of chalcone derivatives: A computational review. *Free Radic. Res.* **2022**, *56*, 378–397. [CrossRef] [PubMed]
12. Constantinescu, T.; Lungu, C.N. Anticancer Activity of Natural and Synthetic Chalcones. *Int. J. Mol. Sci.* **2021**, *22*, 11306. [CrossRef] [PubMed]
13. Michalkova, R.; Mirossay, L.; Gazdova, M.; Kello, M.; Mojzis, J. Molecular Mechanisms of Antiproliferative Effects of Natural Chalcones. *Cancers* **2021**, *13*, 2730. [CrossRef] [PubMed]

https://doi.org/10.3390/biom16010054

14. Hseu, Y.C.; Chiang, Y.C.; Vudhya Gowrisankar, Y.; Lin, K.Y.; Huang, S.T.; Shrestha, S.; Chang, G.R.; Yang, H.L. The In Vitro and In Vivo Anticancer Properties of Chalcone Flavokawain B through Induction of ROS-Mediated Apoptotic and Autophagic Cell Death in Human Melanoma Cells. *Cancers* **2020**, *12*, 2936. [CrossRef]
15. Michalkova, R.; Kello, M.; Kudlickova, Z.; Gazdova, M.; Mirossay, L.; Mojzisova, G.; Mojzis, J. Programmed Cell Death Alterations Mediated by Synthetic Indole Chalcone Resulted in Cell Cycle Arrest, DNA Damage, Apoptosis and Signaling Pathway Modulations in Breast Cancer Model. *Pharmaceutics* **2022**, *14*, 503. [CrossRef]
16. Mendez-Callejas, G.; Pineros-Avila, M.; Celis, C.A.; Torrenegra, R.; Espinosa-Benitez, A.; Pestana-Nobles, R.; Yosa-Reyes, J. Natural 2′,4-Dihydroxy-4′,6′-dimethoxy Chalcone Isolated from Chromolaena tacotana Inhibits Breast Cancer Cell Growth through Autophagy and Mitochondrial Apoptosis. *Plants* **2024**, *13*, 570. [CrossRef]
17. Wang, J.R.; Luo, Y.H.; Piao, X.J.; Zhang, Y.; Feng, Y.C.; Li, J.Q.; Xu, W.T.; Zhang, Y.; Zhang, T.; Wang, S.N.; et al. Mechanisms underlying isoliquiritigenin-induced apoptosis and cell cycle arrest via ROS-mediated MAPK/STAT3/NF-kappaB pathways in human hepatocellular carcinoma cells. *Drug Dev. Res.* **2019**, *80*, 461–470. [CrossRef]
18. Sun, M.; Wang, Y.; Yuan, M.; Zhao, Q.; Zhang, Y.; Yao, Y.; Duan, Y. Angiogenesis, Anti-Tumor, and Anti-Metastatic Activity of Novel alpha-Substituted Hetero-Aromatic Chalcone Hybrids as Inhibitors of Microtubule Polymerization. *Front. Chem.* **2021**, *9*, 766201. [CrossRef]
19. Kwak, A.W.; Lee, M.J.; Lee, M.H.; Yoon, G.; Cho, S.S.; Chae, J.I.; Shim, J.H. The 3-deoxysappanchalcone induces ROS-mediated apoptosis and cell cycle arrest via JNK/p38 MAPKs signaling pathway in human esophageal cancer cells. *Phytomedicine* **2021**, *86*, 153564. [CrossRef]
20. Sasazawa, Y.; Kanagaki, S.; Tashiro, E.; Nogawa, T.; Muroi, M.; Kondoh, Y.; Osada, H.; Imoto, M. Xanthohumol impairs autophagosome maturation through direct inhibition of valosin-containing protein. *ACS Chem. Biol.* **2012**, *7*, 892–900. [CrossRef] [PubMed]
21. Zhang, Y.Q.; Wen, Z.H.; Wan, K.; Yuan, D.; Zeng, X.; Liang, G.; Zhu, J.; Xu, B.; Luo, H. A novel synthesized 3′, 5′-diprenylated chalcone mediates the proliferation of human leukemia cells by regulating apoptosis and autophagy pathways. *Biomed. Pharmacother.* **2018**, *106*, 794–804. [CrossRef] [PubMed]
22. Wan, B.; Zhu, J.; Chang, Q.; Zhou, H.; Shi, Z.; Min, L.; Cai, Y.; Guan, H. Alpha, 2′-dihydroxy-4,4′-dimethoxydihydrochalcone inhibits cell proliferation, invasion, and migration in gastric cancer in part via autophagy. *Biomed. Pharmacother.* **2018**, *98*, 709–718. [CrossRef]
23. Wang, X.; Liang, Y.; Zhang, B.; He, L.; Li, W.; Zhang, W.; Li, C.; Luo, L.; Umar, T.; Feng, H.; et al. 2′-Hydroxychalcone Induces Autophagy and Apoptosis in Breast Cancer Cells via the Inhibition of the NF-kappaB Signaling Pathway: In Vitro and In Vivo Studies. *Nutrients* **2024**, *16*, 514. [CrossRef]
24. Wang, F.W.; Wang, S.Q.; Zhao, B.X.; Miao, J.Y. Discovery of 2′-hydroxychalcones as autophagy inducer in A549 lung cancer cells. *Org. Biomol. Chem.* **2014**, *12*, 3062–3070. [CrossRef]
25. Rossi, M.; Pellegrino, C.; Rydzyk, M.M.; Farruggia, G.; de Biase, D.; Cetrullo, S.; D'Adamo, S.; Bisi, A.; Blasi, P.; Malucelli, E.; et al. Chalcones induce apoptosis, autophagy and reduce spreading in osteosarcoma 3D models. *Biomed. Pharmacother.* **2024**, *179*, 117284. [CrossRef]
26. Niu, Q.; Zhao, W.; Wang, J.; Li, C.; Yan, T.; Lv, W.; Wang, G.; Duan, W.; Zhang, T.; Wang, K.; et al. LicA induces autophagy through ULK1/Atg13 and ROS pathway in human hepatocellular carcinoma cells. *Int. J. Mol. Med.* **2018**, *41*, 2601–2608. [CrossRef]
27. Li, X.; He, S.; Ma, B. Autophagy and autophagy-related proteins in cancer. *Mol. Cancer* **2020**, *19*, 12. [CrossRef]
28. He, Q.; Liu, W.; Sha, S.; Fan, S.; Yu, Y.; Chen, L.; Dong, M. Adenosine 5′-monophosphate-activated protein kinase-dependent mTOR pathway is involved in flavokawain B-induced autophagy in thyroid cancer cells. *Cancer Sci.* **2018**, *109*, 2576–2589. [CrossRef] [PubMed]
29. Wang, J.; Qi, Q.; Zhou, W.; Feng, Z.; Huang, B.; Chen, A.; Zhang, D.; Li, W.; Zhang, Q.; Jiang, Z.; et al. Inhibition of glioma growth by flavokawain B is mediated through endoplasmic reticulum stress induced autophagy. *Autophagy* **2018**, *14*, 2007–2022. [CrossRef]
30. Ao, M.; Hu, X.; Qian, Y.; Li, B.; Zhang, J.; Cao, Y.; Zhang, Y.; Guo, K.; Qiu, Y.; Jiang, F.; et al. Discovery of new chalone adamantyl arotinoids having RXRalpha-modulating and anticancer activities. *Bioorg. Chem.* **2021**, *113*, 104961. [CrossRef] [PubMed]
31. Xie, Z.; Luo, X.; Zou, Z.; Zhang, X.; Huang, F.; Li, R.; Liao, S.; Liu, Y. Synthesis and evaluation of hydroxychalcones as multifunctional non-purine xanthine oxidase inhibitors for the treatment of hyperuricemia. *Bioorg. Med. Chem. Lett.* **2017**, *27*, 3602–3606. [CrossRef]
32. Gore, M.P.; Gore, A.M. Compositions and Methods for Treating Bleeding and Bleeding Disorders. U.S. Patent 12,343,317, 1 July 2025.
33. Ahmad, I.; Thakur, J.P.; Chanda, D.; Saikia, D.; Khan, F.; Dixit, S.; Kumar, A.; Konwar, R.; Negi, A.S.; Gupta, A. Syntheses of lipophilic chalcones and their conformationally restricted analogues as antitubercular agents. *Bioorg. Med. Chem. Lett.* **2013**, *23*, 1322–1325. [CrossRef]

https://doi.org/10.3390/biom16010054

34. Wanka, L.; Iqbal, K.; Schreiner, P.R. The lipophilic bullet hits the targets: Medicinal chemistry of adamantane derivatives. *Chem. Rev.* **2013**, *113*, 3516–3604. [CrossRef]
35. Xu, Y.; Qian, C.; Wang, Q.; Song, L.; He, Z.; Liu, W.; Wan, W. Deacetylation of ATG7 drives the induction of macroautophagy and LC3-associated microautophagy. *Autophagy* **2024**, *20*, 1134–1146. [CrossRef] [PubMed]
36. Das, S.; Shukla, N.; Singh, S.S.; Kushwaha, S.; Shrivastava, R. Mechanism of interaction between autophagy and apoptosis in cancer. *Apoptosis* **2021**, *26*, 512–533. [CrossRef] [PubMed]
37. Xie, Q.; Liu, Y.; Li, X. The interaction mechanism between autophagy and apoptosis in colon cancer. *Transl. Oncol.* **2020**, *13*, 100871. [CrossRef] [PubMed]
38. Moon, D.O.; Kim, M.O.; Choi, Y.H.; Hyun, J.W.; Chang, W.Y.; Kim, G.Y. Butein induces G(2)/M phase arrest and apoptosis in human hepatoma cancer cells through ROS generation. *Cancer Lett.* **2010**, *288*, 204–213. [CrossRef]
39. Slawinska-Brych, A.; Zdzisinska, B.; Czerwonka, A.; Mizerska-Kowalska, M.; Dmoszynska-Graniczka, M.; Stepulak, A.; Gagos, M. Xanthohumol exhibits anti-myeloma activity in vitro through inhibition of cell proliferation, induction of apoptosis via the ERK and JNK-dependent mechanism, and suppression of sIL-6R and VEGF production. *Biochim. Biophys. Acta Gen. Subj.* **2019**, *1863*, 129408. [CrossRef]
40. Kong, W.; Li, C.; Qi, Q.; Shen, J.; Chang, K. Cardamonin induces G2/M arrest and apoptosis via activation of the JNK-FOXO3a pathway in breast cancer cells. *Cell Biol. Int.* **2020**, *44*, 177–188. [CrossRef]
41. Kim, S.Y.; Park, E.J.; Zhao, Y.Z.; Sohn, D.H. Butein suppresses bile acid-induced hepatocyte apoptosis through a JNK-dependent but ERK-independent pathway. *Planta Med.* **2007**, *73*, 777–781. [CrossRef]
42. He, W.; Wang, Q.; Srinivasan, B.; Xu, J.; Padilla, M.T.; Li, Z.; Wang, X.; Liu, Y.; Gou, X.; Shen, H.M.; et al. A JNK-mediated autophagy pathway that triggers c-IAP degradation and necroptosis for anticancer chemotherapy. *Oncogene* **2014**, *33*, 3004–3013. [CrossRef] [PubMed]

https://doi.org/10.3390/biom16010054

MDPI AG
Grosspeteranlage 5
4052 Basel
Switzerland
Tel.: +41 61 683 77 34

Biomolecules Editorial Office
E-mail: biomolecules@mdpi.com
www.mdpi.com/journal/biomolecules

www.ingramcontent.com/pod-product-compliance
Lightning Source LLC
Chambersburg PA
CBHW041016270926
51850CB00024B/750
* 9 7 8 3 7 2 5 8 7 8 1 5 4 *